Quaternion and Octonion Color Image Processing with MATLAB®

Artyom M. Grigoryan and **Sos S. Agaian**

SPIE PRESS
Bellingham, Washington USA

Library of Congress Cataloging-in-Publication Data

Names: Grigoryan, Artyom M., author. | Agaian, S. S., author.
Title: Quaternion and octonion color image processing with MATLAB / Artyom M. Grigoryan and Sos S. Agaian.
Description: Bellingham, Washington : SPIE Press, [2018] | Includes bibliographical references and index.
Identifiers: LCCN 2017013782 | ISBN 9781510611351 (print : alk. paper) | ISBN 1510611355 (print : alk. paper) | ISBN 9781510611368 (PDF) | ISBN 1510611363 (PDF) | ISBN 9781510611375 (ePub) | ISBN 1510611371 (ePub) | ISBN 9781510611382 (Kindle/Mobi) | ISBN 151061138X (Kindle/Mobi)
Subjects: LCSH: Image processing–Digital techniques–Mathematics. | Quaternions. | Color vision. | Color. | MATLAB.
Classification: LCC TA1637.5 .G75 2017 | DDC 621.36/7028553–dc23 LC record available at https://lccn.loc.gov/2017013782

Published by

SPIE
P.O. Box 10
Bellingham, Washington 98227-0010 USA
Phone: +1 360.676.3290
Fax: +1 360.647.1445
Email: books@spie.org
Web: http://spie.org

Printed in the United States of America.
First Printing.
For updates to this book, visit http://spie.org and type "PM279" in the search field.

SPIE.

Contents

Preface

Color image processing has attracted much interest in recent years. The use of color in image processing is motivated by the facts that (1) the human eyes can discern thousands of colors, and image processing is used both for human interaction and computer interpretation; (2) a color image comprises more information than a grayscale image; (3) color features are robust to several image-processing procedures (for example, to the translation and rotation of the regions of interest); (4) color features are efficiently used in many vision tasks, including object recognition and tracking, image segmentation and retrieval, image registration etc.; and (5) color is necessary in many real-life applications such as visual communications, multimedia systems, fashion and food industries, computer vision, entertainment, consumer electronics, production printing and proofing, digital photography, biometrics, digital artwork reproduction, industrial inspection, and biomedical applications. Finally, the enormous number of color images that are constantly uploaded to the Internet require new approaches to visual media creation, retrieval, processing, and applications. This also gives us new opportunities to create a number of large visual data-driven applications. Three independent quantities are used to describe any particular color; the human eyes see all colors as variable combinations of primary colors of red, green, and blue. Many of the methods of modern color image processing are based on dealing with each primary color separately. However, this methodology fails to capture the inherent correlation between the color image components and results in color artifacts. Moreover, it is not clear how to handle and combine the information from the different primary colors. It is natural therefore to ask how to couple the information contained in the given primary colors and how to process the color components as a whole unit without losing the spectral relation that is present in them, or how to develop a mathematical color model that may help to process all of the color image components. The application of the theory of hypercomplex numbers in color imaging may give us the answers to these questions. The quaternions and octonions are numbers that extend the complex numbers to higher dimensions, 4 and 8, respectively.

Recently, the theory of the quaternion algebra has been used in color science and color systems by processing simultaneously the color channels.

The three color channels of the image can be represented as a vector field of quaternion numbers. The first hypercomplex numbers that are quaternions were discovered by the Irish mathematician and physicist William Rowan Hamilton in 1843. Quaternions are currently accepted as one of the most important concepts in modern computer graphics, in both theoretical and applied mathematics, in group theory and topology, quantum mechanics, color image processing, and virtual reality applications.

The main goal of this book is to provide the mathematics of quaternions and octonions and to show how they can be used in emerging areas of color image processing. The book begins with a chapter covering the introductory material and fundamentals of complex and quaternion numbers, multiplication of quaternions, the geometry of rotations, and many functions of quaternions, such as the exponent, logarithm, power, and trigonometric functions. This chapter includes many illustrative examples and MATLAB® codes to demonstrate why they are important and to explore quaternions unencumbered by their mathematical aspects. Chapter 2 is devoted to octonion numbers and the main operations and functions of octonions. Multiplication of octonions is described and illustrated through examples with scripts. In Chapters 3 and 4, different mappings of grayscale and color images into the spaces of quaternions and octonions are described with examples. The operation of multiplication with its parts being the inner and vector products are illustrated on color images. The authors pay more attention to the fast quaternion discrete Fourier transforms (QDFTs) because of the Fourier transform's major impact on the various components of image processing, such as image filtering, image enhancement image analysis, image reconstruction, and image compression. The concept of the QDFT became a very popular topic in color imaging. Fast algorithms of the QDFT are based on representations by combinations of classical 1-D DFTs, which leads to fast numerical implementation with the fast Fourier transform software. Chapter 5 describes the algorithms of the 1-D fast Fourier transform and introduces fast algorithms for the quaternion and octonion discrete Fourier transforms (ODFTs). Many examples and scripts required to calculate the transforms are given. The concepts of 2-D quaternion and octonion DFTs as generalizations of the DFT are described in detail with fast algorithms and scripts in Chapter 6. The tensor representation of images in the quaternion and octonion spaces is also introduced, and the efficiency of the tensor algorithm of the fast left- and right-side 2-D QDFTs is described and compared with existent methods. In addition, a new concept of the 2-D QDFT on a hexagonal lattice is presented. The final chapters discuss the specific applications: state-of-the-art color image enhancement, gradient operators, grayscale and color visibility images, and face recognition and filtering applications. This book provides the theory and methods, many unique tools, and 72 codes and functions written in MATLAB with useful comments. We believe that the presented computer simulations, numerical experiments, and illustrative solutions of real-world problems in color image processing and the given analysis will allow the readers to develop a deeper understanding of both theoretical and practical aspects and advanced concepts of this subject. This book will be useful for upper-level undergraduates and graduate students, researchers, and image processing engineers, helping them to reorganize quaternions and octonions, and apply them in practice. This book is also for the developer, scientist, and engineer working in computer graphics, signal and image processing, multimedia analytics, visualization, or entertainment computing. Finally, this book will be helpful to the people who need an assortment of quaternion utilities, sample MATLAB codes, and practical examples to help them understand the theory involved in quaternion imaging.

We appreciate all who assisted in the preparation of this book in a short eight-month period. We are grateful to Merughan Grigoryan and the reviewers for many suggestions and recommendations.

Artyom Grigoryan and Sos Agaian
March 2018

Chapter 1
Complex and Hypercomplex Numbers

The theory of complex numbers is well developed; complex numbers have been used in science and engineering for a long time and are still being used for solving many new problems. The arithmetic of these numbers generalizes the arithmetic of real numbers in the sense that, together with the operations of addition and multiplication by real numbers, the inverse number and the division are defined. Such a complete arithmetic exists for other numbers, which are called quaternions and octonions. Quaternions were first discovered by Hamilton in 1843 [6]. More recently, quaternions have been employed in bioinformatics, navigation systems [7], and image and video processing [8, 9]. Octonions, which are defined as doubled quaternion numbers [34], have been used in signal and image processing, and we believe that they can also be used effectively for parallel processing many images.

Recently, the theory of quaternion algebra has been used in the application of color science that processes the three color channels simultaneously [1]–[13]. Quaternion numbers found interesting applications in color image processing, such as image enhancement [19, 31], watermarking [32], adaptive filtering [33], and prostate cancer Gleason grading [16]. The quaternion can be considered as a four-dimensional number with one real part and three imaginary parts. The imaginary dimensions are represented as i,j, and k, which are orthogonal to each other and to 1. In many cases, it is useful to transfer the calculations from the real space of signals and images to complex space, analyze and solve problems by using methods of the complex analysis (arithmetic), and then transfer the solution back to the real space. The transformation to the space of quaternions is also promising. Quaternion algebra for color imaging was first used by Pei and led to the description of new tools, such as the quaternion Fourier transforms and correlations for image processing by representing the red, green, and blue values at each pixel in the color image as single pure quaternion-valued pixels [10]. There are a number of studies on quaternions and quaternion operations and systems in color image processing [11]–[19]. These color-processing systems use pure complex quaternion representation, not the complete quaternion components.

It is natural to ask how to use the complete quaternion representation, or more precisely, how to use the "real" scalar number information in different color image-processing applications, or what the advantage is of using the complete representation model over the pure complex quaternion model, particularly in color image processing. The demand by the consumer to run multiple applications on a mobile platform is increasing every day and is presenting a greater challenge to develop efficient computation tools and proficient energy resources [8]. For example, the users want their cell phones to work on different communication standards with multiple office or home-related applications showing real-time performance.

1.1 Complex and Hypercomplex Numbers

In this section, we present the basics of the complex and hypercomplex numbers, namely quaternions, and describe the main operations and properties of such numbers. Complex numbers are considered as points on a two-dimensional plane, where one coordinate is real and another one is measured along the imaginary vertical, as shown in Fig. 1.1 for the point $(3, 2i)$. The first coordinate, 3, of this point is measured along the real axis, and the unit on the vertical is denoted by $1i$, or simply i. This unit measure i is called *the imaginary unit* and will be described in the next sub-section.

1.1.1 Complex arithmetic

We first go back to the sixteenth century, when the formal and not real solution of the simple quadratic equation

$$z^2 + 1 = 0, \quad \text{or} \quad z^2 = -1, \tag{1.1}$$

was denoted by $z = \sqrt{-1}$ by two Italian mathematicians Rafael Bombelli and Gerolamo Cardano [34]. Another solution of this equation is $z = -\sqrt{-1}$.

Figure 1.2(a) shows the graph of the parabola $y = x^2 + 1$, which does not intersect the horizontal line, i.e., there is no solution of Eq. (1.1) in real arithmetic. Solutions to this equation can be found in other arithmetics that

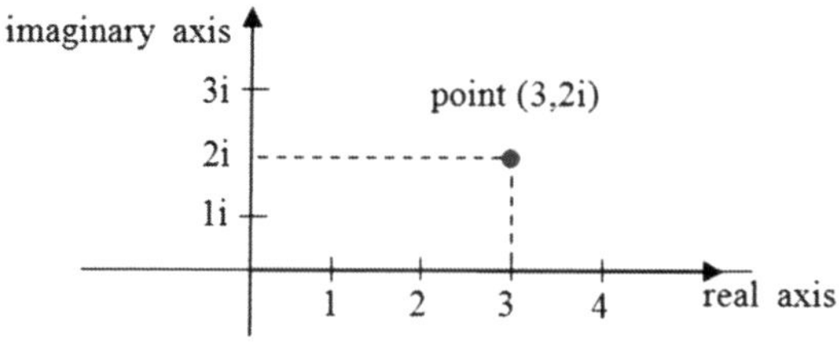

Figure 1.1 The point on a two-dimensional plane.

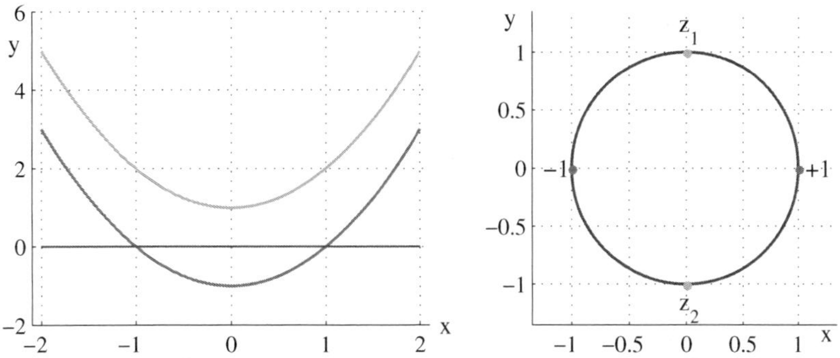

(a) $y=x^2+1$ and $y=x^2-1$ (b) unit circle

Figure 1.2 (a) Parabola and (b) the unit circle with four unit points on it.

will be described in a moment; in fact, these two solutions are on the unit circle shown in Fig. 1.2(b). There are four unit points on this circle. Two points are on the horizontal line, ± 1, and they are the solutions of the equation $z^2 - 1 = 0$, which can be written as $x^2 - 1$ with real numbers x. The graph of the parabola $y = x^2 - 1$ is also shown in Fig. 1.2(a). Two other points of the circle, which are on the vertical line, are the solutions of the equation $z^2 + 1 = 0$. These are the points $z_1 = \sqrt{-1}$ and $z_2 = -\sqrt{-1}$.

Given a real number $y \neq 0$, the solutions of the equation

$$z^2 + y^2 = 0, \quad \text{or} \quad \left(\frac{z}{y}\right)^2 + 1 = 0, \tag{1.2}$$

can be written as $z/y = \pm\sqrt{-1}$, $z = \pm y\sqrt{-1}$, or $z = \pm(\sqrt{-1})y$. If for a given real number x we consider the quadratic equation

$$(z - x)^2 + y^2 = 0, \tag{1.3}$$

then, one can see that $z - x = \pm y\sqrt{-1}$ and the solutions of this equation are the numbers $z = x + y\sqrt{-1} = x + (\sqrt{-1})y$ and $z = x - y\sqrt{-1} = x - (\sqrt{-1})y$.

In the eighteenth century, Euler denoted this imaginary number or symbol $\sqrt{-1}$ by i, i.e., $i = \sqrt{-1}$ and $i^2 = -1$. This symbol represents an imaginary unit, and in the engineering community it is denoted by j, since the letter "i" is used for the electrical current.

The number $z = x + iy = x + yi$ is called *a complex number*. Numbers x and y are real, and x is called the real part of z and y is the imaginary part of z. These two parts are denoted by $x = \mathrm{Re}\,(z)$ and $y = \mathrm{Im}\,(z)$, or $x = \mathfrak{R}(z)$ and $y = \mathfrak{I}(z)$. The concept of the complex number generalizes the real numbers that can be considered as the complex numbers with zero imaginary parts, i.e., when $y = 0$. Arithmetical operations also are generalized in complex arithmetic, and we consider the main operations over complex numbers.

Given complex numbers $z_1 = x_1 + iy_1$ and $z_2 = x_2 + iy_2$, the following properties for operations of addition and multiplication are valid:

1. $z_1 + z_2 = [x_1 + iy_1] + [x_2 + iy_2] = (x_1 + x_2) + i(y_1 + y_2)$,
2. $kz_1 = k[x_1 + iy_1] = (kx_1) + i(ky_1)$,

for any real number k. One can note that $z_1 + z_2 = z_2 + z_1$. The operation of subtraction is defined as $z_1 - z_2 = z_1 + (-1)z_2$:

3. $z_1 - z_2 = [x_1 + iy_1] - [x_2 + iy_2] = (x_1 - x_2) + i(y_1 - y_2)$.

The most important operation of complex numbers is multiplication, $z = z_1 z_2$, which is calculated directly as

$$z_1 z_2 = [x_1 + iy_1][x_2 + iy_2] = x_1 x_2 + ix_1 y_2 + iy_1 x_2 + i^2 y_1 y_2.$$

Considering the definition, $i^2 = -1$, we obtain the following:

4. $z_1 z_2 = (x_1 x_2 - y_1 y_2) + i(x_1 y_2 + y_1 x_2)$.

Thus, the multiplication of two complex numbers z_1 and z_2 is the complex number $z = x + iy$ with the real and imaginary parts defined by $x = x_1 x_2 - y_1 y_2$ and $y = x_1 y_2 + y_1 x_2$, respectively. When the numbers z_1 and z_2 are real, i.e., $y_1 = 0$ and $y_2 = 0$, this operation is reduced to the multiplication of real numbers, $z_1 z_2 = x_1 x_2$. The set of complex numbers is denoted by C.

Example 1.1

If $z_1 = 1 + 2i$ and $z_2 = 2 - 3i$, then the multiplication proceeds as

$$z_1 z_2 = [2 - 2(-3)] + i[-3 + 2(2)] = 8 + i.$$

The operation of multiplication together with the operations of addition and subtraction is commutative, i.e.,

5. $z_2 z_1 = (x_2 x_1 - y_2 y_1) + i(x_2 y_1 + y_2 x_1) = z_1 z_2$.

It is not difficult to see that for any real numbers k_1 and k_2 and complex numbers z_1, z_2, and z_3, the following holds:

5*a*. $(k_1 z_1)(k_2 z_2) = (k_1 k_2)(z_1 z_2)$,

5*b*. $z_1(z_2 + z_3) = z_1 z_2 + z_1 z_3$.

For a complex number $z_1 = x_1 + iy_1$, the number $z_2 = x_1 - iy_1$ is called *the complex conjugate* and denoted by $\bar{z}_1$. It is clear that $\bar{\bar{z}}_1 = z_1$ for any complex number, and $\bar{z}_1 = z_1$, if only the number z_1 is real. The operation of conjugation $z \to \bar{z}$ is a linear operation, namely,

6. $\overline{z_1 + kz_2} = \bar{z}_1 + k\bar{z}_2,$

for any real number k and complex numbers z_1 and z_2.

The module $|z_1|$ of the complex number z_1 is defined as the multiplication $z_1\bar{z}_1$, which according to multiplication can be written as

$$6a.\quad z_1\bar{z}_1 = (x_1x_1 + y_1y_1) + i(-x_1y_1 + y_1x_1) = x_1^2 + y_1^2$$

and denoted as $|z_1|^2$. Therefore, $|z_1| = \sqrt{x_1^2 + y_1^2}$ and it is positive if $z_1 \neq 0$, and it is clear that $|z_1| \geq |x_1|$ and $|z_1| \geq |y_1|$. The number $|z_1|$ also is called *the length* of z_1.

In the general case, the following holds for the complex conjugate of the multiplication:

$$6b.\quad \overline{z_1z_2} = (x_1x_2 - y_1y_2) - i(x_1y_2 + y_1x_2) = \bar{z}_1\bar{z}_2.$$

The following equalities hold for multiplication:

$$|z_1z_2|^2 = (z_1z_2)\overline{(z_1z_2)} = (z_1z_2)(\bar{z}_1\bar{z}_2) = (z_1\bar{z}_1)(z_2\bar{z}_2) = |z_1|^2|z_2|^2.$$

Therefore, the length of the product of two complex numbers equals the product of their lengths:

$$6c.\quad |z_1z_2| = |z_1||z_2|.$$

Example 1.2

For the complex number $z_1 = 3 + 4i$, the length of the number is

$$|z_1| = \sqrt{3^2 + 4^2} = \sqrt{9 + 16} = 5.$$

The complex conjugate number $\bar{z}_1 = 3 - 4i$ has the same length:

$$|\bar{z}_1| = \sqrt{3^2 + (-4)^2} = \sqrt{9 + 16} = 5.$$

The property of triangle inequality holds for the complex numbers

$$6d.\quad |z_1 + z_2| \leq |z_1| + |z_2|.$$

The equality holds for the cases when one of the complex numbers is zero, or $z_1 = kz_2$ when k is real.

When z_1 is an imaginary number, i.e., $x_1 = 0$, the square of the number is $z_1^2 = (iy_1)^2 = -y_1^2$ and therefore,

$$6e.\quad z_1^2 = -|z_1|^2.$$

Example 1.3

For the complex number $z_1 = 4i$, the length of the number is

$$|z_1| = \sqrt{4^2} = 4, \quad |z_1|^2 = 16, \quad \text{and} \quad z_1^2 = (4i)^2 = 16i^2 = -16.$$

From the equation $z_1\bar{z}_1 = |z_1|^2$, the inverse number $1/z_1$ is defined as

$$7.\quad (z_1)^{-1} = \frac{1}{z_1} = \frac{1}{|z_1|^2}\bar{z}_1, \quad \text{when} \quad z_1 \neq 0.$$

Indeed, one can verify that $(z_1)^{-1}(z_1) = (z_1)(z_1)^{-1} = 1$. Therefore, the operation of division of the complex numbers z_2 and z_1, if only $z_1 \neq 0$, is defined as

$$8.\quad \frac{z_2}{z_1} = z_2(z_1)^{-1} = z_2\frac{1}{z_1} = z_2\frac{1}{|z_1|^2}\bar{z}_1 = \frac{1}{|z_1|^2}z_2\bar{z}_1 = \frac{z_2\bar{z}_1}{|z_1|^2}.$$

Example 1.4

For the complex numbers $z_1 = 3 - 4i$ and $z_2 = 2 + i$, we have the following:

$$\frac{1}{z_1} = \frac{1}{3-4i} = \frac{1}{(3^2+4^2)}\overline{(3-4i)} = \frac{1}{25}(3+4i) = \frac{3}{25} + i\frac{4}{25}$$

and

$$\frac{z_2}{z_1} = \frac{2+i}{3-4i} = \frac{1}{25}(2+i)(3+4i) = \frac{1}{25}(2+11i) = \frac{2}{25} + i\frac{11}{25}.$$

In complex arithmetic [35, 36], we can solve Eqs. (1.1) to (1.3). For that, we first consider the complex numbers z on the imaginary line, i.e., $z = iy$. The solution of equation $z^2 + 1 = 0$, which for such complex numbers is $-y^2 + 1 = 0$, are $z = i(+1) = i$ and $z = i(-1) = -i$. Figure 1.3(a) shows a graph of the parabola $z^2 + 1 = (iy)^2 + 1$, when iy runs the interval of imaginary numbers $[-2i, 2i]$. One can see that this parabola intersects the horizontal at points $+i$ and $-i$. Another parabola $(z-1)^2 + 2$ is shown in Fig. 1.3(b). It should be noted that the graph of this parabola is calculated for the complex numbers $z = 1 + iy$ when the imaginary components y run the interval $[-2, 2]$. Therefore, the plot is shown versus these complex numbers $z = 1 + iy$. This parabola intersects the horizontal line at two points, $z_1 = 1 + i\sqrt{2}$ and $z_2 = 1 - i\sqrt{2}$.

1.1.1.1 Geometry of complex numbers

Every complex number $z = x + iy$ can uniquely be presented as the point (x, y) on the real plane R^2. In fact, this is another form of representing the complex number. We denote this point by $P = P(z)$, which has the Cartesian

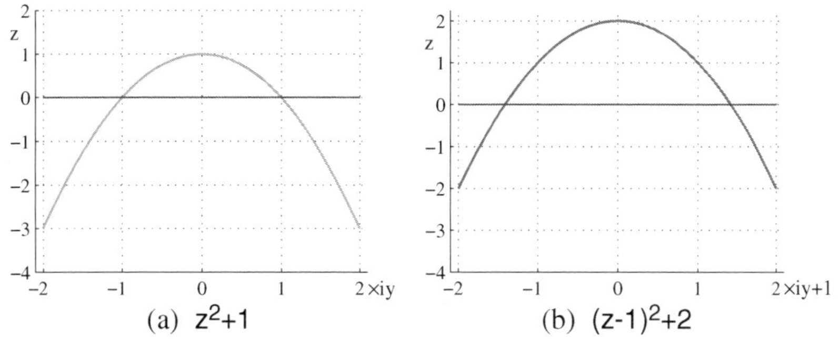

Figure 1.3 Graphs of (a) the parabola z^2+1 and (b) the parabola $(z-1)^2+2$.

coordinates (x, y). The expression $P = x + iy$ relates to complex arithmetic. The distance between the original and P is $d(P) = |z| = \sqrt{x^2+y^2}$. So, the point P lies on the circle of radius $r = d(P)$.

In the specific case where $r = 1$, when point P lies on the unit circle, we can write P as

$$P = P(\theta) = (x, y) = (\cos\,\theta,\ \sin\,\theta) = \cos\,\theta + i\,\sin\,\theta,$$

where the imaginary unit $i^2 = -1$. The cosine and sine functions are calculated by the known Taylor series,

$$\cos(\theta) = 1 - \frac{\theta^2}{2!} + \frac{\theta^4}{4!} - \frac{\theta^6}{6!} + \frac{\theta^8}{8!} - \frac{\theta^{10}}{10!} + \cdots,$$

$$\sin(\theta) = \theta - \frac{\theta^3}{3!} + \frac{\theta^5}{5!} - \frac{\theta^7}{7!} + \frac{\theta^9}{9!} - \frac{\theta^{11}}{11!} + \cdots.$$

Similarly, the point $\bar{P} = P(-\theta)$, which represents the complex conjugate number $\bar{z}$, can be written as

$$\bar{P} = P(-\theta) = (x, -y) = (\cos\,\theta, -\sin\,\theta) = \cos\,\theta - i\,\sin\,\theta.$$

Both points P and $\bar{P}$ are on the same circle of radius $r = d(P) = d(\bar{P})$. Therefore,

$$P(\theta) + P(-\theta) = 2\,\cos\,\theta,$$

$$P(\theta) - P(-\theta) = i2\,\sin\,\theta.$$

The function $P(\theta)$ is denoted by $e^{i\theta}$ in honor of Euler, who used this function and founded the above relations for the real and imaginary parts of $e^{i\theta}$,

$$e^{i\theta} + e^{-i\theta} = 2\,\cos\,\theta, \tag{1.4}$$

$$e^{i\theta} - e^{-i\theta} = i2\,\sin\,\theta. \tag{1.5}$$

The point on the unit circle is completely described by the angle θ. The unit circle intersects the axes at the points

$$1 = i^0, \quad i = i^1, \quad -1 = i^2, \quad -i = (i^2)i = i^3.$$

The points $P(\theta)$ on the unit circle with $\theta = \pi, \pi/2, \pi/3$, and $\pi/4$ are the following:

$$\begin{aligned}
e^{i\pi} &= \cos(\pi) + i\,\sin(\pi) = -1 + i0 = -1,\\
e^{i\pi/2} &= \cos(\pi/2) + i\,\sin(\pi/2) = 0 + i1 = 1i = i,\\
e^{i\pi/3} &= \cos(\pi/3) + i\,\sin(\pi/3) = 0.5 + i\frac{\sqrt{3}}{2} = 0.5 + 0.8660i,\\
e^{i\pi/4} &= \cos(\pi/4) + i\,\sin(\pi/4) = \frac{\sqrt{2}}{2} + i\frac{\sqrt{2}}{2} = 0.7071 + 0.7071i.
\end{aligned}$$

Since $i^{2n} = (-1)^n$ and $i^{2n+1} = (-1)^n i$ for integers $n = 0, 1, 2, \ldots,$ the Taylor series for the complex exponential function $\exp(i\theta)$ can be obtained as

$$\begin{aligned}
\cos(\theta) + i\,\sin(\theta) &= 1 + \frac{(i\theta)^2}{2!} + \frac{(i\theta)^4}{4!} + \frac{(i\theta)^6}{6!} + \frac{(i\theta)^8}{8!} + \frac{(i\theta)^{10}}{10!} + \cdots\\
&\quad + i\theta + \frac{(i\theta)^3}{3!} + \frac{(i\theta)^5}{5!} + \frac{(i\theta)^7}{7!} + \frac{(i\theta)^9}{9!} + \frac{(i\theta)^{11}}{11!} + \cdots =\\
e^{i\theta} &= \sum_{n=0}^{\infty} \frac{(i\theta)^n}{n!}, \quad \text{(considering } 0! = 1).
\end{aligned}$$

In the general case, when point P lies on the circle of radius r, the point can be represented as (r, θ). This is the so-called *polar form* of representation of P (as shown in Fig. 1.4) instead of $P = (x, y) = (r\cos\theta, r\sin\theta)$,

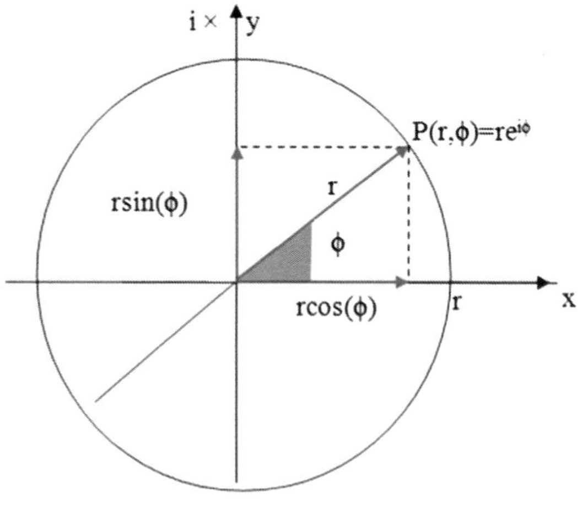

Figure 1.4 Point $P(r, \theta)$ on the complex plane C.

$$P = P(r,\theta) = re^{i\theta},$$

or we can simply write $P = (r,\theta)$. Here, the radius and angle are calculated as

$$r = \sqrt{x^2+y^2}, \quad \text{and} \quad \theta = \arctan\frac{y}{x} = \tan^{-1}\frac{y}{x},$$

and $\theta = \arccos(y) = \pm\pi/2$, if $x = 0$. This angle θ is also called the argument of z and denoted by $\theta = \arg(z)$. If point P lies in the left semi-plane, calculation of angle θ should be corrected by adding angle $\pm\pi$ depending on the sign of the imaginary part y.

Example 1.5

The complex number $z = 3+4i$ as the point $P(z) = (3,4)$ is described in the polar form as $(r,\theta) = (5, 0.9273)$. Indeed,

$$r = \sqrt{3^2+4^2} = 5 \quad \text{and} \quad \theta = \arctan\frac{4}{3} = 0.9273 \text{ rad},$$

or

$$\theta = \frac{180}{\pi} \times \arctan\frac{4}{3} = 53.1301°.$$

The polar form for the complex conjugate $\bar{z} = 3-4i$ is $P(\bar{z}) = (5, \ -0.9273)$,

$$r = \sqrt{3^2+(-4)^2} = 5, \quad \text{and} \quad \theta = \arctan\frac{-4}{3} = -0.9273 \text{ rad},$$

or $\theta = -53.1301°$.

Now, we describe the geometry of the multiplication of two complex numbers $z_1 = x_1 + iy_1$ and $z_2 = x_2 + iy_2$. Considering these numbers as two-dimension vectors

$$z_1 = (x_1, y_1)^T = \begin{pmatrix} x_1 \\ y_1 \end{pmatrix} \quad \text{and} \quad z_2 = (x_2, y_2)^T = \begin{pmatrix} x_2 \\ y_2 \end{pmatrix},$$

the complex number $z = x + iy = z_1 z_2$ can be calculated in vector form as

$$z = \begin{pmatrix} x \\ y \end{pmatrix} = \mathbf{A}\begin{pmatrix} x_2 \\ y_2 \end{pmatrix} = \begin{pmatrix} x_1 & -y_1 \\ y_1 & x_1 \end{pmatrix}\begin{pmatrix} x_2 \\ y_2 \end{pmatrix}.$$

The determinant of the matrix $\mathbf{A}$ is $\det(\mathbf{A}) = x_1^2 + y_1^2$ and we consider the case when $\det(\mathbf{A}) \neq 0$. This matrix multiplication also can be written as

$$z = \begin{pmatrix} x \\ y \end{pmatrix} = \sqrt{x_1^2+y_1^2}\begin{pmatrix} \frac{x_1}{\sqrt{x_1^2+y_1^2}} & -\frac{y_1}{\sqrt{x_1^2+y_1^2}} \\ \frac{y_1}{\sqrt{x_1^2+y_1^2}} & \frac{x_1}{\sqrt{x_1^2+y_1^2}} \end{pmatrix}\begin{pmatrix} x_2 \\ y_2 \end{pmatrix},$$

or

$$z = \begin{pmatrix} x \\ y \end{pmatrix} = |z_1| \begin{pmatrix} \cos\theta_1 & -\sin\theta_1 \\ \sin\theta_1 & \cos\theta_1 \end{pmatrix} \begin{pmatrix} x_2 \\ y_2 \end{pmatrix}.$$

Angle θ_1 is defined from the equations

$$\cos\theta_1 = \frac{x_1}{\sqrt{x_1^2 + y_1^2}} \quad \text{and} \quad \sin\theta_1 = \frac{y_1}{\sqrt{x_1^2 + y_1^2}}.$$

The matrix

$$\mathbf{A} = A(r_1, \theta_1) = r_1 \begin{pmatrix} \cos\theta_1 & -\sin\theta_1 \\ \sin\theta_1 & \cos\theta_1 \end{pmatrix}$$

is the matrix of rotation by angle θ_1 around the circle of radius amplified by a factor of $r_1 = |z_1|$. Thus, when multiplying the complex number z_2 by z_1, the point (x_2, y_2) is rotated by angle θ_1 on a circle of radius $r_2 = |z_2|$ and then is moved along its angle to another circle of radius $r_1 r_2$. The result of multiplication is the same if point (x_2, y_2) is first moved along angle $\theta_2 = \tan^{-1}(y_2/x_2)$ to the circle of radius $r_1 r_2$ and then rotated by angle θ_1 on this new circle.

Example 1.6

Consider the multiplication of two complex numbers $z_1 = 2 + i$ and $z_2 = 3 - 4i$. The product $z = z_1 z_2 = (2 \cdot 3 + 1 \cdot 4) + i(-4 \cdot 2 + 1 \cdot 3) = 10 - 5i$. In this case, $r_1 = \sqrt{2^2 + 1} = \sqrt{5} = 2.2360$, and angle $\theta_1 = 0.4636$ rad (or $26.5651°$). The second complex number is on the circle of radius $r_2 = \sqrt{3^2 + 4^2} = 5$ along the ray at angle $\theta_2 = -0.9273$ rad (or $-53.1301°$) to the horizontal axis. The two points representing complex numbers z_1 and z_2 are shown in Fig. 1.5(a).

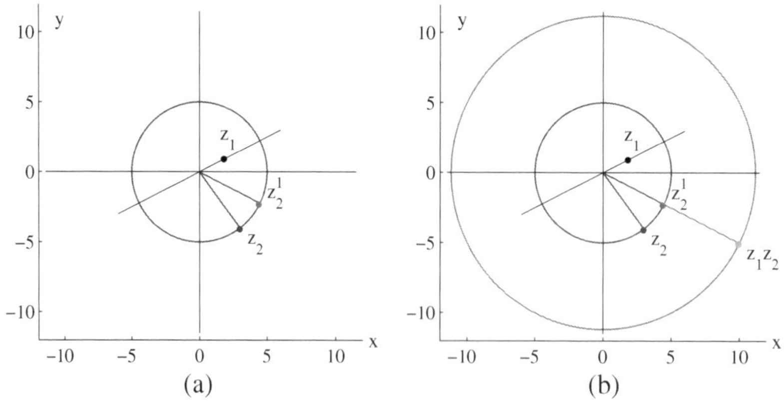

Figure 1.5 Geometry of multiplication of two complex numbers: (a) rotation of the point z_2 and (b) moving to another circle.

The second point with the coordinate $(3, -4)$ is rotated counter-clockwise to the point $(4.4720, -2.2360)$, which is denoted by z_2' in the figure and calculated by

$$\begin{pmatrix} 4.4720 \\ -2.2360 \end{pmatrix} = \begin{pmatrix} \cos\theta_1 & -\sin\theta_1 \\ \sin\theta_1 & \cos\theta_1 \end{pmatrix} \begin{pmatrix} 3 \\ -4 \end{pmatrix} = \begin{pmatrix} 0.8944 & -0.4472 \\ 0.4472 & 0.8944 \end{pmatrix} \begin{pmatrix} 3 \\ -4 \end{pmatrix}.$$

Then, this new point is moved along the angle $\theta = \theta_1 + \theta_2 = -0.4636$ rad (or $-26.5651°$) to the new circle of radius $r = r_1 r_2 = 11.1803$, by multiplying the coordinates of the point by $r_1 = \sqrt{5}$, as shown in Fig. 1.5(b). This new point can be calculated as

$$P(z_1 z_2) = \sqrt{5}\begin{pmatrix} 4.4720 \\ -2.2360 \end{pmatrix} = \begin{pmatrix} 10 \\ -5 \end{pmatrix}.$$

Therefore, $z_1 z_2 = 10 - 5i$ and, indeed, $r = |z_1 z_2| = \sqrt{10^2 + 5^2} = 11.1803$. All numbers in the above calculations are given with a precision of four digits after the point.

Example 1.7

Consider the multiplication of the complex number $z_2 = x_2 + iy_2$ by the complex number $z_1 = x_1 + iy_1 = 1 + i$. In this case, $x_1 = y_1 = 1$ and the complex number $z = z_1 z_2 = x + iy$ can be calculated as

$$z = \begin{pmatrix} x \\ y \end{pmatrix} = \mathbf{A}\begin{pmatrix} x_2 \\ y_2 \end{pmatrix} = \begin{pmatrix} 1 & -1 \\ 1 & 1 \end{pmatrix} \begin{pmatrix} x_2 \\ y_2 \end{pmatrix},$$

with the determinant of the matrix $\mathbf{A}$ equal to $\det(\mathbf{A}) = 2x_1 y_1 = 2$. This binary matrix $\mathbf{A}$ is well-known as a basic 2×2 matrix of the Hadamard and Fourier transformations. In this case, radius $r_1 = \sqrt{2}$ and angle $\theta_1 = \arctan(1) = \pi/4$. The matrix

$$\mathbf{A} = A(\sqrt{2}, \pi/4) = \sqrt{2}\begin{bmatrix} \cos(\pi/4) & -\sin(\pi/4) \\ \sin(\pi/4) & \cos(\pi/4) \end{bmatrix}$$

describes the rotation of point (x_2, y_2) by an angle of $\pi/4$ (in radians) following the translation of the point from the circle of radius $r_2 = \sqrt{x_2^2 + y_2^2}$ to the circle of radius $\sqrt{2} r_2$.

It is not difficult to see that the complex number $z = x + iy = z_1 z_2$ can also be calculated in vector form with another matrix,

$$z = \begin{pmatrix} x \\ y \end{pmatrix} = \mathbf{A}\begin{pmatrix} x_1 \\ y_1 \end{pmatrix} = \begin{pmatrix} x_2 & -y_2 \\ y_2 & x_2 \end{pmatrix} \begin{pmatrix} x_1 \\ y_1 \end{pmatrix}.$$

Here, matrix $\mathbf{A}$ is defined by the second complex number z_2 and can be written as

$$\mathbf{A} = A(r_2, \theta_2) = r_2 \begin{pmatrix} \cos\theta_2 & -\sin\theta_2 \\ \sin\theta_2 & \cos\theta_2 \end{pmatrix},$$

where radius $r_2 = \sqrt{x_2^2 + y_2^2}$ and the angle θ_2 of rotation of point (x_1, y_1) is defined from the equations

$$\cos\theta_2 = \frac{x_2}{\sqrt{x_2^2 + y_2^2}} \quad \text{and} \quad \sin\theta_2 = \frac{y_2}{\sqrt{x_2^2 + y_2^2}}.$$

In the multiplication $z_1 z_2$ by this matrix, point (x_1, y_1) can be rotated by angle θ_2 on a circle of radius $r_1 = |z_1|$ and then moved to another circle of radius $r_2 r_1$.

1.1.2 Quaternion numbers

The operation of multiplication in complex arithmetic allows for division by nonzero numbers. As shown in the previous sub-section, division is possible because of the unique inverse complex number $1/z$, when $z \neq 0$. If we try to generalize the concept of complex numbers and define a new arithmetic, the operation of multiplication must guarantee the existence of such inverse numbers. The operations of multiplication and division are dual, meaning that for any two complex numbers $z_1, z_2 \neq 0$,

$$\frac{z_1}{z_2} = \frac{1}{\frac{1}{z_1} z_2}.$$

Let C be the complex plane with numbers that we denote by $z = a + ib$, where a and b are real numbers, and i is the imaginary unit. We may consider a different imaginary unit j and new numbers q in the form of

$$q = z + jc = (a + ib) + jc = a + bi + jc,$$

where c is real. It is well known that for such numbers with three components (a, b, c), it is not possible to define the multiplication together with division. For example, consider the simple imaginary number $q = i = 0 + 1i + 0j$ and its inverse $q^{-1} = (a + ib + jc)$, if it exists, from the equation

$$(0 + 1i + 0j)(a + ib + jc) = 1.$$

We can write $1 = i(a + ib + jc) = ia + i^2 b + ijc = ia - b + ijc$, or $ia + ijc = 1 + b$ and, therefore, $a = 0$ and $ijc = 1 + b$. If we multiply both parts of this equality by i from the left, we obtain $i(ijc) = i(1 + b)$, or $-jc = i(1 + b)$. This is possible only if $j = -i(1 + b)/c$, but j should be a different imaginary unit, not proportional to i. Thus, the inverse number to i does not exist in the arithmetic of numbers $a + ib + jc$.

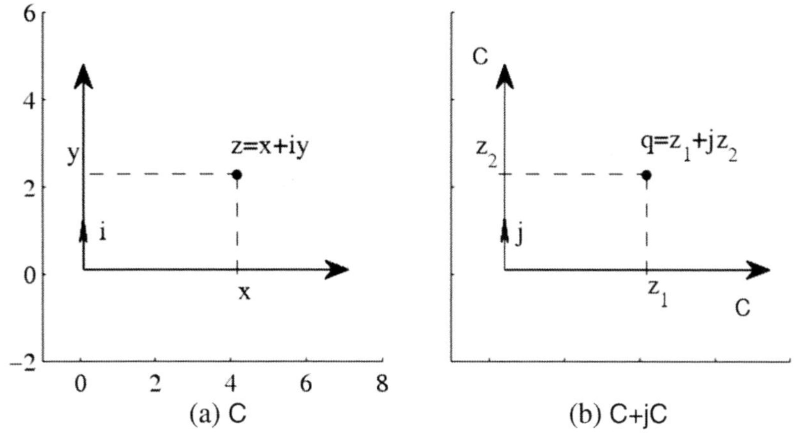

Figure 1.6 (a) The complex plane C and (b) the abstract combination of two complex spaces.

The complex numbers $z = x + iy$ represent the points (x, y) in the complex space C. Figure 1.6(a) shows such a plane. As shown above, adding one line to the complex plane does not result in a full arithmetic with multiplication and division. Instead, we may think of "adding" similar complex planes, formally writing this operation as $C + jC$, as shown in Fig. 1.6(b). For that, we can imagine two complex planes; one complex plane C with numbers $z_1 = x_1 + iy_1$ and another complex plane C with numbers $z_2 = x_2 + iy_2$. If we assume another imaginary unit j, then the following numbers can be considered:

$$q = z_1 + jz_2 = (x_1 + iy_1) + j(x_2 + iy_2). \tag{1.6}$$

We can denote such a doubled complex space by C^2 and call such representation of numbers q *the* $(2, 2)$*-representation.*

Thus, the complex numbers z_1 and z_2 in this construction play the same role as the real numbers x and y in the complex numbers $x + iy$. These new numbers q can be written as

$$q = x_1 + iy_1 + jx_2 + (ji)y_2,$$

where the number (ji) should represent another number, or may be a new imaginary number unit, which will be denoted by k or $-k$, i.e., $k = ji$ or $-k = ji$. These numbers as elements of four (or "quaternion" in Latin) are called *quaternions* and were first described by the Irish mathematician William Hamilton (1805–1865) in Ref. [6].

1.1.3 Rules for multiplication

The new imaginary number j has been considered, and a new number k has been defined as $k = ji$. Now, we need to describe the multiplications among the numbers i, j, and k. We assume that the operation of multiplication of numbers q is associative, i.e., $(q_1 q_2)q_3 = q_1(q_2 q_3)$, for any numbers q_1, q_2, and q_3 of type

$$q = x_1 + iy_1 + jx_2 + ky_2. \tag{1.7}$$

Also we assume that all such numbers q are commutative with real numbers λ, i.e., $q\lambda = \lambda q$.

It is important to note that the multiplication of such numbers q is not commutative, i.e., $q_1 q_2 \neq q_1 q_2$, for many quaternions where $q_1 \neq q_2$. Indeed, we consider the number $q = 1 + i + j$. Then, the following calculations hold:

$$\begin{aligned} qi - qj &= (1 + i + j)i - (1 + i + j)j = (i + i^2 + ji) - (j + ij + j^2) \\ &= (i - 1 + ji) - (j + ij - 1) = i - j + ji - ij, \end{aligned}$$

or

$$q(i - j) = (i - j) + (ji - ij).$$

If $ij = ji$, then $q = 1$, which is not true. Thus, $ij \neq ji$, and it can similarly be shown that $jk \neq kj$ and $ik \neq ki$.

The following properties can also easily be obtained:

Since $j^2 = -1$,

$$\begin{aligned} jk &= j(ji) = (j^2)i = -i, \\ jki &= j(ji)i = (jj)ii = -i^2 = 1, \\ ijk &= ij(ji) = i(j^2)i = -i^2 = 1, \\ ki &= (ji)i = j(i^2) = -j. \end{aligned}$$

It is also true that

$$(ki)^2 = -1 \quad \text{and} \quad (kj)^2 = -1,$$

since

$$\begin{aligned} (ki)(ki) &= (ji)i(ji)i = j(ii)j(ii) = j(-1)j(-1) = jj = -1 \\ (kj)(kj) &= (ji)j(ji)j = (ji)jj(ij) = -(ji)(ij) = -j(ii)j = jj = -1. \end{aligned}$$

It is clear that these equalities do not describe all rules of multiplication from the left and right between the numbers i, j, and k. Since we have not yet defined the multiplication ij, this system of equalities is incomplete; other missing multiplications are ik, kj, and k^2.

We know that $ij \neq ji$, therefore, we can assume the following condition:

$$ij = -ji, \quad \text{or} \quad ij = -k.$$

Then, we obtain the following equalities:

$$k^2 = kk = (ji)(ji) = (-ij)(ji) = -i(jj)i = ii = -1,$$
$$kj = (ji)j = -(ij)j = -i(jj) = i,$$
$$ik = i(ji) = i(-ij) = -(ii)j = j.$$

It is also true that one of these equalities results in the equality $ij = -ji$. For instance, let $k^2 = -1$; then we obtain the following:

$$[jiji = -1] \to [j \cdot jiji = -j] \to [(jj)iji = -j] \to [iji = j] \to [-ji = ij].$$

Thus, under the assumption that $ij = -ji$, we obtain the following full set of multiplication laws:

$$ij = -ji = -k, \quad jk = -kj = -i,$$
$$ki = -ik = -j, \quad i^2 = j^2 = k^2 = kji = jik = -1.$$

Together with multiplication by the real unit number 1, all of these basic multiplications are shown in Table 1.1, which is referred to as the $T(1, j, i, k)$ table; the order of imaginary units j, i, k shows that if we periodically extend this triplet as j, i, k, j, i, then the result of multiplication of each of the two numbers from left to right (right to left) is the next number with the plus (minus) sign.

It should be noted that such a table of multiplications is not unique, and we can use other tables. For instance, if we start with the condition $ji = -k$, we obtain the similar table $T(1, j, i, k)$, which is shown in Table 1.2.

Table 1.1 Table $T(1,j,i,k)$ of the multiplications of the four quaternion unit numbers.

	1	i	j	k
1	1	i	j	k
i	i	-1	$-k$	j
j	j	k	-1	$-i$
k	k	$-j$	i	-1

Table 1.2 Table $T(1,i,j,k)$ of the multiplications of the four quaternion unit numbers

	1	i	j	k
1	1	i	j	k
i	i	-1	k	$-j$
j	j	$-k$	-1	i
k	k	j	$-i$	-1

The three imaginary units i, j, and k satisfy the following multiplication laws:

$$\begin{aligned} ij &= -ji = k, \quad jk = -kj = i, \\ ki &= -ik = j, \quad i^2 = j^2 = k^2 = ijk = -1. \end{aligned} \tag{1.8}$$

In these two tables, the unit i is changed by j and j is changed by i. We will stand on $T(1, i, j, k)$ since this table is most often used in quaternion arithmetic. The multiplication of quaternion numbers, which is based on this table of rules, is called *the left multiplication* of quaternions.

1.1.4 Basic operations

We consider the arithmetical operations with quaternion numbers. Thus, a quaternion number is represented as

$$q = a + bi + cj + dk = a + (bi + cj + dk),$$

where a, b, c, and d are real numbers. The number a is called the real part of q, and $(bi + cj + dk)$ is the "imaginary" part of q. The quaternion q has a three-component imaginary part, which we denote by q', such that

$$q = a + q' = a + (bi + cj + dk) = a + (ib + jc + kd).$$

Other notations for the real and imaginary components are Sq and Vq, respectively, i.e., $Sq = a$, $Vq = q'$, and $q = Sq + Vq$.

In the cases when $q' = ib, jc$, and kd, the numbers $q = a + ib$, $q = a + jc$, and $q = a + kd$ can be considered as complex numbers. The zero quaternion is $0 + 0i + 0j + 0k = 0$, and the unity element is $1 + 0i + 0j + 0k = 1$.

Now, we describe the arithmetic of quaternion numbers, i.e., the main operations that include the sum, multiplication by a real constant, and multiplication and division of quaternion numbers.

The sum of two quaternions $q_1 = a_1 + q_1' = a_1 + (ib_1 + jc_1 + kd_1)$ and $q_2 = a_2 + q_2' = a_2 + (ib_2 + jc_2 + kd_2)$ is defined as

1. $q = q_1 + q_2 = a + q' = (a_1 + a_2) + (q_1' + q_2') = a + (bi + cj + dk),$

and $b = b_1 + b_2$, $c = c_1 + c_2$, and $d = d_1 + d_2$. Therefore, this operation is defined to be component-wise and commutative:

1*a*. $q = q_1 + q_2 = q_2 + q_1.$

For any real number λ,

2. $\lambda q = \lambda a + \lambda q' = \lambda a + (i\lambda b + j\lambda c + k\lambda d).$

If $a = 0$, the quaternion q is called a pure quaternion number:

$$q = 0 + (ib + jc + kd).$$

The multiplication q_1q_2 of two quaternions q_1 and q_2 is calculated according to Table 1.2 of multiplication as

3. $q = q_1q_2 = (a_1 + q_1')(a_2 + q_2') = a_1a_2 + a_1q_2' + a_2q_1' + q_1'q_2'$

and can be expanded as

$$\begin{aligned} q &= a_1a_2 + a_1q_2' + a_2q_1' + (ib_1 + jc_1 + kd_1)(ib_2 + jc_2 + kd_2) \\ &= a_1q_2' + a_2q_1' + a_1a_2 - (b_1b_2 + c_1c_2 + d_1d_2) \\ &\quad + i(c_1d_2 - d_1c_2) - j(b_1d_2 - d_1b_2) + k(b_1c_2 - c_1b_2). \end{aligned}$$

This multiplication $q = q_1q_2$ can also be written as

$$q = [a_1q_2' + a_2q_1'] + a_1a_2 - (b_1b_2 + c_1c_2 + d_1d_2) + \begin{vmatrix} i & j & k \\ b_1 & c_1 & d_1 \\ b_2 & c_2 & d_2 \end{vmatrix}. \tag{1.9}$$

Example 1.8

Consider two quaternions $q_1 = 1 - 2i + 3j + k$ and $q_2 = 1 - i + 4j + 3k$. The multiplication $q = q_1q_2$ is calculated as follows.

(1). The imaginary components of these numbers are $q_1' = -2i + 3j + k$ and $q_2' = -i + 4j + 3k$; therefore,

$$a_1q_2' + a_2q_1' = 1 \cdot (-2i + 3j + k) + 1 \cdot (-i + 4j + 3k) = -3i + 7j + 4k.$$

(2). The real part of the multiplication is

$$a_1a_2 - (b_1b_2 + c_1c_2 + d_1d_2) = 1 - [-2(-1) + 3(4) + 1(3)] = 1 - 17 = -16.$$

(3). The quaternion determinant equals

$$\begin{vmatrix} i & j & k \\ -2 & 3 & 1 \\ -1 & 4 & 3 \end{vmatrix} = i(9 - 4) - j(-6 + 1) + k(-8 + 3) = 5i + 5j - 5k.$$

(4). Therefore, the quaternion number $q = q_1q_2$ equals

$$q = q_1q_2 = (-3i + 7j + 4k) - 16 + (5i + 5j - 5k) = -16 + (-2i + 12j - k).$$

It is not difficult to check that the multiplication of these two numbers is not commutative:

$$q_1q_2 \neq q_2q_1 = -16 + (-8i + 2j + 9k).$$

Below is the script of the MATLAB®-based code "test_example.m" for this example. It uses function "mult2qs_direct" to multiply two quaternions.

```
%------------------------------------------------------------
% test_example.m / with function mult2qs_direct(q1,q2)
% multiplication of two quaternion numbers
   q1=[1,-2,3,1]; q2=[1,-1,4,3];
   q=mult2qs_direct(q1,q2);       % -16 2 12 -1
%------------------------------------------------------------
 function q = mult2qs_direct(q1,q2)
   a1=q1(1); b1=q1(2); c1=q1(3); d1=q1(4);
   a2=q2(1); b2=q2(2); c2=q2(3); d2=q2(4);
   Q1=q1(2:4); Q2=q2(2:4);  % imaginary parts
%
   Q12=a1*Q2 + a2*Q1;           % -3 7 4
   a=a1*a2-Q1*Q2';              % 1-17=-16
   bi=(c1*d2-d1*c2);            % 5
   cj=-(b1*d2-d1*b2);           % 5
   dk=(b1*c2-c1*b2);            % -5
%
   b=Q12(1) + bi;               % 2
   c=Q12(2) + cj;               % 12
   d=Q12(3) + dk;               % -1
   q=[a,b,c,d];                 % -16 2 12 -1
%------------------------------------------------------------
```

4. When q_1 and q_2 are the quaternions with only one and the same imaginary units, for instance, $q_1 = a_1 + jc_1$ and $q_2 = a_2 + jc_2$, the operation of multiplication q_1q_2 is complex multiplication. Indeed, the following holds:

$$q_1' = jc_1, q_2' = jc_2, \quad q_1' \times q_1' = \begin{vmatrix} i & j & k \\ 0 & c_1 & 0 \\ 0 & c_2 & 0 \end{vmatrix} = 0,$$

and

$$q_1q_2 = [a_1(jc_2) + a_2(jc_1)] + a_1a_2 - c_1c_2 = (a_1a_2 - c_1c_2) + j(a_1c_2 + a_2c_1).$$

If $q_1' = 0$ and $q_2' = 0$, then

5. $q_1q_2 = (a_1 + q_1')(a_2 + q_2') = a_1a_2.$

For any real numbers λ_1 and λ_2,

6. $(\lambda_1 q_1)(\lambda_2 q_2) = \lambda_1 \lambda_2 (q_1 q_2)$.

The quaternion multiplication is distributive:

7. $q_1(q_2 + q_3) = q_1 q_2 + q_1 q_3$ and $(q_1 + q_2) q_3 = q_1 q_3 + q_2 q_3$,

and associative by multiplication:

8. $(q_1 q_2) q_3 = q_1 (q_2 q_3)$.

1.1.5 Properties of multiplication of quaternions

In this section we consider a few properties of the operation of multiplication in quaternion arithmetic.

P1. It is important to mention that the property of commutativity does not hold in quaternion algebra; i.e., for quaternions $q_1 = a_1 + q_1'$ and $q_2 = a_2 + q_2'$,

$$q_1 q_2 = q_2 q_1 \quad \text{or} \quad q_1 q_2 \neq q_2 q_1,$$

because of the imaginary units, $ij = -k \neq ji = k$, and so on. It can also be seen from Eq. (1.9) that

$$q_2 q_1 = [a_2 q_1' + a_1 q_2'] + a_2 a_1 - (b_2 b_1 + c_2 c_1 + d_2 d_1) + \begin{vmatrix} i & j & k \\ b_2 & c_2 & d_2 \\ b_1 & c_1 & d_1 \end{vmatrix}$$

and therefore,

$$q_1 q_2 - q_2 q_1 = 2 \begin{vmatrix} i & j & k \\ b_1 & c_1 & d_1 \\ b_2 & c_2 & d_2 \end{vmatrix}. \tag{1.10}$$

The multiplication of a quaternion number q_1 by a real number λ is commutative, i.e., $\lambda q_1 = q_1 \lambda$.

When the quaternion vectors $\mathbf{q}_1'$ and $\mathbf{q}_2'$ are similar, i.e., $\mathbf{q}_2' = \lambda \mathbf{q}_1'$, where λ is a real number, then

$$q_2 q_1 = q_1 q_2 = (a_2 + a_1 \lambda) q_1' + a_2 a_1 - \lambda (b_1^2 + c_1^2 + d_1^2).$$

P2. In the case when $q_1 = q_2$, we obtain the following for the square of the quaternion:

$$q = q_1^2 = [a_1^2 - (b_1^2 + c_1^2 + d_1^2)] + 2 a_1 q_1'. \tag{1.11}$$

The real part of the square is the number $a = a_1^2 - (b_1^2 + c_1^2 + d_1^2)$, and the imaginary part is $2 a_1 q_1'$. If q_1 is a pure quaternion, i.e., $a_1 = 0$, then

$$q_1^2 = (q_1')^2 = -(b_1^2 + c_1^2 + d_1^2) \leq 0. \tag{1.12}$$

Thus, for a quaternion number $q = a + q' = a + (ib + jc + kd)$,

$$q^2 = a^2 - |q'|^2 + 2aq', \tag{1.13}$$

and therefore, the equations of type $q^2 = 4 > 0$ cannot be solved, if $q' \neq 0$. The equations of type $q^2 = -4 < 0$ can be solved in quaternion numbers, since in this case $a = 0$ and we have the condition $q^2 = -|q'|^2 = -4$, or $|q'|^2 = 4$. The number $q = q'$ as a point (b, c, d) is on the sphere of radius 2 in three-dimensional (3-D) space. Such points on this sphere are, for instance, $(2, 0, 0), (0, 2, 0), (0, 0, -2), (\sqrt{2}, \sqrt{2}, 0)$, and $(1, 1, \sqrt{2})$. In other words, the solutions of the equation $q^2 = -4$ include the quaternion numbers $q = 2i + 0j + 0k$, $0i + 2j + 0k$, $0i + 0j - 2k$, $\sqrt{2}(i + j) + 0k$, and $i + j + \sqrt{2}k$.

Example 1.9

Consider the solution of the equation

$$q_1^2 = 1 + 2i + 4j.$$

It directly follows from Eq. (1.11) that the following system of equations should be considered:

$$\begin{cases} a_1^2 - (b_1^2 + c_1^2 + d_1^2) = 1, \\ a_1 q_1' = i + 2j. \end{cases}$$

Therefore, $d_1 = 0$, and this system of equations is reduced to the following one:

$$\begin{cases} a_1^2 - (b_1^2 + c_1^2) = 1, \\ a_1 b_1 = 1, \\ a_1 c_1 = 2, \end{cases}$$

and, therefore, $q = a_1 + (1/a_1)i + (2/a_1)j$. To find a_1, we need to solve the equation

$$a_1^2 - \left(\frac{1}{a_1^2} + \frac{4}{a_1^2}\right) = 1.$$

Two solutions of this equation $(a_1^2)^2 - (a_1^2) - 5 = 0$ are

$$a_1^2 = \frac{1 \pm \sqrt{1 + 20}}{2} = \frac{1 \pm \sqrt{21}}{2},$$

and since we are looking for real a_1, we obtain

$$a_1^2 = \frac{1+\sqrt{21}}{2}, \quad a_1 = \pm\sqrt{\frac{1+\sqrt{21}}{2}}.$$

Thus, the solutions of the above equation are

$$q = a_1 + \frac{i}{a_1} + \frac{2j}{a_1} = a_1 + \frac{1}{a_1}(i+2j)$$
$$= \pm\left[\sqrt{\frac{1+\sqrt{21}}{2}} + \frac{\sqrt{2}}{1+\sqrt{21}}(i+2j)\right] = \pm[1.6707 + 0.5985(i+2j)].$$

If the square q_1^2 is a real number, then $2a_1q_1' = 0$, which means that $a_1 = 0$ or $q_1' = 0$. In the first case, the square is negative:

$$q = q_1^2 = -(b_1^2 + c_1^2 + d_1^2) < 0.$$

The square of any pure quaternion is a negative number. In the second case, we have $q = a_1^2 > 0$. The square of the quaternion number is positive only if the number is real.

Example 1.10

Consider the first equation of this chapter in quaternion arithmetic:

$$q^2 + 1 = 0. \tag{1.14}$$

For a solution $q = a + q' = a + (ib + jc + kd)$ of this equation, we obtain the following:

$$a^2 - |q'|^2 + 1 + 2aq' = 0,$$

where $|q'|^2 = b^2 + c^2 + d^2$. The imaginary part of the expression of this equation is zero; therefore, $a = 0$ and the equation is reduced to the equation $|q'|^2 = 1$. All solutions of this equation $q = q' = (ib + jc + kd)$ as points (b, c, d) are on the unit sphere $b^2 + c^2 + d^2 = 1$. Thus, the solutions of Eq. (1.14) are pure unit quaternion numbers. The case when $c = d = 0$ corresponds to the complex solutions $q = i$ and $q = -i$.

The solutions of the general equation

$$q^2 + y^2 = 0, \tag{1.15}$$

where y is a positive number, can be obtained from the solutions of Eq. (1.14),

$$\left(\frac{q}{y}\right)^2 + 1 = 0.$$

Thus, $q/y = ib + jc + kd$, where $b^2 + c^2 + d^2 = 1$, or we can write $q = iyb + jyc + kyd$. The solutions q are pure quaternions with length $|q| = y$. Presenting such quaternions as points (yb, yc, yd), they are in the sphere of radius y.

Example 1.11

Consider all solutions $q = a + q'$ of the equation

$$q^2 - q + 2 = 0. \tag{1.16}$$

This equation can be written as

$$[2aq' + a^2 + (q')^2] - (a + q') + 2 = 0,$$

or

$$[a^2 - a + (q')^2 + 2] + q'(2a - 1) = 0,$$

and it can reduced to the following system of equations:

$$\begin{cases} q'(2a - 1) = 0, \\ a^2 - a + (q')^2 + 2 = 0. \end{cases}$$

If $q' = 0$, then we do not get a real solution from the equation $a^2 - a + 2 = 0$. Therefore, we consider the $a = 1/2$ case, for which we obtain the condition

$$b^2 + c^2 + d^2 = -(q')^2 = a^2 - a + 2 = \frac{1}{4} - \frac{1}{2} + 2 = \frac{7}{4}.$$

Thus, a solution q in Eq. (1.16) is of the form $q = 1/2 + q'$, where point $q' = (b, c, d)$ is on a sphere of radius $r = \sqrt{7}/2$. Such solutions are, for instance,

$$q_1 = \frac{1}{2} + i\frac{\sqrt{7}}{2}, q_2 = \frac{1}{2} - j\frac{\sqrt{7}}{2}, q_3 = \frac{1}{2} - i + j\frac{\sqrt{2}}{2} + k\frac{1}{2}, q_4 = \frac{1}{2} + (i - j)\frac{\sqrt{2}}{2} + k\frac{\sqrt{3}}{2}.$$

The numbers q_1 and q_2 are only solutions of the equation in complex arithmetic, and q_3 and q_4 are full quaternion numbers. A sphere with the points marked q_2, q_3, and q_4 is shown in Fig. 1.7. This sphere contains all solutions of the equation $q^2 - q + 2 = 0$, which are pure quaternions.

Example 1.12 *Equation of the golden ratio*

Consider all solutions $q = a + q'$ of the following equation:

$$q^2 - q - 1 = 0. \tag{1.17}$$

This equation can be written as

$$[a^2 - a + (q')^2 - 1] + q'(2a - 1) = 0,$$

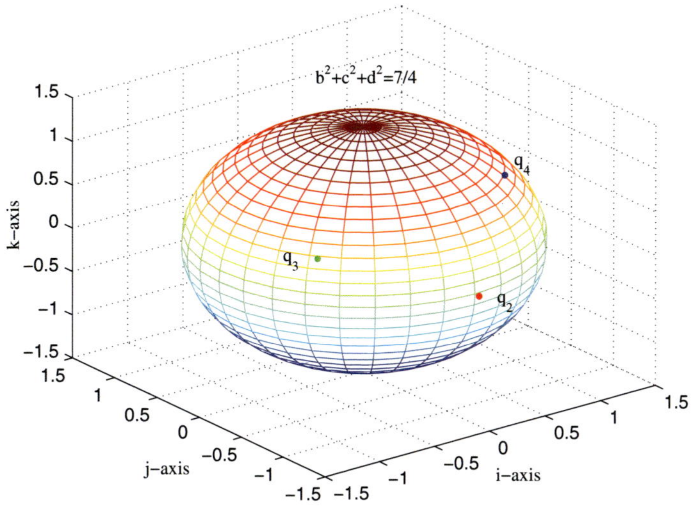

Figure 1.7 A sphere of radius $\sqrt{7}/2$ in the space of quaternion vectors.

and it can reduced to the following system of equations:

$$\begin{cases} q'(2a-1)=0, \\ a^2-a+(q')^2-1=0. \end{cases}$$

We consider two cases, when $q'=0$ and when $q'\neq 0$.

In the first case, we obtain two real solutions of the equation $a^2-a-1=0$:

$$a_1=\Phi=\frac{1+\sqrt{5}}{2} \quad \text{and} \quad a_2=\varphi=\frac{1-\sqrt{5}}{2}.$$

The number $\Phi=1.6180339887498948\ldots$ is known as *the golden ratio*, or *the golden mean*. The second solution is $\varphi=1-\Phi=-0.6180339887498948\ldots$.

Figure 1.8 illustrates these two solutions as the points of intersection of the line $y=x+1$ with the parabola $y=x^2$. The first point (from the left) of the intersection is $\varphi=1-\Phi$, and the second point is Φ.

In the second case when $a=1/2$, we obtain the following "condition" for the imaginary part of q to be considered:

$$|q'|^2=-(q')^2=a^2-a-1=\frac{1}{4}-\frac{1}{2}-1=-\frac{5}{4}.$$

A length of the vector cannot be negative.

Thus, in 3-D space of pure quaternions, the equation $q^2-q-1=0$ is unsolvable. The only solutions of an equation such as $a^2-a-1=0$ are in the real line, and these solutions are related to the golden ratio. This example

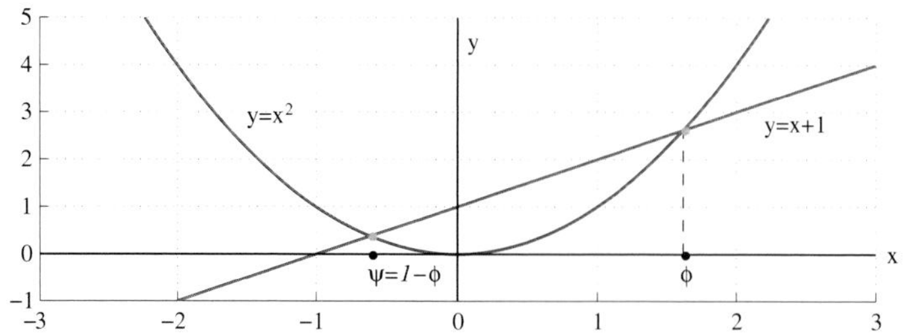

Figure 1.8 The line $y = x + 1$ and the parabola $y = x^2$ in the real plane R^2.

differs from Example 1.11, where the consideration of the equation in quaternion arithmetic results in not only known complex solutions but the entire sphere of radius $\sqrt{7}/2$. The golden ratio equation is unique in the sense that it is solvable only in the real line.

As is well known, the golden ratio had originally been defined as a ratio of two real numbers b and a (when $b > a$), such that (see Fig. 1.9)

$$\frac{b+a}{b} = \frac{b}{a}, \tag{1.18}$$

or $(b+a)b^{-1} = ba^{-1}$. Therefore, we can write this condition as $1 + ab^{-1} = ba^{-1}$, or $ba^{-1} - ab^{-1} = 1$. After denoting $x = ba^{-1}$, we obtain the equation $x - x^{-1} = 1$, or $x^2 - x - 1 = 0$.

In quaternion arithmetic, the ratios $(b+a)/b$ and b/a in Eq. (1.18) can be considered as divisions from the left or divisions from the right. In other words, instead of one equation in the real case, we have two equations for quaternion numbers,

$$\begin{aligned} (b+a)b^{-1} &= ba^{-1} \quad \text{or} \quad ba^{-1} - ab^{-1} = 1, \\ b^{-1}(b+a) &= a^{-1}b, \quad \text{or} \quad a^{-1}b - b^{-1}a = 1, \end{aligned}$$

which are different, since the operation of multiplication of quaternions is not commutative. In both cases, these equations can be reduced to $q^2 - q - 1 = 0$, when $q = ba^{-1}$ and $a^{-1}b$, respectively.

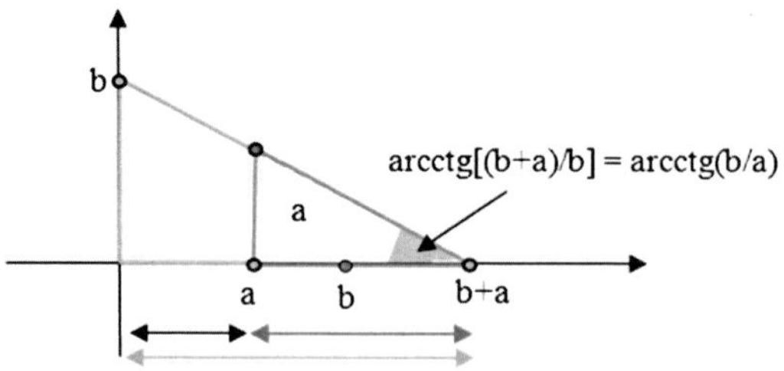

Figure 1.9 Triangles with the golden ratio property.

We consider the two quaternions $q_1 = a_1 + ib_1 + jc_1 + kd_1$ and $q_2 = a_2 + ib_2 + jc_2 + kd_2$ to be in the golden ratio if all components of these numbers have the golden ratio. Namely, let $q_2 > q_1$, meaning that $a_2 > a_1$, $b_2 > b_1$, $c_2 > c_1$, and $d_2 > d_1$. We assume the golden ratio for the components

$$\frac{a_2}{a_1} = \frac{b_2}{b_1} = \frac{c_2}{c_1} = \frac{d_2}{d_1} = \Phi.$$

Then,

$$\frac{q_2 + q_1}{q_2} = \frac{(1+\Phi)q_1}{\Phi q_1} = \frac{(1+\Phi)}{\Phi} = \Phi = \frac{q_2}{q_1},$$

since $1 + \Phi = \Phi^2$.

It should be noted that the golden ratio was first introduced for the real numbers a and b, and the ratio itself concerns the ratio of the lengths $a + b$, b, and a, which are real and positive. This means that we should apply this concept to the lengths of the numbers. The quaternion number $q = a + q'$ has the real and imaginary parts with lengths $|a|$ and $|q'|$, respectively. The length of the quaternion is $|q|$, and we know that $|q| \leq |a| + |q'|$. Let us assume that $|a| < |q'|$ and these two parts are in the golden ratio, i.e., $|q'| = \Phi|a|$. Then,

$$|q| = \sqrt{a^2 + |q'|^2} = \sqrt{(1+\Phi^2)a^2} = |a|\sqrt{1+\Phi^2} = |a|\sqrt{2+\Phi}.$$

Thus, for such a quaternion q, the imaginary part q' can be considered as a point (b, c, d) on a sphere of radius $\Phi|a|$:

$$b^2 + c^2 + d^2 = \Phi^2 a^2, \quad \text{or} \quad \left(\frac{b}{a}\right)^2 + \left(\frac{c}{a}\right)^2 + \left(\frac{d}{a}\right)^2 = \Phi^2.$$

The number $\Phi^2 = \Phi + 1 = 2.6180339887498948\ldots$. The point $(b/a, c/a, d/a)$ is on the sphere of the radius Φ. If the quaternion q is a complex number, i.e., $q' = ib$, then this condition can be written as $b = \pm a\Phi$. Thus,

$$q = a + ia\Phi = a(1+i)\Phi, \quad \text{or} \quad q = a - ia\Phi = a(1-i)\Phi.$$

If $|a| > |q'|$ and these two parts are in the golden ratio, i.e., $|a|/|q'| = \Phi$,

$$|q| = \sqrt{a^2 + |q'|^2} = \sqrt{(1+\varphi^2)a^2} = |a|\sqrt{1+\varphi^2} = |a|\sqrt{2+\varphi}.$$

Here, we consider $1/\Phi = -\varphi$. Thus, the imaginary part q' of the quaternion q can be considered as point (b, c, d) on a sphere of radius $|a\varphi|$:

$$b^2 + c^2 + d^2 = a^2\varphi^2, \quad \text{or} \quad \left(\frac{b}{a}\right)^2 + \left(\frac{c}{a}\right)^2 + \left(\frac{d}{a}\right)^2 = \varphi^2,$$

where $\varphi^2 = \varphi + 1 = 0.3819660112501052\ldots$.

Now, we generalize the results of the above two examples. When considering the polynomial equation

$$q^2 + 2\lambda q + v = 0 \tag{1.19}$$

with real coefficients λ and v, we can separate the two cases when the discriminant of the equation $\Delta = (\lambda^2 - v)$ is negative and non-negative.

Case $\Delta < 0$: The equation

$$z^2 + 2\lambda z + v = 0$$

has two solutions,

$$z_{1,2} = -\lambda \pm \sqrt{\lambda^2 - v} = -\lambda \pm i\sqrt{v - \lambda^2},$$

where i is the complex imaginary unit. The transformation of this equation into the space of quaternion numbers enlarges the set of solutions. Indeed, Eq. (1.19) can be written as

$$q^2 + 2\lambda q + v = (a^2 - |q'|^2 + 2aq') + 2\lambda(a + q') + v = 0$$

and reduced to the system of equations

$$\begin{cases} a^2 - |q'|^2 + 2a\lambda + v = 0 \\ (a + \lambda)q' = 0. \end{cases} \tag{1.20}$$

The $q' = 0$ case is not considered since the equation $a^2 + 2a\lambda + v = 0$ is not solvable in real numbers. Therefore, we consider the $a + \lambda = 0$ case, i.e., when $a = -\lambda$. From the first equation of Eq. (1.20), we obtain

$$|q'|^2 = a^2 + 2a\lambda + v = \lambda^2 - 2\lambda^2 + v = v - \lambda^2 = -\Delta > 0.$$

Thus, the solutions of Eq. (1.19) are all quaternion numbers of type

$$q = -\lambda + q', \quad \text{such that} \quad |q'| = \sqrt{-\Delta}.$$

The imaginary parts of these solutions are on a 3-D sphere of radius $r = \sqrt{-\Delta}$. In Example 7, which concerns the equation $q^2 - q + 2 = 0$, the numbers $\lambda = -1/2$ and $v = 2$. Therefore, $-\Delta = v - \lambda^2 = 2 - 1/4 = 7/4$, and $r = \sqrt{7}/2$.

Case $\Delta \geq 0$: The equation

$$x^2 + 2\lambda x + v = 0$$

has the real solutions

$$x_{1,2} = -\lambda \pm \sqrt{\lambda^2 - v} = -\lambda \pm \sqrt{\Delta}.$$

These two solutions relate to the $q' = 0$ case in Eq. (1.20). The condition $q' \neq 0$, or $a + \lambda = 0$, leads to the equality $|q'|^2 = v - \lambda^2 = -\Delta \leq 0$, which may only be satisfied for $q' = 0$. Thus, as in Example 9, quaternion arithmetic does not give a new solution of the quadratic equation that is solvable in real numbers.

P3. The quaternion conjugate of $q = a + q' = a + (bi + cj + dk)$ is the quaternion

$$\bar{q} = a - (bi + cj + dk),$$

such that

$$q\bar{q} = [a + (bi + cj + dk)][a - (bi + cj + dk)] = a^2 + b^2 + c^2 + d^2 \geq 0.$$

The operation of complex conjugate is linear, i.e.,

$$\overline{(q_1 + q_2)} = \bar{q}_1 + \bar{q}_2.$$

Indeed, the following calculations hold: $q_1 + q_2 = (a_1 + q_1') + (a_2 + q_2') = (a_1 + a_2) + (q_1' + q_2')$; therefore,

$$\overline{q_1 + q_2} = (a_1 + a_2) - (q_1' + q_2') = (a_1 - q_1') + (a_2 - q_2') = \bar{q}_1 + \bar{q}_2.$$

For any real number λ,

$$\overline{\lambda q} = \lambda \bar{q}.$$

It follows from Eq. (1.9) that the quaternion conjugate of multiplication $q_1 q_2$ equals

$$\overline{q_1 q_2} = -[a_1 q_2' + a_2 q_1'] + a_1 a_2 - (b_1 b_2 + c_1 c_2 + d_1 d_2) - \begin{vmatrix} i & j & k \\ b_1 & c_1 & d_1 \\ b_2 & c_2 & d_2 \end{vmatrix} \quad (1.21)$$

and

$$\bar{q}_2 \bar{q}_1 = -[a_1 q_2' + a_2 q_1'] + a_1 a_2 - (b_1 b_2 + c_1 c_2 + d_1 d_2) + \begin{vmatrix} i & j & k \\ b_2 & c_2 & d_2 \\ b_1 & c_1 & d_1 \end{vmatrix}.$$

Thus, $\overline{q_1 q_2} = \bar{q}_2 \bar{q}_1$ and might not equal $\bar{q}_1 \bar{q}_2$.

Example 1.13

Consider two quaternion numbers $q_1 = 1 + i - 2j + k$ and $q_2 = 1 + 2i - j - k$. The following holds for the conjugate of multiplication:

$$\overline{q_1 q_2} = \overline{(1 + i - 2j + k)(1 + 2i - j - k)} = \overline{-2 + 6i + 3k} = -2 - 6i - 3k$$

and $\bar{q}_2\bar{q}_1 = -2 - 6i - 3k$. The multiplication of their conjugate numbers $\bar{q}_1 = 1 - i + 2j - k$ and $\bar{q}_2 = 1 - 2i + j + k$ equals

$$\bar{q}_1\bar{q}_2 = (1 - i + 2j - k)(1 - 2i + j + k) = -2 + 6j + 3k.$$

P4. The module of the quaternion q equals

$$|q| = \sqrt{q\bar{q}} = \sqrt{\bar{q}q} = \sqrt{a^2 + b^2 + c^2 + d^2}.$$

In the particular case, when $q = q'$, the module $|q'| = \sqrt{b^2 + c^2 + d^2}$. In general, $|q| = \sqrt{a^2 + |q'|^2}$ and $|q| \geq |q'|$.

It is not difficult to see that the following holds for the operation of the quaternion conjugate:

$$\bar{\bar{q}} = q,$$

and

$$q + \bar{q} = 2a, \quad q - \bar{q} = 2q' = 2(bi + cj + dk).$$

The following calculations are valid:

$$|q_1q_2|^2 = (q_1q_2)\overline{q_1q_2} = (q_1q_2)(\bar{q}_2\bar{q}_1) = q_1|q_2|^2\bar{q}_1 = |q_2|^2(q_1\bar{q}_1) = |q_2|^2|q_1|^2.$$

Thus,

$$|q_1q_2| = |q_2||q_1| = |q_1||q_2|.$$

It is not difficult to show the property of triangle inequality,

$$|q_1 + q_2| \leq |q_1| + |q_2|.$$

To do so, it is enough to prove that for any quaternion $q = a + (ib + cd + kd)$, the inequality $|q + 1| \leq |q| + 1$ holds. The following calculations hold:

$$\begin{aligned}|q + 1|^2 &= (q + 1)(\bar{q} + 1) = |q|^2 + 1 + (q + \bar{q}) = |q|^2 + 1 + 2a \\ &\leq |q|^2 + 1 + 2|a| \leq |q|^2 + 1 + 2|q| = (|q| + 1)^2.\end{aligned}$$

P5. The inverse q^{-1} to the quaternion $q \neq 0$ is defined as the quaternion number such that $q^{-1}q = 1$. This is the inverse number when multiplying q from the left and is defined (from the definition $\bar{q}q = |q|^2$) as the multiplication of the quaternion q by the real number

$$q^{-1} = \frac{1}{|q|^2}\bar{q}.$$

The inverse from the left and right are the same, i.e., $qq^{-1} = q^{-1}q = 1$. Note that if q is a unit quaternion, i.e., $|q| = 1$, then $q^{-1} = \bar{q}$.

Example 1.14

Consider the quaternion number $q = 1 + 2i - 3j + 4k$. Then, $\bar{q} = 1 - 2i + 3j - 4k$, and the inverse q^{-1} is calculated as

$$(1+2i-3j+4k)^{-1} = \frac{1}{1+2^2+3^2+4^2}(1-2i+3j-4k) = \frac{1}{30}(1-2i+3j-4k).$$

The following property holds for the inverse of multiplication:

$$(q_1q_2)^{-1} = \frac{1}{|q_1q_2|^2}\overline{q_1q_2} = \frac{1}{|q_1|^2|q_2|^2}\bar{q}_2\bar{q}_1 = \frac{\bar{q}_2}{|q_2|^2}\frac{\bar{q}_1}{|q_1|^2} = (q_2)^{-1}(q_1)^{-1}.$$

In the particular case when $q_1 = q_2$, we obtain $(q_1^2)^{-1} = (q_1^{-1})^2$.

P6. Multiplication is not a commutative operation; therefore, the operation of division of one quaternion q_1 by another $q_2 \neq 0$ is considered when solving one of the following two equations:

$$q_1 = q_2q \quad \text{and} \quad q_1 = qq_2.$$

For division from the left, multiplication of the equality $q_1 = q_2q$ by $\bar{q}_2$ from the left results in

$$\bar{q}_2q_2q = \bar{q}_2q_1 \quad \text{or} \quad |q_2|^2q = \bar{q}_2q_1.$$

Therefore, we obtain the solution

$$q = \frac{1}{|q_2|^2}\bar{q}_2q_1.$$

We can denote this division operation from the left as

$$q = q_l = \frac{1}{q_2}q_1.$$

Similarly, after multiplying both sides of equation $q_1 = qq_2$ by $\bar{q}_2$ from the right, we obtain the solution of this equation, which equals

$$q = \frac{1}{|q_2|^2}q_1\bar{q}_2 = q_1\frac{1}{|q_2|^2}\bar{q}_2.$$

We can denote this division from the right by

$$q = q_r = q_1\frac{1}{q_2}.$$

The divisions from the right and left are the same if $\bar{q}_2 q_1 = q_1 \bar{q}_2$, which can also be written as $\bar{q}_1 q_2 = q_2 \bar{q}_1$.

Example 1.15

Consider two quaternions $q_1 = 1 + i + k$ and $q_2 = 1 + 2i - j + k$. The left and right divisions of q_1 by q_2 are calculated as

$$q_l = \frac{1}{|q_2|^2}\bar{q}_2 q_1 = \frac{1}{1+4+1+1}(1 - 2i + j - k)(1 + i + k) = \frac{1}{7}(4 + 2j - k)$$

and

$$q_r = \frac{1}{|q_2|^2} q_1 \bar{q}_2 = \frac{1}{1+4+1+1}(1 + i + k)(1 - 2i + j - k) = \frac{1}{7}(4 - 2i + k).$$

We can see that these two numbers are different.

Below is the script of the MATLAB-based code "test_example10.m" to calculate the division of two quaternion numbers for this example.

```
%-----------------------------------------------
% test_example10.m
    q1=[1,1,0,1];                    % q1=1+i+0j+k
    q2=[1,2,-1,1];                   % q2=1+2i-j+k
    q2c=[q2(1),-q2(2:4)];            % conjugate to q2
    q2m=q2*q2';                      % |q2|^2=7
%-------------- Left Division (ql) -------------
    q2c1= mult2qs_direct(q2c,q1); % 4 0 2 -1
    ql=(1/q2m)*q2c1;             % 0.5714 0 0.2857 -0.1429
%   checking the result
    mult2qs_direct(q2,ql);           % 1 1 0 1 (q1)
%--------------- Right Division (qr) -------------
    q12c=mult2qs_direct(q1,q2c); % 4 -2 0 1
    qr=(1/q2m)*q12c;   % 0.5714 -0.2857 0 0.1429
%   checking the result
    mult2qs_direct(qr,q2);           % 1 1 0 1 (q1)
%-----------------------------------------------
```

P7. It directly follows from Eq. (1.9) that if the numbers q_1 and q_2 are pure quaternions, i.e., $a_1 = a_2 = 0$, then

$$q = q_1 q_2 = -(b_1 b_2 + c_1 c_2 + d_1 d_2) + \begin{vmatrix} i & j & k \\ b_1 & c_1 & d_1 \\ b_2 & c_2 & d_2 \end{vmatrix}. \tag{1.22}$$

When $q_1 = q_2$, we obtain from this Eq. (1.22) the following:

$$(q_1)^2 = -(b_1^2 + c_1^2 + d_1^2) + \begin{vmatrix} i & j & k \\ b_1 & c_1 & d_1 \\ b_1 & c_1 & d_1 \end{vmatrix} = -(b_1^2 + c_1^2 + d_1^2) = -|q_1|^2. \quad (1.23)$$

Therefore, for the pure unit quaternion q_1, i.e., $|q_1| = 1$, its square $q_1^2 = -1$.

P8. For any two pure quaternion numbers q_1 and q_2, the following holds:

$$q_1q_2 - q_2q_1 = 2\begin{vmatrix} i & j & k \\ b_1 & c_1 & d_1 \\ b_2 & c_2 & d_2 \end{vmatrix},$$

and this difference is zero (or $q_1q_2 = q_2q_1$) only if q_1 and q_2 are similar, i.e., $q_2 = Aq_2$, where A is a real number.

P9. We also use the notation $q = q_e + iq_i + jq_j + kq_k$ for the quaternion number $q = a + bi + cj + dk$, i.e., $q_e = a$, $q_i = b$, $q_j = c$, and $q_k = d$. The quaternion q can be considered as a vector

$$\mathbf{q} = (q_e, q_i, q_j, q_k)^T = (a, b, c, d)^T$$

in four-dimensional real space R^4, when the basic vectors of this space are $\mathbf{e} = (1, 0, 0, 0)^T$, $\mathbf{i} = (0, 1, 0, 0)^T$, $\mathbf{j} = (0, 0, 1, 0)^T$, and $\mathbf{k} = (0, 0, 0, 1)^T$. Here, the operation $(\cdot)^T$ stands for the transposition.

The direct calculation of the multiplication q_1q_2 results in the following:

$$\begin{aligned} q_1q_2 &= (a_1 + ib_1 + jc_1 + kd_1)(a_2 + ib_2 + jc_2 + kd_2) \\ &= (a_1a_2 - b_1b_2 - c_1c_2 - d_1d_2) + i(a_1b_2 + a_2b_1 + c_1d_2 - c_2d_1) \\ &\quad + j(a_1c_2 + a_2c_1 - b_1d_2 + b_2d_1) + k(a_1d_2 + a_2d_1 + b_1c_2 - b_2c_1). \end{aligned}$$

This formula can also be obtained in symbolic programming in MATLAB. As an example, a simple script of the code "test_example.m" for symbolic calculation of the multiplication of two quaternions is given below.

```
% mult2q_symbolic.m     / Art Grigoryan, 2015
% the formula for multiplication q1q2 of two quaternions q1
  and q2
% ---------------------------------------------------------
  syms a1 b1 c1 d1 as real;
  syms a2 b2 c2 d2 as real;
  g1= [a1 b1 c1 d1];
  g2= [a2 b2 c2 d2];
  q1q2 =mult2qs(g1,g2)
% ---------------------------------------------------------
```

The script of the code "mult2qs.m" is given in the next script.

Therefore, in matrix form, the multiplication $q = q_1 q_2$ can be written as

$$\mathbf{q} = \mathbf{A}\mathbf{q}_2 = \begin{pmatrix} a_1 & -b_1 & -c_1 & -d_1 \\ b_1 & a_1 & -d_1 & c_1 \\ c_1 & d_1 & a_1 & -b_1 \\ d_1 & -c_1 & b_1 & a_1 \end{pmatrix} \begin{pmatrix} a_2 \\ b_2 \\ c_2 \\ d_2 \end{pmatrix}, \tag{1.24}$$

where the four-dimensional (4-D) vectors are

$$\mathbf{q} = \begin{pmatrix} a \\ b \\ c \\ d \end{pmatrix} \quad \text{and} \quad \mathbf{q}_2 = \begin{pmatrix} a_2 \\ b_2 \\ c_2 \\ d_2 \end{pmatrix}.$$

The quaternion numbers $\mu = \mu_e + i\mu_i + j\mu_j + k\mu_k$ such that $\mu^2 = -1$ are pure quaternions. For instance, the numbers

$$\mu = \frac{i+j+k}{\sqrt{3}}, \quad \mu = \frac{i-j+k}{\sqrt{3}}, \quad \text{and} \quad \mu = \frac{i+j}{\sqrt{2}}$$

are pure unit quaternions. For such quaternions, $\mu_e = 0$ and the imaginary part lies on the unit sphere. It should be noted that for any real triplet $(\mu_i, \mu_j, \mu_k) \neq (0, 0, 0)$, the corresponding number

$$\mu = \frac{i\mu_i + j\mu_j + k\mu_k}{\sqrt{\mu_i^2 + \mu_j^2 + \mu_k^2}}$$

is a pure unit quaternion.

The equation $\mu^2 = -1$ can also be considered in matrix form as

$$\begin{pmatrix} \mu_e & -\mu_i & -\mu_j & -\mu_k \\ \mu_i & \mu_e & -\mu_k & \mu_j \\ \mu_j & \mu_k & \mu_e & -\mu_i \\ \mu_k & -\mu_j & \mu_i & \mu_e \end{pmatrix} \begin{pmatrix} \mu_e \\ \mu_i \\ \mu_j \\ \mu_k \end{pmatrix} = \begin{pmatrix} -1 \\ 0 \\ 0 \\ 0 \end{pmatrix}.$$

It follows that

$$\mu_e^2 - \mu_i^2 - \mu_j^2 - \mu_k^2 = -1,$$
$$2\mu_e\mu_i = 0, \quad 2\mu_e\mu_j = 0, \quad 2\mu_e\mu_k = 0.$$

Therefore, $\mu_e = 0$ and $|\mu|^2 = \mu_i^2 + \mu_j^2 + \mu_k^2 = 1$, i.e., the point (μ_i, μ_j, μ_k) is on the unit sphere.

Below is the script of the MATLAB-based code-function "mult2qs.m" for calculating the multiplication of two quaternions by using the matrix **A**.

```
%-----------------------------------------------------------------
% call: muit2qs.m   / Art Grigoryan, UTSA January 10, 2015
% the matrix multiplication of 4-D quaternion-vectors
% q1= [a1,b1,c1,d1] * q2= [a2,b2,c2,d2]
function q12=mult2qs(q1,q2)
    a1=q1(1);  b1=q1(2);  c1=q1(3);  d1=q1(4);
    if size(q2,1)==1 q2=q2'; end  % q2 to be a column-vector
    A=[ a1 -b1 -c1 -d1
        b1  a1 -d1  c1
        c1  d1  a1 -b1
        d1 -c1  b1  a1];         % See Eq. (1.24)
    q12=A*q2;
%-----------------------------------------------------------------
```

This function as the function "mult2qs_direct.m" uses 16 real multiplications and 12 real additions in the multiplication of two quaternion numbers.

Matrix **A** in Eq. (1.24) is defined by the quaternion q_1, so we can write $\mathbf{A} = \mathbf{A}_{L;q_1}$. The subscript L stands for the vector q_1 on the left of $q_1 q_2$. It is not difficult to see that the same multiplication $q = q_1 q_2$ can be written by a matrix defined by the quaternion q_2 as

$$\mathbf{q} = \mathbf{A}_{R;q_2}\mathbf{q}_1 = \begin{pmatrix} a_2 & -b_2 & -c_2 & -d_2 \\ b_2 & a_2 & d_2 & -c_2 \\ c_2 & -d_2 & a_2 & b_2 \\ d_2 & c_2 & -b_2 & a_2 \end{pmatrix} \begin{pmatrix} a_1 \\ b_1 \\ c_1 \\ d_1 \end{pmatrix}, \tag{1.25}$$

where the 4-D vector $\mathbf{q}_1 = (a_1, b_1, c_1, d_1)^T$. Here, we use the subscript R for multiplication by the matrix that is generated by the quaternion from the right in the multiplication $q_1 q_2$.

Below is the code-function "mult2qsR.m" for calculating the multiplication of two quaternions by using the matrix $\mathbf{A}_{R;q_2}$.

```
%-----------------------------------------------------------------
% call: muit2qsR.m   / Art Grigoryan, UTSA January 10, 2015
% the matrix multiplication of 4-D quaternion-vectors
% q1= [a1,b1,c1,d1] * q2= [a2,b2,c2,d2]
function q12=mult2qsR(q1,q2)
    a2=q2(1);  b2=q2(2);  c2=q2(3);  d2=q2(4);
    if size(q1,1)==1 q1=q1'; end  % q1 to be a column-vector
```

```
    Aq2 = [ a2 -b2 -c2 -d2
            b2  a2  d2 -c2
            c2 -d2  a2  b2
            d2  c2 -b2  a2];          % See Eq. (1.25)
    q12 = Aq2*q1;
%-----------------------------------------------------------
```

We now describe an important property of the inverse matrices to matrices **A**, but first we use a simple script of the calculation of such matrices.

Example 1.16

Consider the quaternion number $q = 1 + 2i + 3j - 4k$ with $|q|^2 = 15$. If we run the code "inver2Amatrices.m," the script of which is given below, we obtain the following multiplication of the matrices with their transposition:

$$\mathbf{A}_{L;q} = \begin{pmatrix} 1 & -2 & -3 & 1 \\ 2 & 1 & 1 & 3 \\ 3 & -1 & 1 & -2 \\ -1 & -3 & 2 & 1 \end{pmatrix}, \quad \mathbf{A}_{L;q}^T = \begin{pmatrix} 1 & 2 & 3 & -1 \\ -2 & 1 & -1 & -3 \\ -3 & 1 & 1 & 2 \\ 1 & 3 & -2 & 1 \end{pmatrix}, \quad \mathbf{A}_{L;q} \cdot \mathbf{A}_{L;q}^T = 15\mathbf{I}_4,$$

where $\mathbf{I}_4$ denotes the 4×4 identity matrix.

```
%-----------------------------------------------------------
% call: inver2Amatrices.m / Artyom Grigoryan, January 2, 2016
% calculate inverse matrices of the left and right
  multiplication
    q = [1,2,3,-1];
    % q=q/norm(q);  % |q|^2 = norm(q)^2 = 15
    a = q(1);  b = q(2);  c = q(3);  d = q(4);
    A1 = [ a -b -c -d
           b  a -d  c
           c  d  a -b
           d -c  b  a];          % See Eq. (1.17)
    A2 = [ a -b -c -d
           b  a  d -c
           c -d  a  b
           d  c -b  a];          % See Eq. (1.18)
d1 = det(A1); d2 = det(A2);      % d1 = d2 = 225 which is 15*15
what_weget = A1'*A1;
what_weget = A2'*A2;
%  15   0   0   0
%   0  15   0   0
%   0   0  15   0
%   0   0   0  15
% if open the command q=q/norm(q), then we get the identity
  matrix
%-----------------------------------------------------------
```

Similarly, for multiplication from the right, we have

$$\mathbf{A}_{R;q} = \begin{pmatrix} 1 & -2 & -3 & 1 \\ 2 & 1 & -1 & -3 \\ 3 & 1 & 1 & 2 \\ -1 & 3 & -2 & 1 \end{pmatrix}, \quad \mathbf{A}_{R;q}^T = \begin{pmatrix} 1 & 2 & 3 & -1 \\ -2 & 1 & 1 & 3 \\ -3 & -1 & 1 & -2 \\ 1 & -3 & 2 & 1 \end{pmatrix}, \quad \mathbf{A}_{R;q} \cdot \mathbf{A}_{R;q}^T = 15\mathbf{I}_4.$$

We note that $|q|^2 = 15$ and have the following inverse matrices:

$$\mathbf{A}_{L;q}^{-1} = \frac{1}{15}\mathbf{A}_{L;q}^T = \frac{1}{15}\begin{pmatrix} 1 & 2 & 3 & -1 \\ -2 & 1 & -1 & -3 \\ -3 & 1 & 1 & 2 \\ 1 & 3 & -2 & 1 \end{pmatrix}$$

and

$$\mathbf{A}_{R;q}^{-1} = \frac{1}{15}\mathbf{A}_{R;q}^T = \frac{1}{15}\begin{pmatrix} 1 & 2 & 3 & -1 \\ -2 & 1 & 1 & 3 \\ -3 & -1 & 1 & -2 \\ 1 & -3 & 2 & 1 \end{pmatrix}.$$

Also, we can see that the determinant of both matrices is $225 = (15)^2$.

In general, we have the following for matrices $\mathbf{A}$:

$$\det(\mathbf{A}_{L;q}) = \det(\mathbf{A}_{R;q}) = |q|^4 \tag{1.26}$$

and

$$\mathbf{A}_{L;q}^{-1} = \frac{1}{|q|^2}\mathbf{A}_{L;q}^T \quad \text{and} \quad \mathbf{A}_{R;q}^{-1} = \frac{1}{|q|^2}\mathbf{A}_{R;q}^T. \tag{1.27}$$

If the quaternion is a unit quaternion, i.e., $|q| = 1$, these inverse matrices are transpose matrices.

1.2 Vector Space and Pure Quaternions

To describe the geometry of the multiplication presented in the previous section, we consider the Cartesian coordinate system in 3-D space, vectors $\mathbf{q}' = [b, c, d]$ with original $(0, 0, 0)$, and the terminal points (b, c, d). Along three perpendicular X,- Y,- and Z-axes in this system, we can measure the unit vectors as

$$\mathbf{i} = [1, 0, 0], \qquad \mathbf{j} = [0, 1, 0], \qquad \text{and} \qquad \mathbf{k} = [0, 0, 1],$$

respectively, as shown in Fig. 1.10. The space of quaternion numbers is 4-D, and the Cartesian model in Fig. 1.10 is shown only to present the space of the

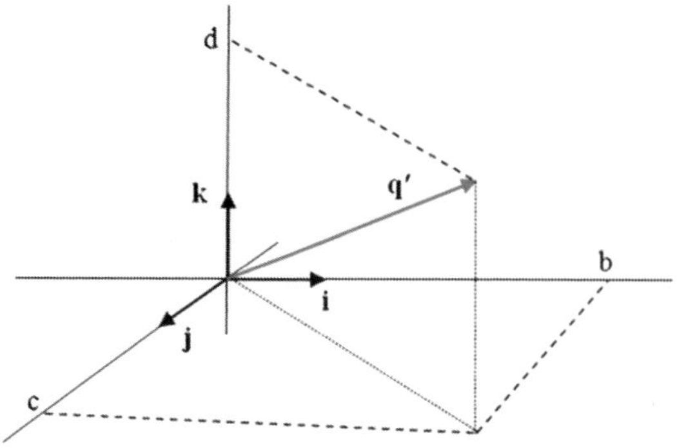

Figure 1.10 The Cartesian coordinate system in the 3-D space.

imaginary components of the quaternion numbers. In general, it is not necessary to draw the *i*-, *j*-, and *k*-axes perpendicular to each other in the space of quaternions.

A pure quaternion number $q' = (b, c, d)$ can be presented by the vector $\mathbf{q}' = [b, c, d]$ in this vector space. The sum of pure quaternion numbers and multiplication by real numbers correspond to similar operations on vectors. Thus, we can associate vectors with pure quaternion numbers and call them *vector quaternions*. Given a quaternion $q = a + q'$, a is the real part of q, and q' is *the vector part* of q.

In vector space, two interesting operations, the inner product and vector product, are described. These operations are used in many applications of vector algebra and mechanics, when calculating the work done by a force, or finding the normal vector to the plane, moment of a force, velocity of a rotating body, etc. We now describe these operations performed on vector quaternions.

1.2.1 Inner product (or dot product)

The operation of the inner product of two vectors $\mathbf{q}'_1 = [b_1, c_1, d_1]$ and $\mathbf{q}'_2 = [b_2, c_2, d_2]$ is defined as

$$(\mathbf{q}'_1, \mathbf{q}'_2) = b_1 b_2 + c_1 c_2 + d_1 d_2. \tag{1.28}$$

Thus, the real part of the multiplication in Eq. (1.22) is the inner product with the sign minus.

In vector space, the inner product is defined by the following rule: for each two elements $\mathbf{q}'_1$ and $\mathbf{q}'_2$ of the space with which a real number is associated, *the inner product* that is denoted as $(\mathbf{q}'_1, \mathbf{q}'_2)$ has the following four properties (axioms):

1. $(\mathbf{q}'_1, \mathbf{q}'_2) = (\mathbf{q}'_2, \mathbf{q}'_1)$.
2. $(\mathbf{q}'_1 + \mathbf{q}'_3, \mathbf{q}'_2) = (\mathbf{q}'_1, \mathbf{q}'_2) + (\mathbf{q}'_3, \mathbf{q}'_2)$.
3. $(\lambda\mathbf{q}'_1, \mathbf{q}'_2) = \lambda(\mathbf{q}'_1, \mathbf{q}'_2)$, for any real number λ.
4. $(\mathbf{q}', \mathbf{q}') \geq 0$, and $(\mathbf{q}', \mathbf{q}') = 0$ if only $\mathbf{q}' = 0$.

The value of $(\mathbf{q}', \mathbf{q}')$ is denoted by $|\mathbf{q}'|^2$, and the positive number $|\mathbf{q}'| = \sqrt{(\mathbf{q}', \mathbf{q}')}$ is called *the length* of vector $\mathbf{q}'$. As in real Euclidean space, the concept of the angle between the vectors can be defined in the vector space of pure quaternions. Indeed, it directly follows from the above axioms, which for given vectors $\mathbf{q}_1'$ and $\mathbf{q}_2'$, the following polynomial is non-negative:

$$p(\lambda) = (\lambda\mathbf{q}_1' - \mathbf{q}_2', \lambda\mathbf{q}_1' - \mathbf{q}_2') = |\lambda\mathbf{q}_1' - \mathbf{q}_2'|^2 \geq 0.$$

Here, λ is a real variable. Thus,

$$\lambda^2(\mathbf{q}_1', \mathbf{q}_1') - 2\lambda(\mathbf{q}_1', \mathbf{q}_2') + (\mathbf{q}_2', \mathbf{q}_2') \geq 0,$$

and the discriminant of this equation

$$\Delta = (\mathbf{q}_1', \mathbf{q}_2')^2 - (\mathbf{q}_1', \mathbf{q}_1')(\mathbf{q}_2', \mathbf{q}_2') = (\mathbf{q}_1', \mathbf{q}_2')^2 - |\mathbf{q}_1'|^2|\mathbf{q}_2'|^2 \leq 0.$$

Therefore, $(\mathbf{q}_1', \mathbf{q}_2')^2/(|\mathbf{q}_1'|^2|\mathbf{q}_2'|^2) \leq 1$, or

$$-1 \leq \frac{(\mathbf{q}_1', \mathbf{q}_2')}{|\mathbf{q}_1'||\mathbf{q}_2'|} \leq 1.$$

The fraction in this expression can be considered as a cosine of angle ϑ:

$$\cos(\vartheta) = \frac{(\mathbf{q}_1', \mathbf{q}_2')}{|\mathbf{q}_1'||\mathbf{q}_2'|}, \quad \text{and} \quad \sin(\vartheta) = \sqrt{1 - \frac{(\mathbf{q}_1', \mathbf{q}_2')^2}{|\mathbf{q}_1'|^2|\mathbf{q}_2'|^2}}.$$

This angle is called the angle between vectors $\mathbf{q}_1'$ and $\mathbf{q}_2'$. Since

$$(\mathbf{q}_1', \mathbf{q}_2') = |\mathbf{q}_1'||\mathbf{q}_2'|\cos(\vartheta), \tag{1.29}$$

we can call two vectors *perpendicular* if $(\mathbf{q}_1', \mathbf{q}_2') = 0$. This condition can be denoted as $\mathbf{q}_1' \perp \mathbf{q}_2'$.

It directly follows from Eq. (1.22) that if vectors $\mathbf{q}_1'$ and $\mathbf{q}_2'$ are perpendicular, then the following holds for the pure quaternions $q_1 = q_1'$ and $q_2 = q_1'$:

$$q_1q_2 = \begin{vmatrix} i & j & k \\ b_1 & c_1 & d_1 \\ b_2 & c_2 & d_2 \end{vmatrix} = -\begin{vmatrix} i & j & k \\ b_2 & c_2 & d_2 \\ b_1 & c_1 & d_1 \end{vmatrix} = -q_2q_1. \tag{1.30}$$

Also, if $q_1q_2 = -q_2q_1$, then q_1 and q_2 are pure quaternions, and the corresponding vectors $\mathbf{q}_1'$ and $\mathbf{q}_2'$ are perpendicular.

Since

$$q_1'q_2' = -(\mathbf{q}_1', \mathbf{q}_2') + \begin{vmatrix} i & j & k \\ b_1 & c_1 & d_1 \\ b_2 & c_2 & d_2 \end{vmatrix},$$

the real part of $q_1'q_2'$ is zero if vectors $\mathbf{q}_1'$ and $\mathbf{q}_2'$ are perpendicular.

Example 1.17

Consider three vector quaternions $\mathbf{q}_1' = (1, 2-3)$, $\mathbf{q}_2' = (-1, 5, 3)$, and $\mathbf{q}_3' = (2, 1, 2)$. The first two vectors are orthogonal,

$$(\mathbf{q}_1', \mathbf{q}_2') = 1(-1) + 2(5) - 3(3) = 0,$$

and

$$q_1'q_2' = (1 + 2j - 3k)(-i + 5j + 3k) = 21i + 7k.$$

One can verify this result by using the calculation of the determinant:

$$\begin{vmatrix} i & j & k \\ b_1 & c_1 & d_1 \\ b_2 & c_2 & d_2 \end{vmatrix} = \begin{vmatrix} i & j & k \\ 1 & 2 & -3 \\ -1 & 5 & 3 \end{vmatrix} = i(6+15) - j(3-3) + k(5+2) = 21i + 7k.$$

These two vectors are not perpendicular to vector $\mathbf{q}_3'$. Indeed,

$$(\mathbf{q}_1', \mathbf{q}_3') = 1(2) + 2(1) - 3(2) = -2,$$
$$(\mathbf{q}_2', \mathbf{q}_3') = -1(2) + 5(1) + 3(2) = 9.$$

The angle $\vartheta_{1,3}$ between vectors $\mathbf{q}_1'$ and $\mathbf{q}_3'$ is calculated by

$$\cos(\vartheta_{1,3}) = \frac{(\mathbf{q}_1', \mathbf{q}_3')}{|\mathbf{q}_1'||\mathbf{q}_3'|} = \frac{-2}{\sqrt{1+4+9}\sqrt{4+1+4}} = \frac{-2}{3\sqrt{14}} = -0.1782,$$

$$\vartheta_{1,3} = \arccos\left(-\frac{2}{3\sqrt{14}}\right) = 1.7499 \text{ rad, or } 100.2634°.$$

The angle $\vartheta_{2,3}$ between vectors $\mathbf{q}_2'$ and $\mathbf{q}_3'$ is calculated by

$$\cos(\vartheta_{2,3}) = \frac{(\mathbf{q}_2', \mathbf{q}_3')}{|\mathbf{q}_2'||\mathbf{q}_3'|} = \frac{9}{\sqrt{1+25+9}\sqrt{4+1+4}} = \frac{3}{\sqrt{35}} = 0.5071,$$

$$\vartheta_{2,3} = \arccos\left(\frac{3}{\sqrt{35}}\right) = 1.0390 \text{ rad, or } 59.5296°.$$

1.2.2 Vector product

In vector space, we can consider the symbolic determinate

$$\begin{vmatrix} \mathbf{i} & \mathbf{j} & \mathbf{k} \\ b_1 & c_1 & d_1 \\ b_2 & c_2 & d_2 \end{vmatrix} = \mathbf{i}(c_1 d_2 - d_1 c_2) - \mathbf{j}(b_1 d_2 - d_1 b_2) + \mathbf{k}(b_1 c_2 - c_1 b_2). \quad (1.31)$$

According to Eq. (1.22), the vector with the terminal point

$$[(c_1 d_2 - d_1 c_2), -(b_1 d_2 - d_1 b_2), (b_1 c_2 - c_1 b_2)]$$

describes the whole imaginary part of the multiplication $q = q_1 q_2$, when $a_1 = a_2 = 0$. We also can say that this point corresponds to $(q_1' q_2')$. This vector is denoted by $\mathbf{q}_1' \times \mathbf{q}_2'$ or $[\mathbf{q}_1', \mathbf{q}_2']$ and is called *the vector product* of vectors $\mathbf{q}_1'$ and $\mathbf{q}_2'$.

It is not difficult to see that

$$\begin{vmatrix} \mathbf{i} & \mathbf{j} & \mathbf{k} \\ b_2 & c_2 & d_2 \\ b_1 & c_1 & d_1 \end{vmatrix} = \mathbf{i}(c_2 d_1 - d_2 c_1) - \mathbf{j}(b_2 d_1 - d_2 b_1) + \mathbf{k}(b_2 c_1 - c_2 b_1) = - \begin{vmatrix} \mathbf{i} & \mathbf{j} & \mathbf{k} \\ b_1 & c_1 & d_1 \\ b_2 & c_2 & d_2 \end{vmatrix},$$

and, therefore,

$$\mathbf{q}_2' \times \mathbf{q}_1' = -\mathbf{q}_1' \times \mathbf{q}_2'.$$

The vector product is perpendicular to vectors $\mathbf{q}_1'$ and $\mathbf{q}_2'$. To verify this property, it is sufficient to show that the real parts of vectors $\mathbf{q}_1'(\mathbf{q}_1' \times \mathbf{q}_2')$ and $\mathbf{q}_2'(\mathbf{q}_1' \times \mathbf{q}_2')$ are zero. The real part of the first multiplication can be calculated as

$$b_1(c_1 d_2 - d_1 c_2) - c_1(b_1 d_2 - d_1 b_2) + d_1(b_1 c_2 - c_1 b_2) = 0.$$

The real part of the second multiplication is similarly calculated and equals zero, too. The three perpendicular vectors $\mathbf{q}_1' \times \mathbf{q}_2'$, $\mathbf{q}_1'$, and $\mathbf{q}_2'$ are oriented in space similarly to the way unit vectors **i**, **j**, and **k** are oriented. The orientation of these unit vectors can be considered in the chosen right- or left-handed Cartesian coordinate system in 3-D space.

The length of the vector product can be calculated by

$$\begin{aligned} |\mathbf{q}_1' \times \mathbf{q}_2'|^2 &= (c_1 d_2 - d_1 c_2)^2 + (b_1 d_2 - d_1 b_2)^2 + (b_1 c_2 - c_1 b_2)^2 \\ &= (c_1^2 d_2^2 - 2c_1 c_2 d_1 d_2 + d_1^2 c_2^2) + (b_1^2 d_2^2 - 2b_1 b_2 d_1 d_2 + d_1^2 b_2^2) \\ &\quad + (b_1^2 c_2^2 - 2b_1 b_2 c_1 c_2 + c_1^2 b_2^2) \\ &= (b_1^2 + c_1^2 + d_1^2)(b_2^2 + c_2^2 + d_2^2) - c_1^2 c_2^2 - b_1^2 b_2^2 - d_1^2 d_2^2 \\ &\quad - 2c_1 c_2 d_1 d_2 - 2b_1 b_2 d_1 d_2 - 2b_1 b_2 c_1 c_2 \\ &= (b_1^2 + c_1^2 + d_1^2)(b_2^2 + c_2^2 + d_2^2) - (b_1 b_2 + c_1 c_2 + d_1 d_2)^2. \end{aligned}$$

Because

$$
\begin{aligned}
|\mathbf{q}_1'|^2 &= b_1^2 + c_1^2 + d_1^2, \\
|\mathbf{q}_2'|^2 &= b_2^2 + c_2^2 + d_2^2, \\
(\mathbf{q}_1', \mathbf{q}_2') &= |\mathbf{q}_1'||\mathbf{q}_2'|\cos(\vartheta) = b_1 b_2 + c_1 c_2 + d_1 d_2,
\end{aligned} \tag{1.32}
$$

where ϑ is the angle between the vector quaternions, we obtain

$$|\mathbf{q}_1' \times \mathbf{q}_2'|^2 = |\mathbf{q}_1'|^2|\mathbf{q}_2'|^2 - |\mathbf{q}_1'|^2|\mathbf{q}_2'|^2\cos^2(\vartheta) = |\mathbf{q}_1'|^2|\mathbf{q}_2'|^2\sin^2(\vartheta).$$

Thus, the length of the vector products equals

$$|\mathbf{q}_1' \times \mathbf{q}_2'| = |\mathbf{q}_1'||\mathbf{q}_2'|\sin(\vartheta), \tag{1.33}$$

when considering $\sin(\vartheta)$ to be non-negative. It is the area of the parallelogram composed on vectors $\mathbf{q}_1'$ and $\mathbf{q}_2'$. If the vectors are perpendicular, then $\vartheta = \pi/2$, $\sin(\vartheta) = 1$, and

$$|\mathbf{q}_1' \times \mathbf{q}_2'| = |\mathbf{q}_1'||\mathbf{q}_2'|. \tag{1.34}$$

Now, we can define the inner product and vector product of quaternions as

$$(q_1, q_2) = (\mathbf{q}_1', \mathbf{q}_2') \quad \text{and} \quad (q_1 \times q_2) = (\mathbf{q}_1' \times \mathbf{q}_2'),$$

and consider pure quaternions as *vector quaternions*. The multiplication of pure quaternion numbers unites the inner and vector products:

$$q = q_1 q_2 = -(q_1, q_2) + (q_1 \times q_2). \tag{1.35}$$

One can notice that

$$q_2 q_1 = -(q_2, q_1) + (q_2 \times q_1) = -(q_1, q_2) - (q_1 \times q_2) = q_1 q_2 - 2(q_1 \times q_2).$$

In the general case of quaternions, the multiplication of two quaternions in Eq. (1.9) can be written as

$$q = q_1 q_2 = a_1 a_2 - (q_1, q_2) + [a_1 q_2' + a_2 q_1'] + (q_1 \times q_2). \tag{1.36}$$

The real part of this vector is the minus inner product of the imaginary parts of the quaternions,

$$a = a_1 a_2 - (q_1, q_2),$$

and the imaginary part of the multiplication is the sum of vector quaternions,

$$q' = [a_1 q_2' + a_2 q_1'] + (q_1 \times q_2).$$

The first vector in the square brackets presents a vector in the "$(\mathbf{q}'_1) - (\mathbf{q}'_2)$" plane, and the second vector is the vector product of the imaginary parts of the quaternions, which is perpendicular to this plane.

Example 1.18

Consider two quaternions $q_1 = 1 + i - j + 2k$ and $q_2 = 1 + 2i + j - 3k$. The vector quaternions are $\mathbf{q}'_1 = (1, -1, 2)$ and $\mathbf{q}'_2 = (2, 1, -3)$. The inner product equals

$$(q'_1, q'_2) = 1 \cdot 2 + (-1) \cdot 1 + 2 \cdot (-3) = -5,$$

and $a = 1 - (-5) = 6$. The vector product is calculated as

$$(q'_1 \times q'_2) = \begin{vmatrix} \mathbf{i} & \mathbf{j} & \mathbf{k} \\ 1 & -1 & 2 \\ 2 & 1 & -3 \end{vmatrix} = \mathbf{i}(3-2) - \mathbf{j}(-3-4) + \mathbf{k}(1+2) = 1\mathbf{i} + 7\mathbf{j} + 3\mathbf{k}$$

and can be written as the vector with the terminal point $(1, 7, 3)$. The imaginary part of $q_1 q_2$ is

$$\begin{aligned} q' &= [a_1 q'_2 + a_2 q'_1] + (q'_1 \times q'_2) = [(2i + j - 3k) + (i - j + 2k)] + (i + 7j + 3k) \\ &= [3i - k] + (i + 7j + 3k) = 4i + 7j + 2k; \end{aligned}$$

therefore, $q = a + q' = 6 + 4i + 7j + 2k$.

1.3 Quaternion Multiplication and Rotation

Quaternion multiplication allows one to describe a rotation of one vector in 3-D space around any other vector. In this section, we show a few examples of rotations by multiplication. First, we consider the multiplication of two quaternions, one of which is a pure quaternion. The geometry of such multiplications with rotations in the 3-D subspace of vectors presenting quaternion vectors is described by using the matrix representation of multiplication as well as the polar form of quaternions.

1.3.1 Multiplication and sum of elementary rotations

Consider the quaternions in the space R^4 as vectors. When multiplying the quaternion $\mathbf{q}_1 = (a_1, b_1, c_1, d_1)^T$ by the vector quaternion $\mathbf{q}_2 = (0, b_2, c_2, d_2)^T$, the real component of the multiplication is $-(b_1 b_2 + c_1 c_2 + d_1 d_2)$. It follows from Eq. (1.24),

$$\mathbf{q} = \mathbf{A}\mathbf{q}_2 = \begin{pmatrix} a_1 & -b_1 & -c_1 & -d_1 \\ b_1 & a_1 & -d_1 & c_1 \\ c_1 & d_1 & a_1 & -b_1 \\ d_1 & -c_1 & b_1 & a_1 \end{pmatrix} \begin{pmatrix} 0 \\ b_2 \\ c_2 \\ d_2 \end{pmatrix}, \tag{1.37}$$

that the matrix of multiplication in the 4-D real space $\mathbf{A}$ can be reduced to a (3×3) matrix:

$$\mathbf{B} = \begin{pmatrix} a_1 & -d_1 & c_1 \\ d_1 & a_1 & -b_1 \\ -c_1 & b_1 & a_1 \end{pmatrix},$$

when calculating the vector component of the multiplication $q_1 q_2$,

$$\begin{pmatrix} b \\ c \\ d \end{pmatrix} = \mathbf{B} \begin{pmatrix} b_2 \\ c_2 \\ d_2 \end{pmatrix}.$$

The point (b_2, c_2, d_2) is on a circle of radius $r_{2,j} = \sqrt{b_2^2 + c_2^2 = d_2^2}$.

This matrix can be represented as

$$\mathbf{B} = \mathbf{B}\Big(\frac{a_1}{2}, 0, -c_1\Big) + \mathbf{B}\Big(\frac{a_1}{2}, d_1, 0\Big) + \mathbf{B}\Big(0, \frac{a_1}{2}, b_1\Big),$$

where the matrices

$$\mathbf{B}_j = \mathbf{B}\Big(\frac{a_1}{2}, 0, -c_1\Big) = \begin{pmatrix} \frac{a_1}{2} & 0 & c_1 \\ 0 & 0 & 0 \\ -c_1 & 0 & \frac{a_1}{2} \end{pmatrix} \sim \mathbf{B}'_j = \begin{pmatrix} \frac{a_1}{2} & c_1 \\ -c_1 & \frac{a_1}{2} \end{pmatrix}, \ \det(\mathbf{B}'_j) = \frac{a_1^2}{4} + c_1^2,$$

$$\mathbf{B}_k = \mathbf{B}\Big(\frac{a_1}{2}, d_1, 0\Big) = \begin{pmatrix} \frac{a_1}{2} & -d_1 & 0 \\ d_1 & \frac{a_1}{2} & 0 \\ 0 & 0 & 0 \end{pmatrix} \sim \mathbf{B}'_k = \begin{pmatrix} \frac{a_1}{2} & -d_1 \\ d_1 & \frac{a_1}{2} \end{pmatrix}, \ \det(\mathbf{B}'_k) = \frac{a_1^2}{4} + d_1^2,$$

$$\mathbf{B}_i = \mathbf{B}\Big(0, \frac{a_1}{2}, b_1\Big) = \begin{pmatrix} 0 & 0 & 0 \\ 0 & \frac{a_1}{2} & -b_1 \\ 0 & b_1 & \frac{a_1}{2} \end{pmatrix} \sim \mathbf{B}'_i = \begin{pmatrix} \frac{a_1}{2} & -b_1 \\ b_1 & \frac{a_1}{2} \end{pmatrix}, \ \det(\mathbf{B}'_i) = \frac{a_1^2}{4} + b_1^2.$$

Each of these three matrices describes the rotation in the corresponding 2-D plane. Matrix $\mathbf{B}_j$ describes the rotation of the projection $(b_2, 0, d_2)$ of the vector (b_2, c_2, d_2) in the $(i-k)$-plane around the j-axis by the angle

$$\phi_j = -\arctan\Big(\frac{2c_1}{a_1}\Big).$$

Following this rotation, the point moves along the radial line from its circle of radius $r_{2,j} = \sqrt{b_2^2 + d_2^2}$ to another circle of radius amplified by a factor of $\lambda_j = \sqrt{\det(\mathbf{B}'_j)}$. A similar rotation for complex numbers was described in Example 1.6 with Fig. 1.5.

In the first stage, matrix $\mathbf{B}_k$ describes the rotation in the $(i-j)$-plane, and $\mathbf{B}_i$ describes the rotation in the $(j-k)$-plane of projections $(b_2, c_2, 0)$ and $(0, b_2, c_2)$ by angles

$$\phi_k = \arctan\left(\frac{2d_1}{a_1}\right) \quad \text{and} \quad \phi_i = \arctan\left(\frac{2b_1}{a_1}\right),$$

respectively. Then, the rotated points move to circles with radii amplified by the factors of $\lambda_k = \sqrt{\det(\mathbf{B}'_k)}$ and $\lambda_i = \sqrt{\det(\mathbf{B}'_i)}$, respectively. Thus, matrix $\mathbf{B}$ is the sum of the three matrices of the elementary, or Givens rotations, as shown in Fig. 1.11. If q is the quaternion vector, i.e., $a_1 = 0$, all of these angles are considered to be $\pi/2$.

Statement 1.1 *Multiplication of the quaternion number $q_1 = a_1 + ib_1 + jc_1 + kd_1$ by a pure quaternion q_2 describes the rotation of the sum of three rotations of the vector quaternion q_2 around the i-, j-, and k-axes by the angles*

$$\phi_i = \arctan\left(\frac{2b_1}{a_1}\right), \quad \phi_j = \arctan\left(\frac{2c_1}{a_1}\right), \quad \textit{and} \quad \phi_k = \arctan\left(\frac{2d_1}{a_1}\right),$$

respectively, or all $\pi/2$ if $a_1 = 0$.

If q_2 is not a pure quaternion, $q_2 = a_2 + ib_2 + jc_2 + kd_2$ and $a_2 \neq 0$, we can write the multiplication $q_1 q_2$ in the following matrix form:

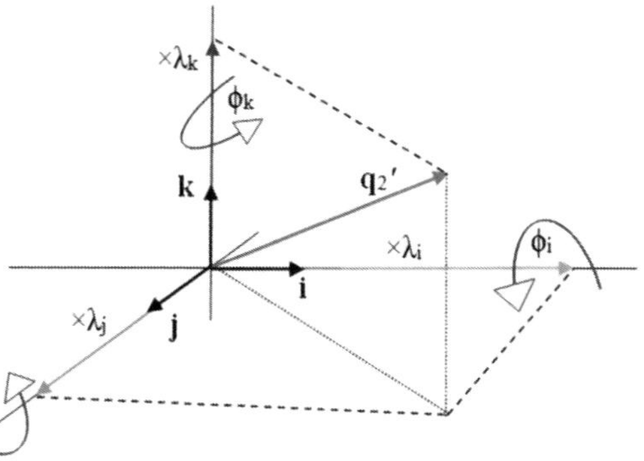

Figure 1.11 Three rotations around the axes in the vector subspace of quaternions.

$$\mathbf{A}\mathbf{q}_2 = \mathbf{A}\begin{pmatrix} a_2 \\ 0 \\ 0 \\ 0 \end{pmatrix} + \mathbf{A}\begin{pmatrix} 0 \\ b_2 \\ c_2 \\ d_2 \end{pmatrix} = a_2\mathbf{q}_1 + \begin{bmatrix} -(q_1', q_2') \\ \mathbf{B}\begin{pmatrix} b_2 \\ c_2 \\ d_2 \end{pmatrix} \end{bmatrix},$$

or we can write the multiplication as

$$\mathbf{A}\mathbf{q}_2 = \begin{bmatrix} a_2a_1 - (q_1', q_2') \\ a_2\mathbf{q}_1' + \mathbf{B}\mathbf{q}_2' \end{bmatrix}. \tag{1.38}$$

In the second line of matrix Eq. (1.38), the amplified vector $\mathbf{q}_1'$ is added to the combined rotation of vector $\mathbf{q}_2'$ by matrix $\mathbf{B}$. This is a geometrical interpretation of the quaternion multiplication in the general case.

Example 1.19

Consider two quaternions $q_1 = 2 + 1i + 3j + 4k$ and $q_2 = 0 + 2i + j - k$, where the numbers $a_1 = 2$, $b_1 = 1$, $c_1 = 3$, and $d_1 = 4$. The multiplication q_1q_2 of these two quaternions can be calculated as

$$\mathbf{q} = \mathbf{A}\mathbf{q}_2 = \begin{pmatrix} 2 & -1 & -3 & -4 \\ 1 & 2 & -4 & 3 \\ 3 & 4 & 2 & -1 \\ 4 & -3 & 1 & 2 \end{pmatrix}\begin{pmatrix} 0 \\ 2 \\ 1 \\ -1 \end{pmatrix} = \begin{pmatrix} -1 \\ -3 \\ 11 \\ -7 \end{pmatrix}.$$

Thus, the number $q = q_1q_2 = -1 - 3i + 11j - 7k$.

Figure 1.12 shows two vectors $\mathbf{q}_1' = (1, 3, 4)$ and $\mathbf{q}_2' = (2, 1, -1)$ and the vector $\mathbf{q}_{12}' = (-3, 11, -7)$. It can be shown that vector $\mathbf{q}_{12}'$ describes the rotation of vector $\mathbf{q}_2'$ around vector $\mathbf{q}_1'$ with a subsequent change in the length of vector $\mathbf{q}_2'$.

We now consider the rotations of the vector $\mathbf{q}_2' = (2, 1, -1)$ when calculating the imaginary part $q' = (-3i + 11j - 7k)$ of q. Matrix $\mathbf{B}$ is

$$\mathbf{B} = \begin{pmatrix} 2 & -4 & 3 \\ 4 & 2 & -1 \\ -3 & 1 & 2 \end{pmatrix} = \begin{pmatrix} 1 & 0 & 3 \\ 0 & 0 & 0 \\ -3 & 0 & 1 \end{pmatrix} + \begin{pmatrix} 1 & -4 & 0 \\ 4 & 1 & 0 \\ 0 & 0 & 0 \end{pmatrix} + \begin{pmatrix} 0 & 0 & 0 \\ 0 & 1 & -1 \\ 0 & 1 & 1 \end{pmatrix}.$$

Here, the matrix components and their determinants equal

$$\mathbf{B}_j = \mathbf{B}(1, 0, -3) = \begin{pmatrix} 1 & 0 & 3 \\ 0 & 0 & 0 \\ -3 & 0 & 1 \end{pmatrix} = \sqrt{10}\begin{pmatrix} \frac{1}{\sqrt{10}} & 0 & \frac{3}{\sqrt{10}} \\ 0 & 0 & 0 \\ -\frac{3}{\sqrt{10}} & 0 & \frac{1}{\sqrt{10}} \end{pmatrix}$$

and $\det(\mathbf{B}_j) = 1 + 9 = 10$;

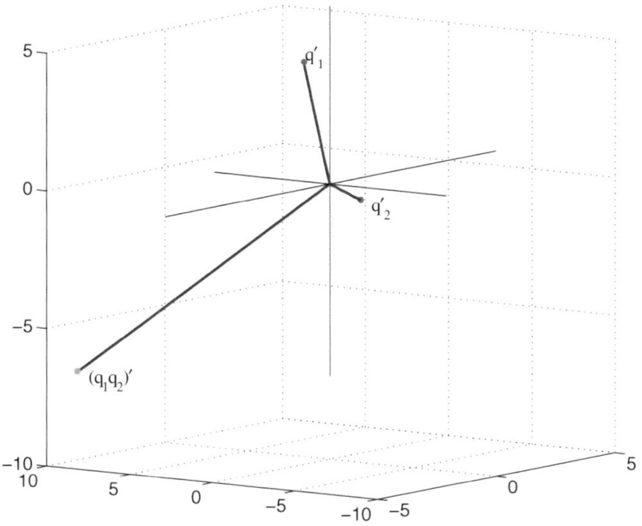

Figure 1.12 Vector $\mathbf{q}'_{12}$: vector $\mathbf{q}'_2$ rotation around vector $\mathbf{q}'_1$ with a subsequent change in length.

$$\mathbf{B}_k = \mathbf{B}(1,4,0) = \begin{pmatrix} 1 & -4 & 0 \\ 4 & 1 & 0 \\ 0 & 0 & 0 \end{pmatrix} = \sqrt{17}\begin{pmatrix} \frac{1}{\sqrt{17}} & -\frac{4}{\sqrt{17}} & 0 \\ \frac{4}{\sqrt{17}} & \frac{1}{\sqrt{17}} & 0 \\ 0 & 0 & 0 \end{pmatrix},$$

and $\det(\mathbf{B}_k) = 1 + 16 = 17$; and

$$\mathbf{B}_i = \mathbf{B}(0,1,1) = \begin{pmatrix} 0 & 0 & 0 \\ 0 & 1 & -1 \\ 0 & 1 & 1 \end{pmatrix} = \sqrt{2}\begin{pmatrix} 0 & 0 & 0 \\ 0 & \frac{1}{\sqrt{2}} & -\frac{1}{\sqrt{2}} \\ 0 & \frac{1}{\sqrt{2}} & \frac{1}{\sqrt{2}} \end{pmatrix},$$

and $\det(\mathbf{B}_i) = 1 + 1 = 2$.

Matrix $\mathbf{B}_j$ first rotates the projection vector $(2, 0, -1)$ along the circle of radius $r_{2,j} = \sqrt{5}$ around the j-axis by the angle

$$\phi_j = -\arctan\left(\frac{2c_1}{a_1}\right) = -\arctan(3) = 1.2490\,\text{rad},$$

or 71.5651° (see Fig. 1.13). The rotated projection is calculated as

$$\begin{pmatrix} \frac{1}{\sqrt{10}} & 0 & \frac{3}{\sqrt{10}} \\ 0 & 0 & 0 \\ -\frac{3}{\sqrt{10}} & 0 & \frac{1}{\sqrt{10}} \end{pmatrix}\begin{pmatrix} 2 \\ 0 \\ -1 \end{pmatrix} = \frac{1}{\sqrt{10}}\begin{pmatrix} -1 \\ 0 \\ -7 \end{pmatrix}.$$

Then, matrix $\mathbf{B}_j$ moves the rotated projection to the vector $(-1, 0, -7)$,

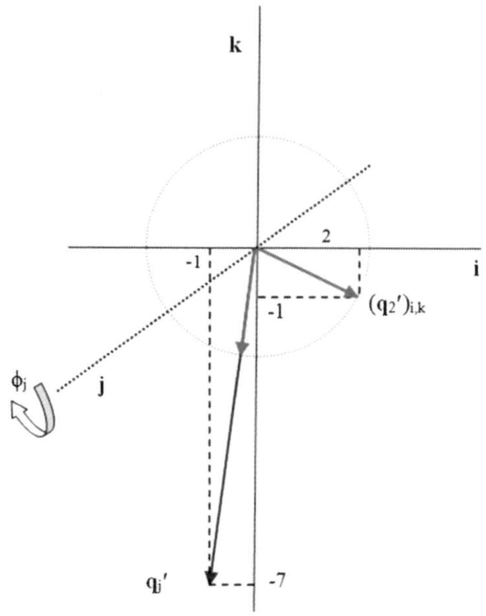

Figure 1.13 The rotation around the j-axis.

$$\mathbf{q}_j' = \sqrt{10}\cdot\begin{pmatrix} \frac{1}{\sqrt{10}} & 0 & \frac{3}{\sqrt{10}} \\ 0 & 0 & 0 \\ -\frac{3}{\sqrt{10}} & 0 & \frac{1}{\sqrt{10}} \end{pmatrix}\begin{pmatrix} 2 \\ 0 \\ -1 \end{pmatrix} = \begin{pmatrix} -1 \\ 0 \\ -7 \end{pmatrix}.$$

We can assume that the projection is first amplified by a factor of $\lambda_j = \sqrt{\det(\mathbf{B}_j)} = \sqrt{10}$, and then rotated by the angle ϕ_j,

$$\mathbf{q}_j' = \begin{pmatrix} \frac{1}{\sqrt{10}} & 0 & \frac{3}{\sqrt{10}} \\ 0 & 0 & 0 \\ -\frac{3}{\sqrt{10}} & 0 & \frac{1}{\sqrt{10}} \end{pmatrix}\left[\sqrt{10}\begin{pmatrix} 2 \\ 0 \\ -1 \end{pmatrix}\right] = \begin{pmatrix} -1 \\ 0 \\ -7 \end{pmatrix},$$

along a circle of radius $r = \lambda_j r_{2,j} = \sqrt{50}$.

Matrix $\mathbf{B}_k$ rotates and moves the projection vector $(2, 1, 0)$ to the vector $(-2, 9, 0)$,

$$\mathbf{q}_k' = \begin{pmatrix} 1 & -4 & 0 \\ 4 & 1 & 0 \\ 0 & 0 & 0 \end{pmatrix}\begin{pmatrix} 2 \\ 1 \\ 0 \end{pmatrix} = \begin{pmatrix} -2 \\ 9 \\ 0 \end{pmatrix},$$

around the k-axis by the angle

$$\phi_k = -\arctan\left(\frac{2d_1}{a_1}\right) = -\arctan(2) = 1.1071\,\text{rad},$$

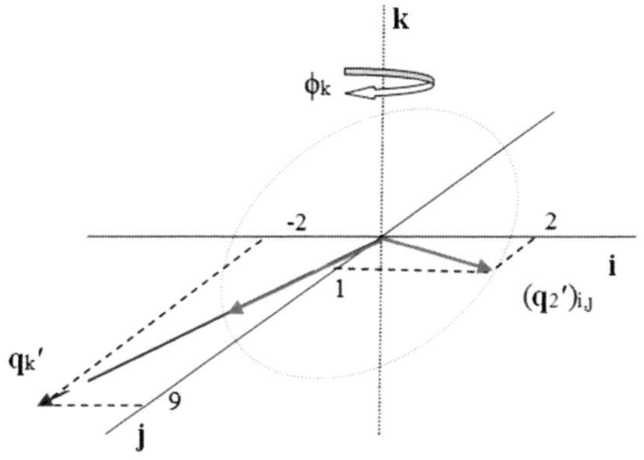

Figure 1.14 The rotation around the k-axis.

or 63.4349° (see Fig. 1.14). The rotated projection is calculated as

$$\begin{pmatrix} \frac{1}{\sqrt{17}} & -\frac{4}{\sqrt{17}} & 0 \\ \frac{4}{\sqrt{17}} & \frac{1}{\sqrt{17}} & 0 \\ 0 & 0 & 0 \end{pmatrix} \begin{pmatrix} 2 \\ 1 \\ 0 \end{pmatrix} = \frac{1}{\sqrt{17}} \begin{pmatrix} -2 \\ 9 \\ 0 \end{pmatrix}.$$

Then, matrix $\mathbf{B}_k$ moves the rotated projection to the vector $(-2, 9, 0)$ by multiplying the rotation by a factor of $\lambda_k = \sqrt{\det(\mathbf{B}_k)} = \sqrt{17}$.

Matrix $\mathbf{B}_i$ rotates and moves the vector $(0, 1, -1)$ to the vector $(-1, 0, -7)$. First, it rotates the vector as

$$\mathbf{q}_i' = \begin{pmatrix} 0 & 0 & 0 \\ 0 & \frac{1}{\sqrt{2}} & -\frac{1}{\sqrt{2}} \\ 0 & \frac{1}{\sqrt{2}} & \frac{1}{\sqrt{2}} \end{pmatrix} \begin{pmatrix} 0 \\ 1 \\ -1 \end{pmatrix} = \frac{1}{\sqrt{2}} \begin{pmatrix} 0 \\ 2 \\ 0 \end{pmatrix}$$

around the i-axis by the angle

$$\phi_i = -\arctan\left(\frac{2b_1}{a_1}\right) = -\arctan\left(\frac{1}{2}\right) = 0.4636\,\text{rad},$$

or 26.5651° (see Fig. 1.15). The rotated projection then is moved to the vector $(0, 2, 0)$, after multiplying by a factor of $\lambda_i = \sqrt{\det(\mathbf{B}_i)} = \sqrt{2}$.

We can verify the above calculations by

$$\mathbf{q}_j' + \mathbf{q}_k' + \mathbf{q}_i' = \begin{pmatrix} -1 \\ 0 \\ -7 \end{pmatrix} + \begin{pmatrix} -2 \\ 9 \\ 0 \end{pmatrix} + \begin{pmatrix} 0 \\ 2 \\ 0 \end{pmatrix} = \begin{pmatrix} -3 \\ 11 \\ -7 \end{pmatrix} = \mathbf{q}'.$$

The decomposition of the matrix $\mathbf{B}$ shows that vector $\mathbf{q}_{12}'$ can be obtained by a linear combination of three elementary rotations of the projections of the vector $\mathbf{q}_2'$ around the three axes of the Cartesian system.

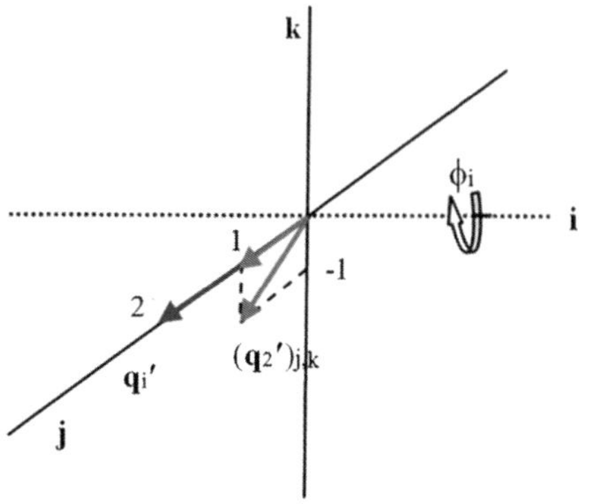

Figure 1.15 The rotation around the *i*-axis.

1.3.2 Rotation: Multiplication by a perpendicular vector

We consider the unit quaternion number (not pure) $q = a + q'$ and multiplication of this number by vector $\mathbf{s}'$, which is perpendicular to $\mathbf{q}'$.

Such a number can be represented as

$$q = \cos(\phi) + \frac{q'}{|q'|}\sin(\phi),$$

where the angle θ is defined from the condition $|q|^2 = a^2 + |q'|^2 = 1$ as

$$a = \cos(\phi) \quad \text{and} \quad |q'| = \sin(\phi).$$

The unit vector $\mathbf{p}' = \mathbf{q}'/|q'|$ is in the direction of the vector $\mathbf{q}'$.

Example 1.20

Consider the unit quaternion

$$q = \frac{1}{\sqrt{15}}(1 + i - 2j + 3k) = \frac{1}{\sqrt{15}} + \frac{i - 2j + 3k}{\sqrt{15}},$$

which can be written as

$$q = \frac{1}{\sqrt{15}} + \frac{i - 2j + 3k}{\sqrt{14}}\frac{\sqrt{14}}{15}.$$

In this case,

$$a = \frac{1}{\sqrt{15}}, \quad \mathbf{q}' = \frac{\mathbf{i} - 2\mathbf{j} + 3\mathbf{k}}{\sqrt{15}}, \quad |q'| = \frac{\sqrt{14}}{15},$$

and the pure unit quaternion vector is

$$\mathbf{p}' = \frac{\mathbf{i} - 2\mathbf{j} + 3\mathbf{k}}{\sqrt{14}} = 0.2673\mathbf{i} - 0.5345\mathbf{j} + 0.8018\mathbf{k}.$$

The angle θ is calculated from the equations

$$\cos(\phi) = \frac{1}{\sqrt{15}} = 0.2582 \quad \text{and} \quad \sin(\phi) = \frac{\sqrt{14}}{15} = 0.9661,$$

i.e., from the interval $[0, \pi/2]$ as

$$\phi = \arccos\left(\frac{1}{\sqrt{15}}\right) = 1.3096\,\text{rad},$$

or 75.0368°.

An illustration of the point (a, q') and vectors $\mathbf{p}'$ and $\mathbf{q}'$ are given in Fig. 1.16. The unit quaternion number q is presented as the point $(a, |q'|)$.

In the general case, when the module $|q|$ may be not 1, the quaternion q can be represented as

$$q = |q|[\cos(\phi) + p'\sin(\phi)], \quad \phi \in [0, \pi).$$

Thus, the number q is described as a triplet $(|q|, p', \phi)$, which is the polar form of the quaternion q. Below is the script of the MATLAB-based code "test_qinpolform.m" for Example 1.20, with the function "qinpolform.m" to calculate the polar form of q.

```
%------------------------------------------------------------
% test_qinpolform   / Art Grigoryan, November 10, 2015
   q=[1,1,-2,3]/sqrt(15);       % for Example 1.20
   [q_abs,p_vector,phi] = qinpolform(q);
```

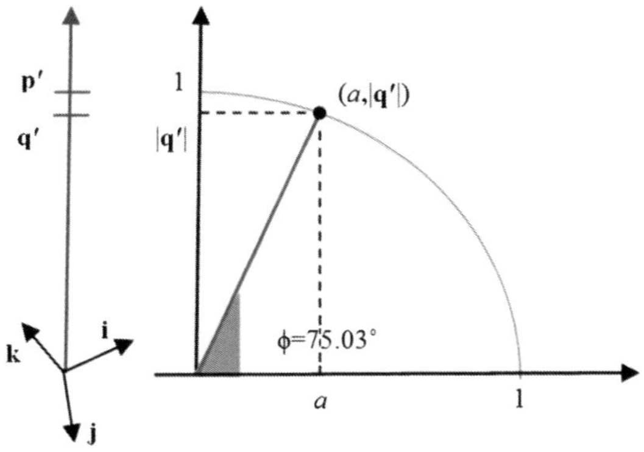

Figure 1.16 Components of the unit quaternion number $q = a + q'$.

```
   % The result is: 1, (0.2673, −0.5345,0.8018), and
     1.3096 rad/s
%----------------------------------------------------------------------
% call: qinpolform.m
% To calculate the polar form of a quaternion number
%      q=|q|[cos(phi) + p'sin(phi)]
%      as the triplet (|q|,p',phi).
  function [q_abs,p_vector,phi] = qinpolform(q)
   q_abs=norm(q);                       % is |q|
   qn=q/q_abs;                          % q/|q|
   qim=qn(2:4);                         % imaginary part of qn
   qim_abs=norm(qim);                   % is |qim|
   p_vector =qim/qim_abs;               % unit vector p'
   phi=acos(qn(1));                     % angle phi
%----------------------------------------------------------------------
```

The multiplication $q\mathbf{s}'$ equals

$$q\mathbf{s}' = [\cos(\phi) + \mathbf{p}' \sin(\phi)]\mathbf{s}' = \mathbf{s}' \cos(\phi) + (\mathbf{p}'\mathbf{s}') \sin(\phi).$$

Here, the multiplication of the vectors is calculated as

$$\mathbf{p}'\mathbf{s}' = -(\mathbf{p}', \mathbf{s}') + \mathbf{p}' \times \mathbf{s}' = \mathbf{p}' \times \mathbf{s}'.$$

We could also come to this result by writing the multiplication $q\mathbf{s}'$ as

$$\begin{aligned} q\mathbf{s}' &= a\mathbf{s}' - (\mathbf{q}', \mathbf{s}') + \mathbf{q}' \times \mathbf{s}' = a\mathbf{s}' + \mathbf{q}' \times \mathbf{s}' = a\mathbf{s}' + (|\mathbf{q}'|\mathbf{p}') \times \mathbf{s}' \\ &= a\mathbf{s}' + |\mathbf{q}'|(\mathbf{p}' \times \mathbf{s}') = \mathbf{s}' \cos(\phi) + (\mathbf{p}' \times \mathbf{s}') \sin(\phi). \end{aligned}$$

As shown above, the vector product $\mathbf{p}' \times \mathbf{s}'$ is a vector that is perpendicular to both vectors $\mathbf{p}'$ and $\mathbf{s}'$. According to Eq. (1.34), the length of the vector product equals

$$|\mathbf{p}' \times \mathbf{s}'| = |\mathbf{p}'||\mathbf{s}'| = |\mathbf{s}'|.$$

Denoting the vector product $\mathbf{p}' \times \mathbf{s}'$ by $\mathbf{v}'$, we can write $q\mathbf{s}' = \mathbf{s}' \cos(\phi) + \mathbf{v}' \sin(\phi)$. Both vectors $\mathbf{s}'$ and $\mathbf{v}'$ have the same length of $|\mathbf{s}'|$. Therefore, this equation describes the rotation of vector $\mathbf{s}'$ around vector $\mathbf{q}'$, or around $\mathbf{p}'$ by the angle θ as shown in Fig. 1.17. Vector $\mathbf{v}'$ is perpendicular to $\mathbf{s}'$, and we may say that $\mathbf{v}'$ is the result of the rotation of $\mathbf{s}'$ by 90° around the vector $\mathbf{p}'$. Thus, the following can be stated.

Statement 1.2 *The multiplication $q\mathbf{s}'$ describes the rotation of the perpendicular vector $\mathbf{s}'$ around the unit vector $\mathbf{p}'$ by the angle ϕ:*

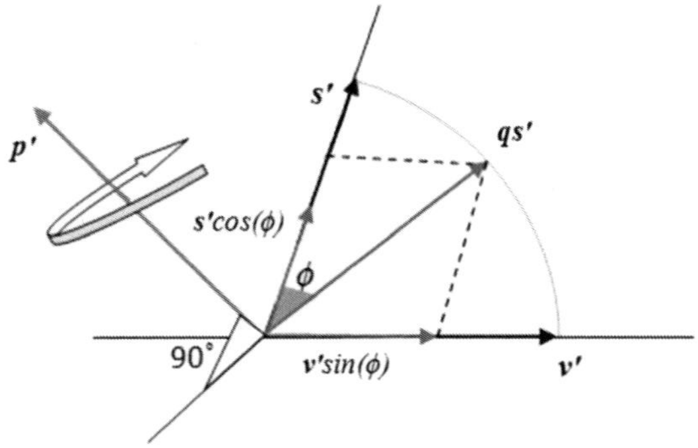

Figure 1.17 Rotation of vector $\mathbf{s}'$ around the unit vector $\mathbf{p}'$ by the angle ϕ.

$$q\mathbf{s}' = \mathbf{s}'\cos(\phi) + v'\sin(\phi). \tag{1.39}$$

Example 1.21

Consider the unit quaternion number

$$q = a + q' = \frac{1}{\sqrt{15}} + \frac{1}{\sqrt{15}}(i - 2j + 3k) = \cos(\phi) + p'\sin(\phi).$$

As shown in Example 1.20, the pure unit quaternion vector $\mathbf{p}'$ and angle ϕ are

$$\mathbf{p}' = \frac{\mathbf{i} - 2\mathbf{j} + 3\mathbf{k}}{\sqrt{14}} \quad \text{and} \quad \phi = \arccos\left(\frac{1}{\sqrt{15}}\right) = 1.3096\,\text{rad}.$$

Let us assume that vector $\mathbf{s}'$ to be rotated around the vector $\mathbf{p}'$ by angle ϕ is

$$\mathbf{s}' = 4\mathbf{i} - \mathbf{j} - 2\mathbf{k} \quad (|s'| = \sqrt{21}).$$

This vector is perpendicular to $\mathbf{p}'$, since

$$(p', s') = \frac{1}{\sqrt{15}}[1(4) + (-2)(-1) + 3(-2)] = 0.$$

Vector $\mathbf{s}'$ is rotated to vector $q\mathbf{s}'$, which can be calculated as

$$\begin{aligned} q\mathbf{s}' &= a\mathbf{s}' + \mathbf{q}' \times \mathbf{s}' = \tfrac{1}{\sqrt{15}}(4\mathbf{i} - \mathbf{j} - 2\mathbf{k}) + \tfrac{1}{\sqrt{15}}\begin{vmatrix} \mathbf{i} & \mathbf{j} & \mathbf{k} \\ 1 & -2 & 3 \\ 4 & -1 & -2 \end{vmatrix} \\ &= \tfrac{1}{\sqrt{15}}[(4\mathbf{i} - \mathbf{j} - 2\mathbf{k}) + (7\mathbf{i} + 14\mathbf{j} + 7\mathbf{k})] = \tfrac{1}{\sqrt{15}}(11\mathbf{i} + 13\mathbf{j} + 5\mathbf{k}). \end{aligned}$$

We also can calculate this rotated vector by using multiplication in matrix form, as described in Eq. (1.37),

$$qs' = As' = \frac{1}{\sqrt{15}}\begin{pmatrix} 1 & -1 & 2 & -3 \\ 1 & 1 & -3 & -2 \\ -2 & 3 & 1 & -1 \\ 3 & 2 & 1 & 1 \end{pmatrix}\begin{pmatrix} 0 \\ 4 \\ -1 \\ -2 \end{pmatrix} = \frac{1}{\sqrt{15}}\begin{pmatrix} 0 \\ 11 \\ 13 \\ 5 \end{pmatrix}.$$

Here, vector $\mathbf{s}' = (0, s') = (0, 4, -1, -2)^T$. We can verify that the length of vector $q\mathbf{s}'$ equals $|s'| = \sqrt{21}$ by

$$|q\mathbf{s}'| = \frac{\sqrt{11^2 + 13^2 + 5^2}}{\sqrt{15}} = \frac{\sqrt{315}}{\sqrt{15}} = \frac{\sqrt{15 \cdot 21}}{\sqrt{15}} = \sqrt{21}.$$

1.3.3 The 2nd rotation: Multiplication by a perpendicular vector

The above described rotation by the angle ϕ is defined by the linear operator

$$q \to L_q(\mathbf{s}') = q\mathbf{s}', \tag{1.40}$$

for a given unit quaternion q, which is considered in the form of $q = \cos(\phi) + p' \sin(\phi)$ with $(p')^2 = -1$. Here, vectors $\mathbf{s}'$ are perpendicular to the imaginary part of q, i.e., $s' \perp \mathbf{q}'$, or $\mathbf{s}' \perp \mathbf{p}'$. To release this constraint and describe the rotation by any vector $\mathbf{s}'$ by operations of multiplication, we consider a modification of the operator L_q.

In quaternion algebra, the following operator of rotation is used:

$$q \to L_{q,\bar{q}}(\mathbf{s}') = q\mathbf{s}'q^{-1}, \tag{1.41}$$

where it is not necessary for vector $\mathbf{s}'$ to be perpendicular to $\mathbf{q}'$. We first consider the operator for a perpendicular vector $\mathbf{s}'$ and then use the fact that any vector $\mathbf{s}'$ can be presented as the sum of two vectors, one of which is perpendicular to $\mathbf{q}'$.

The operator $L_{q,\bar{q}}$ as the operator $L_{q,\bar{q}}$ preserves the length of the vector $\mathbf{s}'$. Indeed, one can see that

$$|L_{q,\bar{q}}(\mathbf{s}')| = |q\mathbf{s}'q^{-1}| = |q||s'||q^{-1}| = |s'|.$$

Since $|q| = 1$, the inverse quaternion q^{-1} is $\bar{q}$, i.e.,

$$q^{-1} = \bar{q} = \cos(\phi) - p' \sin(\phi). \tag{1.42}$$

Therefore,

$$\begin{aligned} L_{q,\bar{q}}(\mathbf{s}') &= (\mathbf{s}' \cos(\phi) + [\mathbf{p}' \times \mathbf{s}'] \sin(\phi))(\cos(\phi) - \mathbf{p}' \sin(\phi)) \\ &= \mathbf{s}' \cos^2(\phi) - \mathbf{s}'\mathbf{p}' \cos(\phi) \sin(\phi) + [\mathbf{p}' \times \mathbf{s}'] \sin(\phi) \cos(\phi) \\ &\quad - [\mathbf{p}' \times \mathbf{s}']\mathbf{p}' \sin^2(\phi). \end{aligned}$$

Vectors $\mathbf{s}'$ and $\mathbf{p}'$ are perpendicular; therefore, $\mathbf{s}'\mathbf{p}' = -[\mathbf{p}' \times \mathbf{s}']$, and vectors $[\mathbf{p}' \times \mathbf{s}']$ and $\mathbf{p}'$ are also perpendicular. This simplifies the above calculations as

$$\begin{aligned} L_{q,\bar{q}}(\mathbf{s}') &= \mathbf{s}'\cos^2(\phi) + 2[\mathbf{p}' \times \mathbf{s}']\sin(\phi)\cos(\phi) + (\mathbf{s}'\mathbf{p}')\mathbf{p}'\sin^2(\phi) \\ &= \mathbf{s}'\cos^2(\phi) + 2[\mathbf{p}' \times \mathbf{s}']\sin(\phi)\cos(\phi) + \mathbf{s}'(\mathbf{p}'\mathbf{p}')\sin^2(\phi) \\ &= \mathbf{s}'\cos^2(\phi) + 2[\mathbf{p}' \times \mathbf{s}']\sin(\phi)\cos(\phi) - \mathbf{s}'\sin^2(\phi) \\ &= \mathbf{s}'\cos(2\phi) + [\mathbf{p}' \times \mathbf{s}']\sin(2\phi), \end{aligned} \quad (1.43)$$

since $(\mathbf{p}')^2 = -1$. Thus, we obtain the rotation of vector $\mathbf{s}'$ around vector $\mathbf{p}'$, which is similar to the rotation in Eq. (1.39) but by the angle 2ϕ,

$$L_{q,\bar{q}}(\mathbf{s}') = q\mathbf{s}'\bar{q} = \mathbf{s}'\cos(2\phi) + \mathbf{v}'\sin(2\phi). \quad (1.44)$$

Such a rotation is illustrated in Fig. 1.18.

Example 1.22

We describe the rotation of the quaternion vector $\mathbf{s}' = \mathbf{i} + 2\mathbf{j} - 3\mathbf{k}$ around the unit vector

$$\mathbf{p}' = \frac{1}{3\sqrt{5}}(2\mathbf{i} + 5\mathbf{j} + 4\mathbf{k})$$

by the angle $\phi = 30°$. The required vector $\mathbf{v}'$ in Eq. (1.44) is calculated by

$$\begin{aligned} \mathbf{v}' = \mathbf{p}' \times \mathbf{s}' &= \frac{1}{3\sqrt{5}}\begin{vmatrix} \mathbf{i} & \mathbf{j} & \mathbf{k} \\ 2 & 5 & 4 \\ 1 & 2 & -3 \end{vmatrix} = \frac{1}{3\sqrt{5}}[\mathbf{i}(-15-8) - \mathbf{j}(-6-4) + \mathbf{k}(4-5)] \\ &= \frac{1}{3\sqrt{5}}(-23\mathbf{i} + 10\mathbf{j} - \mathbf{k}). \end{aligned}$$

Therefore, the rotated vector $\mathbf{s}' \to \tilde{\mathbf{s}}'$ is calculated by

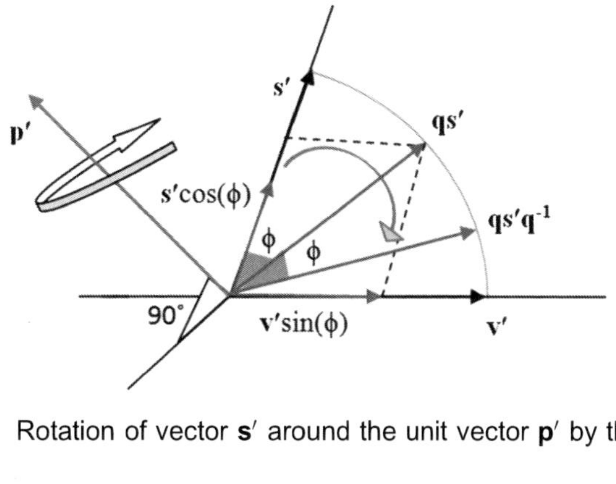

Figure 1.18 Rotation of vector $\mathbf{s}'$ around the unit vector $\mathbf{p}'$ by the angle 2ϕ.

$$\tilde{\mathbf{s}}' = (\mathbf{i} + 2\mathbf{j} - 3\mathbf{k})\cos(30°) + \frac{1}{3\sqrt{5}}(-23\mathbf{i} + 10\mathbf{j} - \mathbf{k})\sin(30°)$$
$$= \mathbf{i}\left(\frac{\sqrt{3}}{2} - \frac{23}{6\sqrt{5}}\right) + \mathbf{j}\left(\sqrt{3} + \frac{\sqrt{5}}{3}\right) - \mathbf{k}\left(\frac{3\sqrt{3}}{2} + \frac{1}{6\sqrt{5}}\right)$$
$$= -0.8483\mathbf{i} + 2.4774\mathbf{j} - 2.6726\mathbf{k},$$

considering that $\cos(30°) = \sqrt{3}/2$ and $\sin(30°) = 1/2$.

The vector $\tilde{\mathbf{s}}'$ is $q\mathbf{s}'\bar{q}$, but q has not been used directly in the above calculations. The quaternion q can be calculated by

$$q = \cos\left(\frac{\varphi}{2}\right) + p'\sin\left(\frac{\varphi}{2}\right) = \cos(15°) + p'\sin(15°)$$
$$= \cos(15°) + \frac{\sin(15°)}{3\sqrt{5}}(2i + 5j + 4k) = \frac{\sqrt{2+\sqrt{3}}}{2} + \frac{\sqrt{2-\sqrt{3}}}{2}\frac{2i + 5j + 4k}{3\sqrt{5}}$$
$$= \frac{\sqrt{2+\sqrt{3}}}{2} + \frac{\sqrt{2-\sqrt{3}}}{6\sqrt{5}}(2i + 5j + 4k) = 0.9659 + 0.0386(2i + 5j + 4k).$$

1.3.4 Multiplication: Rotation of any vector

The operation of rotation by the angle 2ϕ can also be described for the general case of vector $\mathbf{s}'$. Indeed, let $\mathbf{s}'$ be the vector that is not perpendicular to $\mathbf{p}'$. It can be presented as the sum of two vectors

$$\mathbf{s}' = \mathbf{s}'_\perp + \mathbf{s}'_\sim,$$

where vector $\mathbf{s}'_\perp$ is perpendicular to $\mathbf{p}'$, and $\mathbf{s}'_\sim$ is proportional to $\mathbf{p}'$; i.e., $\mathbf{s}'_\sim = \lambda\mathbf{p}'$ for a real number λ.

It is not difficult to notice that the multiplication $q(\lambda\mathbf{s}')$ is commutative. Indeed, the following calculations are valid:

$$q(\lambda\mathbf{p}') = (a + \mathbf{p}'|q'|)(\lambda\mathbf{p}') = \lambda(a\mathbf{p}' + \mathbf{p}'\mathbf{p}'|p'|) = \lambda\mathbf{p}'(a + p'|p'|) = (\lambda\mathbf{p}')q.$$

Therefore, vector $\mathbf{s}'_1$ remains invariable when applying the operator $L_{q,\bar{q}}$,

$$L_{q,\bar{q}}(\mathbf{s}'_\sim) = L_{q,\bar{q}}(\lambda\mathbf{p}') = q(\lambda\mathbf{p}')\bar{q} = (\lambda\mathbf{p}')q\bar{q} = (\lambda\mathbf{p}')|q|^2 = (\lambda\mathbf{p}') = \mathbf{s}'_\sim.$$

The operator $L_{q,\bar{q}}$ is linear; therefore, we obtain the following:

$$L_{q,\bar{q}}(\mathbf{s}') = L_{q,\bar{q}}(\mathbf{s}'_\perp) + L_{q,\bar{q}}(\mathbf{s}'_\sim) = L_{q,\bar{q}}(\mathbf{s}'_\perp) + \mathbf{s}'_\sim. \tag{1.45}$$

The component $\mathbf{s}'_\perp$ is rotated around vector $\mathbf{p}'$, and another component remains invariable, which means that vector $\mathbf{s}'$ is rotated around $\mathbf{p}'$.

Statement 1.3 *Given pure unit quaternion q and vector $\mathbf{s}'$, the equation*

$$L_{q,\bar{q}}(\mathbf{s}') = q\mathbf{s}'\bar{q} = [\mathbf{s}'_{\perp}\cos(2\phi) + (p'\mathbf{s}'_{\perp})\sin(2\phi)] + \mathbf{s}'_{\sim} \tag{1.46}$$

describes the rotation of vector $\mathbf{s}'$ around vector $\mathbf{p}'$ by angle 2ϕ.

1.3.5 Matrix representation of rotation

We consider the operator of rotation $L_{q,\bar{q}}(\mathbf{s}') = q\mathbf{s}'\bar{q}$ in matrix form. The multiplication $q\mathbf{s}'$ can be calculated by matrix $\mathbf{A}_{L;q}$, i.e.,

$$q\mathbf{s}' = \mathbf{A}_{L;q}\mathbf{s}' = \begin{pmatrix} a & -b & -c & -d \\ b & a & -d & c \\ c & d & a & -b \\ d & -c & b & a \end{pmatrix} \begin{pmatrix} 0 \\ s_i \\ s_j \\ s_k \end{pmatrix}. \tag{1.47}$$

Here, vector $\mathbf{s}'$ is composed of its components $\mathbf{s}' = 0 + \mathbf{i}s_i + \mathbf{j}s_j + \mathbf{k}s_k$, and, similarly, $q\mathbf{s}'$ is a vector of $q\mathbf{s}'$.

We denote $\tilde{\mathbf{s}}' = (q\mathbf{s}')\bar{q}$ and the corresponding column vector by $\tilde{\mathbf{s}}'$. The multiplication of the number $q\mathbf{s}'$ from the right by $\bar{q}$ can be accomplished by matrix $\mathbf{A}_{R;\bar{q}}$, i.e.,

$$\tilde{\mathbf{s}}' = \mathbf{A}_{R;\bar{q}}(q\mathbf{s}') = \begin{pmatrix} a & b & c & d \\ -b & a & -d & c \\ -c & d & a & -b \\ -d & -c & b & a \end{pmatrix} (q\mathbf{s}').$$

Therefore, the rotation can be written as

$$\tilde{\mathbf{s}}' = \mathbf{A}_{R;\bar{q}}(q\mathbf{s}') = \mathbf{A}_{R;\bar{q}}(\mathbf{A}_{L;q}\mathbf{s}') = (\mathbf{A}_{R;\bar{q}}\mathbf{A}_{L;q})\mathbf{s}'. \tag{1.48}$$

The operator of rotation $L_{q,\bar{q}}$ is described by the matrix

$$\mathbf{A}_{\bar{q},q} = \mathbf{A}_{R;\bar{q}}\mathbf{A}_{L;q} = \begin{pmatrix} a & b & c & d \\ -b & a & -d & c \\ -c & d & a & -b \\ -d & -c & b & a \end{pmatrix} \begin{pmatrix} a & -b & -c & -d \\ b & a & -d & c \\ c & d & a & -b \\ d & -c & b & a \end{pmatrix}$$
$$= \begin{bmatrix} 1 & 0 & 0 & 0 \\ 0 & a^2+b^2-c^2-d^2 & 2(bc-ad) & 2(ac+bd) \\ 0 & 2(ad+bc) & a^2-b^2+c^2-d^2 & 2(cd-ab) \\ 0 & 2(bd-ac) & 2(ab+cd) & a^2-b^2-c^2+d^2 \end{bmatrix}, \tag{1.49}$$

considering that $a^2 + b^2 + c^2 + d^2 = |q|^2 = 1$.

We can also write this matrix of rotation as

$$\mathbf{A}_{\bar{q},q} = \mathbf{I}_4 + 2\begin{bmatrix} 0 & 0 & 0 & 0 \\ 0 & -(c^2+d^2) & (bc-ad) & (ac+bd) \\ 0 & (ad+bc) & -(b^2+d^2) & (cd-ab) \\ 0 & (bd-ac) & (ab+cd) & -(b^2+c^2) \end{bmatrix}, \tag{1.50}$$

where $\mathbf{I}_4$ is the 4×4 identity matrix.

In the particular case when q is a pure quaternion number, i.e., $a = 0$, this matrix is symmetric and equals

$$\mathbf{A}_{\bar{q},q} = \mathbf{I}_4 + 2\begin{bmatrix} 0 & 0 & 0 & 0 \\ 0 & -(c^2+d^2) & bc & bd \\ 0 & bc & -(b^2+d^2) & cd \\ 0 & bd & cd & -(b^2+c^2) \end{bmatrix}, \tag{1.51}$$

or

$$\mathbf{A}_{\bar{q},q} = \mathbf{I}_4 + 2\begin{pmatrix} 0 & 0 & 0 & 0 \\ 0 & 1-b^2 & bc & bd \\ 0 & bc & 1-c^2 & cd \\ 0 & bd & cd & 1-d^2 \end{pmatrix}, \tag{1.52}$$

since $|q|^2 = b^2 + c^2 + d^2 = 1$. One can notice that $\mathbf{A}_{\bar{q},q} = \mathbf{A}_{q,\bar{q}}$ and the determinant of this matrix $\det(\mathbf{A}_{\bar{q},q}) = 1$.

Example 1.23

Consider the rotation related to vector $\mathbf{s}' = 4.5\mathbf{i} - 2\mathbf{j} + 3.5\mathbf{k}$ and the quaternion

$$q = \frac{1}{\sqrt{15}} + p'\frac{\sqrt{14}}{15} = \frac{1}{\sqrt{15}} + \frac{i - 2j + 3k}{\sqrt{14}}\frac{\sqrt{14}}{15} = \frac{1}{\sqrt{15}}(1 + i - 2j + 3k).$$

Vector $\mathbf{s}'$ can be presented as $\mathbf{s}' = \mathbf{s}'_\perp + \mathbf{s}'_\sim$, where

$$\mathbf{s}'_\perp = 4\mathbf{i} - \mathbf{j} + 2\mathbf{k} \quad \text{and} \quad \mathbf{s}'_\sim = 0.5\mathbf{i} - \mathbf{j} + 1.5\mathbf{k} = 0.5(\mathbf{i} - 2\mathbf{j} + 3\mathbf{k}).$$

To calculate the rotated vector $L_{q,\bar{q}}(\mathbf{s}') = q\mathbf{s}'\bar{q}$, we compose the matrix of rotation $\mathbf{A}_{\bar{q},q}$. For a given q, we have $a = 1$, $b = 1$, $c = -2$, and $d = 3$ all up to the factor of $1/\sqrt{15}$. The matrix can be written as

$$\mathbf{A}_{\bar{q},q} = \frac{1}{15}\begin{pmatrix} 15 & 0 & 0 & 0 \\ 0 & -11 & -10 & 2 \\ 0 & 2 & -5 & -14 \\ 0 & 10 & -10 & 5 \end{pmatrix}, \quad [\det(\mathbf{A}_{\bar{q},q}) = 1]. \tag{1.53}$$

Therefore, the rotated vector is calculated as

$$\tilde{\mathbf{s}}' = \mathbf{A}_{\bar{q},q}\begin{pmatrix} 0 \\ 4 \\ -1 \\ 2 \end{pmatrix} + \mathbf{A}_{\bar{q},q}\begin{pmatrix} 0 \\ 0.5 \\ -1 \\ 1.5 \end{pmatrix} = \begin{pmatrix} 0 \\ -2 \\ -1 \\ 4 \end{pmatrix} + \begin{pmatrix} 0 \\ 0.5 \\ -1 \\ 1.5 \end{pmatrix} = \begin{pmatrix} 0 \\ -1.5 \\ -2 \\ 5.5 \end{pmatrix}.$$

Below is the script of the MATLAB-based code "example_one23.m" for this example with the function "rotateVarroundP.m," which calculates the matrix $\mathbf{A}_{\bar{q},q}$ and rotated vector $\tilde{\mathbf{s}}'$.

```
%-------------------------------------------------------------------
% example_one23.m   / Art Grigoryan, December 30, 2015
  q= [1,1, -2,3]/sqrt(15);
  s1= [0,4, -1,2];
  s2= [0,1, -2,3]/2;
  s=s1+s2; % 0   4.5 -2  3.5
  [qsq,matrixR] = rotateVarroundP(q,s);
  % qsq =     [ 0  -1.5 -2  5.5]
  [qsq1,matrixR] = rotateVarroundP(q,s1);
  % qsq1 =    [ 0  -2    -1    4]
  [qsq2,matrixR] = rotateVarroundP(q,s2);
  % qsq2 =    [ 0   .5   -1  1.5]
  % ----------------- qsq = qsq1 + qsq2 -----------------
  R=round(15*matrixR)
  % 15   0    0    0
  %  0  -11  -10   2
  %  0    2   -5 -14
  %  0   10  -10   5
%-------------------------------------------------------------------
% call: rotateVarroundP / Art Grigoryan, December 31, 2015
% Calculate the matrix of rotation and rotated vector s
% around the unit vector p' by the doubled angle 2*phi,
% where phi = acos(a) and q = a + p' sin(phi).
  function [qsq,matrixR] = rotateVarroundP(q,s)
    a=q(1); b=q(2); c=q(3); d=q(4);
    if size(s,1) ==1 s=s'; end  % s to be a column-vector
    % 1st rotation
```

```
    A1=[a -b -c -d
        b  a -d  c
        c  d  a -b
        d -c  b  a];
    % 2nd rotation
    b=-b; c=-c; d=-d;   % for conjugate of quaternion q
    A2=[a -b -c -d
        b  a  d -c
        c -d  a  b
        d  c -b  a];
    % Matrix and Rotation by the doubled angle of arcos(a)
    matrixR=A2*A1;
    qsq=matrixR*s;            % as qs=A1*s; qsq=A2*qs;
%-----------------------------------------------------------
```

We also can formulate the task in the following way. Given unit vector $\mathbf{p}'$, rotate vector $\mathbf{s}'$ around $\mathbf{p}'$ by an angle 2ϕ. To use the above matrix of rotation, first the quaternion q should be calculated as

$$q = \cos(\phi) + p' \sin(\phi),$$

and then the matrix $\mathbf{A}_{\bar{q},q}$ can be constructed and used to calculate the rotated vector $L_{q,\bar{q}}(\mathbf{s}')$.

Example 1.24

Let vector $\mathbf{s}'$ be $4.5\mathbf{i} - 2\mathbf{j} + 3.5\mathbf{k}$ and rotated around the vector

$$\mathbf{p}' = \frac{1}{\sqrt{14}}(\mathbf{i} - 2\mathbf{j} + 3\mathbf{k})$$

by the angle 2ϕ, where

$$\phi = \arccos(1/\sqrt{15}) = 1.3096\,\text{rad},$$

or $75.0368°$. The quaternion q is calculated as

$$q = \cos(\phi) + p' \sin(\phi) = \frac{1}{\sqrt{15}} + \frac{i - 2j + 3k}{\sqrt{15}}.$$

As we know from Example 1.23, rotation of vector $\mathbf{s}'$ results in the vector

$$L_{q,\bar{q}}(\mathbf{s}') = -1.5\mathbf{i} - 2\mathbf{j} + 5.5\mathbf{k}.$$

Below is the script of the demo code "example_one24.m" for this example with the function "qsprotation.m," which calculates the rotation of the vector s' around the vector p' by a given angle.

```
%------------------------------------------------------------
% example_one24.m / Art Grigoryan, December 31, 2015
  s = [0, 4.5, -2, 3.5];
  p = [1,-2,3]/sqrt(14);
  phi_doubled=2*acos(1/sqrt(15));
  qsq = qsprotation(s,p,phi_doubled)
  % result: 0 -1.5 -2.0 5.5
%------------------------------------------------------------
% qsprotation.m
% rotate a vector s around a unit vector p' by angle phi
% by using the operator with multiplication by quaternion
%    q = cos(phi) + p'sin(phi), (phi in radians).
  function qsq = qsprotation(s,p,phi_doubled)
    phi=phi_doubled/2;
    q=zeros(1,4);
    q(1)=cos(phi);
    q(2:4)=sin(phi)*p;
    qsq=rotateVarroundP(q,s);
%------------------------------------------------------------
```

For this example, we have

$$\mathbf{A}_{q,\bar{q}} = \frac{1}{15}\begin{pmatrix} 15 & 0 & 0 & 0 \\ 0 & -11 & 2 & 10 \\ 0 & -10 & -5 & -10 \\ 0 & 2 & -14 & 5 \end{pmatrix}, = \mathbf{A}^T_{\bar{q},q},$$

and this property holds for all matrices $\mathbf{A}_{q,\bar{q}}$.

One can note that two quaternion numbers $q = a + ib + jc + kd$ and $-q = -a - ib - jc - kd$ define the same matrix, i.e.,

$$\mathbf{A}_{\bar{q},q} = \mathbf{A}_{-\bar{q},-q}.$$

Thus, up to the sign, the number q can be restored from the matrix $\mathbf{A} = \mathbf{A}_{\bar{q},q}$. We denote the elements of this matrix by $a_{n,m}$, $n, m = 0, 1, 2, 3$. If we add all elements of this matrix along the main diagonal, which is the trace, $\mathrm{tr}(\mathbf{A})$, we obtain the following (see Eq. 1.49):

$$\mathrm{tr}(\mathbf{A}) = a_{0,0} + a_{1,1} + a_{2,2} + a_{3,3} = 1 + 3a^2 - b^2 - c^2 - d^2 = 4a^2.$$

Thus, then the real part of q can be calculated by

$$a = \pm\frac{1}{2}\sqrt{\mathrm{tr}(\mathbf{A})}. \tag{1.54}$$

Three components of the imaginary part of q can be defined from other elements of the matrix as

$$a_{2,1} - a_{1,2} = 2ad, \quad a_{1,3} - a_{3,1} = 2ac, \quad a_{3,2} - a_{2,3} = 2ab,$$

and therefore

$$d = \frac{a_{2,1} - a_{1,2}}{2a}, \quad c = \frac{a_{1,3} - a_{3,1}}{2a}, \quad b = \frac{a_{3,2} - a_{3,2}}{2a}. \tag{1.55}$$

If the quaternion is pure, i.e., $a = 0$, we can use the second matrix-summand in Eq. (1.52), the elements of which we denote by $s_{n,m}$,

$$\mathbf{S}_{\bar{q},q} = (\{s_{n,m}\}_{n,m=0}^{3}) = 2\begin{pmatrix} 0 & 0 & 0 & 0 \\ 0 & 1-b^2 & bc & bd \\ 0 & bc & 1-c^2 & cd \\ 0 & bd & cd & 1-d^2 \end{pmatrix}. \tag{1.56}$$

Three components of the imaginary part of q can be defined from the diagonal of this matrix as

$$s_{1,1} = 2(1-b^2), \quad s_{2,2} = 2(1-c^2), \quad a_{3,3} = 2(1-d^2).$$

Thus, up to the sign $\alpha = \pm 1$, we obtain the imaginary components:

$$b = \frac{\alpha\sqrt{1 - s_{1,1}}}{2}, \quad c = \frac{\alpha\sqrt{1 - s_{2,2}}}{2}, \quad d = \frac{\alpha\sqrt{1 - s_{3,3}}}{2}. \tag{1.57}$$

1.3.6 Addition of rotations in 3-D space

We have used the property of linearity of the operation of rotation,

$$L_{q,\bar{q}}(\mathbf{s}' + \mathbf{v}') = L_{q,\bar{q}}(\mathbf{s}') + L_{q,\bar{q}}(\mathbf{v}')$$

for arbitrary vectors $\mathbf{s}'$ and $\mathbf{v}'$.

We also can consider the composition of the rotations, when a vector is rotated around an axis with unit vector $\mathbf{p}_1'$ and then around a different axis characterized by unit vector $\mathbf{p}_2'$ as

$$\mathbf{s}' \to L_{q_1,\bar{q}_1}(\mathbf{s}') \to L_{q_2,\bar{q}_2}[L_{q_1,\bar{q}_1}(\mathbf{s}')]. \tag{1.58}$$

Here, q_1 and q_2 are unit quaternions, which that can be written as

$$q_1 = \cos(\phi_1) + p_1' \sin(\phi_1) \quad \text{and} \quad q_2 = \cos(\phi_2) + p_2' \sin(\phi_2).$$

The quaternion multiplication is distributive, and therefore,

$$L_{q_2,\bar{q}_2}[L_{q_1,\bar{q}_1}(\mathbf{s}')] = L_{q_2,\bar{q}_2}(q_1\mathbf{s}'\bar{q}_1) = q_2(q_1\mathbf{s}'\bar{q}_1)\bar{q}_2 = (q_2q_1)\mathbf{s}'(\bar{q}_1\bar{q}_2),$$

and we know that $(\bar{q}_1\bar{q}_2) = \overline{q_2q_1}$. Therefore, denoting the quaternion q_2q_1 by

$$q = \cos(\phi) + p' \sin(\phi), \tag{1.59}$$

we obtain the following statement.

Statement 1.4 *The consequence of performing two rotations equivalent to the rotation around vector* $\mathbf{p}'$ *by angle* 2ϕ*, where both of these values can be easily calculated from the quaternion multiplication* $q = q_2 q_1$*, is*

$$L_{q_2,\bar{q}_2}[L_{q_1,\bar{q}_1}(\mathbf{s}')] = q\mathbf{s}'\bar{q} = L_{q,\bar{q}}(\mathbf{s}'). \tag{1.60}$$

Example 1.25

Consider two unit quaternions

$$q_1 = a_1 + q_1' = \frac{1}{\sqrt{7}} + \frac{i - j + 2k}{\sqrt{7}} \quad \text{and} \quad q_2 = a_2 + q_2' = \frac{1}{3} + \frac{2i - 2j}{3},$$

and the corresponding unit vectors

$$\mathbf{p}_1' = \frac{\mathbf{q}_1'}{|q_1'|} = \frac{\mathbf{i} - \mathbf{j} + 2\mathbf{k}}{\sqrt{6}} \quad \text{and} \quad \mathbf{p}_2' = \frac{\mathbf{q}_2'}{|q_2'|} = \frac{2\mathbf{i} - 2\mathbf{j}}{\sqrt{8}} = \frac{\mathbf{i} - \mathbf{j}}{\sqrt{2}}.$$

Let us assume that a vector $\mathbf{s}'$ is rotated first around vector $\mathbf{p}_1'$ and then around $\mathbf{p}_2'$ by angles $2\phi_1$ and $2\phi_2$, where

$$\phi_1 = \arccos\left(\frac{1}{\sqrt{7}}\right) = 67.7923° \quad \text{and} \quad \phi_2 = \arccos\left(\frac{1}{3}\right) = 70.5288°,$$

respectively. To find the resulting rotation, we first calculate the quaternion $q = q_2 q_1$. The real part of this number is

$$a = a_1 a_2 - (q_1', q_2') = \frac{1}{3\sqrt{7}}[1 - (2 + 2)] = -\frac{1}{\sqrt{7}},$$

and the imaginary part is

$$q' = (a_1 q_2' + a_2 q_1') + (q_1' \times q_2') = \tfrac{1}{3\sqrt{7}}[(2i - 2j) + (i - j + 2k)] - \tfrac{1}{3\sqrt{7}}\begin{vmatrix} i & j & k \\ 1 & -1 & 2 \\ 2 & -2 & 0 \end{vmatrix}$$
$$= \tfrac{1}{3\sqrt{7}}(3i - 3j + 2k) - \tfrac{1}{3\sqrt{7}}(4i + 4j) = \tfrac{1}{3\sqrt{7}}(-i - 7j + 2k).$$

The vector $\mathbf{q}'$ can be written as

$$\mathbf{q}' = \frac{-\mathbf{i} - 7\mathbf{j} + 2\mathbf{k}}{3\sqrt{7}} = \frac{-\mathbf{i} - 7\mathbf{j} + 2\mathbf{k}}{3\sqrt{6}} \cdot \sqrt{\frac{6}{7}}.$$

Therefore, the corresponding quaternion is written as

$$q = q_2 q_1 = -\frac{1}{\sqrt{7}} + \frac{-i - 7j + 2k}{3\sqrt{7}} = \frac{-3 - i - 7j + 2k}{3\sqrt{7}},$$

the angle of rotation is 2ϕ, where

$$\phi = \arccos(a) = \arccos\left(-\frac{1}{\sqrt{7}}\right) = 112.2077°,$$

and vector $\mathbf{p}'$ around which vector $\mathbf{s}'$ is to be rotated is

$$\mathbf{p}' = \frac{-\mathbf{i} - 7\mathbf{j} + 2\mathbf{k}}{3\sqrt{6}}.$$

Thus, two consecutive rotations of vector $\mathbf{s}'$ by angles $2 \times 67.7923°$ and $2 \times 70.5288°$ around vectors $\mathbf{p}'_1$ and $\mathbf{p}'_2$ can be accomplished by one rotation of vector $\mathbf{s}'$ around vector $\mathbf{p}'$ by angle $2 \times 112.2077°$.

Example 1.26

In Example 1.25, the angles of rotations $2\phi_1$ and $2\phi_2$, as well as the summary angle 2ϕ, are large. Let us consider the consequence of rotating vector $\mathbf{s}'$ around the same unit vectors

$$\mathbf{p}'_1 = \frac{\mathbf{i} - 7\mathbf{j} + 2\mathbf{k}}{\sqrt{6}} \quad \text{and} \quad \mathbf{p}'_2 = \frac{\mathbf{i} - \mathbf{j}}{\sqrt{2}}$$

by 30° and 60°, respectively. Our task is to calculate the quaternion $q = q_2 q_1$ that describes this rotation.

Since

$$\cos(15°) = \frac{\sqrt{1 + \cos(30°)}}{2} = \frac{\sqrt{1 + \frac{\sqrt{3}}{2}}}{2} = \frac{\sqrt{2 + \sqrt{3}}}{2}$$

and

$$\sin(15°) = \frac{\sqrt{1 - \cos(30°)}}{2} = \frac{\sqrt{1 - \frac{\sqrt{3}}{2}}}{2} = \frac{\sqrt{2 - \sqrt{3}}}{2},$$

the quaternion $q_1 = a_1 + q'_1$ is calculated by

$$q_1 = \cos(15°) + p'_1 \sin(15°) = \frac{\sqrt{2 + \sqrt{3}}}{2} + p'_1 \frac{\sqrt{2 - \sqrt{3}}}{2}.$$

The second quaternion $q_2 = a_2 + q_2'$ is

$$q_2 = \cos(30°) + p_2' \sin(30°) = \frac{\sqrt{3}}{2} + p_2' \frac{1}{2}.$$

To calculate the vector $q = q_2 q_1$, we first consider its real part

$$a = a_1 a_2 - (q_1', q_2') = a_1 a_2 - \frac{\sqrt{2-\sqrt{3}}}{2}\frac{1}{2}(p_1', p_2')$$
$$= \frac{\sqrt{2+\sqrt{3}}}{2}\frac{\sqrt{3}}{2} - \frac{\sqrt{2-\sqrt{3}}}{4}\frac{1}{\sqrt{6}\sqrt{2}}(1+1+0) = \frac{\sqrt{6+3\sqrt{3}}}{4} - \frac{\sqrt{6-3\sqrt{3}}}{12},$$

or $a = 0.7618$. This number defines the doubled angle 2ϕ of rotation, where

$$\phi = \arccos(a) = 0.7047\,\text{rad},$$

or 40.3767°.

The imaginary part of $q = q_2 q_1$ can be calculated by

$$q' = (a_2 q_1' + a_1 q_2') + q_2' \times q_1' = (a_2 q_1' + a_1 q_2') + \frac{\sqrt{2-\sqrt{3}}}{2}\frac{1}{2}\frac{1}{\sqrt{2}}\frac{1}{\sqrt{6}}\begin{vmatrix} i & j & k \\ 1 & -1 & 0 \\ 1 & -1 & 2 \end{vmatrix}$$
$$= \left[\frac{\sqrt{3}}{2}\frac{\sqrt{2-\sqrt{3}}}{2} \cdot \frac{i-j+2k}{\sqrt{6}} + \frac{\sqrt{2+\sqrt{3}}}{2} \cdot \frac{i-j}{2\sqrt{2}}\right] + \frac{\sqrt{2-\sqrt{3}}}{8\sqrt{3}}(-2i-2j)$$
$$= \left[\frac{\sqrt{4-2\sqrt{3}}}{8}(i-j+2k) + \frac{\sqrt{4+2\sqrt{3}}}{8}(i-j)\right] - \frac{\sqrt{6-3\sqrt{3}}}{12}(i+j).$$

The components of this vector are

$$b = \frac{\sqrt{4-2\sqrt{3}} + \sqrt{4+2\sqrt{3}}}{8} - \frac{\sqrt{6-3\sqrt{3}}}{12} = 0.3583,$$
$$c = -\frac{\sqrt{4-2\sqrt{3}} + \sqrt{4+2\sqrt{3}}}{8} - \frac{\sqrt{6-3\sqrt{3}}}{12} = -0.5077,$$
$$d = \frac{\sqrt{4-2\sqrt{3}}}{4} = 0.1830.$$

Thus, the required quaternion is

$$q = a + q' = 0.7618 + (0.3583i - 0.5077j + 0.1830k), \tag{1.61}$$

and it can be verified that $|q|^2 = 1$. From this number, we find the unit vector

$$p' = \frac{q'}{|q'|} = \frac{0.3583i - 0.5077j + 0.1830k}{\sqrt{1-a^2}} = \frac{0.3583i - 0.5077j + 0.1830k}{0.6478}.$$

Of course, it is difficult to make such calculations by hand, and we can write the code to calculate all required data. As an example, a simple script of the code "example_one26.m" is given below with the input data of our example.

```
%---------------------------------------------------------------
% example_one26.m   / Art Grigoryan, January 1, 2016
% Rotation of the vector around two unit vectors p1 and p2
% by the given angles phi_1 and phi_2, respectively.
% 1. Given data:
  s= [0, 4.5, -2, 3.5];
  p1= [1, -1, 2]/sqrt(6); phi_1=30/180*pi;
  p2= [1, -1, 0]/sqrt(2); phi_2=60/180*pi;
% 2. Calculation of the quaternions q1, q2, and q=q2*q1:
  q1= [cos(phi_1/2),sin(phi_1/2)*p1]; % 0.9659 0.1057
  -0.1057 0.2113
  q2= [cos(phi_2/2),sin(phi_2/2)*p2]; % 0.8660 0.3536
  -0.3536 0
  q = mult2qs_direct(q2,q1);           % 0.7618 0.3583
  -0.5077 0.1830
% 3. Calculation of the vector p, angle, matrix, and rotated
  vector:
  [q_abs,p,phi]=qinpolform(q);          % 1, 0.5531 -0.7838
  0.2825, 0.7047
  [qsq,matrixR]=rotateVarroundP(q,s);
  %    qsq = [ -0.0000 0.9153 -4.2961 4.1480]
  % matrixR = [1        0        0        0
  %            0   0.4174  -0.6427  -0.6424
  %            0  -0.0850   0.6763  -0.7317
  %            0   0.9047   0.3601   0.2277]
%---------------------------------------------------------------
```

The results of this program are the following:

$$\begin{aligned}
q_1 &= 0.9659 + 0.1057i - 0.1057j + 0.2113k,\\
q_2 &= 0.8660 + 0.3536i - 0.3536j,\\
q &= q_2 q_1 = 0.7618 + 0.3583i - 0.5077j + 0.1830k,\\
p' &= 0.5531i - 0.7838j + 0.2825k,\\
\phi &= 0.7047 \text{ rad}.
\end{aligned} \tag{1.62}$$

Here, q coincides with the quaternion number in Eq. (1.61). The matrix of vector $\mathbf{s}'$ rotation around vector $\mathbf{p}'$ by the angle $\phi = 0.7047$, or 40.3767°, is

$$\mathbf{A}_{\bar{q},q} = \begin{pmatrix} 1 & 0 & 0 & 0 \\ 0 & 0.4174 & -0.6427 & -0.6424 \\ 0 & -0.0850 & 0.6763 & -0.7317 \\ 0 & 0.9047 & 0.3601 & 0.2277 \end{pmatrix}, \quad [\det(\mathbf{A}_{\bar{q},q}) = 1],$$

and the rotated vector is

$$L_{q,\bar{q}}(\mathbf{s}') = 0.9153\mathbf{i} - 4.2961\mathbf{j} + 4.1480\mathbf{k}.$$

One can verify that, for this example, $|L_{q,\bar{q}}(\mathbf{s}')| = |\mathbf{s}'| = 6.0415$.

1.4 The Quaternion Exponential Function

The above-considered module $|q|$ of the quaternion is an example of the non-negative function over the quaternions $q = a + q'$. Other functions are the real component a and imaginary, or vector, component q' of the quaternion q, and the quaternion conjugate $\bar{q}$. In this section, we consider the concept of the exponential function in quaternion arithmetic, which is important in understanding rotations in the vector space of a quaternion, as well as in studying of the application of quaternion Fourier transforms, which will be described in the next chapters.

P1. We consider the Taylor series for the exponential function in the real case: [36]

$$e^x = \exp(x) = 1 + x + \frac{x^2}{2!} + \frac{x^3}{3!} + \frac{x^4}{4!} + \frac{x^5}{5!} + \cdots$$

together with two series

$$\cos(x) = 1 + \sum_{n=1}^{\infty} (-1)^n \frac{x^{2n}}{(2n)!},$$
$$\sin(x) = \sum_{n=0}^{\infty} (-1)^n \frac{x^{2n+1}}{(2n+1)!}.$$

The first series for the imaginary number ix is

$$
\begin{aligned}
e^{ix} = \exp(ix) &= 1 + (ix) + \frac{(ix)^2}{2!} + \frac{(ix)^3}{3!} + \frac{(ix)^4}{4!} + \frac{(ix)^5}{5!} + \cdots \\
&= 1 + ix - \frac{x^2}{2!} - i\frac{x^3}{3!} + \frac{x^4}{4!} + i\frac{x^5}{5!} + \cdots \\
&= \left[1 - \frac{x^2}{2!} + \frac{x^4}{4!} - \cdots\right] + i\left[x - \frac{x^3}{3!} + \frac{x^5}{5!} - \cdots\right] \\
&= \cos(x) + i\,\sin(x).
\end{aligned}
$$

The Taylor series can also be used for the quaternions as

$$
e^q = \exp(q) = 1 + \sum_{n=1}^{\infty} \frac{q^n}{n!}. \tag{1.63}
$$

The convergence of this series can be shown by considering two properties of the modulus operation: (1) the triangle inequality $|q_1 + q_2| \leq |q_1| + |q_2|$ and (2) modulus of the power $|q^n| = |q|^n$, for positive integer n. Indeed, the following inequalities hold:

$$
|e^q| = \left|1 + \sum_{n=1}^{\infty} \frac{q^n}{n!}\right| \leq 1 + \sum_{n=1}^{\infty} \left|\frac{q^n}{n!}\right| = 1 + \sum_{n=1}^{\infty} \frac{|q^n|}{n!} = 1 + \sum_{n=1}^{\infty} \frac{|q|^n}{n!} = e^{|q|}.
$$

Let us first describe the case when $q = \mu\vartheta$, where ϑ is a real number as an angle, and μ is a pure unit quaternion ($\mu^2 = -1$). One can note that

$$
\mu^3 = (\mu^2)\mu = -\mu,\ \mu^4 = (\mu^2)^2 = 1,\ \mu^5 = (\mu^4)\mu = \mu,\ \mu^6 = (\mu^2)^3 = -1, \ldots
$$

or

$$
\mu^{2n} = (-1)^n, \quad \mu^{2n+1} = (-1)^n\mu, \quad n = 0, 1, 2, \ldots.
$$

Therefore, the quaternion number $\exp(\mu\vartheta)$ can be written as

$$
\begin{aligned}
e^{\mu\vartheta} &= 1 + \sum_{n=1}^{\infty} \frac{(\mu\vartheta)^n}{n!} = 1 + \sum_{n=1}^{\infty} \frac{(\mu)^n(\vartheta)^n}{n!} \\
&= \left[1 + \sum_{n=1}^{\infty}(-1)^n \frac{(\vartheta)^{2n}}{(2n)!}\right] + \mu\left[\sum_{n=0}^{\infty}(-1)^n \frac{(\vartheta)^{2n+1}}{(2n+1)!}\right] \\
&= \cos(\vartheta) + \mu\,\sin(\vartheta).
\end{aligned}
$$

Any pure quaternion q can be written as

$$
q = q' = \frac{q'}{|q'|}|q'| = \mu\vartheta, \quad \text{where } \mu = \frac{q'}{|q'|}, \vartheta = |q'| > 0,
$$

and therefore,

$$e^q = \cos(\vartheta) + \mu \sin(\vartheta) = \cos(|q'|) + \frac{q'}{|q'|}\sin(|q'|).$$

In general, when $q = a + q'$, the real number a is commutative with q', i.e., $aq' = q'a$. Therefore, we can define the exponential function as

$$e^q = e^{a+q'} = e^a e^{q'} = e^a[\cos(\vartheta) + \mu \sin(\vartheta)], \tag{1.64}$$

or

$$e^q = e^a\left[\cos(|q'|) + \frac{q'}{|q'|}\sin(|q'|)\right].$$

We consider the exponent of the imaginary part of q to be

$$e^{q'} = \cos(\vartheta) + \mu \sin(\vartheta) = \cos(\vartheta) + i\mu_i \sin(\vartheta) + j\mu_j \sin(\vartheta) + k\mu_k \sin(\vartheta).$$

Therefore, the quaternion exponent can be written as

$$e^{q'} = [1 - (\mu_i + \mu_j + \mu_k)]\cos(\vartheta) + \mu_i e^{i\vartheta} + \mu_j e^{j\vartheta} + \mu_k e^{k\vartheta}.$$

It should be noted that, for a given pure quaternion μ, the number

$$Q = \cos(\vartheta) + \mu \sin(\vartheta)$$

is a quaternion number, and this number is the exponent $Q = e^{\mu\vartheta}$ only if μ is a pure unit quaternion. Otherwise, we can express the number

$$Q = (1 - |\mu|)\cos(\vartheta) + |\mu|\left[\cos(\vartheta) + \frac{\mu}{|\mu|}\sin(\vartheta)\right] = (1 - |\mu|)\cos(\vartheta) + |\mu|e^{\frac{\mu}{|\mu|}\vartheta}$$

as the arithmetic sum of the cosine and exponent.

Example 1.27

Consider the quaternion $q = 1 + i - 2j + k$, which can be written as

$$q = 1 + \mu\vartheta = 1 + \frac{i - 2j + k}{\sqrt{6}}\sqrt{6}.$$

Then, the exponential number $\exp(q)$ is calculated as

$$e^q = e^1\left[\cos(\sqrt{6}) + \frac{i - 2j + k}{\sqrt{6}}\sin(\sqrt{6})\right].$$

The real part of the exponent is

$$(e^q)_e = e^1 \cos(\sqrt{6}) = -2.0928,$$

and the components of the imaginary part equal

$$(e^q)_i = e^1 \frac{1}{\sqrt{6}} \sin(\sqrt{6}) = 0.7082,$$

$$(e^q)_j = e^1 \frac{-2}{\sqrt{6}} \sin(\sqrt{6}) = -1.4164,$$

$$(e^q)_k = e^1 \frac{1}{\sqrt{6}} \sin(\sqrt{6}) = 0.7082.$$

The exponent of q equals

$$e^q = -2.0928 + (0.7082i - 1.4164j + 0.7082k).$$

The length of this number $|e^q| = 2.7183$ and $e^{|q|} = e^{2.6458} = 14.0940$. One can note that, if the imaginary part has equal components, then the corresponding components of the imaginary part of the exponent $\exp(q)$ are equal, too. For instance, if $q_i = q_k$, then $(e^q)_i$ equals $(e^q)_k$; and if $q_i = q_j$, then $(e^q)_i = (e^q)_j$.

Below is the script of the MATLAB-based code "test_expofq.m" for this example with the function "expofq.m" to calculate the quaternion exponential function. We can also check that, for the numbers $q_1 = 1 + i$, $q_2 = 1 - 2j$, and $q_3 = 1 + k$, the exponential function equals

$$e^{1+i} = 1.4687 + 2.2874i,\ e^{1+k} = 1.4687 + 2.2874k,$$
$$e^{1-2j} = -1.1312 - 2.4717j,$$

and the complex exponential function $e^{1+i} = 1.4687 + 2.2874i$ and $e^{1-2i} = -1.1312 - 2.4717i$.

```
%-------------------------------------------------------
% test_expof.m
  q=[1,1,-2,1];
  qexp = expofq(q);      % -2.0928 0.7082 -1.4164 0.7082
% Checking the function
  qexp=expofq([1,0,-2,0]);      % -1.1312 0 -2.4717 0
  zexp=exp(1-2*i);              % -1.1312 -2.4717i
  %
  qexp=expofq([1,1,0,0]);       % 1.4687 2.2874 0 0
  zexp=exp(1+i);                % 1.4687 + 2.2874i
  qexp=expofq([1,0,0,1]);       % 1.4687 0 0 2.2874
```

```
%-----------------------------------------------------------
% expofq.m / Artyom Grigoryan, January 8, 2016
% calculate the quaternion exponent function
  function qexp=expofq(q)
    qim=q(2:4);                % imaginary part of q
    theta=norm(qim);           % angle theta=|qim|=sqrt
                                 (b^2+c^2+d^2)
    mu_vector=qim/theta;  % unit \mu
    e1=exp(q(1));
    qexp(1)=e1*cos(theta);
    qexp(2:4)=(e1*sin(theta))*mu_vector; % See Eq. 1.64
%-----------------------------------------------------------
```

P2. It is known that the cosine, sine, and complex exponential functions are periodic with the fundamental period $T = 2\pi$,

$$\cos(x) = \cos(x + 2\pi n), \quad \sin(x) = \sin(x + 2\pi n), \quad e^{ix} = e^{i(x+2\pi n)},$$

for any integer n. In the definition of the quaternion exponential function [Eq. (1.64)], the pure unit vector $\mu = q'/|q'|$ is used. This vector is directed as the vector q'; both vectors are on the same radial ray. Let us consider new vectors

$$q'_n = q' + \mu 2\pi n = q' + \frac{q'}{|q'|}2\pi n, \quad n = 0, \pm 1, \pm 2, \ldots.$$

One can notice that, if two numbers q_1 and q_2 are commutative, i.e., $q_1 q_2 = q_2 q_1$, then $\exp(q_1 + q_2) = \exp(q_1)\exp(q_2)$. In our case, vectors q' and $(q'/|q'|)2\pi n$ are commutative; therefore,

$$e^{q'_n} = e^{q' + \frac{q'}{|q'|}2\pi n} = e^{q'} e^{\frac{q'}{|q'|}2\pi n} = e^{q'}\left[\cos(2\pi n) + \frac{q'}{|q'|}\sin(2\pi n)\right] = e^{q'}.$$

We also could write the following:

$$e^{q' + \frac{q'}{|q'|}2\pi n} = e^{\frac{q'}{|q'|}(|q'| + 2\pi n)} = \cos(|q'| + 2\pi n) + \frac{q'}{|q'|}\sin(|q'| + 2\pi n)$$

$$= \cos(|q'|) + \frac{q'}{|q'|}\sin(|q'|) = e^{q'}.$$

P3. Consider a pure unit quaternion number μ and the exponential number

$$\exp(\mu\vartheta) = \cos(\vartheta) + \mu \sin(\vartheta).$$

When the number $\mu = i, j$, or k, this exponential function is the traditional complex exponential function. Any unit quaternion number $q = a + q'$ can be written in such a form. Indeed, let $|q| = 1$, and μ be the pure unit quaternion $\mu = q'/|q'|$, i.e., $q' = \mu|q'|$. Because of the equality

$$a^2 + |q'|^2 = 1,$$

there exists an angle $\vartheta \in [0, \pi]$, such that

$$\cos(\vartheta) = a \quad \text{and} \quad \sin(\vartheta) = |q'|.$$

Therefore, we can write

$$q = a + \mu|q'| = \cos(\vartheta) + \mu \sin(\vartheta) = e^{\mu\vartheta}. \tag{1.65}$$

The unit vector μ is in the direction of vector q'.

P4. We now consider the inverse q^{-1} to the quaternion $q = \cos(\vartheta) + \mu \sin(\vartheta)$, i.e., such that $qq^{-1} = 1$. In complex arithmetic, the inverse to $z = \cos(\vartheta) + i \sin(\vartheta)$ is the complex conjugate $z^{-1} = \bar{z} = \cos(\vartheta) - i \sin(\vartheta)$. For the quaternion exponential function, a similar property holds:

$$q^{-1} = \bar{q} = \cos(\vartheta) - \mu \sin(\vartheta) = \exp(-\mu\vartheta). \tag{1.66}$$

Indeed, $q(\bar{q}) = |q|^2 = 1$, or we can perform the direct calculations as

$$[\cos(\vartheta) + \mu \sin(\vartheta)][\cos(\vartheta) - \mu \sin(\vartheta)] = \cos^2(\vartheta) - \mu^2 \sin^2(\vartheta) = 1,$$

where $\mu^2 = -|\mu|^2 = -1$.

Now we consider the inverse number to e^q, for any quaternion $q = a + \mu\vartheta$, where $\mu^2 = -1$. The following calculations hold:

$$\begin{aligned}(e^q)^{-1} &= (e^a e^{\mu\vartheta})^{-1} = e^{-a}(e^{\mu\vartheta})^{-1} \\ &= e^{-a}\frac{e^{-\mu\vartheta}}{|e^{\mu\vartheta}|^2} = e^{-a}e^{-\mu\vartheta} = e^{-a-\mu\vartheta} = e^{-q}.\end{aligned}$$

Thus,

$$(e^q)^{-1} = e^{-q} = e^{-a}[\cos(\vartheta) - \mu \sin(\vartheta)]. \tag{1.67}$$

P5. Consider two quaternions, $q_1 = 1 + 2i - j + k$ and $q_2 = 2 - i + 4j - 2k$. The exponents of these numbers and their sum are the following:

$$e^{q_1} = -2.0928 + (1.4164i - 0.7082j + 0.7082k),$$

$$e^{q_2} = -0.9565 + (1.5989i - 6.3954j + 3.1977k),$$

and

$$e^{(q_1+q_2)} = -19.7786 + (-1.0546i - 3.1638j + 1.0546k).$$

The product of two the exponents equals

$$e^{q_1}e^{q_2} = -7.0565 + (-2.4363i + 10.6650j - 15.2956k).$$

All of these numbers were calculated by the code "test_expof2.m," a script of which is given below.

```
%-----------------------------------------------------------
% test_expof2.m
  q1 = [1, 2, - 1, 1];
  q2 = [2, - 1 ,4, - 2];
  q1exp = expofq(q1);        % −2.0928 1.4164 −0.7082 0.7082
  q2exp = expofq(q2);        % −0.9565 1.5989 −6.3954 3.1977
  q12exp = expofq(q1 + q2); % −19.7786 −1.0546 −3.1638
                              1.0546
  qexp12=mult2qs_direct(q1exp,q2exp);
                            % −7.0565 −2.4363 10.6650
                              − 15.2956
%-----------------------------------------------------------
```

One can see that, for these quaternions,

$$e^{(q_1+q_2)} \neq e^{q_1}e^{q_2}.$$

This statement is also valid in the general case when $q_1 \neq kq_2$, where k is a real number.

P6. It is not difficult to see that, for any angles ϑ_1 and ϑ_2, the following holds:

$$\begin{aligned}\exp(\mu\vartheta_1)\exp(\mu\vartheta_2) &= [\cos(\vartheta_1) + \mu\ \sin(\vartheta_1)][\cos(\vartheta_2) + \mu\ \sin(\vartheta_2)]\\ &= [\cos(\vartheta_1)\ \cos(\vartheta_2) + \mu^2 \sin(\vartheta_1)\ \sin(\vartheta_2)]\\ &\quad + \mu[\cos(\vartheta_1)\ \sin(\vartheta_2) + \sin(\vartheta_1)\ \cos(\vartheta_2)]\\ &= \cos(\vartheta_1 + \vartheta_2) + \mu\ \sin(\vartheta_1 + \vartheta_2) = \exp[\mu(\vartheta_1 + \vartheta_2)].\end{aligned}$$

In the $\vartheta_1 = \vartheta_2$ case, we obtain

$$(e^{\mu\vartheta_1})^2 = e^{2\mu\vartheta_1};$$

therefore, $(e^q)^2 = e^{2q}$ for any quaternion number.

For any pure unit quaternions μ and ν,

$$\exp(\mu\vartheta_1)\exp(\nu\vartheta_1) = [\cos^2(\vartheta_1) + \mu\nu\ \sin^2(\vartheta_1)] + \frac{\mu+\nu}{2}\sin(2\vartheta_1);$$

therefore, if $\mu \neq \nu$,

$$\exp(\mu\vartheta_1)\exp(\nu\vartheta_1) \neq \exp[(\mu+\nu)\vartheta_1].$$

P7. The quaternion also can be represented in a polar form as

$$q = |q|\exp(\mu\vartheta),$$

where μ is a pure unit quaternion $\mu = i\mu_i + j\mu_j + k\mu_k$, such that $|\mu| = 1$, $\mu^2 = -1$, and ϑ is a real angle in the interval $[0, \pi]$ Indeed,

$$q = a + q' = |q|\left(\frac{a}{|q|} + \frac{q'}{|q|}\right) = |q|\left(\frac{a}{|q|} + \frac{q'}{|q'|} + \frac{|q'|}{|q|}\right)$$
$$= |q|\left(\frac{a}{|q|} + \mu\frac{|q'|}{|q|}\right) = |q|[\cos(\vartheta) + \mu\,\sin(\vartheta)],$$

where the pure unit quaternion μ and the angle ϑ are calculated by

$$\mu = \frac{q'}{|q'|} \quad \text{and} \quad \vartheta = \arccos\left(\frac{a}{|q|}\right). \tag{1.68}$$

It should be noted that the sine of this angle is not negative, $\sin(\vartheta) = |q'|/|q|$ therefore, the angles ϑ in the polar form of the quaternions are from the interval $[0, \pi]$.

The polar form is unique, as the classic polar form of complex numbers. Indeed, consider the following equation for the unit quaternion q with $|q| = 1$:

$$q = qe + ib + jc + kd = \cos(\vartheta) + (i\mu_i + j\mu_j + k\mu_k)\,\sin(\vartheta),$$

from which it follows that if $|q_e| \neq 1$; then,

$$\cos(\vartheta) = a, \mu_i = b/\sin(\vartheta), \mu_j = c/\sin(\vartheta), \mu_k = d/\sin(\vartheta).$$

In the $|a| = 1$ case, $b = c = d = 0$; therefore, $\mu_i = \mu_j = \mu_k = 0$ and $\vartheta = \pi$.

P8. Given a real number x, the power of quaternion q is defined as

$$q^x = [|q|\exp(\mu\vartheta)]^x = |q|^x[\exp(\mu\vartheta)]^x$$
$$= |q|^x[\exp(\mu x\vartheta)] = |q|^x[\cos(x\vartheta) + \mu\,\sin(x\vartheta)].$$

We can verify this definition for the square of q as follows:

$$|q|^2[\cos(2\vartheta)] + \mu\,\sin(2\vartheta)] = |q|^2[\cos^2(\vartheta) - \sin^2(\vartheta) + 2\mu\,\sin(\vartheta)\,\cos(\vartheta)]$$
$$= |q|^2[\cos(\vartheta) + \mu\,\sin(\vartheta)]^2 = \{|q|[\cos(\vartheta) + \mu\,\sin(\vartheta)]\}^2 = (q)^2.$$

The $x = 1/2$ case corresponds to the calculation of the square root of the quaternion q,

$$\sqrt{q} = q^{1/2} = \sqrt{|q|}\left[\cos\left(\frac{\vartheta}{2}\right) + \mu \sin\left(\frac{\vartheta}{2}\right)\right].$$

Example 1.28

Consider the square root, square, and cube of the quaternion $q = 1 - i + 2j + 3k$ with the length $|q| = \sqrt{1+1+4+9} = \sqrt{15}$, which can be written as

$$q = \sqrt{15}\left(\frac{1}{\sqrt{15}} + \frac{-i + 2j + 3k}{\sqrt{15}}\right),$$

and whose imaginary part is

$$q' = -i + 2j + 3k = \frac{-i + 2j + 3k}{\sqrt{14}} \cdot \sqrt{14}.$$

The angle ϑ is calculated by

$$\vartheta = \arccos\left(\frac{1}{\sqrt{15}}\right) = 75.0368°,$$

and the vector is

$$\mu = \frac{q'}{|q'|} = \frac{-i + 2j + 3k}{\sqrt{14}}.$$

The following calculations hold for the square root:

$$\begin{aligned}\sqrt{q} &= \sqrt{|q|}\left[\cos\left(\frac{\vartheta}{2}\right) + \mu \sin\left(\frac{\vartheta}{2}\right)\right]\\ &= \sqrt{\sqrt{15}}\left[\cos(37.5184°) + \frac{-i + 2j + 3k}{\sqrt{14}}\sin(37.5184°)\right]\\ &= 1.5609 + 0.3203(-i + 2j + 3k).\end{aligned}$$

The square of q can be calculated by

$$\begin{aligned}q^2 &= |q|^2[\cos(2\vartheta) + \mu \sin(2\vartheta)] = 15[\cos(2\vartheta) + \mu \sin(2\vartheta)]\\ &= 15\left[\cos(150.0736°) + \frac{-i + 2j + 3k}{\sqrt{14}}\sin(150.0736°)\right]\\ &= -13 + 2(-i + 2j + 3k).\end{aligned}$$

The cube of q can be calculated similarly;

$$\begin{aligned} q^3 &= |q|^3[\cos(3\vartheta) + \mu \sin(3\vartheta)] \\ &= (\sqrt{15})^3\left[\cos(225.1103°) + \frac{-i+2j+3k}{\sqrt{14}}\sin(225.1103°)\right] \\ &= -41 - 11(-i+2j+3k). \end{aligned}$$

P9. Using the polar form of the quaternion $q = a + (ib + jc + kd)$, we can define the quaternion logarithm function. Indeed, we can write q as

$$q = |q|e^{\mu\vartheta} = e^{\ln|q|}e^{\mu\vartheta} = e^{\ln|q|+\mu\vartheta}.$$

Then, the logarithm of q can be defined as

$$\ln q = \ln|q| + \mu\vartheta = \ln|q| + \frac{ib+jc+kd}{\sqrt{b^2+c^2+d^2}}\arccos\left(\frac{a}{|q|}\right). \tag{1.69}$$

This is a quaternion number, which we denote by

$$\ln q = a_1 + ib_1 + jc_1 + kd_1$$

with the components

$$\begin{aligned} a_1 &= \ln|q| = \frac{1}{2}\ln(\sqrt{a^2+b^2+c^2+d^2}), \\ b_1 &= \frac{b}{\sqrt{b^2+c^2+d^2}}\arccos\left(\frac{a}{|q|}\right), \\ c_1 &= \frac{c}{\sqrt{b^2+c^2+d^2}}\arccos\left(\frac{a}{|q|}\right), \\ d_1 &= \frac{d}{\sqrt{b^2+c^2+d^2}}\arccos\left(\frac{a}{|q|}\right). \end{aligned} \tag{1.70}$$

Thus, the components b_1, c_1, and d_1 are proportional to the components b, c, and d, respectively:

$$\frac{b_1}{b} = \frac{c_1}{c} = \frac{d_1}{d} = \frac{1}{|q'|}\arccos\left(\frac{a}{|q|}\right).$$

Example 1.29

Consider the quaternion $q = 1 + 2i - 3j + k$. The logarithm of this number is

$$\ln q = 1.3540 + 0.700i - 1.0500j + 0.3500k.$$

Indeed, $|q| = \sqrt{1+4+9+1} = \sqrt{15} = 3.8730$, $a_1 = \ln 3.8730 = 1.3540$, and the imaginary components are

$$b_1 = \frac{2}{\sqrt{14}} \arccos\left(\frac{1}{\sqrt{15}}\right) = \frac{2}{\sqrt{14}} \cdot 1.3096 = 0.7000,$$

$$c_1 = \frac{-3}{\sqrt{14}} \arccos\left(\frac{1}{\sqrt{15}}\right) = -b_1\frac{3}{2} = -1.0500,$$

$$d_1 = \frac{1}{\sqrt{14}} \arccos\left(\frac{1}{\sqrt{15}}\right) = \frac{b_1}{2} = 0.3500.$$

Below is the script of the MATLAB-based code for this example, "test_logofq.m," which uses the function "logofq.m" to calculate the logarithm by Eq. (1.70). The command "qexp = expofq(qq)" is used to verify that the exponent of the logarithm of q is q.

```
%--------------------------------------------------------
% test_logofq.m
  q=[1,2,-3,1];
  qq=logofq(q);           % 1.3540 0.7000 -1.0500 0.3500
  qexp=expofq(qq);        % is q=[1 2 -3 1];
%--------------------------------------------------------
% call: logofq.m / Art Grigoryan, January 9, 2016
% calculate the quaternion exponent function
%--------------------------------------------------------
  function qln=logofq(q)
    q_modul=norm(q);
    a1=log(q_modul);
    qim=q(2:4);                % imaginary part of q
    qm=norm(qim);              % qm=sqrt(b^2+c^2+d^2)
    phi=acos(q(1)/q_modul);
    k=phi/qm;
    bcd1=k*q(2:4);             % imaginary part [b1 c1 d1]
    qln=[a1 bcd1];             % ln(q)=[a1 b1 c1 d1]
%--------------------------------------------------------
```

To express the components of the number q by the components of the number $\ln q$, we can represent $q = a + bi + cj + dk$ as the logarithm in the exponent

$$q = e^{\ln q} = e^{a_1+ib_1+jc_1+kd_1} = e^{a_1}e^{ib_1+jc_1+kd_1}.$$

We consider all components of the quaternion exponent,

$$q = e^{a_1} e^{ib_1+jc_1+kd_1} = e^{a_1} e^{\frac{ib_1+jc_1+kd_1}{\sqrt{b_1^2+c_1^2+d_1^2}}\sqrt{b_1^2+c_1^2+d_1^2}}$$

$$= e^{a_1}\left[\cos\left(\sqrt{b_1^2+c_1^2+d_1^2}\right) + \frac{ib_1+jc_1+kd_1}{\sqrt{b_1^2+c_1^2+d_1^2}} \sin\left(\sqrt{b_1^2+c_1^2+d_1^2}\right)\right].$$

Let $\varphi = \sqrt{b_1^2 + c_1^2 + d_1^2}$. Then, we can write that the real part of q is

$$a = e^{a_1}\cos(\varphi), \text{ and } a_1 = \ln|q|. \tag{1.71}$$

To express the imaginary components of the logarithm by the number q, we use the following equalities:

$$b = \frac{b_1|q|}{\varphi}\sin(\varphi), \quad c = \frac{c_1|q|}{\varphi}\sin(\varphi), \quad d = \frac{d_1|q|}{\varphi}\sin(\varphi).$$

Using the sinc function

$$\sin(\varphi) = \frac{\sin(\varphi)}{\varphi}, \quad \text{sinc}(0) = 1,$$

we can write

$$b = b_1|q|\text{sinc}(\varphi), \quad c = c_1|q|\text{sinc}(\varphi), \quad d = d_1|q|\text{sinc}(\varphi). \tag{1.72}$$

Therefore, the components of the quaternion logarithm can also be calculated by

$$\begin{aligned} b_1 &= \frac{b}{|q|}[\text{sinc}(\varphi)]^{-1}, \\ c_1 &= \frac{c}{|q|}[\text{sinc}(\varphi)]^{-1}, \quad \text{or} \quad c_1 = c\frac{b_1}{b} \text{ if } b \neq 0, \\ d_1 &= \frac{d}{|q|}[\text{sinc}(\varphi)]^{-1}, \quad \text{or} \quad d_1 = d\frac{b_1}{b} \text{ if } b \neq 0. \end{aligned} \tag{1.73}$$

Below is the script of the code "test_logofq2.m" for the calculation of the logarithm of the quaternion $q = 1 + 2i - 3j + k$. In this script, the function "logofq2.m" is used to calculate the logarithm of the quaternion by Eq. (1.73).

```
%--------------------------------------------------------------
% test_logofq2.m
  q=[1,2,-3,1];
  qq=logofq2(q);          % 1.3540 0.7000 -1.0500 0.3500
  qexp = expofq(qq);      % is q = [1 2 -3 1];
%--------------------------------------------------------------
```

```
% call: logofq2.m / Art Grigoryan, January 9, 2016
% calculate the quaternion exponent function
  function qln = logofq2(q)
    q_modul=norm(q);
    a1=log(q_modul);
    phi=acos(q(1)/q_modul);
    k=q_modul*sinc(phi/pi);          % See Eq. 1.72
    bcd1 = q(2:4)/k;                 % imaginary part [b1 c1 d1]
    qln=[a1 bcd1];                   % ln(q) = [a1 b1 c1 d1]
%-----------------------------------------------------------
```

Thus, two functions "logofq(q)" and "logofq2(q)" can be used to calculate the logarithm of a quaternion number q. We note here that the command "sinc" in MATLAB calculates sinc(φ) as $\sin(\pi\varphi)/[\pi(\varphi)]$.

1.5 Quaternion Trigonometric and Hyperbolic Functions

The quaternion exponential function allows for extending the concept of the cosine, sine, and tangent functions to the quaternion space. Given a pure unit quaternion number μ, the exponential function of angle ϑ is defined as

$$e^{\mu\vartheta} = \cos(\vartheta) + \mu \sin(\vartheta),$$

and $e^{-\mu\vartheta} = \cos(\vartheta) - \mu \sin(\vartheta)$. Therefore, the real cosine and sine functions can be written as

$$\cos(\vartheta) = \frac{1}{2}(e^{\mu\vartheta} + e^{-\mu\vartheta}),$$

and

$$\sin(\vartheta) = \frac{1}{2\mu}(e^{\mu\vartheta} - e^{-\mu\vartheta}) = \frac{\bar{\mu}}{2}(e^{\mu\vartheta} - e^{-\mu\vartheta}).$$

To extend these concepts to the quaternion functions, we define the quaternion cosine and sine in a way that is similar to the complex case:

$$\cos(q) = \frac{1}{2}(e^{\mu q} + e^{-\mu q}), \quad \sin(q) = \frac{1}{2\mu}(e^{\mu q} - e^{-\mu q}).$$

Here, the unit μ can be considered to be the number from the representation of the quaternion $q = a_1 + \mu\vartheta$.

The quaternion hyperbolic cosine and sine can be defined as

$$\cosh(q) = \frac{1}{2}(e^{q} + e^{-q}), \quad \sinh(q) = \frac{1}{2}(e^{q} - e^{-q}).$$

Example 1.30

Consider the quaternion $q = 1 + 2i + \sqrt{3}j - 3k$ and write it as

$$q = 1 + \frac{2i + \sqrt{3}j - 3k}{\sqrt{4+3+9}} \cdot 4.$$

When representing this quaternion as $q = a + \mu\vartheta$, we have $a = 1, \vartheta = 4$, and

$$\mu = \frac{2i + \sqrt{3}j - 3k}{\sqrt{4}} = \frac{1}{2}i + \frac{\sqrt{3}}{4}j - \frac{3}{4}k, \quad (\mu^2 = -1).$$

Therefore, the exponent $\exp(q)$ can be written as

$$e^q = e^a e^{\mu\vartheta} = e^{-1}\left[\cos(4) + \left(\frac{1}{2}i + \frac{\sqrt{3}}{4}j - \frac{3}{4}k\right)\sin(4)\right],$$

and

$$e^{-q} = e^{-a}e^{-\mu\vartheta} = e^{-1}\left[\cos(4) + \left(\frac{1}{2}i + \frac{\sqrt{3}}{4}j - \frac{3}{4}k\right)\sin(4)\right].$$

The number μq is $\mu(a + \mu\vartheta) = -\vartheta + a\mu$; therefore,

$$\begin{aligned} e^{\mu q} &= e^{-\vartheta}e^{a\mu} = e^{-\vartheta}[\cos(a) + \mu \sin(a)] \\ &= e^4\left[\cos(1) + \left(\frac{1}{2}i + \frac{\sqrt{3}}{4}j - \frac{3}{4}k\right)\sin(1)\right], \end{aligned}$$

and

$$e^{-\mu q} = e^{-\vartheta}e^{-a\mu} = e^4\left[\cos(1) - \left(\frac{1}{2}i + \frac{\sqrt{3}}{4}j - \frac{3}{4}k\right)\sin(1)\right].$$

Therefore, the cosine of the quaternion q can be calculated as follows:

$$\cos(q) = \frac{1}{2}(e^{\mu q} + e^{-\mu q}) = \cos(1) \cdot \frac{e^{-4} + e^4}{2} + \frac{e^{-4} - e^4}{2}\left(\frac{1}{2}i + \frac{\sqrt{3}}{4}j - \frac{3}{4}k\right)\sin(1).$$

This cosine can also be written as

$$\cos(q) = \cos(1)\cosh(4) - \mu \sinh(4)\sin(1).$$

It also not difficult to see that the quaternion sine can be calculated by

$$\sin(q) = \sin(1)\cosh(4) + \mu \sinh(4)\cos(1).$$

Indeed, since $\mu^2 = -1$, we can write

$$\begin{aligned}\frac{1}{2\mu}(e^{\mu q} - e^{-\mu q}) &= \cos(1)\cdot\frac{e^{-4}-e^{4}}{2\mu} + \frac{e^{-4}+e^{4}}{2\mu}\left(\frac{1}{2}i + \frac{\sqrt{3}}{4}j - \frac{3}{4}k\right)\sin(1)\\ &= \cos(1)\cdot\frac{e^{-4}-e^{4}}{2\mu} + \frac{e^{-4}+e^{4}}{2}\sin(1)\\ &= \cos(1)\cdot\frac{e^{-4}-e^{4}}{-2}\mu + \frac{e^{-4}+e^{4}}{2}\sin(1)\\ &= \cos(1)\sinh(4)\mu + \cosh(4)\sin(1).\end{aligned}$$

The quaternion hyperbolic cosine can be calculated by

$$\cosh(q) = \frac{1}{2}(e^{q} + e^{-q}) = \cos(4)\cdot\frac{e^{1}+e^{-1}}{2} + \frac{e^{1}-e^{-1}}{2}\left(\frac{1}{2}i + \frac{\sqrt{3}}{4}j - \frac{3}{4}k\right)\sin(4),$$

or

$$\cosh(q) = \cos(4)\cosh(1) + \left(\frac{1}{2}i + \frac{\sqrt{3}}{4}j - \frac{3}{4}k\right)\sinh(1)\sin(4).$$

The quaternion hyperbolic sine can similarly be calculated as

$$\sinh(q) = \frac{1}{2}(e^{q} - e^{-q}) = \cos(4)\cdot\frac{e^{1}-e^{-1}}{2} + \frac{e^{1}+e^{-1}}{2}\left(\frac{1}{2}i + \frac{\sqrt{3}}{4}j - \frac{3}{4}k\right)\sin(4),$$

or

$$\sinh(q) = \cos(4)\sinh(1) + \left(\frac{1}{2}i + \frac{\sqrt{3}}{4}j - \frac{3}{4}k\right)\cosh(1)\sin(4).$$

Summarizing the above calculations, we can write that the quaternion cosine and sine and hyperbolic cosine and sine can be represented as

$$\begin{aligned}\cos(q) &= \cos(a)\cosh(\vartheta) - \mu\sinh(\vartheta)\sin(a),\\ \sin(q) &= \sin(a)\cosh(\vartheta) + \mu\sinh(\vartheta)\cos(a),\\ \cosh(q) &= \cos(\vartheta)\cdot\cosh(a) + \mu\sinh(a)\sin(\vartheta),\\ \sinh(q) &= \cos(\vartheta)\cdot\sinh(a) + \mu\cosh(a)\sin(\vartheta),\end{aligned}\tag{1.74}$$

This representation is valid for any quaternion q.

The hyperbolic tangent is defined as

$$\tanh(q) = \frac{\sinh(q)}{\cosh(q)} = \frac{e^{q}-e^{-q}}{e^{q}+e^{-q}} = \frac{e^{q}-e^{-q}}{e^{q}+e^{-q}}\cdot\frac{e^{q}}{e^{q}} = \frac{e^{2q}-1}{e^{2q}+1},$$

since according to Eq. (1.67), $e^{-q} = (e^{q})^{-1}$, or $e^{-q}e^{q} = 1$.

Here, we note that in the division of the hyperbolic sine, the left and right divisions by the hyperbolic cosine are the same, i.e.,

$$\frac{\sinh(q)}{\cosh(q)} = \frac{1}{\cosh(q)}\sinh(q) = \sinh(q)\frac{1}{\cosh(q)}.$$

Indeed, as mentioned in sub-section 1.1.5 (property P4), when dividing quaternions q_1 by q_2, the divisions from the right and left are the same if $\overline{q}_2 q_1 = q_1 \overline{q}_2$. In our case, this requirement holds:

$$\overline{(e^q + e^{-q})}(e^q - e^{-q}) = (e^{-q} + e^q)(e^q - e^{-q}) = e^{2q} - e^{12q}$$

and

$$(e^q - e^{-q})\overline{(e^q + e^{-q})} = (e^q - e^{-q})(e^{-q} + e^q) = e^{2q} - e^{-2q}.$$

The hyperbolic cotangent is similarly defined as

$$\coth(q) = \frac{\cosh(q)}{\sinh(q)} = \frac{e^q + e^{-q}}{e^q - e^{-q}} = \frac{e^q + e^{-q}}{e^q - e^{-q}} \cdot \frac{e^q}{e^q} = \frac{e^{2q} + 1}{e^{2q} - 1}.$$

1.6 Quaternion-Type Numbers

In 4-D space, numbers other than quaternion numbers can be considered and applied in signal and image processing. We mention three such types of numbers, which are called *pseudo-quaternions*, *degenerate quaternions*, and *degenerate pseudo-quaternions* [34].

- Pseudo-quaternions $q = (a, b, c, d)$ are defined as the numbers

$$q = a + ib + ec + fd,$$

where i, e, and f are unit numbers with multiplications given in Table 1.3.

The conjugate number $\overline{q}$ is defined as

$$\overline{q} = a - ib - ec - fd;$$

therefore, the module of the pseudo-quaternion number q is defined by

Table 1.3 Table $T(1, i, e, f)$ of multiplications of pseudo-quaternion unit numbers.

	1	i	e	f
1	1	i	e	f
i	i	-1	f	$-e$
e	e	$-f$	1	$-i$
f	f	e	i	1

$$|q|^2 = |q\overline{q}| = a^2 - b^2 - c^2 - d^2.$$

- The degenerate quaternions $q = (a, b, c, d)$ are defined as the numbers

$$q = a + ib + \varepsilon c + \eta d,$$

where i, ε, and η are the numbers with multiplications given in Table 1.4.
The conjugate number $\overline{q}$ is defined as

$$\overline{q} = a - ib - \varepsilon c - \eta d.$$

The module of the degenerate quaternion number q is defined by

$$|q|^2 = |q\overline{q}| = a^2 + b^2.$$

- The degenerate pseudo-quaternions $q = (a, b, c, d)$ are defined as the numbers

$$q = a + eb + \varepsilon c + \zeta d,$$

where e, ε, and ζ are the numbers with multiplications given in Table 1.5.
The conjugate number $\overline{q}$ is defined as

$$\overline{q} = a - ib - \varepsilon c - \xi d.$$

The module of the degenerate quaternion number q is defined by

$$|q|^2 = |q\overline{q}| = a^2 - b^2.$$

Table 1.4 Table $T(1, i, \varepsilon, \eta)$ of multiplications of degenerate quaternion unit numbers.

	1	i	ε	η
1	1	i	ε	η
i	i	-1	η	$-\varepsilon$
ε	ε	$-\eta$	0	0
η	η	ε	0	0

Table 1.5 Table $T(1, e, \varepsilon, \zeta)$ of multiplications of degenerate pseudo-quaternion unit numbers.

	1	e	ε	ζ
1	1	e	ε	ζ
e	e	1	ζ	ε
ε	ε	$-\zeta$	0	0
ζ	ζ	$-\varepsilon$	0	0

References

[1] S. Sangwine, "Fourier transforms of colour images using quaternion or hypercomplex numbers," *Electronic Letters*, **32**(21), 1979–1980 (1996).

[2] Ö. N. Subakan and B. C. Vemuri, "A quaternion framework for color image smoothing and segmentation," *Int. Journal Comput. Vis.*, **91**, 233–250 (2011).

[3] F.-Y. Lang et al., "Quaternion and color image edge detection," *Computer Science*, **34**(11), 212–216 (2007).

[4] P. Denis, P. Carre, and C. Fernandez-Maloigne, "Spatial and spectral quaternionic approaches for colour images," *Comput. Vis. Image Und.*, **107**, 74–87 (2007).

[5] A. M. Grigoryan and S. S. Agaian, "Retooling of color imaging in the quaternion algebra," *Applied Mathematics and Sciences: An International Journal (MathSJ)*, **1**(3), 23–39, (2014).

[6] W. R. Hamilton, *Lectures on Quaternions: Containing a Systematic Statement of a New Mathematical Method*, Hodges and Smith, Dublin (1853).

[7] J. B. Kuipers, *Quaternions and Rotation Sequences: A Primer with Applications to Orbits, Aerospace and Virtual Reality*, Princeton University Press, Princeton (2002).

[8] H. J. Trussell, E. Saber, and M. Vrhel, "Color image processing," *IEEE Signal Processing Magazine*, **22**(1), 14–22 (2005).

[9] A. Tremeau, S. Tominaga, and K. N. Plataniotis, "Color in image and video processing: Most recent trends and future research directions," *EURASIP Journal on Image and Video Processing*, **2008**, 1–26 (2008).

[10] S. C. Pei, J. J. Ding, and J. H. Chang, "Efficient implementation of quaternion Fourier transform, convolution and correlation by 2-D complex FFT," *IEEE Trans. Signal Processing*, **49**, 2783–2797 (2001).

[11] S. J. Sangwine, "Fourier transforms of colour images using quaternion, or hypercomplex, numbers," *Electronics Letters*, **32**(21), 1979–1980 (1996).

[12] S. C. Pei and C. M. Cheng, "A novel block truncation coding of color images by using quaternion-moment-preserving principle," *IEEE International Symposium on Circuits and Systems*, **2**, 684–687 (1996).

[13] T. Bulow, "Hypercomplex spectral signal representations for the processing and analysis of images." Ph.D. Dissertation, Christian-Albrechts University, Kiel (1999).

[14] A. Kolaman and O. Yadid-Pecht, "Quaternion structural similarity: A new quality index for color images," *IEEE Transactions on Image Processing*, **21**(4), 1526–1536 (2012).

[15] Z. Yang and S.-I. Kamata, "Hypercomplex polar Fourier analysis for color image," *Proc. 18th International Conference on Image Processing (ICIP)* (2011).

[16] A. Greenblatt, C. Mosquera-Lopez, and S. S. Agaian, "Quaternion neural networks applied to prostate cancer Gleason grading," *Proc. Man and Cybernetics, IEEE International Conference*, Manchester, 1144–1149 (2013).

[17] H.-Y. Gao and K.-M. Lam, "From quaternion to octonion: Feature-based image saliency detection," *ICASSP, IEEE International Conference on Acoustics, Speech and Signal Processing*, 2808–2812 (2014).

[18] L. Guo, M. Dai, and M. Zhu, "Quaternion moment and its invariants for color object classification," *Information Sciences*, **273**, 132–143 (2014).

[19] A. M. Grigoryan and S. S. Agaian, "Alpha-rooting method of color image enhancement by discrete quaternion Fourier transform," *Proc. SPIE* , **9019**, 901904 (2014) [doi: 10.1117/12.2040596].

[20] J. S. Walker, *Fast Fourier Transforms*, CRC Press, Boca Raton, Florida (1991).

[21] A. M. Grigoryan and S. S. Agaian, "Split manageable efficient algorithm for Fourier and Hadamard transforms, *IEEE Trans. Signal Processing*, **48**, 172–183 (2000).

[22] A. M. Grigoryan, "2-D and 1-D multi-paired transforms: Frequency-time type wavelets," *IEEE Trans. on Signal Processing*, **49**(2), 344–353 (2001).

[23] A. M. Grigoryan and S. S. Agaian, *Multidimensional Discrete Unitary Transforms: Representation, Partitioning and Algorithms*, Marcel Dekker Inc., New York (2003).

[24] A. M. Grigoryan and M. M. Grigoryan, *Brief Notes in Advanced DSP: Fourier Analysis with MATLAB*, CRC Press, Taylor and Francis Group, Boca Raton, Florida (2009).

[25] L. Rabiner and B. Gold, *Theory and Application of Digital Signal Processing*, Prentice-Hall Inc. (1975).

[26] S. S. Agaian, "Advances and problems of the fast orthogonal transforms for signal-images processing applications (Part 1)," *Pattern Recognition, Classification, Forecasting. Yearbook, The Russian Academy of Sciences, Nauka*, Moscow, **3**, 146–215 (1990).

[27] S. S. Agaian, K. Panetta, and A. M. Grigoryan, "Transform-based image enhancement algorithms," *IEEE Trans. on Image Processing*, **10**(3), 367–382 (2001).

[28] A. Grigoryan and S. Agaian, "Tensor form of image representation: enhancement by image-signals," *Proc. SPIE* , **5014** (2003) [doi: 10.1117/12.473063].

[29] A. M. Grigoryan and S. S. Agaian, "Transform-based image enhancement algorithms with performance measure," *Advances in Imaging and Electron Physics*, **130**, Academic Press, 165–242 (2004).

[30] F. T. Arslan and A. M. Grigoryan, "Fast splitting α-rooting method of image enhancement: Tensor representation," *IEEE Trans. on Image Processing*, **15**(11), 3375–3384 (2006).

[31] A. M. Grigoryan, J. Jenkinson, and S. S. Agaian, "Quaternion Fourier transform based alpha-rooting method for color image measurement and enhancement," *SIGPRO-D-14-01083R1, Signal Processing*, **109**, 269–289 (2015).

[32] B. Chen, G. Coatrieux, G. Chen, J.-L. Coatrieux, and H. S. Full, "4-D quaternion discrete Fourier transform based watermarking for color images," *Digital Signal Processing*, **28**, 106–119 (2014).

[33] B. C. Ujang, C. C. Took, and D. P. Mandic, "On quaternion analyticity: Enabling quaternion-valued nonlinear adaptive filtering," *Proc. ICASSP, IEEE International Conference on Acoustics, Speech and Signal Processing* (2012).

[34] I. L. Kantor and A. S. Solodovnikov, *Hypercomplex Numbers*, Nauka, Moskow (1973).

[35] I. M. Yaglom, *Complex Numbers and Their Applications in Geometry*, Fizmatgiz, Nauka, Moscow (1963).

[36] E. Kreyszig, *Advanced Engineering Mathematics, Equations*, 10th Edition, John Wiley & Sons Inc. (2011).

Chapter 2
Octonion Numbers

It is known that the number of imaginary components that could be added to the real numbers to define new arithmetics with operations of addition, multiplication, and division can be only 1, 3, and 7. The first two cases relate to the complex and quaternion numbers [4], respectively. In this chapter, we consider the arithmetic of numbers with a seven-component imaginary part, which generalizes the quaternion numbers and allows for adding, multiplying, and dividing the numbers [5]–[12]. The properties, examples, and functions on such numbers are described. These numbers are called *octonions* or *Cayley numbers* and were discovered independently by the Irish mathematician John Graves (1806–1870) and the English mathematician Arthur Cayley (1821–1895).

2.1 Octonion Arithmetic

Quaternion numbers q were introduced in Eq. (1.6) in Section 1.1.2 as numbers from the doubled complex plane C^2, which are defined as $q = z_1 + jz_2 = (x_1 + iy_1) + j(x_2 + iy_2) = x_1 + iy_1 + jx_2 + (ji)y_2$. The quaternions also can be defined as $q = z_1 + z_2 j = (x_1 + iy_1) + (x_2 + iy_2)j = x_1 + iy_1 + jx_2 + (ij)y_2$, where the product (ij) is denoted by k, and $(ij) = -(ji) = k$.

Such $(2, 2)$-representation was applied to quaternions and new numbers, which were first called "double-quaternions" or "octives" by John Graves, and then called "octonions" [1]–[3]. The arithmetic with octonions includes the operations of addition, multiplication, and division. Let $q_1 = a_1 + q'_1$ and $q_2 = a_2 + q'_2$ be quaternion numbers, and let o be the double-quaternion, i.e., the number of type

$$o = (q_1, q_2) = q_1 + q_2 E = (a_1 + q'_1) + (a_2 + q'_2)E, \tag{2.1}$$

where E is a symbol to be defined for a unit that commutates with the real numbers. Considering the imaginary parts of the quaternions, $q'_1 = ib_1 + jc_1 + kd_1$ and $q'_2 = ib_2 + jc_2 + kd_2$, we obtain

$$o = a_1 + ib_1 + jc_1 + kd_1 + (a_2 + ib_2 + jc_2 + kd_2)E$$
$$= a_1 + ib_1 + jc_1 + kd_1 + a_2E + (iE)b_2 + (jE)c_2 + (kE)d_2.$$

We denote the numbers iE, jE, and kE by I, J, and K, respectively. This octonion number is the eight component number,

$$o = (a_1, b_1, c_1, d_1, a_2, b_2, c_2, d_2)$$
$$= a_1 + ib_1 + jc_1 + kd_1 + a_2E + Ib_2 + Jc_2 + Kd_2.$$

Thus, we can write octonions in the form of

$$o = a + ib + jc + kd + AE + IB + JC + KD, \tag{2.2}$$

where a, b, c, and d and A, B, C, and D are real numbers. The first component, a, is considered to be the real part of the octonion, and the remaining seven component together compose the "imaginary" part of o. The operation of addition of octonions $o_1 = q_1 + q_2E$ and $o_2 = r_1 + r_2E$, is performed component-wise,

$$o_1 + o_2 = (q_1 + q_2E) + (r_1 + r_2E) = (q_1 + r_1) + (q_2 + r_2)E.$$

The conjugate of the octonion $o = q_1 + q_2E$ is defined as

$$\begin{aligned}\bar{o} &= a - (ib + jc + kd + AE + IB + JC + KD)\\ &= (a - ib - jc - kd) - (AE + IB + JC + KD)\\ &= \bar{q}_1 - q_2E.\end{aligned} \tag{2.3}$$

All seven numbers (i, j, k, E, I, J, and K) are called *imaginary units.* The multiplication table, or rules for multiplications of all triplets of these units can be derived from the octonion multiplication of numbers $o_1 = q_1 + q_2E$ and $o_2 = r_1 + r_2E$, which is defined as

$$o_1o_2 = (q_1 + q_2E)(r_1 + r_2E) = (q_1r_1 - \bar{r}_2q_2) + (r_2q_1 + q_2\bar{r}_1)E, \tag{2.4}$$

where q_1, q_2, r_1, and r_2 are arbitrary quaternion numbers. The multiplication is defined in the way to preserve the multiplication of quaternion numbers and to satisfy the property of "equality for eight squares," which was found by Arthur Cayley [2],

$$|o_1o_2|^2 = |o_1|^2|o_2|^2. \tag{2.5}$$

Here, the module o_1 of the octonion $o_1 = q_1 + q_2E$ is defined as $\sqrt{o_1\bar{o}_1}$ and

$$\begin{aligned} o_1\bar{o}_1 &= (q_1 + q_2E)(\bar{q}_1 - q_2E) \\ &= (q_1\bar{q}_1 + \bar{q}_2q_2) + (-q_2q_1 + q_2q_1)E = |q_1|^2 + |q_2|^2. \end{aligned} \quad (2.6)$$

Therefore, $\bar{o}_1 o_1 = |q_1|^2 + |q_2|^2 = o_1\bar{o}_1$.

It also follows from the definition in Eq. (2.4) that for imaginary units $u_1, u_2 = i, j, k$, the following properties hold:

$$u_1(u_2)E = (u_1u_2)E, \quad (2.7)$$

if only $u_1 = u_2$, and

$$u_1(u_2)E = -(u_1u_2)E, \quad (2.8)$$

if $u_1 \neq u_2$. For instance, $iI = i(iE) = (i^2)E = -E$ and $iJ = i(jE) = -(ij)E = -kE = -K$. The second property shows that the multiplication of octonions is not associative. We also can note that for the imaginary units $U_2 = u_2E = I, J$, and K, the following property holds:

$$U_2u_1 = (u_2E)u_1 = (u_2\bar{u}_1)E = -(u_2u_1)E. \quad (2.9)$$

For instance,

$$\begin{aligned} Ii &= (iE)i = -(ii)E = E = -iI, \\ Ji &= (jE)i = -(ji)E = -(-k)E = K = -iJ, \\ Jk &= (jE)k = -(jk)E = -iE = -I = -kJ. \end{aligned}$$

Also, the following multiplications hold between the last four imaginary units:

$$\begin{aligned} EI &= E(iE) = -(EE)i = i = -IE, \\ EJ &= E(jE) = -(EE)j = j = -JE, \\ EK &= E(kE) = -(EE)k = k = -KE, \\ EE &= II = JJ = KK = -1. \end{aligned}$$

Together with the multiplication by the real number 1, all basic multiplications of imaginary units are shown in Table 2.1.

The multiplication of different imaginary units can also be illustrated in the diagram. Figure 2.1 shows the diagram of 7 cyclic rotations in all triplets $\{u_1, u_2, u_3\}$ of the imaginary units, where the units $u_1 \neq u_2 \neq u_3$ are among i, j, k, E, I, J, and K. Each arrow shows the order of the units in the multiplication with the sign plus, the multiplication in the opposite direction result the next units in the triplet with the sign minus. Thus, if we take the triplet $\{u_1, u_2, u_3\}$ in the order with the direction of the arrow, and extend it as $\{u_1, u_2, u_3, u_1, u_2\}$, then $u_1u_2 = u_3$, $u_2u_3 = u_1$, and $u_3u_1 = u_2$, but $u_2u_1 = -u_3$, $u_3u_2 = -u_1$, and $u_1u_3 = -u_2$. For the symmetry in the diagram, the rotation of units in the triplet jJE is shown twice.

Table 2.1 Table $T(1, i, j, k, E, I, J, K)$ of multiplications of the octonion unit numbers.

	1	***i***	***j***	***k***	***E***	***I***	***J***	***K***
1	1	i	j	k	E	I	J	K
i	i	-1	k	$-j$	I	$-E$	$-K$	J
j	j	$-k$	-1	i	J	K	$-E$	$-I$
k	k	j	$-i$	-1	K	$-J$	I	$-E$
E	E	$-I$	$-J$	$-K$	-1	i	j	k
I	I	E	$-K$	J	$-i$	-1	$-k$	j
J	J	K	E	$-I$	$-j$	k	-1	$-i$
K	K	$-J$	I	E	$-k$	$-j$	i	-1

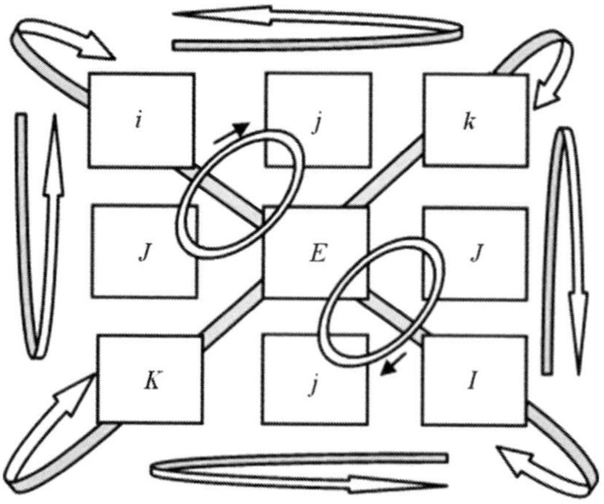

Figure 2.1 Diagram of 7 rotations in triplets of the imaginary units.

2.1.1 Properties of the octonion multiplication

The multiplication defined in Eq. (2.4) is a complicate operation and it is difficult to describe this operation in an elegant form with inner and vector products, as for the quaternions. A few properties of the multiplication and division of octonions are described below.

First, we consider the operation of multiplication of octonions in matrix form. The direct calculation of the multiplication of two octonions, $o = o_1 o_2$, can be written as follows:

$$\mathbf{o} = \mathbf{A}\mathbf{o}_2 = \begin{pmatrix} a_1 & -b_1 & -c_1 & -d_1 & -A_1 & -B_1 & -C_1 & -D_1 \\ b_1 & a_1 & -d_1 & c_1 & -B_1 & A_1 & D_1 & -C_1 \\ c_1 & d_1 & a_1 & -b_1 & -C_1 & -D_1 & A_1 & B_1 \\ d_1 & -c_1 & b_1 & a_1 & -D_1 & C_1 & -B_1 & A_1 \\ A_1 & B_1 & C_1 & D_1 & a_1 & -b_1 & -c_1 & -d_1 \\ B_1 & -A_1 & D_1 & -C_1 & b_1 & a_1 & d_1 & -c_1 \\ C_1 & -D_1 & -A_1 & B_1 & c_1 & -d_1 & a_1 & b_1 \\ D_1 & C_1 & -B_1 & -A_1 & d_1 & c_1 & -b_1 & a_1 \end{pmatrix} \begin{pmatrix} a_2 \\ b_2 \\ c_2 \\ d_2 \\ A_2 \\ B_2 \\ C_2 \\ D_2 \end{pmatrix}, \tag{2.10}$$

where the octonions o and o_2 are presented by the eight-dimensional (8-D) column-vectors $\mathbf{o} = (a, b, c, d, A, B, C, D)'$ and $\mathbf{o}_2 = (a_2, b_2, c_2, d_2, A_2, B_2, C_2, D_2)'$, respectively. The matrix **A** can be obtained from the 8×8 subtable in Table 2.1. Indeed, one can note that the first column of **A** is obtained from the first column of this table, by the following substitution:

$$1 \to a_1, i \to b_1, j \to c_1, k \to d_1, E \to A_1, I \to B_1, J \to C_1, K \to D_1.$$

The remaining 8×7 part of **A** can be obtained from the subtable by the substitution

$$1 \to -a_1, i \to -b_1, j \to -c_1, k \to -d_1, E \to -A_1, I \to -B_1, J \to -C_1, K \to -D_1.$$

Below is the script of the function "mult2octs.m" to calculate the multiplication of two octonions, by using the above matrix **A**.

```
%-----------------------------------------------------------------
% call: mult2octs.m / Artyom Grigoryan, 06/02/2015
% calculation of multiplication of two octonions o1 and o2.
function o12 = mult2octs(o1,o2)
    a1=o1(1); b1=o1(2); c1=o1(3); d1=o1(4);
    A1=o1(5); B1=o1(6); C1=o1(7); D1=o1(8);
    if size(o2,1)==1 o2=o2'; end
    A=[ a1 -b1 -c1 -d1 -A1 -B1 -C1 -D1
        b1  a1 -d1  c1 -B1  A1  D1 -C1
        c1  d1  a1 -b1 -C1 -D1  A1  B1
        d1 -c1  b1  a1 -D1  C1 -B1  A1
        A1  B1  C1  D1  a1 -b1 -c1 -d1
        B1 -A1  D1 -C1  b1  a1  d1 -c1
        C1 -D1 -A1  B1  c1 -d1  a1  b1
        D1  C1 -B1 -A1  d1  c1 -b1  a1];
    o12=(A*o2)';
%-----------------------------------------------------------------
```

Example 2.1

Consider two octonions

$$o_1 = 1 + 2i + 3j + 4k + E + 2I + J - K,$$
$$o_2 = 2 - i + j + 2k + 3E - 4I + J + 2K.$$

After running the three command lines

```
>> o1 = [1  2 3 4 1  2 1 -1];
>> o2 = [2 -1 1 2 3 -4 1  2];
>> o12 = mult2octs(o1,o2);     % -1 -8 -3 14 -1 2 34 -7
```

we obtain the result of the multiplication of these octonions,

$$o_1o_2 = -1 - 8i - 3j + 14k - E + 2I + 34J - 7K.$$

Here for comparison, we note that the multiplication o_2o_1 by the command

```
>> o21=mult2octs(o2,o1);     % -1 14 17 6 11 -2 -28 7
```

gives the following octonion that is different from o_1o_2:

$$o_2o_1 = -1 + 14i + 17j + 6k + 11E - 2I - 28J + 7K.$$

We can verify the mentioned in Eq. (2.5) equality for eight squares for the octonions o_1 and o_2. After running the command lines

```
>> o12in2 = o12*o12';    % 1480
>> o1in2 = o1*o1';       % 37
>> o2in2 = o2*o2';       % 40
```

we do obtain such equality, $|o_1o_2|^2 = 1480 = 37 \cdot 40 = |o_1|^2 \cdot |o_2|^2$.
The matrix $\mathbf{A}$ of multiplication o_1o_2 is

$$\mathbf{A} = \mathbf{A}_{\mathbf{o}_1} = \begin{pmatrix} 1 & -2 & -3 & -4 & -1 & -2 & -1 & 1 \\ 2 & 1 & -4 & 3 & -2 & 1 & -1 & -1 \\ 3 & 4 & 1 & -2 & -1 & 1 & 1 & 2 \\ 4 & -3 & 2 & 1 & 1 & 1 & -2 & 1 \\ 1 & 2 & 1 & -1 & 1 & -2 & -3 & -4 \\ 2 & -1 & -1 & -1 & 2 & 1 & 4 & -3 \\ 1 & 1 & -1 & 2 & 3 & -4 & 1 & 2 \\ -1 & 1 & -2 & -1 & 4 & 3 & -2 & 1 \end{pmatrix}.$$

This matrix is generated by the first octonion o_1 that lies on the first column of the matrix. The determinant of this matrix is

$$\det(\mathbf{A}) = 1874200 = |o_1|^8 = (37)^4.$$

The matrix is orthogonal, which means that the sums of point-wise product of any two different rows (columns) are zero. The inverse matrix $\mathbf{A}^{-1}$ is the transpose matrix to $\mathbf{A}$ up to the coefficient $|o_1|^2 = 37$,

$$\mathbf{A}^{-1} = \frac{1}{|o_1|^2}\mathbf{A}^T = \frac{1}{37}\mathbf{A}^T.$$

The multiplication of octonions can also be calculated by using the function "mult2qs_direct.m" or "mult2qs.m" that were described above for the multiplication of quaternions. Indeed, as follows from the definition in Eq. (2.4)

$$o = o_1 o_2 = (q_1 r_1 - \bar{r}_2 q_2) + (r_2 q_1 + q_2 \bar{r}_1)E,$$

this operation requires four operations of quaternion multiplication. The first two multiplications in $(q_1 r_1 - \bar{r}_2 q_2)$ determine the first four components of the multiplication o, namely, the real part and i-, j-, and k-components of the imaginary part of o. The next two quaternion multiplications in $(r_2 q_1 + q_2 \bar{r}_1)$ determine the E-, I-, J-, and K-components of the imaginary part of o.

Below is the script of the MATLAB®-based code "test_octavionmult.m" to calculate the multiplication of two octonions o_1 and o_2 from Example 2.1. and verify the result of multiplication. The multiplication of two octonions is accomplished by the function "mult2octs_direct.m," which uses the funcion "mult2qs_direct.m" for the quaternion multiplication.

```
%---------------------------------------------------------------
% test_octavionmult.m
    o1=[1  2 3 4 1  2 1 -1];
    o2=[2 -1 1 2 3 -4 1  2];
    o12=mult2octs_direct(o1,o2) % -1 -8 -3 14 -1
         2 34 -7
%---------------------------------------------------------------
% call mult2octs_direct.m / Artyom Grigoryan, 06/02/2015
% calculation of multiplication of two octonions o1 and o2.
function o12=mult2octs_direct(o1,o2)
    q1=o1(1:4); q2=o1(5:8);
    r1=o2(1:4); r2=o2(5:8);
    r1c=[r1(1),-r1(2:4)];
    r2c=[r2(1),-r2(2:4)];
    % 4 quaternion multiplications
    c1=mult2qs_direct(q1,r1);
    c2=mult2qs_direct(r2c,q2);
    c3=mult2qs_direct(r2,q1);
    c4=mult2qs_direct(q2,r1c);
    o12=[c1-c2,c3+c4];
%---------------------------------------------------------------
```

P1. According to the definition of the imaginary units, $E^2 = I^2 = J^2 = K^2 = -1$, and therefore we obtain the following in Eq. (2.6):

$$|o|^2 = a^2 + b^2 + c^2 + d^2 + E^2 + I^2 + J^2 + K^2. \tag{2.11}$$

For example, for the octonion $o = 1 - 2i + 3J - 2K$, the square of its module $|o|^2 = 1 + 2^2 + 3^2 + 2^2 = 18$.

P2. The operation of multiplication of octonions is not associative, i.e., not for all octonions o_1, o_2, and o_3, the equality $(o_1 o_2)o_3 = o_1(o_2 o_3)$ holds.

Here, we remind that such operation for quaternions is associative. The associative property holds for any pair of octonions, i.e., when $o_3 = o_2$ and $o_3 = \bar{o}_2$,

$$(o_1o_2)o_2 = o_1(o_2o_2) \quad \text{and} \quad o_1(o_1o_2) = (o_1o_1)o_2,$$

and

$$(o_1o_2)\bar{o}_2 = o_1(o_2\bar{o}_2) \quad \text{and} \quad \bar{o}_1(o_1o_2) = (\bar{o}_1o_1)o_2.$$

Example 2.2

Consider two octonions

$$o_1 = 1 + 2i + 3j + 4k + E + 2I + J - K,$$
$$o_2 = 2 - i + j + 2k + 3E - 4I + J + 2K,$$

and the products $(o_1o_2)o_2$ and $o_1(o_2o_2)$. To avoid complex calculations, we can use the above functions for octonion multiplication. After running the command lines

```
o1=[1  2 3 4 1  2 1 −1];
o2=[2 −1 1 2 3 −4 1  2];
o12=mult2octs(o1,o2);    %  −1  −8  −3  14  −1  2  34  −7
o122=mult2octs(o12,o2); % −44 −112 −84 −104 −44 −72 96 12
o22=mult2octs(o2,o2);    % −32 −4    4   8   12 −16 4   8
o122=mult2octs(o1,o22); % −44 −112 −84 −104 −44 −72 96 12
```

we obtain the following results:

$$o_1o_2 = -1 - 8i - 3j + 14k - E + 2I + 34J - 7K,$$
$$(o_1o_2)o_2 = -44 - 112i - 84j - 104k - 44E - 72I + 96J + 12K,$$

and

$$o_2o_2 = -32 - 4i + 4j + 8k + 12E - 16I + 4J + 8K,$$
$$o_1(o_2o_2) = -44 - 112i - 84j - 104k - 44E - 72I + 96J + 12K.$$

Thus, $(o_1o_2)o_2 = o_1(o_2o_2)$.

One can check that the equality $o_1(o_1o_2) = (o_1o_1)o_2$ is also hold. Indeed, after running the next three command lines

```
o122=mult2octs(o1,o12);  % −76 21 −43 −46 −113 152 31 −88
o11=mult2octs(o1,o1);    % −35  4   6   8    2   4  2  −2
o112=mult2octs(o11,o2);  % −76 21 −43 −46 −113 152 31 −88
```

we obtain the following results:

$$\begin{aligned} o_1(o_1o_2) &= -76 + 21i - 43j - 46k - 113E + 152I + 31J - 88K, \\ o_1o_1 &= -35 + 4i + 6j + 8k + 2E + 4I + 2J - 2K, \\ (o_1o_1)o_2 &= -76 + 21i - 43j - 46k - 113E + 152I + 31J - 88K. \end{aligned}$$

We can also verify the equality $\bar{o}_2(o_2o_1) = (\bar{o}_2o_2)o_1$, by running the next five command lines

```
o2c=[o2(1) -o2(2:8)];          % the conjugate to octonion o2
o21=mult2octs(o2,o1);          % -1 14 17    6 11 -2 -28   7
o2c21=mult2octs(o2c,o21);      % 40 80 120 160 40 80   40 -40
o22=mult2octs(o2c,o2);         % 40  0   0   0  0  0    0   0
o221=mult2octs(o22,o1);        % 40 80 120 160 40 80   40 -40
```

we obtain the following results:

$$\begin{aligned} o_2o_1 &= -1 + 14i + 17j + 6k + 11E - 2I - 28J + 7K, \\ \bar{o}_2(o_2o_1) &= 40 + 80i + 120j + 160k + 40E + 80I + 40J - 40K, \\ \bar{o}_2o_2 &= 40, \\ (\bar{o}_2o_2)o_1 &= 40 + 80i + 120j + 160k + 40E + 80I + 40J - 40K. \end{aligned}$$

P3. Consider the square of the octonion $o = q_1 + q_2E$, where $q_1 = a_1 + q'_1$ and $q_2 = a_2 + q'_2$. As directly follows from Eq. (2.4), the square can be written as

$$\begin{aligned} o^2 &= oo = (q_1 + q_2E)(q_1 + q_2E) \\ &= (q_1q_1 - \bar{q}_2q_2) + (q_2q_1 + q_2\bar{q}_1)E = q_1^2 - |q_2|^2 + q_2(q_1 + \bar{q}_1)E \\ &= q_1^2 - |q_2|^2 + 2a_1q_2E = (a_1^2 - |q'_1|^2 + 2a_1q'_1) - |q_2|^2 + 2a_1q_2E \\ &= (a_1^2 - a_2^2) - (|q'_1|^2 + |q'_2|^2) + 2a_1(q'_1 + q_2E) \\ &= (a_1^2 - a_2^2) - (|q'_1|^2 + |q'_2|^2) + 2a_1(o - a_1) \\ &= -(a_1^2 + a_2^2) - (|q'_1|^2 + |q'_2|^2) + 2a_1o = -|o|^2 + 2a_1o. \end{aligned}$$

If we consider that $o = a_1 + o'$, where the imaginary part of the octonion is $o' = q'_1 + q_2E$, then the square of the octonion can be written as

$$o^2 = 2a_1o - |o|^2 = 2a_1(a_1 + o') - (a_1^2 + |o'|^2) = a_1^2 - |o'|^2 + 2a_1o'. \quad (2.12)$$

If the octonion is pure imaginary, i.e., $a_1 = 0$, then the square $o^2 = -|o'|^2$ is a negative number. A similar result was received for quaternion numbers, $q^2 = a^2 - |q'|^2 + 2aq'$ in Eq. (1.13).

As an example, for the octonion

$$o_1 = 1 + o'_1 = 1 + (2i + 3j + 4k + E + 2I + J - K),$$

we obtain the square

$$o_1^2 = o_1 o_1 = -35 + 4i + 6j + 8k + 2E + 4I + 2J - 2K.$$

In this case, $a_1 = 1$, $o'_1 = 2i + 3j + 4k + E + 2I + J - K$, and

$$|o'_1|^2 = 2^2 + 3^2 + 4^2 + 1 + 2^2 + 1 + 1 = 36.$$

Therefore, the real part of o_1^2 is $a_1^2 - |o'_1|^2 = 1 - 36 = -35$.

P4. We now describe the case when the octonion $o = a_1 + o'$ is unit, i.e., $|o|^2 = 1$, and its square $o^2 = -1$. As follows from Eq. (2.12), we have

$$a_1^2 - |o'|^2 + 2a_1 o' = -1.$$

This equality can be solved under the condition that $2a_1 o' = 0$, which means that $a_1 = 0$ and therefore $o = o'$. Then, $|o|^2 = 1$ and $o^2 = -1$, i.e., the octonion is a pure unit octonion (with its real part equals zero). The examples of pure unit octonions are i, j, k, E, I, J, and K, and

$$o_1 = \frac{i + j + k + E + I + J + K}{\sqrt{7}}, \quad o_2 = \frac{i + j + I + J}{2}.$$

Example 2.3

Consider the solution of the following equation in the octonion arithmetic:

$$o_1^2 = 1 - 2i + 6j - 4d - 2E + 4I + 2J. \tag{2.13}$$

We are looking for the solution of the equation

$$a_1^2 - |o'_1|^2 + 2a_1 o' = 1 - 2i + 6j - 4d - 2E + 4I + 2J.$$

The first condition for a solution of this equation is

$$a_1^2 - |o'_1|^2 = 1, \quad \text{or} \quad |o'_1|^2 = a_1^2 - 1.$$

Here, $o_1 = a_1 + o'_1 = a_1 + (ib_1 + jc_1 + kd_1 + EA_1 + IB_1 + JC_1 + KD_1)$.

The second condition is $2a_1 o'_1 = -2i + 6j - 4d - 2E + 4I + 2J$, or

$$a_1 o'_1 = -i + 3j - 2d - E + 2I + J.$$

Since,

$$a_1 o'_1 = a_1(ib_1 + jc_1 + kd_1 + EA_1 + IB_1 + JC_1 + KD_1)$$

the following system of equations should be considered:

$$\begin{cases} a_1b_1 = -1, \\ a_1c_1 = 3, \\ a_1d_1 = -2, \\ a_1A_1 = -1, \\ a_1B_1 = 2, \\ a_1C_1 = 1, \\ a_1D_1 = 0. \end{cases}$$

Thus, $b_1 = -1/a_1$, $c_1 = 3/a_1$, $d_1 = -2/a_1$, $A_1 = -1/a_1$, $B_1 = 2/a_1$, $C_1 = 1/a_1$, and $D_1 = 0$. The required condition $|o'|^2 = a_1^2 - 1$ can be written as

$$a_1^2 - 1 = \frac{1}{a_1^2} + \frac{9}{a_1^2} + \frac{4}{a_1^2} + \frac{1}{a_1^2} + \frac{4}{a_1^2} + \frac{1}{a_1^2} = \frac{20}{a_1^2}.$$

We obtain the quadratic equation

$$(a_1^2)^2 - a_1^2 - 20 = 0$$

with the solution

$$a_1^2 = \frac{1 + \sqrt{1 + 80}}{2} = 5.$$

Thus, we have two solutions $a_1 = \sqrt{5}$ and $a_1 = -\sqrt{5}$. For the first case,

$$b_1 = A_1 = -C_1 = -\frac{1}{\sqrt{5}} = -\frac{\sqrt{5}}{5},$$

$$c_1 = \frac{3}{\sqrt{5}} = \frac{3\sqrt{5}}{5},$$

$$d_1 = -B_1 = -\frac{2}{\sqrt{5}} = -\frac{2\sqrt{5}}{5},$$

and $D_1 = 0$. Therefore, the octonion o_1 equals

$$o_1 = \sqrt{5} + \frac{\sqrt{5}}{5}(-i + 3j - 2k - E + 2I + J).$$

For the $a_1 = -\sqrt{5}$ case, the solution equals the above octonion o_1 with the sign minus.

To verify the solutions, we can run two command lines in MATLAB

```
>> o1=[sqrt(5) 1/sqrt(5)*[-1, 3, -2, -1, 2, 1, 0]];
>> o11=mult2octs_direct(o1,o1);    % 1 -2 6 -4 -2 4 2 0
```

and obtain the result $o11 = [1, -2, 6, -4, -2, 4, 2, 0]$, which is for the octonion $1 - 2i + 6j - 4d - 2E + 4I + 2J$ in the right side of Eq. (2.13).

Example 2.4

Consider the solutions $o_1 = a_1 + o_1'$ of the following quadratic equation:

$$o_1^2 - o_1 + 1 = 0. \tag{2.14}$$

This equation can be written as

$$(a_1^2 - |o_1'|^2 + 2a_1 o_1') - (a_1 + o_1') + 1 = 0.$$

Zeroing the real and imaginary parts of this equation, we obtain

$$a_1^2 - a_1 - |o_1'|^2 + 1 = 0, \quad \text{and} \quad (2a_1 - 1)o_1' = 0.$$

We are looking for not real solution and therefore considering only the $o_1' \neq 0$ case. Assuming $a_1 = 0.5$, we obtain from the first equation the following condition for the imaginary part of the octonion $o' = ib_1 + jc_1 + kd_1 + EA_1 + IB_1 + JC_1 + KD_1$:

$$\frac{1}{4} - \frac{1}{2} - |o_1'|^2 + 1 = 0, \quad \text{or} \quad |o_1'|^2 = \frac{3}{4}.$$

Thus, the imaginary part of the octonion presented as the seven-dimensional (7-D) vector-point

$$\mathbf{o}_1' = (b_1, c_1, d_1, A_1, B_1, C_1, D_1)$$

is one of the points in the 7-D sphere of radius $\sqrt{3}/2$,

$$b_1^2 + c_1^2 + d_1^2 + A_1^2 + B_1^2 + C_1^2 + D_1^2 = \frac{3}{4}.$$

For instance, such points are

$$\mathbf{o}_1' = (1/2, 1/2, 0, 1/2, 0, 0, 0) \quad \text{and} \quad \mathbf{o}_1' = (0, 0, 0, 1/2, 0, 1/2, 1/2),$$

and the corresponding solutions o_1 of Eq. (2.14) are the octonions

$$o_1 = \frac{1}{2} + \frac{1}{2}(i + j + E) \quad \text{and} \quad o_1 = \frac{1}{2} + \frac{1}{2}(E + J + K).$$

P5. In the octonion arithmetic, we separate the operations of division of one quaternion o_1 by another $o_2 \neq 0$ from the left and right, by solving the equations $o_1 = o_2 o$ and $o_1 = o o_2$, respectively.

For the division from the left, the multiplication of both sides of the equality $o_1 = o_2 o$ by $\bar{o}_2$ from the left results in the following: $\bar{o}_2 o_2 o = \bar{o}_2 o_1$. Therefore, $|o_2|^2 o = \bar{o}_2 o_1$ and we obtain the solution o of the equation $o_1 = o_2 o$,

$$o = \frac{1}{|o_2|^2} \bar{o}_2 o_1.$$

We denote this division operation from the left as

$$o = o_l = \frac{1}{o_2} o_1.$$

The solution of the equation $o_1 = o o_2$,

$$o = o_r = o_1 \frac{1}{o_2},$$

which we also call the division from the right, can similarly be obtained after multiplying by $\bar{o}_2$ (from the right) both sides of the equation $o_1 = o o_2$,

$$o_1 \bar{o}_2 = o|o_2|^2 \rightarrow o = \frac{1}{|o_2|^2} o_1 \bar{o}_2 = o_1 \frac{1}{|o_2|^2} \bar{o}_2.$$

Example 2.5

Consider the division operations

$$o_l = \frac{1}{o_2} o_1 \quad \text{and} \quad o_r = o_1 \frac{1}{o_2},$$

for octonions

$$o_1 = 1 - i + 2j - 3d - 2E + I + 3J - K,$$
$$o_2 = 2 + 3i + j + d - E + 3I - J + 2K.$$

The square of the module

$$|o_2|^2 = 2^2 + 3^2 + 1 + 1 + 1 + 3^2 + 1 + 2^2 = 30.$$

The inverse $1/o_2$ equals

$$\frac{1}{o_2} = \frac{\bar{o}_2}{|o_2|^2} = \frac{2 - 3i - j - d + E - 3I + J - 2K}{30}.$$

To calculate the product $\bar{o}_2 o_1$, we can run the following script in MATLAB

```
o1 =  [1 −1 2 −3 −2 1  3 −1];
o2 =  [2  3 1  1 −1 3 −1  2];
o2c =[o2(1) -o2(2:8)];
mo2 = o2*o2';      % 30, the square of the module |o2|^2
o2c1 = mult2octs_direct(o2c,o1)  % −2 −10 −5 −9 13 1 18 14
% verify the answer
o1is = mult2octs_direct(o2,o2c1)  % 30 −30 60 −90 −60 30
90 −30
o1is = o1is/mo2;     % 1 −1 2 −3 −2 1 3 −1 (is o1)
```

and obtain the product

$$\bar{o}_2 o_1 = -2 - 10i - 5j - 9k + 13E + I + 18J + 14K.$$

Therefore,

$$o_l = \frac{1}{|o_2|^2}\bar{o}_2 o_1 = \frac{1}{30}(-2 - 10i - 5j - 9k + 13E + I + 18J + 14K).$$

To calculate the division from the right $o_r = o_1 \frac{1}{o_2}$ for the solution o of the equation $o_1 = o o_2$, we can run the next few commands

```
o12c=mult2octs_direct(o1,o2c)  % −2 0 1 −5 −19 −3 −4 −22
% verify the result
o1is=mult2octs_direct(o12c,o2)  % 30 −30 60 −90 −60 30
90 −30
o1is=o1is/mo2;    % 1 −1 2 −3 −2 1 3 −1
```

and obtain the following results:

$$o_1 \bar{o}_2 = -2 + j - 5k - 19E - 3I - 4J - 22K.$$

Therefore, the solution is

$$o_r = \frac{1}{|o_2|^2} o_1 \bar{o}_2 = \frac{1}{30}(-2 + j - 5k - 19E - 3I - 4J - 22K).$$

P6. For any octonions o_1 and o_2, the following holds:

$$\overline{o_1 o_2} = \bar{o}_2 \bar{o}_1.$$

Example 2.6

Consider two octonions

$$o_1 = 1 + 2i + 3j + 4k + E + 2I + J - K,$$
$$o_2 = 2 - i + j + 2k + 3E - 4I + J + 2K.$$

Then,

$$o_1 o_2 = -1 - 8i - 3j + 14k - E + 2I + 34J - 7K,$$

and

$$\bar{o}_2 \bar{o}_1 = -1 + 8i + 3j - 14k + E - 2I - 34J + 7K,$$

which is $\overline{o_1 o_2}$.

2.2 Functions on Octonions

Many functions described in Section 1.4 on quaternions can similarly be defined and described for octonion numbers. We first consider the exponential function and representation of octonions in a polar form.

2.2.1 Octonion exponential function

The exponential function on octonions o is defined by the Taylor series as

$$e^o = \exp(o) = 1 + \sum_{n=1}^{\infty} \frac{o^n}{n!}. \tag{2.15}$$

Consider a unit pure octonion, which we denote by

$$\lambda = \lambda' = i\lambda_i + j\lambda_j + k\lambda_k + E\lambda_E + I\lambda_I + J\lambda_J + K\lambda_K,$$

and a real number ϑ, which will stand for the variable angle. Since $\lambda^2 = -1$, we can define the exponential function

$$\exp(\lambda\vartheta) = \cos(\vartheta) + \lambda\sin(\vartheta) \tag{2.16}$$

and its conjugate function $\exp(-\lambda\vartheta) = \cos(\vartheta) - \lambda\sin(\vartheta)$.

P1. Any octonion can be written as $o = a + \lambda\vartheta$, where the λ is a pure octonion and ϑ is a real number. Therefore we can write the following general formula for the octonion exponential function

$$e^o = e^{a+\lambda\vartheta} = e^a e^{\lambda\vartheta} = e^a[\cos(\vartheta) + \lambda\sin(\vartheta)]. \tag{2.17}$$

Example 2.7

Consider the octonion $o = 2 + 2i - j + 2k + E + I - 2J + K$, which can be written as

$$o = 2 + \lambda\vartheta = 2 + \frac{2i - j + 2k + E + I - 2J + K}{4}4.$$

For this octonion, $a = 2$ and the angle

$$\vartheta = |o'| = \sqrt{2^2 + 1 + 2^2 + 1 + 1 + 2^2 + 1} = \sqrt{16} = 4,$$

and the pure unit octonion

$$\lambda = \frac{2i - j + 2k + E + I - 2J + K}{4}.$$

The exponential number $\exp(o)$ is calculated by

$$e^o = e^2\left[\cos(4) + \frac{2i - j + 2k + E + I - 2J + K}{4}\sin(4)\right].$$

Denoting $C = e^2/4 \approx 1.8473$, we can write the octonion exponent as

$$e^o = e^2\cos(4) + C\sin(4)(2i - j + 2k + E + I - 2J + K).$$

Below is the script of the code "test_expofoct.m" for this example with the function "expofoct.m" to calculate the quaternion exponential function.

```
%-------------------------------------------------------
% test_expofoct.m
  oct=[2,2,-1,2,1,1,-2,1];  % o=2+2i-j+2k+E+I-2J+K
  octexp = expofoct(oct);
% -4.8298 -2.7960 1.3980 -2.7960 -1.3980 -1.3980 2.7960
  -1.3980
%-------------------------------------------------------
% call: expofoct.m / Artyom Grigoryan, February 14, 2016
% calculate the octonion exponent function
  function octexp = expofoct(oct)
    octim=oct(2:8);     % imaginary part of the octonion
    theta=norm(octim); % angle |octim|=sqrt(b^2+...+K^2)
    lambda_vector =octim/theta;  % imaginary unit \lambda
    e1=exp(oct(1));
    octexp(1) =e1*cos(theta);
    octexp(2:8)=(e1*sin(theta))*lambda_vector;    % See
    Eq. 2.17
%-------------------------------------------------------
```

The calculation by this code results in the octonion

$$e^0 = -4.8298 - 2.7960i + 1.3980j - 2.7960k - 1.3980E - 1.3980I + 2.7960J - 1.3980K.$$

P2. We can verify that the module of the octonion $\cos(\vartheta) + \lambda\sin(\vartheta)$ is 1 for any pure unit octonion λ and angle ϑ. In other words, for any pure octonion o we have $|e^o| = 1$. For example, if

$$o = 0 + 2i - j + 2k + E + I - 2I + K,$$

then after running the commands given below

```
%-----------------------------------------------------------
  oct=[0,2,-1,2,1,1,-2,1];
  octexp = expofoct(oct);
% -0.6536 -0.3784 0.1892 -0.3784 -0.1892 -0.1892 0.3784
  -0.1892
  octexp*octexp'    % 1
%-----------------------------------------------------------
```

we obtain

$$e^o = -0.6536 - 0.3784i + 0.1892j - 0.3784k - 0.1892E - 0.1892I + 0.3784J - 0.1892K$$

and $|e^o|^2 = 1$.

P3. For the octonion $o = e^{\lambda\vartheta} = \cos(\vartheta) + \lambda\sin(\vartheta)$, the inverse octonion o^{-1}, such that $oo^{-1} = 1 = o^{-1}o$, is the conjugate octonion

$$o^{-1} = e^{-\lambda\vartheta} = \cos(\vartheta) - \lambda\sin(\vartheta).$$

P4. The octonion $o = a + o'$, as the quaternion and complex numbers, can be represented in the polar form as

$$o = |o|\exp(\lambda\vartheta),$$

where λ is a pure unit quaternion and ϑ is a real angle in the interval $[0, \pi]$. Indeed,

$$o = |o|\frac{o}{|o|} = |o|\left(\frac{a}{|o|} + \frac{o'}{|o'|}\frac{|o'|}{|o|}\right) = |o|\left(\frac{a}{|o|} + \lambda\frac{|o'|}{|o|}\right)$$
$$= |o|[\cos(\vartheta) + \lambda\sin(\vartheta)],$$

where the pure unit octonion λ and the angle ϑ are calculated by

$$\lambda = \frac{o'}{|o'|} \quad \text{and} \quad \vartheta = \arccos\left(\frac{a}{|o|}\right). \tag{2.18}$$

The sine of this angle is not negative, $\sin(\vartheta) = |o'|/|o| > 0$. Therefore, the angles ϑ in the polar form of octonions are in the interval $[0, \pi]$.

2.2.2 Octonion power and logarithm

For octonions, many functions can be defined similarly to the functions described in Section 1.4 for quaternions. Here, we consider the power and logarithm functions on octonions and describe their properties. Both functions are based on the polar form of representation of octonions, $o = e^{\lambda\vartheta}$.

2.2.3 Power function

It should be noted that for any angles ϑ_1 and ϑ_2, the following holds:

$$\exp(\lambda\vartheta_1)\exp(\lambda\vartheta_2) = \exp[\lambda(\vartheta_1 + \vartheta_2)],$$

when λ is a pure octonion. In the $\vartheta_1 = \vartheta_2$ case, we obtain

$$(e^{\lambda\vartheta_1})^2 = e^{2\lambda\vartheta_1}, \quad (e^{\lambda\vartheta_1})^3 = e^{3\lambda\vartheta_1}, \quad (e^{\lambda\vartheta_1})^4 = e^{4\lambda\vartheta_1}, \ldots,$$

and therefore, for any octonion o, the following is valid:

$$o^n = (|o|e^{\lambda\vartheta})^n = |o|^n e^{n\lambda\vartheta},$$

where $n > 1$ is integer.

In general case, when x is a real number, the power of the octonion o is defined as

$$\begin{aligned} o^x &= [|o|\exp(\lambda\vartheta)]^x = |o|^x[\exp(\lambda\vartheta)]^x = |o|^x[\exp(\lambda x\vartheta)] \\ &= |o|^x[\cos(x\vartheta) + \lambda\sin(x\vartheta)]. \end{aligned} \tag{2.19}$$

For instance, when $x = 1/2$ and $1/3$, the corresponding roots of the octonion o are calculated by

$$\begin{aligned} o^{\frac{1}{2}} &= |o|^{\frac{1}{2}}[\cos(\vartheta/2) + \lambda\sin(\vartheta/2)], \\ o^{\frac{1}{3}} &= |o|^{\frac{1}{3}}[\cos(\vartheta/3) + \lambda\sin(\vartheta/3)]. \end{aligned} \tag{2.20}$$

Example 2.8

Consider the octonion $o = 1 + 3i + E - J + 2K$ and its powers o^x, for $x = 3, 1/2$, and $1/3$. In the polar form, the module of the octonion is $|o| = \sqrt{1+9+1+1+4} = 4$, and the angle is calculated by

$$\vartheta = \arccos\left(\frac{a}{|o|}\right) = \arccos(1/4) = 1.3181\ (\text{rad, or } 75.5225^\circ).$$

The imaginary part of o is $o' = 3i + E - J + 2K$ and the module is $\sqrt{15} = 3.8730$. According to Eq. (2.18), the pure unit octonion is calculated by

$$\lambda = \frac{o'}{|o'|} = \frac{3i + E - J + 2K}{\sqrt{15}} = 0.7746i + 0.2582E - 0.2582J + 0.5164K.$$

The square root of the octonion is calculated by

$$\begin{aligned} o^{\frac{1}{2}} &= 2\cos(0.6591) + 2\lambda\sin(0.6591) = 1.5811 + \lambda 1.2447 \\ &= 1.5811 + 0.9487i + 0.3162E - 0.3162J + 0.6325K. \end{aligned}$$

For the 1/3-power, $4^{\frac{1}{3}} = 1.5874$ and therefore

$$\begin{aligned} o^{\frac{1}{3}} &= 1.5874[\cos(0.4394) + \lambda\sin(0.4394)] = 1.4366 + \lambda 0.6752 \\ &= 1.4366 + 0.5230i + 0.1743E - 0.1743J + 0.3487K. \end{aligned}$$

The square of the octonion equals

$$\begin{aligned} o^2 &= |o|^2[\cos(2\vartheta) + \lambda\sin(2\vartheta)] \\ &= 16[\cos(2.6362) + \lambda\sin(2.6362)] = 16(-0.8750 + \lambda 0.4841) \\ &= -14 + 6i + 2E - 2J + 4K. \end{aligned}$$

The powers were calculated by using the code with the script "test_ octpower.m" given below. All calculations given above are verified in this code as well. For instance, after calculating the power $o^{\frac{1}{3}}$ of the octonion, the cube $(o^{\frac{1}{3}})^3$ of this power is calculated to check that the result is the original octonion o.

```
%----------------------------------------------------------
  % call: test_octpower.m
  o=[1 3 0 0 1 0 -1 2];     % o=1+3i+E-J+2K
  oinx = octpower(o,2);     % -14 6 0 0 2 0 -2 4
  o2 = mult2octs_direct(o,o);  % to verify the same result, o^2
  oinx = octpower(o,0.5);     % power in 1/2
  % 1.5811 0.9487 0 0 0.3162   0 -0.3162 0.6325
```

```
   o2 = mult2octs_direct(oinx,oinx);     % 1 3 0 0 1 0 -1 2
   oinx = octpower(o,1/3);      % power in 1/3
   % 1.4366 0.5230 0 0 0.1743     0 -0.1743 0.3487
   ois = octpower(oinx,3);      % 1 3 0 0 1 0 -1 2
%----------------------------------------------------------
```

This code uses the function "octpower.m," which calculates the power o^x of the octonion o by Eq. (2.19), for any real $x > 0$.

```
%----------------------------------------------------------
% call: octpower.m / Artyom Grigoryan, February 15, 2016
% to calculate the power (o^x) of the octonion o, when r>0.
function oinx = octpower(o,x)
   o_modul=norm(o);  % sqrt(a^2+b^2+c^2+...+J^2+K^2)
   theta=acos(o(1)/o_modul);     % the angle in Eq. 2.15
   oim=o(2:8);     % imaginary part of the octonion
   oim_modul=norm(oim);  % sqrt(b^2+c^2+d^2+...
    +J^2+K^2)
   lambda_vector =oim/oim_modul;  % the pure unit octonion
   % caclulation of the power
   thetax=x*theta;
   o_modulx=o_modul^x;
   oinx(1) = o_modulx*cos(thetax);
   oinx(2:8) = o_modulx*lambda_vector*sin(thetax);
%----------------------------------------------------------
```

2.2.4 Octonion logarithm

The formula for calculating the logarithm of the octonion $o = a + o'$ with the imaginary part $ib + jc + kd + EA + IB + JC + KD$ can be derived by using the polar form of the octonion, when writing it as

$$o = |o|e^{\mu\vartheta} = e^{\ln|o|}e^{\lambda\vartheta} = e^{\ln|o|+\lambda\vartheta}.$$

Here,

$$\lambda = \frac{o'}{|o'|} \quad \text{and} \quad \cos(\vartheta) = \frac{a}{|o|}, \quad \sin(\vartheta) = \frac{o'}{|o|}.$$

The logarithm of the octonion o therefore is defined as

$$\ln o = \ln|o| + \lambda\vartheta = \ln|o| + \frac{ib + jc + kd + EA + IB + JC + KD}{\sqrt{b^2 + c^2 + d^2 + A^2 + B^2 + C^2 + D^2}}\vartheta. \quad (2.21)$$

If we denote this octonion by

$$\ln o = a_1 + ib_1 + jc_1 + kd_1 + EA_1 + IB_1 + JC_1 + KD_1,$$

its components can be calculated as

$$\begin{aligned}
a_1 &= \ln|o| = \frac{1}{2}\ln(|o|^2),\\
b_1 &= \frac{b}{|o'|}\vartheta, \quad c_1 = \frac{c}{|o'|}\vartheta, \quad d_1 = \frac{d}{|o'|}\vartheta,\\
E_1 &= \frac{E}{|o'|}\vartheta, \quad B_1 = \frac{B}{|o'|}\vartheta, \quad C_1 = \frac{C}{|o'|}\vartheta, \quad D_1 = \frac{D}{|o'|}\vartheta,
\end{aligned} \tag{2.22}$$

where,

$$\vartheta = \arccos\left(\frac{a}{|o|}\right),$$
$$|o'| = \sqrt{b^2 + c^2 + d^2 + A^2 + B^2 + C^2 + D^2}, \quad |o| = \sqrt{a^2 + |o'|^2}.$$

The components b_1, c_1, d_1, and A_1, B_1, C_1, D_1 up to the constant $T = \vartheta/|o'|$ are proportional to the components b, c, d, and A, B, C, D, respectively. In other words,

$$\ln(o) = \frac{1}{2}\ln(|o|^2) + To', \qquad \left(T = \frac{\vartheta}{|o'|}\right). \tag{2.23}$$

Example 2.9

Consider the octonion $o = 2 - 2i + j + k - 2E + I - 2J + K$. The logarithm of this octonion is

$$\ln o = 1.4979 + 0.2768(-2i + j + k - 2E + I - 2J + K).$$

Indeed, we have $a = 2$, and $|o|^2 = 4 + 4 + 1 + 1 + 4 + 1 + 4 + 1 = 20$ and $|o'|^2 = |o|^2 - a^2 = 20 - 4 = 16$. The real part of the logarithm $\ln(o)$ is

$$a_1 = \frac{1}{2}\ln 20 = 1.4979.$$

The required angle ϑ is calculated by

$$\vartheta = \arccos\left(\frac{a}{|o|}\right) = \arccos\left(\frac{2}{\sqrt{20}}\right) = \arccos\left(\frac{1}{\sqrt{5}}\right) = 1.1071 \text{ (rad)},$$

or $\vartheta = 63.4349^\circ$ in degrees, and the coefficient of proportionality is

$$T = \frac{\vartheta}{|o'|} = \frac{1.1071}{4} = 0.2768.$$

Therefore, the imaginary part of the logarithm equals

$$(\ln o)' = To' = 0.2768(-2i + j + k - 2E + I - 2J + K).$$

Below is the script of the code "test_logofoct.m" for this example, which uses the function "logofoct.m" to calculate the logarithm by Eq. (2.23). The last command in the code "octexp = expofoct(oln)" is used to verify that the exponent of the $\ln(o)$ is o.

```
%---------------------------------------------------------------
% call: test_logofoct.m
    o=[o=[2 -2 1 1 -2 1 -2 1];
    noR=norm(o);                % 4.4721 = sqrt(20)
    noI=norm(o(2:8));           % 4 = sqrt(16)
    oln = logofoct(o);
% 1.4979 -0.5536 0.2768 0.2768 -0.5536 0.2768 -0.5536
  0.2768
% verify the calculation by using the exponent of this
  logarithm
  octexp = expofoct(oln)     % 2 -2 1 1 -2 1 -2 1
%---------------------------------------------------------------
% call: logofoct.m / Art Grigoryan, February 15, 2016
% calculate the octonion logarithm function ln(o).
  function oln = logofoct(o)
    o_modul=norm(o);       % the module of the octonion
    a1=log(o_modul);       % the real part of the logarithm
    oim=o(2:8);       % imaginary part of octonion o
    theta=acos(o(1)/o_modul);       % angle
    om=norm(oim);      % om=sqrt(b^2+c^2+d^2+...+K^2)
    T=theta/om;       % coefficient of proportionality
    bcdABCD1 = T*o(2:8);       % b1 c1 d1 E1 I1 J1 K1
    oln=[a1 bcdABCD1];      % a1 b1 c1 d1 E1 I1 J1 K1
%---------------------------------------------------------------
```

It should be noted that all given above functions on octonions can be used for quaternions, since the quaternions $q = a + ib + jc + kd$ are the octonions $o = a + ib + jc + kd + EA + IB + JC + KD$, when four imaginary components $A = B = C = D = 0$.

For example, the logarithm of the octonion quaternion $q = 2 - 2i + j + 3k$ can be calculated by the command "logofoct(o)" where

o is the vector $o = [2, -2, 1, 3, 0, 0, 0, 0]$ representing the octonion $o = 2 - 2i + j + 3k$, as shown below.

```
%---------------------------------------------------------
    o=[2,-2,1,3,0,0,0,0];
    oln=logofoct(o) % 1.4452 -0.5772 0.2886 0.8659 0 0 0 0
    qln=logofq(o(1:4)); % 1.4452 -0.5772 0.2886 0.8659
%---------------------------------------------------------
```

The result of calculation is the vector $[1.4452, -0.5772, 0.2886, 0.8659, 0, 0, 0, 0]$ representing the logarithm is

$$\ln(o) = 1.4452 - 0.5772i + 0.2886j + 0.8659k.$$

The first four components of this vector can be used for the quaternion logarithm. The same result of the logarithm is obtained by the command "logofq(q)" where the quaternion $q = 2 - 2i + j + 3k$ is represented by the vector $q = [2, -2, 1, 3]$.

To calculate the exponent of this quaternion q, we also can use the function "expofoct" and compare the result with calculation by the function "expofq" as shown below.

```
%---------------------------------------------------------
    octex=expofoct(o); % -6.0982 2.2303 -1.1152 -3.3455 0
    0 0 0
    qex=expofq(o(1:4)); % -6.0982 2.2303 -1.1152 -3.3455
%---------------------------------------------------------
```

For the power of the quaternion, q^x, where $x > 0$ is a real number, we can use the above power function "octpower" written for octonions.

```
%---------------------------------------------------------
    ox=octpower(o,1/4); % 1.3832 -0.2046 0.1023 0.3069 0 0 0 0
    q=ox(1:4)
    q=qpower(o(1:4),1/4) % 1.3832 -0.2046 0.1023 0.3069
    qpower(q,4) % 2.0000 -2.0000 1.0000 3.0000
%---------------------------------------------------------
```

One can see in this example that the power $q^{\frac{1}{4}}$ calculated by the function "octpower" equals

$$q_1 = (2 - 2i + j + 3k)^{\frac{1}{4}} = 1.3832 - 0.2046i + 0.1023j + 0.3069k.$$

Here, we also use the function of the power, which is written only for quaternions, by simplifying the above function "octpower." The results of the calculation were verified by calculating the 4th power of the quaternion q_1 by the function "qpower" to get the right answer, $q_1^4 = q$.

```
%-------------------------------------------------------------
% call: qpower.m / Artyom Grigoryan, February 17, 2016
% to calculate the power (q^x) of the quaternion q.
  x is real >0.
function qinx = qpower(q,x)
    q_modul=norm(q);     % sqrt(a^2+b^2+c^2)
    theta=acos(q(1)/q_modul);     % the angle in Eq. 1.68
    qim=q(2:4);     % imaginary part of the quaternion
    qim_modul=norm(qim);     % sqrt(b^2+c^2+d^2)
    lambda_vector =qim/qim_modul;   % the pure unit
    quaternion
    % caclulation of the power
    thetax=x*theta;
    q_modulx=q_modul^x;
    qinx(1) = q_modulx*cos(thetax);
    qinx(2:4) = q_modulx*lambda_vector*sin(thetax);
%-------------------------------------------------------------
```

2.2.5 Octonion trigonometric functions

The cosine, sine, and tangent functions are defined by using the standard procedure with the exponential function in the octonion space. For a given pure unit octonion number λ, the pair of octonion conjugate exponential functions of angle ϑ is defined as

$$e^{\lambda\vartheta} = \cos(\vartheta) + \lambda \sin(\vartheta), \quad e^{-\lambda\vartheta} = \cos(\vartheta) - \lambda \sin(\vartheta).$$

The real cosine and sine functions can be calculated from the octonion exponential functions as

$$\cos(\vartheta) = \frac{1}{2}(e^{\lambda\vartheta} + e^{-\lambda\vartheta}), \quad \text{and} \quad \sin(\vartheta) = \frac{1}{2\lambda}(e^{\lambda\vartheta} - e^{-\lambda\vartheta}).$$

The octonion cosine and sine functions can be defined by the Taylor series

$$\cos(o) = 1 + \sum_{n=1}^{\infty}(-1)^n \frac{o^{2n}}{(2n)!},$$

$$\sin(o) = \sum_{n=0}^{\infty}(-1)^n \frac{o^{2n+1}}{(2n+1)!}.$$

If o is a pure octonion, it can be presented as $o = \lambda\vartheta$, where λ is a pure unite octonion ($\lambda^2 = -1$) and ϑ is real, as the angle. Therefore, the function $\cos(o)$ is real and can be written as

$$\cos(o) = \cos(\lambda\vartheta) = 1 + \sum_{n=1}^{\infty} \frac{\vartheta^{2n}}{(2n)!} = 1 + \sum_{n=1}^{\infty}(-1)^n \frac{(i\vartheta)^{2n}}{(2n)!} = \cos(i\vartheta),$$

since $(-1)^n(\lambda\vartheta)^{2n} = \vartheta^{2n}$, for integers $n > 0$. Here i is the complex imaginary, for instance. In other words, in this case $\cos(o)$ does not depend on λ. For the sine function, we can write

$$\sin(o) = \sin(\lambda\vartheta) = \sum_{n=1}^{\infty}(-1)^n \frac{\lambda\lambda^{2n}\vartheta^{2n+1}}{(2n+1)!} = \lambda\sum_{n=1}^{\infty}(-1)^n \frac{i^{2n}\vartheta^{2n+1}}{(2n+1)!} = \frac{\lambda}{i}\sin(i\vartheta),$$

since $(-1)^n(\lambda\vartheta)^{2n+1} = \lambda\vartheta^{2n+1}$, for integers $n > 0$. One can note the following equality:

$$\cos(o) + \lambda\sin(o) = \cos(i\vartheta) + \frac{\lambda^2}{i}\sin(i\vartheta) = \cos(i\vartheta) + i\sin(i\vartheta).$$

In the general case when $o = a_1 + \lambda\vartheta$, we can extend these concepts to the octonion functions, by defining the cosine and sine as

$$\cos(o) = \frac{1}{2}(e^{\lambda o} + e^{-\lambda o}), \quad \sin(o) = \frac{1}{2\lambda}(e^{\lambda o} - e^{-\lambda o}).$$

The octonion hyperbolic cosine and sine can be defined as

$$\cosh(o) = \frac{1}{2}(e^{o} + e^{-o}), \quad \sinh(o) = \frac{1}{2}(e^{o} - e^{-o}).$$

Similar to the quaternion hyperbolic functions, the octonion cosine and sine and hyperbolic cosine and sine can be written as follows:

$$\begin{aligned}
\cos(o) &= \cos(a)\cosh(\vartheta) - \lambda\sinh(\vartheta)\sin(a),\\
\sin(o) &= \sin(a)\cosh(\vartheta) + \lambda\sinh(\vartheta)\cos(a),\\
\cosh(o) &= \cos(\vartheta)\cdot\cosh(a) + \lambda\sinh(a)\sin(\vartheta),\\
\sinh(o) &= \cos(\vartheta)\cdot\sinh(a) + \lambda\cosh(a)\sin(\vartheta).
\end{aligned} \tag{2.24}$$

The octonion tangent and cotangent are defined as

$$\tanh(o) = \frac{\sinh(o)}{\cosh(o)} = \frac{e^{o} - e^{-o}}{e^{o} + e^{-o}}$$

and

$$\coth(o) = \frac{\cosh(o)}{\sinh(o)} = \frac{e^{o} + e^{-o}}{e^{o} - e^{-o}}.$$

References

[1] J. Graves, "On a connection between the general theory of normal couples and the theory of complete quadratic functions of two variables," *Philos. Mag.*, **26**, 315–320 (1845).
[2] A. Cayley, "On Jacobi's elliptic functions, in reply to the Rev..; and on quaternions," *Philosophical Magazine*, **26**, 208–211 (1845), Appendix reprinted in The Collected Mathematical Papers, p. 127, Johnson Reprint Co., New York (1963).
[3] W. R. Hamilton, "Note, by Sir W. R. Hamilton, respecting the researches of John T. Graves, Esq.," *Transactions of the Royal Irish Academy*, **21**, 338–341 (1848).
[4] W. R. Hamilton, *Lectures on quaternions: Containing a systematic statement of a new mathematical method*, Hodges and Smith, Dublin (1853).
[5] H. Freudenthal, "Oktaven, ausnahmegruppen und oktavengeometrie," *Geom. Dedicata*, **19**(1), 7–63 (1985).
[6] F. van der Blij, "History of the octaves," *Simon Stevin*, **34**, 106–125 (1961).
[7] I. L. Kantor and A. S. Solodovnikov, *Hypercomplex Numbers*, Nauka, Moskow (1973).
[8] G. M. Dixon, *Division Algebras: Octonions, Quaternions, Complex Numbers and the Algebraic Design of Physics*, Kluwer Academic Publishers, Berlin (1994).
[9] R. Ablamowicz, P. Lounesto, and J. M. Parra, *Clifford Algebras with Numeric and Symbolic Computations*, Birkhäuser, Switzerland (1996).
[10] L. V. Sabinin, L. Sbitneva, and I. P. Shestakov, "§ 17.2 Octonion algebra and its regular bimodule representation," *Non-Associative Algebra and its Applications*, p. 235, CRC Press, Boca Raton, Florida (2006).
[11] J. C. Baez, "The Octonions," *Bulletin of the American Mathematical Society*, **39**(2), 145–205 (2002).
[12] J. H. Conway and D. A. Smith, *On Quaternions and Octonions: Their Geometry, Arithmetic, and Symmetry*, A.K. Peters, Ltd. (2003).
[13] A. M. Grigoryan and S. S. Agaian, "Retooling of color imaging in the quaternion algebra," *Applied Mathematics and Sciences: An International Journal (MathSJ)*, **1**(3), 23–39 (2014).

Chapter 3
Quaternions and Color Images

Many ways can be found to apply the theory of quaternion numbers in imaging. In this chapter, we consider the grayscale and color images and their mappings to the quaternion space from the well-known color models [1]–[7]. Different models of mapping one of a few grayscale and color images into the quaternion space are considered. The presentation of the real image in the quaternion space, allows for developing new methods of image representation and processing, which include image enhancement, filtration, and restoration.

3.1 Digital Grayscale Images

The digital grayscale image $f_{n,m}$, or 2-D image can be considered as a table or array of real values that may represent brightness or intensities of the image at spatial points (n, m). The informal term brightness can relate to the stream of light energy that falls to the image plane. These points are usually defined at knots of a regular discrete lattice, for instance the Cartesian or hexagonal lattice. In the 8-bit image format, the values of the image, which also are called gray level values, are integers and range from 0 (for the black color) to 255 (for the white color), with the remaining 254 intermediate values are shades of gray. Figure 3.1 illustrates the digital image, which is composed of a finite number of image elements [picture elements (pels, or pixels)]. Each image element has a location in a 2-D space (matrix) and value. The number at each pixel relates to the amount of intensity of the image. We may think that the value at one pixel in the digital image represents a small region in the continuous image being digitized. This region (that contributes to the pixel value) is called *sampling aperture*. To display the image, the value of each pixel is converted into a grayscale, with a total of 256 gray-levels. This figure shows the original "tree" image in Fig. 3.1(a). The part of the image in the domain $[110, 125] \times [120, 135]$ is shown in Fig. 3.1(b), and the 16×16 matrix of intensities of this part in Fig. 3.1(c).

Such digital images represent real images $f(x, y)$ with the spatial coordinates x and y, which may be captured by visual sensors in cameras, microscopes, telescopes and other imaging systems. The light is not the only

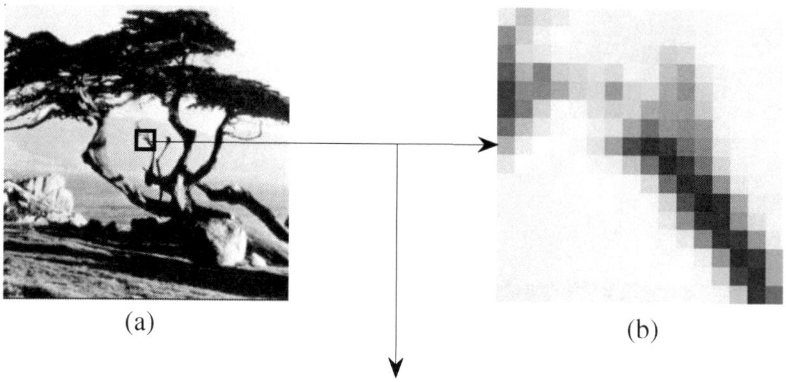

(a) (b)

190	164	182	188	200	198	198	201	200	199	201	201	201	201	204	204
132	174	198	206	200	198	197	198	199	197	198	201	202	202	201	201
120	184	189	200	198	197	199	198	192	174	180	201	204	201	202	203
98	158	122	169	177	183	180	199	197	164	150	190	201	202	203	201
80	129	126	148	171	177	137	175	182	151	133	177	200	201	202	201
85	122	175	201	197	182	171	176	151	144	140	170	201	204	202	200
109	165	190	200	202	191	184	148	111	137	128	140	199	206	205	202
156	197	207	206	199	184	142	89	65	78	112	129	190	201	204	204
187	207	205	201	202	197	192	166	111	56	74	117	180	201	203	201
203	202	203	202	205	202	203	204	181	107	60	68	134	191	201	202
201	202	202	202	201	202	203	205	203	177	89	52	63	148	196	200
199	202	201	200	201	198	202	202	202	189	129	65	60	134	185	199
199	200	200	200	199	199	199	202	203	201	190	127	65	74	138	192
200	198	200	199	200	199	199	200	203	201	202	192	134	58	71	148
199	199	199	200	200	200	201	200	203	203	203	199	170	79	59	148
197	197	197	200	198	200	199	199	199	201	201	201	198	163	62	105

(c)

Figure 3.1 (a) The tree image of size 256×256, (b) the part of the image in the area $[110 : 125, 120 : 135]$, and (c) the table of intensities of this part.

parameter used in scientific imaging. For example, images can be visualized after reconstructing from data obtained by other types of sensors or systems, such as tomographic images that are calculated from the data obtained by a complex X-ray set and detectors measuring the roentgen radiation interaction with the body [8].

3.2 Model 1: (2 × 2)-Mapping

The grayscale image $f_{n,m}$ of size $N \times M$ can be divided by four parts with even-even, even-odd, odd-even, and odd-odd indices as follows:

$$
\begin{array}{cccccc}
\cdot & \cdot & \cdot & \cdot & \cdot & \cdot \\
\cdot & f_{2n,2m} & f_{2n,2m+1} & f_{2n,2m+2} & f_{2n,2m+3} & \cdot \\
\cdot & f_{2n+1,2m} & f_{2n+1,2m+1} & f_{2n+1,2m+2} & f_{2n+1,2m+3} & \cdot \\
\cdot & f_{2n+2,2m} & f_{2n+2,2m+1} & f_{2n+2,2m+2} & f_{2n+2,2m+3} & \cdot \\
\cdot & f_{2n+3,2m} & f_{2n+3,2m+1} & f_{2n+3,2m+2} & f_{2n+3,2m+3} & \cdot \\
\cdot & \cdot & \cdot & \cdot & \cdot & \cdot
\end{array}
$$

$$f_e = \{(f_e)_{n,m}\} = \{f_{2n,2m}\},$$
$$f_i = \{(f_i)_{n,m}\} = \{f_{2n,2m+1}\},$$
$$f_j = \{(f_j)_{n,m}\} = \{f_{2n+1,2m}\},$$
$$f_k = \{(f_k)_{n,m}\} = \{f_{2n+1,2m+1}\},$$

where $n = 0 : (N/2 - 1)$, $m = 0 : (M/2 - 1)$. It is assumed that the numbers N and M are even, if not, the borders of indexes should be corrected. As an example, Fig. 3.2 shows the grayscale "peppers" image $f_{n,m}$. These four parts can be considered as components of the quaternion matrix, which we call the quaternion image and denote by $q(f)$. The first part, the matrix of size $N/2 \times M/2$ is denoted by f_e as the e-component, or real component of $q(f)$. Other three parts can be considered as the components of the imaginary part of the quaternion image. Therefore, they can be denoted as f_i, f_j, and f_k for the i,j,- and k-components, respectively.

Figure 3.3 shows the real part $(f_e)_{n,m}$ in part (a), and the components $(f_i)_{n,m}$, $(f_j)_{n,m}$, and $(f_k)_{n,m}$ in parts (b), (c), and (d), respectively.

Thus, we obtain the full quaternion image with non-zero real part. The components of imaginary part can be mapped into one of the color spaces, such as the RGB or XYZ space [1].

3.3 Color Models

In this section, we briefly describe a few color models that are widely used in color imaging and can be used for mapping grayscale images into color images.

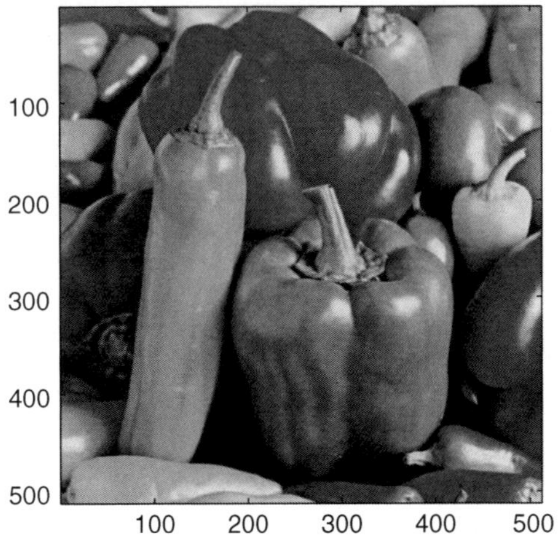

Figure 3.2 The grayscale "pepper" image of size 512 × 512.

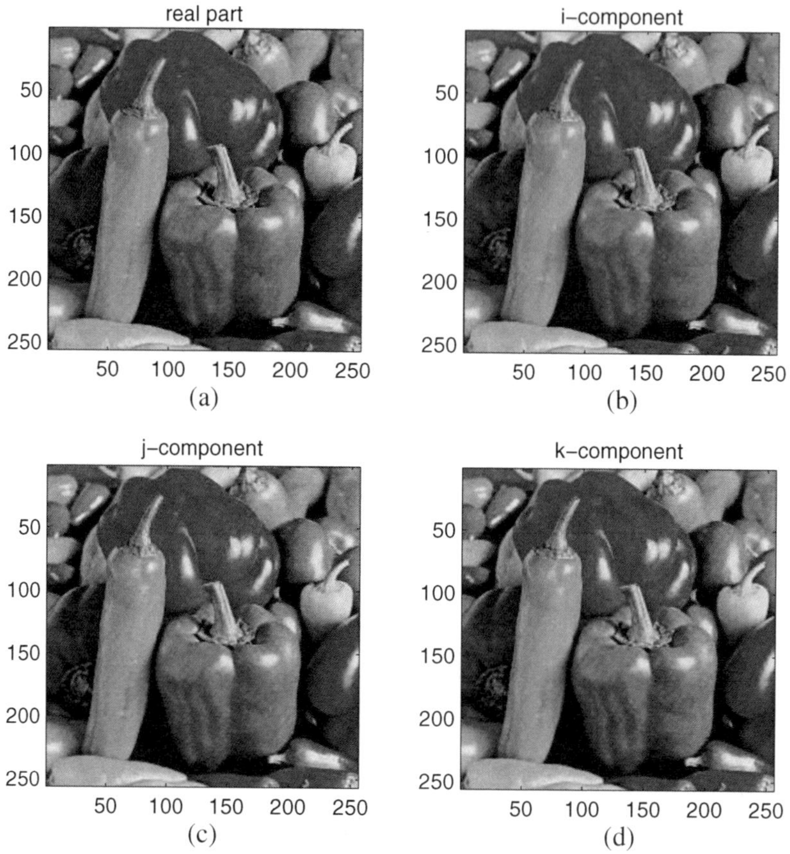

Figure 3.3 The division of the grayscale image into four parts of size 256 × 256 each. (a) The real part and (b) the *i*-component, (c) the *j*-component, and (d) the *k*-component of the quaternion image.

3.3.1 Family of RGB models

In the 19th century, Thomas Young and Herman Helmholtz in the publication "Theory of trichromatic color vision" and James Maxwell's (introduced the color triangle) described the RGB color model, which is a convenient model for computer graphics. The name of the model is associated with the initials of the three primary colors, namely the red (of wavelength 700 nm), green (of wavelength 546.1 nm), and blue (of wavelength 435.8 nm) colors. Each of these colors stimulates one of the three types of the human eye's color receptors with as little stimulation as possible of the other two [1]–[4]. In the television industry, the RGB model is able to match the human eye perception reasonably well. This color model is used for recording colors in digital cameras and video cameras. The RGB model is the most common way to encode color in computing. This color model is illustrated in Fig. 3.4(a), where red (R), green (G), and blue (B) are primary colors, and the color is expressed as the triplet (r, g, b),

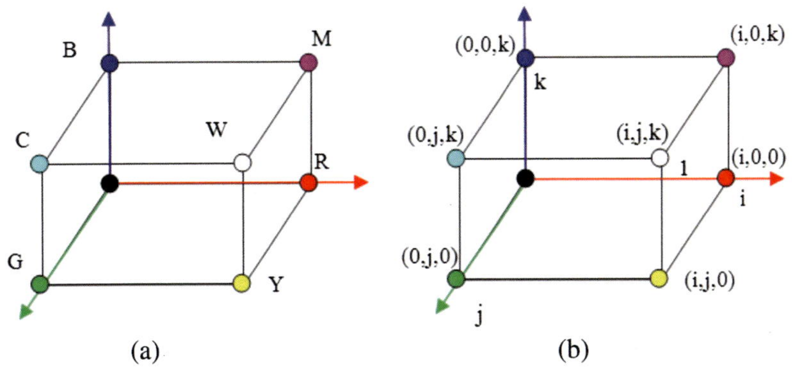

Figure 3.4 The color cube in (a) the RGB color model and (b) in the quaternion subspace.

each component of which can vary from zero to a defined maximum value, for instance 1. The triplets $R = (1, 0, 0)$, $G = (0, 1, 0)$, and $B = (0, 0, 1)$ define the red, green, and blue colors, respectively. Any other color as a triplet (r, g, b) is defined as a linear combination of these primaries,

$$(r, g, b) = r(1, 0, 0) + g(0, 1, 0) + b(0, 0, 1),$$

where the numbers r, g, b are from the unit interval $[0, 1]$.

It is an additive model, in which colors are produced by adding components $C = rR + gG + bB$ and reproduce a broad range of colors. The integer range of each color can be considered in the standard $[0, 255]$ interval of integers. For example, the triplet $(255, 0, 0)$ expresses the red color, the triplet $(0, 255, 0)$ expresses the green color, and the triplet $(0, 0, 255)$ expresses the blue color. If the triplet (r, g, b) is $(255, 0, 255)$ the result is the magenta color $\{M = R + B = 255[(1, 0, 0) + (0, 0, 1)] = 255(1, 0, 1)\}$; if all components are at zero, the result is black; if all components are at maximum, the result is the white color. Together the red and green lights result in the yellow color, $Y = (255, 255, 0)$. Figure 3.5 shows three triangles of primary colors and other secondary colors as the result of overlapping the primary colors. The required number of bits for each color is 8, and 24 for the triplet of colors (r, g, b). Therefore the total number of different colors in the 24-bit RGB

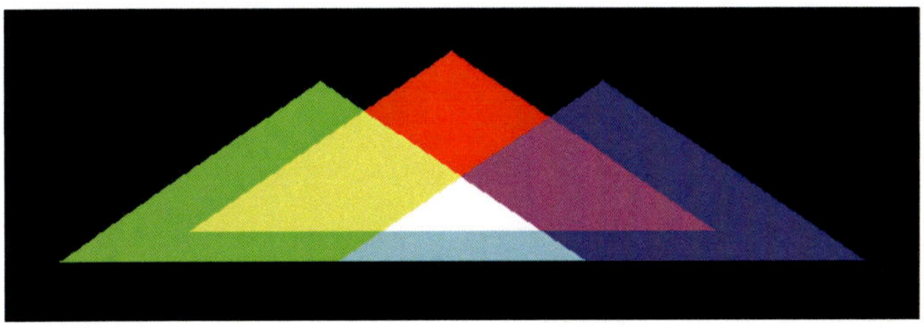

Figure 3.5 Triangles of the primary colors in the RGB model.

model is $2^{24} = 1677216$. However, this number of linear combinations of the three primary colors does not cover all visible colors.

It should be noted that the human sensation of color is subjective. For majority of people, the brightness in the color image is perceived by 30% of the red color, 59% the green color, and 11% of the blue color. Therefore, the brightness at each pixel of the color image, or the global brightness can be calculated by

$$x = x(r, g, b) = 0.30r + 0.59b + 0.11g.$$

In the subspace with pure quaternions, this RGB color cube can be represented in the cube with imaginary units, where each color triplet (r, g, b) is presented as the quaternion (ir, jg, kb), as shown in Fig. 3.4(b).

When transforming the color image into the quaternion subspace, we may assume that the component f_i is set to the red color channel, and f_j and f_k to green and blue channels, respectively, as shown in Table 3.1.

Here, the letter **E** stands for the gray levels, **R** for the red color, **G** for the green color, and **B** for the blue color. In such a mapping, the imaginary part of the quaternion image $q(f)$ can be illustrated as a color image. One can notice from this table that in such a model, each set of four pixels in the moving 2×2 window are considered as the e, i, j, and k-components and mapped to the colors as

$$\begin{pmatrix} \bullet & \bullet \\ \bullet & \bullet \end{pmatrix} \rightarrow \begin{pmatrix} e & i \\ j & k \end{pmatrix} \rightarrow \begin{pmatrix} E & R \\ G & B \end{pmatrix}. \tag{3.1}$$

Other similar 2×2-models can also be considered, for instance the model with mapping

$$\begin{pmatrix} \bullet & \bullet \\ \bullet & \bullet \end{pmatrix} \rightarrow \begin{pmatrix} e & i \\ j & k \end{pmatrix} \rightarrow \begin{pmatrix} E & B \\ R & G \end{pmatrix}.$$

Based on the 2×2-model described by Eq. (3.1), Fig. 3.6(a) shows the real part $(f_e)_{n,m}$ of the quaternion image, and the imaginary part, or color image is shown in Fig. 3.6(b). These two images, namely the grayscale image and color image of size $N/2 \times M/2$ each, represent the original grayscale "peppers" image of size $N \times M$ in the quaternion space.

Table 3.1 Mapping of the grayscale image into the quaternion image.

.	.	.	.	.	.
.	**E**:$f_{2n,2m}$	**R**:$f_{2n,2m+1}$	**E**:$f_{2n,2m+2}$	**R**:$f_{2n,2m+3}$	.
.	**G**:$f_{2n+1,2m}$	**B**:$f_{2n+1,2m+1}$	**G**:$f_{2n+1,2m+2}$	**B**:$f_{2n+1,2m+3}$	.
.	**E**:$f_{2n+2,2m}$	**R**:$f_{2n+2,2m+1}$	**E**:$f_{2n+2,2m+2}$	**R**:$f_{2n+2,2m+3}$	.
.	**G**:$f_{2n+3,2m}$	**B**:$f_{2n+3,2m+1}$	**G**:$f_{2n+3,2m+2}$	**B**:$f_{2n+3,2m+3}$	.
.	.	.	.	.	.

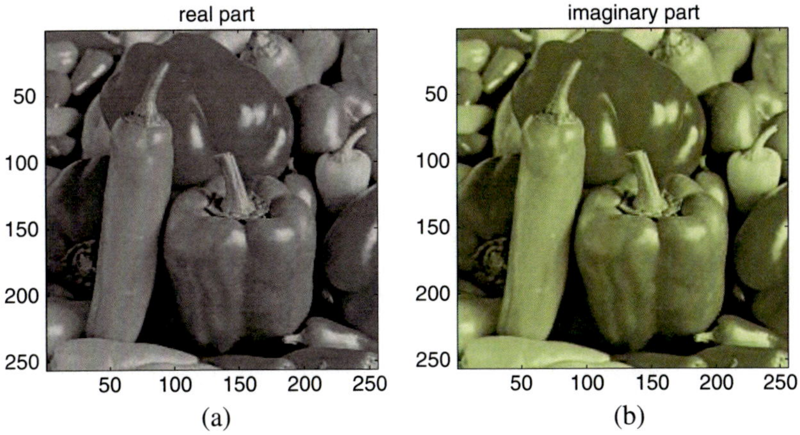

Figure 3.6 The quaternion image of size 256 × 256: (a) the real part and (b) the imaginary part in the RGB color model.

3.3.1.1 Script for the 2 × 2-mapping

Below is the script of the MATLAB®-based code "test_2pepper.m" for calculating the described above operations for the (2 × 2) model of mapping of the "peppers" image. The code works on the original color image "peppers.tiff" which is the image "4.2.07.tiff" from the USC-SIPI Image Database and can be found at the address "http://sipi.usc.edu/database/database.php"

```
%=========================================================
%   call: test_1pepper.m / Artyom Grigoryan, 2016
%1  Read the "peppers" image
    X1=imread('peppers.tiff','tiff');
    [N2 M2 L] = size(X1);
    N2=4*floor(N2/4); M2=4*floor(M2/4);
%2  average grayscale image calculation
    Y=double(X1);
    Y1=Y(:,:,1); Y2=Y(:,:,2); Y3=Y(:,:,3);
    X=round((Y1+Y2+Y3)/3);
    figure;            colormap(gray(255));
    subplot(2,1,2);   image(X);   axis image;     % Fig. 3.2
%3  calculate the model of image in quaternion space
    Qe=X(1:2:N2,1:2:M2);    % to be the real part
    Qi=X(2:2:N2,1:2:M2);    % to be the i-component
    Qj=X(1:2:N2,2:2:M2);    % to be the j-component
    Qk=X(2:2:N2,2:2:M2);    % to be the k-component
    figure;    colormap(gray(255));                % Fig. 3.3
    subplot(2,2,1); image(Qe); axis image;
    title('real part');     xlabel('(a)');
    subplot(2,2,2); image(Qi); axis image;
```

```
    title('i-component');          xlabel('(b)');
    subplot(2,2,3); image(Qj); axis image;
    title('j-component');          xlabel('(c)');
    subplot(2,2,4); image(Qk); axis image;
    title('k-component');          xlabel('(d)');
%4  display the real part and imaginary part as a color image
    figure;     colormap(gray(255));              % Fig. 3.6
    subplot(1,2,1); image(Qe); axis image;
    title('real part');            xlabel('(a)');
    [N3,M3]=size(Qi);                             % 510 x 170
    Q=zeros(N3,M3,3);
    Q(:,:,1)=1.1*Qi; Q(:,:,2)=1.25*Qj; Q(:,:,3)=0.75*Qk;
    Q=uint8(Q);
    subplot(1,2,2);   image(Q);   axis image;
    title('imaginary part'); xlabel('(b)');
%=========================================================
```

3.3.2 Model 2: (3 × 3)-mapping

We consider the model of grayscale image representation in quaternion space in the form of a three-component pure quaternion image. Such an image can be presented and processed in the RGB and other color models.

In Table 3.2, the first 3 × 3 matrix is shown with the quaternion imaginary units, where the rules of multiplications are written from left to right and down to up. This matrix is transformed to the color space as shown in the second 3 × 3 matrix of this table.

Thus, we divide the grayscale image $f = f_{n,m}$ into three parts as shown in Table 3.3. For simplicity of calculation, we consider the case when the size of the original grayscale image is divided by 3, i.e., N and M are multiple to 3. As an example, Fig. 3.7 shows the same grayscale "peppers" image $f_{n,m}$ of the size 510 × 510, considering that 510 = 170(3).

Table 3.2 Table of mapping the quaternion imaginary components into the primary colors.

i	j	k	→	R	G	B
k	i	j	→	B	R	G
j	k	i	→	G	B	R

Table 3.3 Mapping for the RGB color model.

.	.	.	.	.	.
.	**R**:$f_{2n,2m}$	**G**:$f_{2n,2m+1}$	**B**:$f_{2n,2m+2}$	**R**:$f_{2n,2m+3}$	.
.	**B**:$f_{2n+1,2m}$	**R**:$f_{2n+1,2m+1}$	**G**:$f_{2n+1,2m+2}$	**B**:$f_{2n+1,2m+3}$	.
.	**G**:$f_{2n+1,2m}$	**B**:$f_{2n+2,2m+1}$	**R**:$f_{2n+2,2m+2}$	**G**:$f_{2n+2,2m+3}$	.
.	**R**:$f_{2n+1,2m}$	**G**:$f_{2n+3,2m+1}$	**B**:$f_{2n+3,2m+2}$	**R**:$f_{2n+3,2m+3}$	.
.	.	.	.	.	.

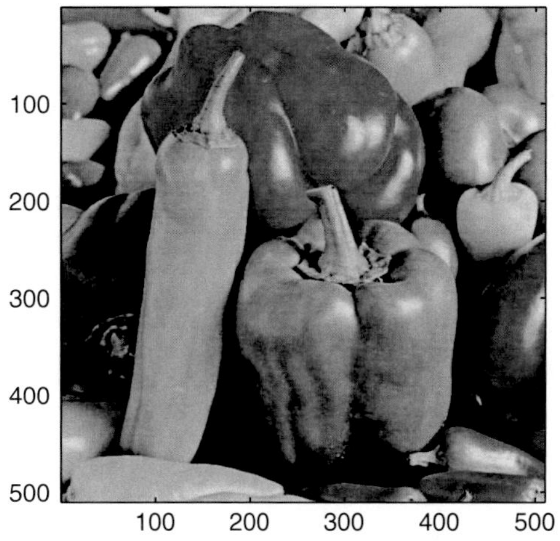

Figure 3.7 The grayscale "peppers" image of size 510 × 510.

Three parts of the image can be composed as follows:

$$f_i = \{f_{n,m}; (n+m) = 0 \bmod 3\},$$
$$f_j = \{f_{n,m}; (n+m) = 1 \bmod 3\},$$
$$f_k = \{f_{n,m}; (n+m) = 2 \bmod 3\},$$

where $n = 0 : (N/3 - 1)$, $m = 0 : (M/3 - 1)$. These parts are denoted as f_i, f_j, and f_k as the *i*-, *j*-, and *k*-components, respectively. The pure quaternion image composed from these components is denoted by $q(f)'$. For the above "peppers" image, Fig. 3.8 shows the parts $(f_i)_{n,m}$, $(f_j)_{n,m}$, and $(f_k)_{n,m}$ in parts (a), (b), and (c), respectively.

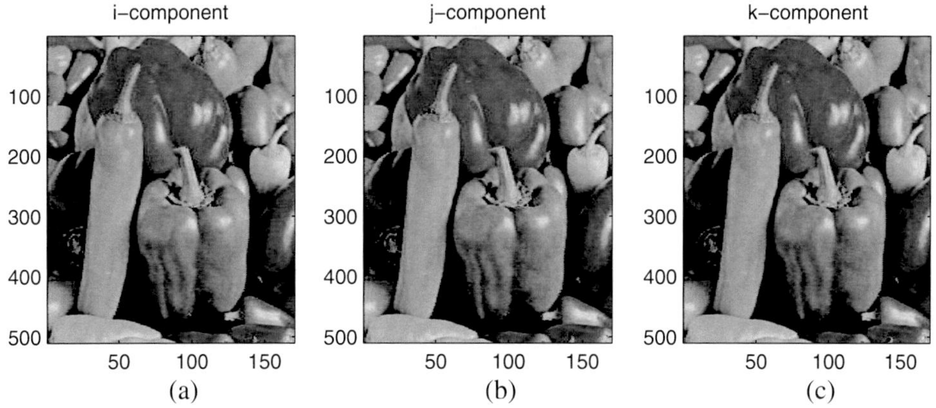

Figure 3.8 Presentation of the image by three parts of size 510 × 170 each.

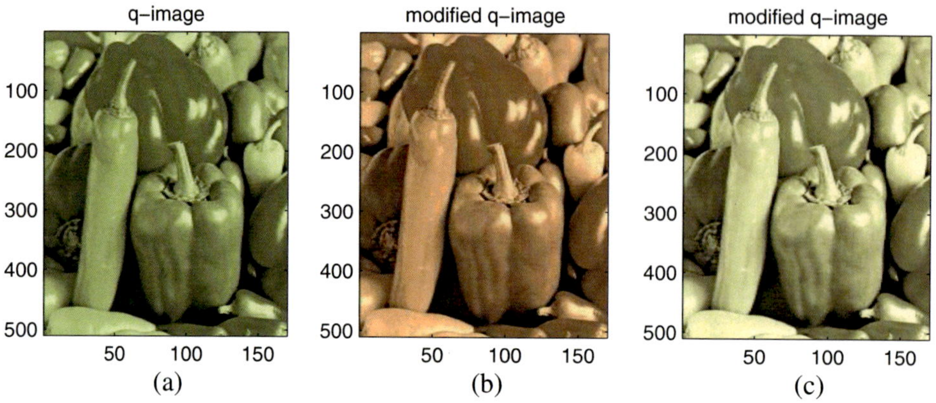

Figure 3.9 (a) The quaternion image of size 510 × 170. (b) and (c) The modified quaternion images in the RGB color model.

These three images can be mapped into the color space. Figure 3.9 shows the quaternion image $q(f)'$ as a color image in the RGB color model in part (a). In Fig. 3.9(b), before composing the color quaternion image, the red channel was amplified by the factor of 1.75, and the green channel by the factor of 1.5 in Fig. 3.9(c). Thus, in the described above second model, the grayscale image f is represented as a pure quaternion image $q(f)'$, which can be visualized as a color image.

3.3.2.1 Script for the 3 × 3-mapping

Below is the script of the MATLAB-based code "test_2pepper.m" for calculating all operations for the (3 × 3) model of mapping the "peppers" image.

```
%==================================================
%   call: test_2pepper.m      / Artyom Grigoryan, 2016
%I  Read the color "pepper" image of size 512x512:
      X1=imread('peppers.tiff','tiff');
      [N2 M2 L] = size(X1);
      N2=3*floor(N2/3);   M2=3*floor(M2/3);       % 510x510
      figure;   colormap(gray(255));              % Fig. 3.7
      Y=double(X1(1:N2,1:M2,:));
      Y1=Y(:,:,1);   Y2=Y(:,:,2);   Y3=Y(:,:,3);
      X=round((Y1+Y2+Y3)/3);
      subplot(2,1,2);   image(X);   axis image;
%II Calculate the model of the image in the quaternion space
   NM2=N2*M2/3;
   Xi=zeros(1,NM2);   % compose the i-component of the image
   n1=1;
   for n=0:N2-1
```

```
        for m=0:M2-1
            if mod(n+m,3)==0
                Xi(n1)=X(n+1,m+1);
                n1=n1+1;
            end
        end
    end
    Qi=reshape(Xi,N2/3,M2);   Qi=Qi';
    Xj=zeros(1,NM2);   % compose the j-component of the image
    n1=1;
    for n=0:N2-1
        for m=0:M2-1
            if mod(n+m,3)==1
                Xj(n1)=X(n+1,m+1);
                n1=n1+1;
            end
        end
    end
    Qj=reshape(Xj,N2/3,M2); Qj=Qj';
    Xk=zeros(1,NM2);   % compose the k-component of the image
    n1=1;
    for n=0:N2-1
        for m=0:M2-1
            if mod(n+m,3)==2
                Xk(n1)=X(n+1,m+1);
                n1=n1+1;
            end
        end
    end
    Qk=reshape(Xk,N2/3,M2); Qk=Qk';
%   Display the imaginary components of the image
    figure;    colormap(gray(255));                %Fig. 3.8
    subplot(2,3,4); image(Qi); axis normal;
    title('i-component'); xlabel('(a)');
    subplot(2,3,5); image(Qj); axis normal;
    title('j-component'); xlabel('(b)');
    subplot(2,3,6); image(Qk); axis normal;
    title('k-component'); xlabel('(c)');
%   Compose the quaternion image and display its modifica-
    tions
    figure;    colormap(gray(255));                %Fig. 3.9
    [N3, M3] = size(Qi);
    Q=zeros(N3,M3,3);
```

```
    Q(:,:,1)=Qi; Q(:,:,2)=Qj; Q(:,:,3)=Qk;
    % calculate and display the modifications of the color
    image
    Q(:,:,1)=1.1*Qi; Q(:,:,2)=1.25*Qj; Q(:,:,3)=0.75*Qk;
    Q=uint8(Q);
    subplot(2,3,4); image(Q); axis normal;
    title('q-image');  xlabel('(a)');
    % 2nd modification
    Q=zeros(N3,M3,3);
    Q(:,:,1)=1.5*Qi; Q(:,:,2)=1.1*Qj; Q(:,:,3)=0.75*Qk;
    Q=uint8(Q);
    subplot(2,3,5);  image(Q);  axis normal;
    title('modified q-image');  xlabel('(b)');
    % 3rd modification
    Q=zeros(N3,M3,3);
    Q(:,:,1)=1.5*Qi; Q(:,:,2)=1.5*Qj; Q(:,:,3)=Qk;
    Q=uint8(Q);
    subplot(2,3,6); image(Q); axis normal;
    title('modified q-image'); xlabel('(c)');
%=====================================================
```

3.3.3 RGBY model

In multi-primary displays, four and more primaries are considered [1, 3] and the RGBY is one of the color models that are used for such devices. The RGBY color model is the RGB model with the addition yellow color. The color image is the four-component image. Therefore, the transformation of the grayscale image into the color and quaternion images is straightforward and the moving 2×2 window can be used, as shown in Table 3.4. Here, the letter **Y** stands for the yellow color.

In this model, four pixels in the moving 2×2 window are considered as the e, i, j, and k-components and mapped to the colors as

$$\begin{pmatrix} \bullet & \bullet \\ \bullet & \bullet \end{pmatrix} \rightarrow \begin{pmatrix} e & i \\ j & k \end{pmatrix} \rightarrow \begin{pmatrix} R & G \\ B & Y \end{pmatrix}. \tag{3.2}$$

Table 3.4 The grayscale image mapping to the RGBY color model.

.	.	.	.	.	.
.	**R**:$f_{2n,2m}$	**G**:$f_{2n,2m+1}$	**R**:$f_{2n,2m+2}$	**G**:$f_{2n,2m+3}$	.
.	**B**:$f_{2n+1,2m}$	**Y**:$f_{2n+1,2m+1}$	**B**:$f_{2n+1,2m+2}$	**Y**:$f_{2n+1,2m+3}$	.
.	**R**:$f_{2n+2,2m}$	**G**:$f_{2n+2,2m+1}$	**R**:$f_{2n+2,2m+2}$	**G**:$f_{2n+2,2m+3}$	.
.	**B**:$f_{2n+3,2m}$	**Y**:$f_{2n+3,2m+1}$	**B**:$f_{2n+3,2m+2}$	**Y**:$f_{2n+3,2m+3}$	.
.	.	.	.	.	.

The color image is composed as

$$\tilde{f}_{n,m} = (f_{2n,2m}, g_{2n,2m+1}, b_{2n+1,2m}, y_{2n+1,2m+1})$$

and the quaternion image is

$$q_{n,m} = f_{2n,2m} + ig_{2n,2m+1} + jb_{2n+1,2m} + ky_{2n+1,2m+1},$$

where $n = 0 : (N/2 - 1)$ and $m = 0 : (M/2 - 1)$.

3.3.3.1 Other RGB models

Different RGB based color models also are used in imaging, including the normalized RGB, ProPhoto RGB, sRGB, scRGB, CIE RGB, AdobeRGB, and standard RGB. For instance, the normalized RGB model is described by the following colors:

$$r = \frac{R}{R+G+B}, \quad g = \frac{G}{R+G+B}, \quad b = \frac{B}{R+G+B}.$$

The de-correlated RGB model is described by the following three components [6]:

$$I_1 = \frac{R+G+B}{3}, \quad I_2 = \frac{R-B}{2}, \quad I_2 = \frac{(G-R)+(G-B)}{4}.$$

The first component is the intensity of the color image, and two other components are defined by differences between the primary colors. This model is a linear transformation of the RGB model.

We mention another linear transformation of the RGB model, which is called the opponent color model (OPP) and described in the normalized form by the following components:

$$\text{R-minus-G} = \frac{R-G}{\sqrt{2}}, \quad \text{B-minus-Y} = \frac{2B-(R+G)}{\sqrt{6}}, \quad \text{I} = \frac{R+G+B}{\sqrt{3}}.$$

This model also is used in the following logarithmic form [7]:

$$\begin{aligned} \text{R-minus-G} &= \log(R) - \log(G), \\ \text{B-minus-Y} &= [2\log(B) - \log(R) + \log(G)]/2, \\ \text{I} &= \log(G). \end{aligned}$$

The RGB model after rotations by 45° degrees in the RB-plane and then in the RG-plane is described by in the new ZXY system of coordinates by

$$Z = \frac{R + \sqrt{2}G + B}{2}, \quad Y = \sqrt{2}(B - R), \quad X = \frac{\sqrt{2}G - (R + B)}{2}.$$

3.3.4 CMY model

The CMY model uses the primary colors of pigment, cyan (C), magenta (M), and yellow (Y), which are considered to be the main colors in color printing [1]. The modification of this model, which is called the CMYK color model, uses the additional key color (K). The primary colors from the RGB color space are transfered to the CMYK space by the simple operations:

$$C = 1 - R, \quad M = 1 - G, \quad Y = 1 - B$$

and then adding the forth color (which usually is the black color) as $K = \min(C, M, Y)$ with the following change of colors: $C = C - K$, $M = M - K$, and $Y = Y - K$. Figure 3.10 shows the three triangles of primary colors in the CMY model and other secondary colors as the result of overlapping the primary colors. The secondary colors are red, green, blue, and black.

The primary colors from the RGB color space can be reconstructed from the CMY space as follows: $R = 1 - C$, $G = 1 - M$, and $B = 1 - Y$. The reconstruction of these colors from the CMYK space can be calculated by

$$R = 1 - C + K, \quad G = 1 - M + K, \quad B = 1 - Y + K.$$

3.3.5 XYZ model

In the XYZ model [1], a mathematical formula is used to convert the RGB data to a system of positive integers as values X, Y, and Z. These values are linear combinations of the colors R, G, and B; they do not directly correspond to the red, green, and blue, but are approximately so. To obtain the XYZ values from the primary colors R, G, and B, the following multiplication in matrix form is used:

$$\begin{bmatrix} X \\ Y \\ Z \end{bmatrix} = \frac{1}{0.17697} \begin{bmatrix} 0.49 & 0.31 & 0.2 \\ 0.17697 & 0.8124 & 0.01063 \\ 0 & 0.01 & 0.99 \end{bmatrix} \begin{bmatrix} R \\ G \\ B \end{bmatrix}. \tag{3.3}$$

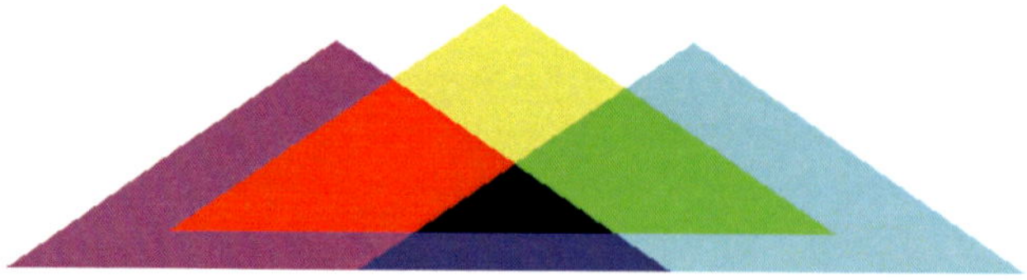

Figure 3.10 Triangles of the primary colors in the CMY model.

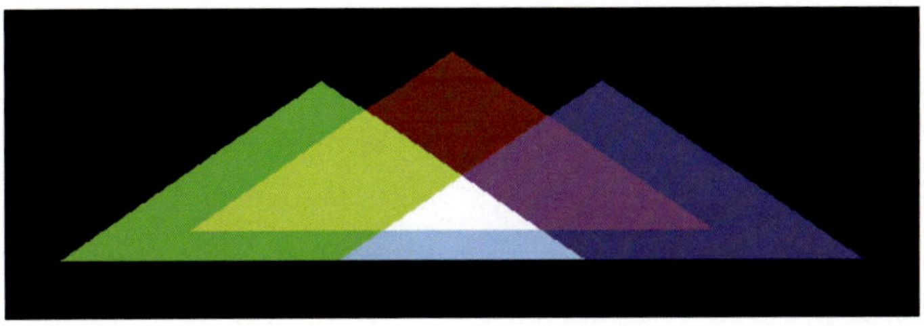

Figure 3.11 Image of color triangles of the RGB model in the XYZ color model.

As an example, Fig. 3.11 shows the image of three triangles of primary colors in the RGB model shown in Fig. 3.5, after transforming to the XYZ model that is described by the matrix in Eq. (3.3). Along each row, the sum of coefficients of this not normalized matrix is 1. The transformation of values X, Y, and Z into the quaternion space is similar to the RGB color model, i.e., $(X, Y, Z) \to 0 + (iX + jY + kZ)$. The values of the X, Y, and Z components for a few colors are given in Table 3.5.

The transformations of the primary colors from the XYZ model to the RGB model can approximately be calculated by

$$\begin{bmatrix} R \\ G \\ B \end{bmatrix} = \begin{bmatrix} 0.418466 & -0.158661 & -0.082835 \\ -0.091169 & 0.252431 & 0.015708 \\ 0.000921 & -0.002550 & 0.178599 \end{bmatrix} \begin{bmatrix} X \\ Y \\ Z \end{bmatrix}. \tag{3.4}$$

The XYZ values can also be calculated without the normalizing matrix by the coefficient 0.17697,

$$\begin{bmatrix} X \\ Y \\ Z \end{bmatrix} = \begin{bmatrix} 0.49 & 0.31 & 0.2 \\ 0.17697 & 0.8124 & 0.01063 \\ 0 & 0.01 & 0.99 \end{bmatrix} \begin{bmatrix} R \\ G \\ B \end{bmatrix}, \tag{3.5}$$

with the inverse transform

$$\begin{bmatrix} R \\ G \\ B \end{bmatrix} = \begin{bmatrix} 2.364614 & -0.896541 & -0.468073 \\ -0.515166 & 1.426408 & 0.088758 \\ 0.005204 & -0.014408 & 1.009204 \end{bmatrix} \begin{bmatrix} X \\ Y \\ Z \end{bmatrix}. \tag{3.6}$$

Table 3.5 The X,Y, and Z components for six colors in the XYZ model.

colors	red	yellow	green	cyan	blue	magenta
(r, g, b)	$(1, 0, 0)$	$(1, 1, 0)$	$(0, 1, 0)$	$(0, 1, 1)$	$(0, 0, 1)$	$(1, 0, 1)$
X	2.768831	4.520540	1.751709	2.881844	1.130135	3.898966
Y	1	5.590609	4.590609	4.650675	0.060067	1.060067
Z	0	0.056507	0.056507	5.650675	5.594169	5.594169

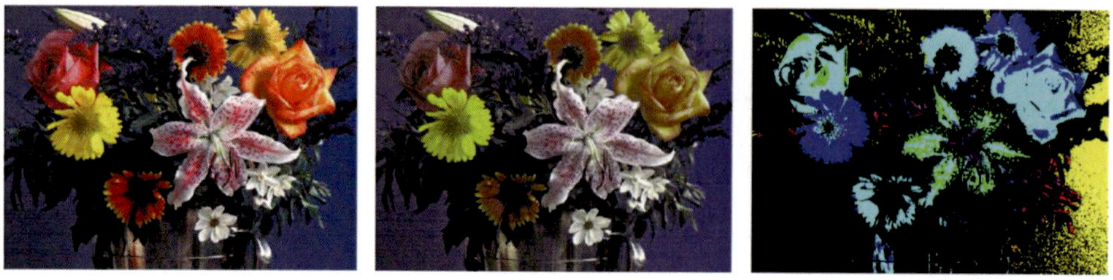

(a) RGB (b) XYZ (c) CMYK

Figure 3.12 The "flowers" image in (a) the RGB, (b) the XYZ, and (c) the CMYK color models.

Figure 3.12 shows the color "flowers" image in the RGB color space in part a, in the XYZ color space in b, and in the CMYK colors space in c.

3.3.6 HSI color model

The Hue-Saturation-Intensity (HSI) color model is a non-linear transformation of the RGB color space [1]. The hue component defines the pure color, the saturation determines the amount of the white in the color, and the intensity of the color is referred to as the gray value of the red, green, and blue colors. The hue is calculated as the angle in the circle of 360°. In this model, the brightness of the image in each point is calculated together with such characteristics of the color as the hue and saturation. The hue and saturation of colors respond closely to the way humans perceive the color, and thus this model is well suited for interactive manipulation of color images.

The transformation of colors R, G, and B into the corresponding H, S, and I values in this model are calculated as follows [1]:

$$
\begin{aligned}
H &= \begin{cases} \vartheta, & \text{if } B \le G; \\ 360 - \vartheta, & \text{if } B > G, \end{cases} \\
S &= 1 - \frac{\min\{R, G, B\}}{I}, \\
I &= \text{Gray-value of primary colors} = (R + G + B)/3,
\end{aligned}
\tag{3.7}
$$

where the angle (in degrees) is calculated by

$$
\vartheta = \vartheta(R, G, B) = \arccos\left[\frac{1}{2}\frac{(R-G)+(R-B)}{\sqrt{(R-G)^2+(R-B)(G-B)}}\right]. \tag{3.8}
$$

According to the definition, the angle ϑ is the function of the differences of primaries, namely, the differences between the read-and-green and red-and-blue. If we denote $\Delta_{R,G} = R - G$, $\Delta_{R,B} = R - B$, and $\Delta_{G,B} = \Delta_{R,B} - \Delta_{R,G} = G - B$, then the angle can be written as

$$\vartheta = \vartheta(\Delta_{R,G}, \Delta_{R,B}) = \arccos\left(\frac{1}{2}\frac{\Delta_{R,G} + \Delta_{R,B}}{\sqrt{\Delta_{R,G}^2 + \Delta_{R,B}\Delta_{G,B}}}\right). \tag{3.9}$$

Thus, the angles for the colors (R, G, B) and $(R + a, G + a, B + a)$ are the same for any constant a such that $0 \leq R + a, G + a, B + a \leq 1$.

It is not difficult to see that the angle can also be defined by the ratio of differences

$$\Delta_{(R,G)/(R,B)} = \frac{\Delta_{R,G}}{\Delta_{R,B}} = \frac{R - G}{R - B}. \tag{3.10}$$

Indeed, it follows from Eq. (3.9) that the angle can be written as

$$\vartheta = \arccos\left(\frac{1}{2}\frac{\Delta_{R,G} + \Delta_{R,B}}{\sqrt{\Delta_{R,G}^2 + \Delta_{R,B}^2 - \Delta_{R,B}\Delta_{R,G}}}\right),$$

and therefore,

$$\vartheta = \arccos\left\{\frac{1}{2}\frac{1 + \Delta_{(R,G)/(R,B)}}{\sqrt{1 + [\Delta_{(R,G)/(R,B)}]^2 - \Delta_{(R,G)/(R,B)}}}\right\}. \tag{3.11}$$

The hue component, H, is defined by the angle ϑ and its few values together with the values of the saturation and intensity for the main colors are given in Table 3.6.

The hue component, saturation, and intensity for a few other colors are given in Table 3.7.

One can rewrite Eq. (3.8), by using the rations of green-to-red and blue-to-red. Indeed, let α_1 and α_2 be the parameters $\alpha_1 = G/R$ and $\alpha_2 = B/G$, if $R \neq 0$. Then, the equation can be written as

Table 3.6 The angle, hue, saturation, and intensity for a few colors in the HSI model.

colors	red	yellow	green	cyan	blue	magenta
(r, g, b)	$(1, 0, 0)$	$(1, 1, 0)$	$(0, 1, 0)$	$(0, 1, 1)$	$(0, 0, 1)$	$(1, 0, 1)$
angle	0°	60°	120°	180°	120°	60°
hue	0°	60°	120°	180°	240°	300°
saturation	1	1	1	1	1	1
intensity	1/3	2/3	1/3	2/3	1/3	2/3

Table 3.7 Hue, saturation, and intensity for a few colors in the HSI models.

colors	maroon	brown	gold	orange	chocolate
(r, g, b)	$(0.5, 0, 0)$	$(0.65, 0.16, 0.16)$	$(1, 0.84, 0)$	$(1, 0.65, 0)$	$(0.82, 0.41, 0.12)$
hue	0°	$8.54(10^{-7})$°	51.6145°	39.6374°	24.5036°
saturation	1	0.4940	1	1	0.7391
intensity	0.1667	0.3255	0.6144	0.5490	0.4510

$$\vartheta = \vartheta(R, G, B) = \arccos\left[\frac{1 - \frac{\alpha_1+\alpha_2}{2}}{\sqrt{(1-\alpha_1)^2 + (1-\alpha_2)(\alpha_1 - \alpha_2)}}\right]. \tag{3.12}$$

Thus, for any red color, R, we can write that

$$\vartheta(R, G, B) = \vartheta(1, \alpha_1, \alpha_2) = \vartheta\left(1, \frac{G}{R}, \frac{B}{R}\right).$$

For instance,

$$\vartheta\left(\frac{1}{2}, \frac{1}{8}, \frac{1}{4}\right) = \vartheta\left(\frac{1}{4}, \frac{1}{16}, \frac{1}{8}\right) = \vartheta\left(\frac{1}{8}, \frac{1}{32}, \frac{1}{16}\right) = \vartheta\left(1, \frac{1}{4}, \frac{1}{2}\right).$$

It means that all these sets of four triplets of colors have the same hue, as well the the saturation, which can also be calculated by these rations,

$$S = 1 - \frac{\min\{R, \alpha_1 R, \alpha_2 R\}}{R(1+\alpha_1+\alpha_1)\frac{1}{3}} = 1 - \frac{3}{(1+\alpha_1+\alpha_1)}\min\{1, \alpha_1, \alpha_2\} \geq 0.$$

The intensity can be expressed by the red color as

$$I = \frac{R+G+B}{3} = \frac{1+\alpha_1+\alpha_1}{3}R.$$

It should also be mentioned that, for any blue color, $(0, 0, b)$, when $b \in (0, 1]$, the hue is 240°. Indeed, it follows from Eq. (3.8) that

$$\vartheta(0, 0, b) = \arccos\left[\frac{1}{2}\frac{(0-0)+(0-b)}{\sqrt{(0-0)^2+(0-b)(0-b)}}\right]$$
$$= \arccos\left(\frac{1}{2}\frac{-b}{\sqrt{b^2}}\right) = \arccos\left(-\frac{1}{2}\right) = 120°;$$

therefore, $H = H(0, 0, b) = 360 - \vartheta = 240$. For the red and green colors, we obtain $\vartheta(r, 0, 0) = 0$ and $\vartheta(0, g, 0) = 120$, where $r, g \in [0, 1]$, respectively.

Below is the script of the function "u4huesi.m" for calculating values of the hue, saturation, intensity, and the angle ϑ for any triplet of colors $(r, g, b) \neq (0, 0, 0)$.

```
%-------------------------------------------------------
function [hueI,Is,In,theta]=u4huesi(r,g,b)
 u=0.5*(2*r-g-b)/sqrt((r-g)^2+(r-b)*(g-b));
 theta = acos(u)/pi*180;
 if b<=g
     hueI=theta;
 else
     hueI=360-theta;
 end
 In=(r+g+b)/3;
 Is=1-min([r,g,b])/In;
%-------------------------------------------------------
```

It also follows from Eq. (3.8) that the hue of the yellow color with any blue, $(1, 1, b)$, is 60°, the hue of the cyan color with any red, $(r, 1, 1)$, where $g \in (0, 1]$, is 180°, and the hue of the magenta color with any green, $(1, g, 1)$, where $g \in (0, 1]$, is 300°. Therefore, the above table can be written in a more general way as in Table 3.8.

Figure 3.13 shows the hue as a function of the colors (r, g, b) for the $b = 0.5$ case. The function changes much at the point $r = 0.5$ when $g = 0.5$; this is the point of the gray-color $(r, g, b) = (0.5, 0.5, 0.5)$. The function hue as the function of the red color increases when $g < 0.5$ and decreases when $g < 0.5$.

Table 3.8 Hue for a few colors in the HSI model, where $r, g, b \in [0, 1]$.

colors	red	yellow	green	cyan	blue	magenta
(r, g, b)	$(r, 0, 0)$	$(1, 1, b)$	$(0, g, 0)$	$(r, 1, 1)$	$(0, 0, b)$	$(1, g, 1)$
hue	0°	60°	120°	180°	240°	300°

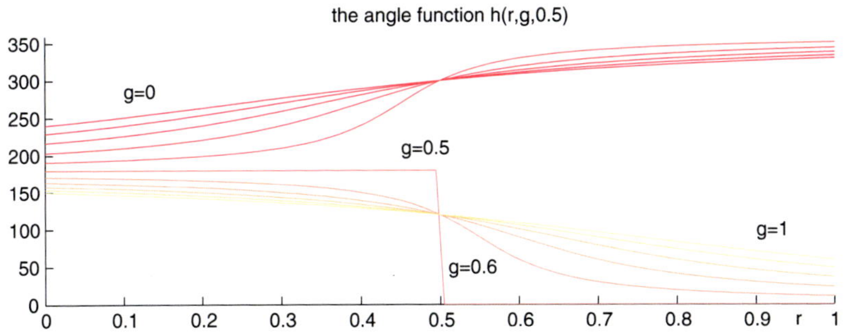

Figure 3.13 Hue functions of colors $(r, g, 0.5)$, when $r \in [0, 1]$, and $g = 0, 0.1, 0.2, 0, 3, \ldots, 1$.

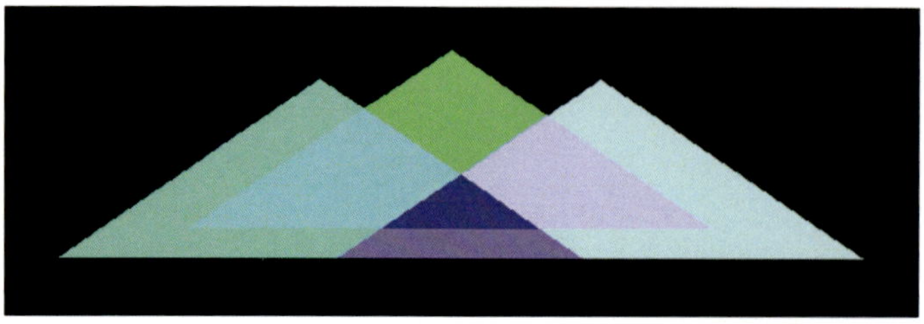

Figure 3.14 Image of color triangles of the XYZ model in the HSI color model.

As an example, Fig. 3.14 shows the image of the XYZ model shown in Fig. 3.11 after transforming to the HSI model.

Given values H, S, and I, the corresponding triplet (R, G, B) can be calculated by the following equations:

$$\begin{aligned} x &= I(1-S), \\ y &= I\left[1+\frac{S\cos(\phi)}{\cos(\phi-60^{\circ})}\right], \\ z &= 3I-(x+y), \end{aligned} \tag{3.13}$$

where

$$(x, y, z, \phi) = \begin{cases} (B, R, G, H), & \text{if } 0^{\circ} \leq H < 120^{\circ}; \\ (R, G, B, H-120^{\circ}), & \text{if } 120^{\circ} \leq H < 240^{\circ}; \\ (G, B, R, H-240^{\circ}), & \text{if } 240^{\circ} \leq H < 360^{\circ}. \end{cases} \tag{3.14}$$

In quaternion space, these three components of the HSI model are defined in the following way [5]. The value (I) component is referred as the norm of the quaternion vector q on the gray axis (axis of real part of q), which is $(q \cdot \mu)\mu$, where for instance $\mu = (1+j+k)/\sqrt{3}$. The saturation is referred to as the angle between the vectors corresponding to numbers q and μ. The hue is defined by a reference vector v, which is orthogonal to μ, for instance, a vector in red color direction.

$$\begin{aligned} H &= \tan^{-1}(q-\mu v q v \mu)/(|q-vqv|), \\ I &= |q-\mu q \mu|/2, \\ S &= |q+\mu q \mu|/2. \end{aligned} \tag{3.15}$$

3.3.7 HSV color model

The Hue-Saturation-Value color model is similar to the HSI model; instead of I(intensity), the V(value) component is used, which shows the darkness of the color [1]. All H, S, and V components are calculated as follows. For each color (r, g, b) with values of components in the unit interval $[0, 1]$, we define three numbers

$$V = \max(r, g, b), \quad m = \min(r, g, b), \quad \text{and} \quad C = V - m.$$

The saturation value is calculated as

$$S = \frac{C}{V}, \quad \text{if } V \neq 0,$$

and $S = 0$ otherwise. The hue component for the color (r, g, b) is defined only when $V \neq 0$ as the angle (in degrees) calculated as follows:

$$H = \begin{cases} 60 \cdot \frac{g-b}{C} \text{mod} 360, & \text{if } r = V, \\ 60 \cdot \frac{b-r}{C} + 120, & \text{if } g = V, \\ 60 \cdot \frac{r-g}{C} + 240, & \text{if } b = V. \end{cases} \tag{3.16}$$

In this model, the hue component H is also defined by the differences of the primary colors. A few values of H together with the values of the saturation and intensity for the main colors are given in Table 3.9.

The hue, saturation, and value components for a few other colors are given in Table 3.10.

3.3.8 YCbCr color model

In video imaging and digital photography, the YCbCr (8 bits per sample) color model is used. This model is calculated from the RGB color model by mapping the full range of colors $[0, 255]$ into its sub-interval $[16, 240]$. Here, Y stands for the light intensity, namely for "luma," which is the weighted sum of gamma-compressed, or gamma-corrected primary RGB colors. The symbols

Table 3.9 The angle, hue, saturation, and intensity for a few colors in the HSI model.

colors	red	yellow	green	cyan	blue	magenta
(r, g, b)	$(1, 0, 0)$	$(1, 1, 0)$	$(0, 1, 0)$	$(0, 1, 1)$	$(0, 0, 1)$	$(1, 0, 1)$
hue	0°	60°	120°	180°	240°	300°
saturation	1	1	1	1	1	1
value	1	1	1	1	1	1

Table 3.10 Hue, saturation, and value for a few colors in the HSV model.

colors	maroon	brown	gold	orange	chocolate
(r, g, b)	$(0.5, 0, 0)$	$(0.65, 0.16, 0.16)$	$(1, 0.84, 0)$	$(1, 0.65, 0)$	$(0.82, 0.41, 0.12)$
hue	50.5882°	38.8235°	50.5882°	38.8235°	25°
saturation	1	0.7455	1	1	0.8571
intensity	0.5	0.6471	1	1	0.8235

Cb and Cr are calculated as specific differences between primary colors, blue and (red plus green) and red and (green plus blue), respectively.

To calculate the YCbCr model values from the primary colors R, G, and B, the following matrix of transformation is used:

$$\begin{bmatrix} Y \\ C_b \\ C_r \end{bmatrix} = \begin{bmatrix} 16 \\ 128 \\ 128 \end{bmatrix} + \begin{bmatrix} 65.4810 & 128.5530 & 24.996 \\ -37.7970 & -74.2030 & 112.000 \\ 112.000 & -93.7860 & -18.214 \end{bmatrix} \begin{bmatrix} R \\ G \\ B \end{bmatrix}, \quad (3.17)$$

where the values of colors R, G, and B are in the interval $[0, 1]$. Here, the sum of coefficients along the rows of the matrix equals $[219.03, 0, 0]$. Therefore, the white color $(1, 1, 1)$ is transformed to $(Y, C_b, C_r) = (235.03, 128, 128)$. For the primary red, green, and blue colors, the values of (Y, C_b, C_r) equal $(81.4810, 90.2030, 240)$, $(144.5530, 53.7970, 34.2140)$, and $(40.9960, 240, 109.7860)$ respectively.

As an example, Fig. 3.15 shows the image of the RGB color model shown in Fig. 3.5 after transforming to the YCbCr color model.

The values of this color space are calculated from its analog version, or the so-called YPbPr space, whose components are calculated by

$$\begin{bmatrix} Y \\ P_b \\ P_r \end{bmatrix} = \begin{bmatrix} 0.299 & 0.587 & 0.114 \\ -0.168736 & -0.331264 & 0.50 \\ 0.50 & -0.418688 & -0.081312 \end{bmatrix} \begin{bmatrix} R \\ G \\ B \end{bmatrix}, \quad (3.18)$$

by the transformation

$$[Y, C_b, C_r] = [16, 128, 128] + [219Y, 224P_b, 224P_r].$$

To original components of the RGB model can be calculated (approximately) from the YCbCr model values by

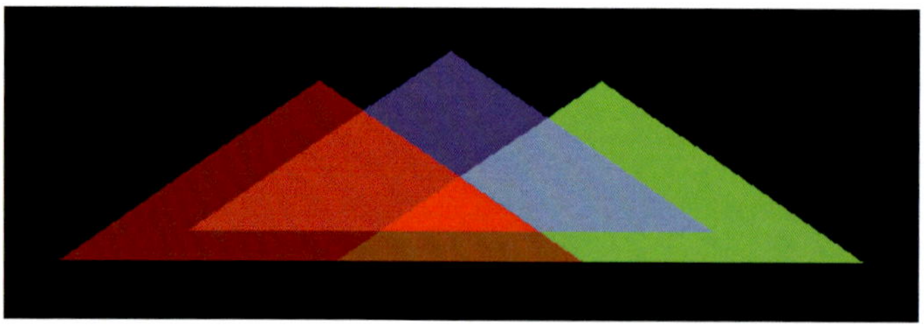

Figure 3.15 Image of color triangles of the RGB model in the YCbCr color model.

$$\begin{bmatrix} R \\ G \\ B \end{bmatrix} = \begin{bmatrix} 0.00456558 & 0.00456558 & 0.00456558 \\ -0.00000108 & -0.00153740 & 0.00790963 \\ 0.00625893 & -0.00318811 & 0.00000001 \end{bmatrix} \begin{bmatrix} Y-16 \\ C_b-128 \\ C_r-128 \end{bmatrix}. \quad (3.19)$$

Here, the inverse matrix is given with precision of 8 digits after the point.

3.4 Image Models with Octonions

In this section, we consider a few models for transferring the grayscale discrete image into the octonion space. Since the octonion is a number with eight components, there are many ways for packing the information in pixels into octonions. We consider the discrete grayscale image $f_{n,m}$ in the Cartesian lattice $X_{N,M} = \{(n,m); n = 0:(N-1), m = 0:(M-1)\}$ and a few models shown in Fig. 3.16.

3.4.1 Model 1 with the 3 × 3 cross window

In the first model, it is proposed to pack the discrete image at every five points into octonions in the following way:

$$\begin{bmatrix} \cdot & f_{n-1,m} & \cdot \\ f_{n,m-1} & \underline{f_{n,m}} & f_{n,m+1} \\ \cdot & f_{n+1,m} & \cdot \end{bmatrix} . \times \begin{bmatrix} \cdot & i & \cdot \\ j & \underline{1} & k \\ \cdot & E & \cdot \end{bmatrix} \to o_{n_1,m_1}. \quad (3.20)$$

The octonion at the corresponding point (n_1, m_1) is calculated by

$$o_{n_1,m_1} = f_{n,m} + if_{n-1,m} + jf_{n,m-1} + kf_{n,m+1} + Ef_{n+1,m}.$$

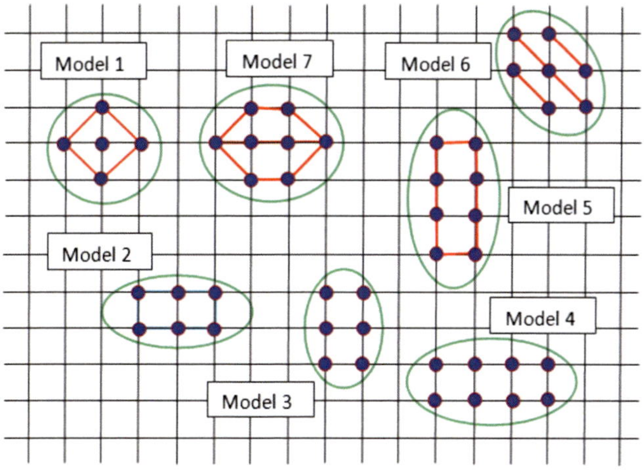

Figure 3.16 Seven models of the grayscale image transformation into octonions.

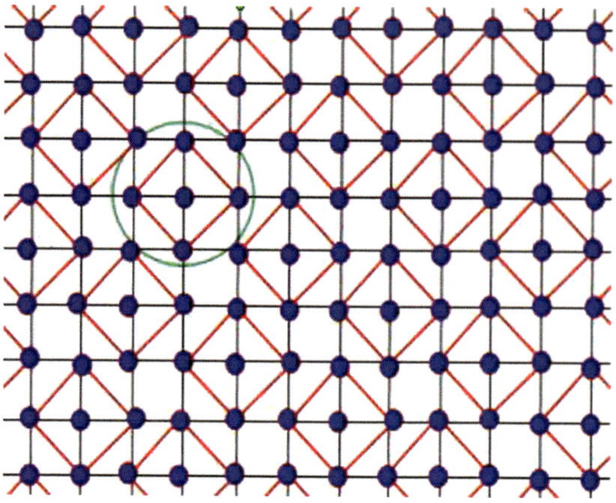

Figure 3.17 Covering of the Cartesian lattice by the 3 × 3 cross-window.

The simple cross window of size (3×3) is used in this model. The covering of the lattice by such windows is shown in Fig. 3.17.

In definition of the octonion image, instead of units $1, i, j, k, E$ any five different units from eight units $1, i, j, k, E, I, J, D$ can be used. For instance, the octonion image can be composed as

$$o_{n_1,m_1} = f_{n,m} + if_{n-1,m} + jf_{n,m-1} + Jf_{n,m+1} + If_{n+1,m}.$$

3.4.2 Model 2 with the 2 × 3 window

We consider the octonion image composition by using six neighbor pixels in the 2×3 window in the following way:

$$\begin{bmatrix} f_{n-1,m-1} & f_{n-1,m} & f_{n-1,m+1} \\ f_{n,m-1} & \underline{f_{n,m}} & f_{n,m+1} \end{bmatrix} . \times \begin{bmatrix} I & J & K \\ i & \underline{j} & k \end{bmatrix} \rightarrow o_{n_1,m_1}. \tag{3.21}$$

The mask shows the imaginary units assigned to each component of the image during the octonion image composition,

$$o_{n_1,m_1} = if_{n,m-1} + jf_{n,m} + kf_{n,m+1} + If_{n-1,m-1} + Jf_{n-1,m} + Kf_{n-1,m+1}.$$

If mapping starts from the point $(1, 1)$, the correspondence between points (n, m) and (n_1, m_1) is calculated by

$$(n, m) = (2n_1 + 1, 3m_1 + 1), \quad n = 0 : N/2, \quad m = 0 : M/3.$$

The octonion image has the size rounded to $(N/2) \times (M/3)$.

In the above mask, instead of units i, j, k, I, J, K, any other six units from eight units $1, i, j, k, E, I, J, D$ can be used.

3.4.3 Model 3 with the 3 × 2 window

We consider the following octonion image composition by six neighbor pixels in the 3×2 window:

$$\begin{bmatrix} f_{n-1,m-1} & f_{n-1,m} \\ f_{n,m-1} & \underline{f_{n,m}} \\ f_{n+1,m-1} & f_{n+1,m} \end{bmatrix} . \times \begin{bmatrix} J & j \\ I & \underline{i} \\ K & k \end{bmatrix} \rightarrow o_{n_1,m_1}. \tag{3.22}$$

The octonion image is calculated by

$$o_{n_1,m_1} = if_{n,m} + jf_{n-1,m} + kf_{n+1,m} + If_{n,m-1} + Jf_{n-1,m-1} + Kf_{n+1,m-1}.$$

If the mapping starts from the point $(1, 1)$, the correspondence between points (n, m) and (n_1, m_1) is calculated by

$$(n, m) = (3n_1 + 1, 2m_1 + 1), \quad n_1 = 0 : N/3, \quad m_1 = 0 : M/2.$$

The octonion image has the size rounded to $(N/3) \times (M/2)$. Instead of units i, j, k, I, J, K, any other six units from eight units $1, i, j, k, E, I, J, D$ can also be used.

3.4.4 Model 4 with the 2 × 4 window

In this model, the octonion image composition at each point from eight neighbor pixels in the 2×4 window is accomplished in the following way:

$$\begin{bmatrix} f_{n-1,m-1} & f_{n-1,m} & f_{n-1,m+1} & f_{n-1,m+2} \\ f_{n,m-1} & \underline{f_{n,m}} & f_{n,m+1} & f_{n,m+2} \end{bmatrix} . \times \begin{bmatrix} I & E & J & K \\ i & \underline{1} & j & k \end{bmatrix} \rightarrow o_{n_1,m_1}. \tag{3.23}$$

The mask shows the imaginary units assigned to each component of the image during the octonion image composition,

$$\begin{aligned} o_{n_1,m_1} = & f_{n,m} + if_{n,m-1} + jf_{n,m+1} + kf_{n,m+2} \\ & + Ef_{n-1,m} + If_{n-1,m-1} + Jf_{n-1,m+1} + Kf_{n-1,m+2}. \end{aligned}$$

If the mapping starts from the point $(1, 1)$, the correspondence between points (n, m) and (n_1, m_1) is calculated by

$$(n, m) = (2n_1 + 1, 4m_1 + 1), \quad n_1 = 0 : N/2, \quad m_1 = 0 : M/4.$$

The octonion image has the size rounded to $(N/2) \times (M/4)$. All eight units are used for each pixel in the octonion image o_{n_1,m_1}, but their order in the mask can be changed.

3.4.5 Model 5 with the 4 × 2 window

This model is similar to Model 4. The octonion image is composed by using values of the grayscale image at eight neighbor pixels in the 4 × 2 window in the following way:

$$\begin{bmatrix} f_{n-1,m-1} & f_{n-1,m} \\ f_{n,m-1} & \underline{f_{n,m}} \\ f_{n+1,m-1} & f_{n+1,m} \\ f_{n+2,m-1} & f_{n+2,m} \end{bmatrix} . \times \begin{bmatrix} I & i \\ E & \underline{1} \\ J & j \\ K & k \end{bmatrix} \rightarrow o_{n_1,m_1}. \tag{3.24}$$

The octonion image is calculated by

$$\begin{aligned} o_{n_1,m_1} = & f_{n,m} + if_{n-1,m} + jf_{n+1,m} + kf_{n+2,m} \\ & + Ef_{n,m-1} + If_{n-1,m-1} + Jf_{n+1,m-1} + Kf_{n+2,m-1}. \end{aligned}$$

If the mapping starts from the point $(1, 1)$, the correspondence between points (n, m) and (n_1, m_1) is calculated by

$$(n, m) = (4n_1 + 1, 2m_1 + 1), \quad n_1 = 0 : (N/4 - 1), \quad m_1 = 0 : (M/2 - 1).$$

The octonion image is of the size rounded to $(N/4) \times (M/2)$. In this model also, all eight units can be used in different orders in the mask.

3.4.6 Model 6 with the 3 × 3 window

The following seven values of the image at the pixel (n, m) are used to compose the seven-component octonion images in the proposed mapping,

$$\begin{bmatrix} f_{n-1,m-1} & f_{n-1,m} & \cdot \\ f_{n,m-1} & \underline{f_{n,m}} & f_{n,m+1} \\ \cdot & f_{n+1,m} & f_{n+1,m+1} \end{bmatrix} . \times \begin{bmatrix} K & J & \cdot \\ i & \underline{E} & j \\ \cdot & I & k \end{bmatrix} \rightarrow o_{n_1,m_1}. \tag{3.25}$$

In this model, the pure octonion image is composed as follows:

$$\begin{aligned} o_{n_1,m_1} = & if_{n,m-1} + jf_{n,m+1} + kf_{n+1,m+1} \\ & + Ef_{n,m} + If_{n+1,m} + Jf_{n-1,m} + Kf_{n-1,m-1}. \end{aligned}$$

We also can use the real unit as one of the coefficients of the mask and map the grayscale image into the octonion image with the real part and six-component imaginary part, for instance as follows:

$$\begin{bmatrix} f_{n-1,m-1} & f_{n-1,m} & \cdot \\ f_{n,m-1} & \underline{f_{n,m}} & f_{n,m+1} \\ \cdot & f_{n+1,m} & f_{n+1,m+1} \end{bmatrix} . \times \begin{bmatrix} K & J & \cdot \\ i & \underline{1} & j \\ \cdot & I & k \end{bmatrix} \rightarrow o_{n_1,m_1}. \tag{3.26}$$

The octonion image is calculated by

$$\begin{aligned} o_{n_1,m_1} = & f_{n,m} + if_{n,m-1} + jf_{n,m+1} + kf_{n+1,m+1} \\ & + If_{n+1,m} + Jf_{n-1,m} + Kf_{n-1,m-1}. \end{aligned}$$

3.4.7 Model 7 with the 3 × 4 full hexagon

Together with the rectangular lattice, the hexagonal lattices are important for solving many problems in image processing in the spatial and frequency domains [9]–[17]. It is known that the vision system relates best to the regular hexagonal tessellation, which has a lower number of neighbors than the rectangular lattice. Therefore, we consider a model with the hexagonal window of samples. This model uses eight values of the image at the pixel (n, m) in the hexagonal window to compose the eight-component octonion images by the mapping

$$\begin{bmatrix} \cdot & f_{n-1,m} & f_{n-1,m+1} & \cdot \\ f_{n,m-1} & \underline{f_{n,m}} & f_{n,m+1} & f_{n,m+2} \\ \cdot & f_{n+1,m} & f_{n+1,m+1} & \cdot \end{bmatrix} . \times \begin{bmatrix} \cdot & I & E & \cdot \\ i & \underline{1} & j & k \\ \cdot & J & K & \cdot \end{bmatrix} \rightarrow o_{n_1,m_1}. \tag{3.27}$$

The covering of the lattice by such hexagonal windows is shown in Fig. 3.18.

In this mapping, the octonion image is composed as follows:

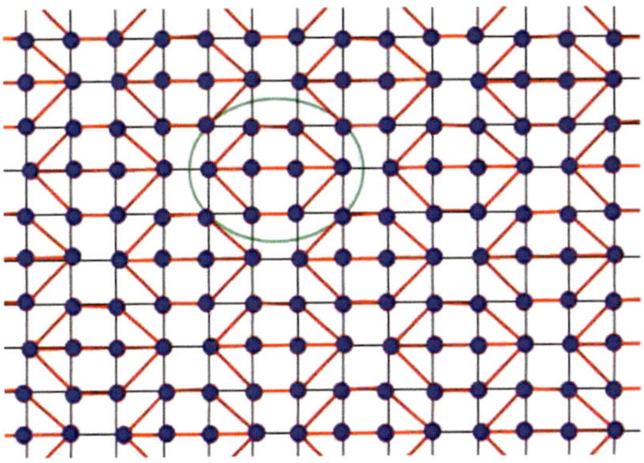

Figure 3.18 Covering of the Cartesian lattice by the 3 × 4 hexagonal window.

$$o_{n_1,m_1} = f_{n,m} + if_{n,m-1} + jf_{n,m+1} + kf_{n,m+2} \\ + Ef_{n-1,m+1} + If_{n-1,m} + Jf_{n+1,m} + Kf_{n+1,m+1}.$$

A similar model, with the rotated by 90° full hexagon can also be considered.

3.4.8 Models with a sequence of images or videos

In all seven models described above, the discrete octonion image is composed from the single grayscale image $f_{n,m}$, by packing the information of neighbor pixels in one union that is an octonion number. Here, we consider three simple models of composing the octonion image from different images.

Model 8A. If we working with a few images, it is possible to pack point-wise maximum of eight images into one octonion image,

$$\{(f_1)_{n,m}, (f_2)_{n,m}, (f_3)_{n,m}, \ldots, (f_8)_{n,m}\} \to \{o_{n,m}\}.$$

The octonion image can be calculated as

$$o_{n,m} = (f_1)_{n,m} + i(f_2)_{n,m} + j(f_3)_{n,m} + k(f_4)_{n,m} \\ + E(f_5)_{n,m} + I(f_6)_{n,m} + J(f_7)_{n,m} + K(f_8)_{n,m}. \tag{3.28}$$

Model 8B. We can pack four different images at the pair of pixels, let say the neighbor pixels $(n, 2m)$ and $(n, 2m+1)$, in one octonion image,

$$\begin{matrix} \{(f_1)_{n,2m}, (f_1)_{n,2m+1}\} \\ \{(f_2)_{n,2m}, (f_2)_{n,2m+1}\} \\ \{(f_3)_{n,2m}, (f_3)_{n,2m+1}\} \\ \{(f_4)_{n,2m}, (f_4)_{n,2m+1}\} \end{matrix} \to \{o_{n,m}\},$$

where $n = 0 : (N-1)$ and $m = 0 : (M/2-1)$. In this mapping, the octonion image can be calculated as

$$o_{n,m} = (f_1)_{n,2m} + i(f_1)_{n,2m+1} + j(f_2)_{n,2m} + k(f_2)_{n,2m+1} \\ + E(f_3)_{n,2m} + I(f_3)_{n,2m+1} + J(f_4)_{n,2m} + K(f_4)_{n,2m+1}. \tag{3.29}$$

Model 8C. Two different images at four neighbor pixels, let say $(2n, 2m)$, $(2n, 2m+1)$, $(2n+1, 2m)$, and $(2n+1, 2m+1)$, can be mapped to one octonion image,

$$\begin{matrix} \{(f_1)_{2n,2m}, (f_1)_{2n,2m+1}, (f_1)_{2n+1,2m}, (f_1)_{2n+1,2m+1}\} \\ \{(f_2)_{2n,2m}, (f_2)_{2n,2m+1}, (f_2)_{2n+1,2m}, (f_2)_{2n+1,2m+1}\} \end{matrix} \to \{o_{n,m}\},$$

where $n = 0 : (N/2 - 1)$ and $m = 0 : (M/2 - 1)$. In this mapping, the octonion image can be calculated as

$$\begin{aligned} o_{n,m} = {} & (f_1)_{2n,2m} + i(f_1)_{2n,2m+1} + j(f_1)_{2n+1,2m} + k(f_1)_{2n+1,2m+1} \\ & + E(f_2)_{2n,2m} + I(f_2)_{2n,2m+1} + J(f_2)_{2n+1,2m} + K(f_2)_{2n+1,2m+1}. \end{aligned} \tag{3.30}$$

References

[1] R. C. Gonzalez and R. E. Woods, *Digital Image Processing*, 2nd Edition, Prentice Hall, Upper Saddle River, New Jersey (2002).

[2] H. J. Trussell, E. Saber, and M. Vrhel, "Color image processing," *IEEE Signal Processing Magazine*, **22**(1), 14–22 (2005).

[3] Y. Murakumi et al., "Color conversion method for multi-primary display for spectral color reproduction," *Journal of Electronic Imaging*, **13**(4), 701–708 (2004).

[4] A. Tremeau, S. Tominaga, and K. N. Plataniotis, "Color in image and video processing: Most recent trends and future research directions," *EURASIP Journal on Image and Video Processing*, p. 26 (2008).

[5] P. Denis, P. Carre, and C. F. Maloigne, "Spatial and spectral quaternionic approaches for color images," *Computer Vision and Image understanding*, **107**, 74–87 (2007).

[6] M. D. Fairchild, *Color Appearance Models*, Addison-Wesley, Boston (1998).

[7] H. Palus, *Color Spaces*, Chapmann and Hall (1998).

[8] G. T. Herman, *Image Reconstruction From Projections. The Fundamentals of Computerized Tomography*, Academic Press, New York (1980).

[9] D. E. Dudgeon and R. M. Mersereau, *Multidimensional Digital Signal Processing*, Prentice Hall, Englewood Cliffs (1984).

[10] R. C. Staunton, "Hexagonal sampling in image processing," *in Advances in Imaging and Electron Physics*, **107**, 231–307 (1999).

[11] R. M. Mersereau, "The processing of hexagonally sampled two-dimensional signals," *Proc. IEEE*, **67**(6), 930–949 (1979).

[12] D. H. Hubel, *Eye, Brain, and Vision*, Scientific American Books, New York (1988).

[13] A. L. Nel, "Hexagonal image processing," *Proc. Communications and Signal Processing*, Southern African Conference on, COMSIG, 109–113 (1989).

[14] R. M. Mersereau, *A two-dimensional FFT for hexagonally sampled data*, in Digital Signal Processing, ed. V. Cappellini and A.G. Constantinides, Academic Press, 93–101 (1980).

[15] I. Her, "Geometric transformations on the hexagonal grid," *IEEE Trans. Image Processing*, **4**(9), 1213–1222 (1995).

[16] J. L. Zapata and G. X. Ritter, "Fast Fourier transform for hexagonal aggregates," *Journal of Mathematical Imaging and Vision*, **12**(3), 183–197 (2000).

[17] A. M. Grigoryan and S. S. Agaian, *Multidimensional Discrete Unitary Transforms: Representation, Partitioning and Algorithms*, Marcel Dekker Inc., New York (2003).

Chapter 4
Color Images as 2-D Grayscale Images

In this chapter, we consider different transformations of the color image into one grayscale image, which allow for processing the three color channels of the image as one unit in the spatial and frequency domains. We consider the RGB model [1], and other color image models can also be used [2]–[8].

In the RGB color model, we consider the color image $f_{n,m}$ with its red, green, and blue color components

$$f_{n,m} = (r_{n,m}, g_{n,m}, b_{n,m}), \qquad n = 0:(N-1), m = 0:(M-1).$$

As an example, Fig. 4.1 shows the color "crane" image $f_{n,m}$ of size 748×1000 in the RGB color model.

The grayscale image presenting the brightness of this image is calculated by

$$i_{n,m} = 0.3r_{n,m} + 0.59g_{n,m} + 0.11b_{n,m}. \tag{4.1}$$

The brightness $i_{n,m}$ of this color image is shown in Fig 4.2(a), and the red, green, and blue color channels in parts (b), (c), and (d), respectively.

This four-component image can be mapped to the grayscale image of size twice larger, $(2N) \times (2M)$, by using the scheme shown in Table 4.1. Here, the letter I stands for the brightness, R for the red color, G for the green color, and B for the blue color. At each point (n, m) with even coordinates (starting with $(0, 0)$), the new grayscale image is composed as

$$\begin{aligned} a_{n,m} &= i_{n,m}, & a_{n,m+1} &= r_{n,m}, \\ a_{n+1,m} &= b_{n,m}, & a_{n+1,m+1} &= g_{n,m}. \end{aligned} \tag{4.2}$$

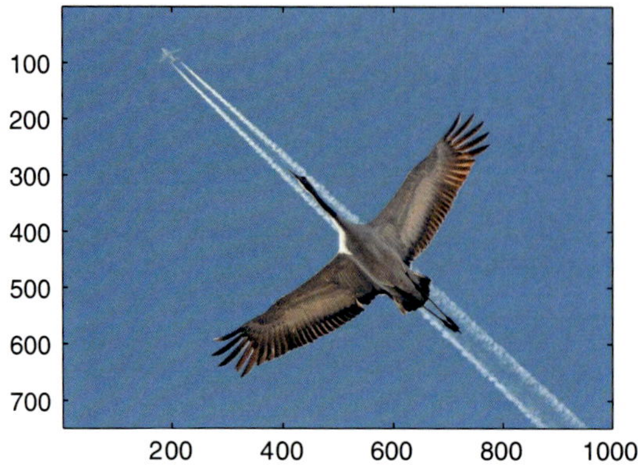

Figure 4.1 The color "crane" image of size 748 × 1000 in the RGB color model.

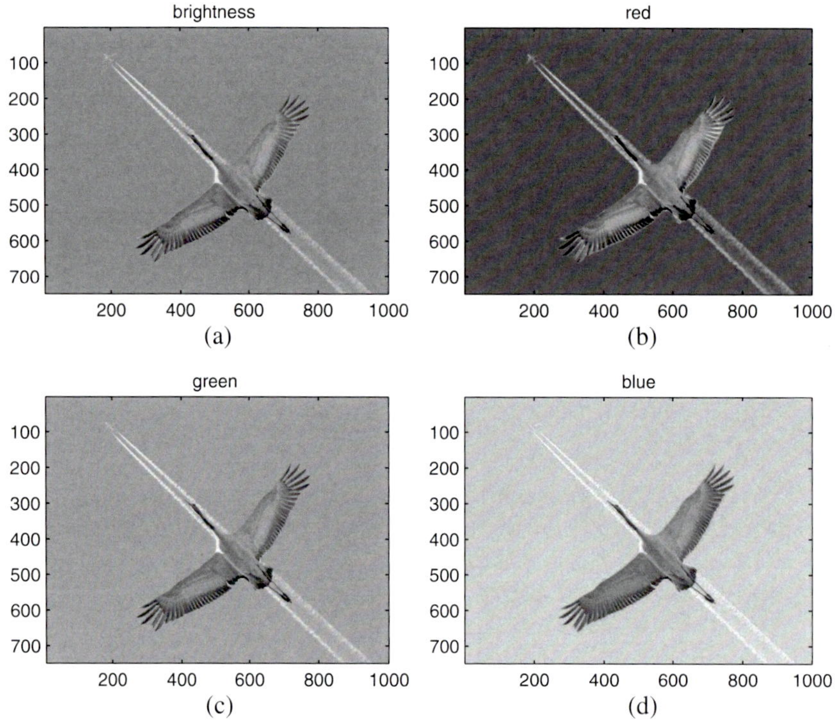

Figure 4.2 (a) The intensity and the color channels of the "crane" image: (b) red, (c) green, and (d) blue.

Table 4.1 Table of mapping the quaternion imaginary components into the primary colors.

I	R	$\rightarrow$	(n, m)	$(n, m+1)$
B	G	$\rightarrow$	$(n+1, m)$	$(n+1, m+1)$

Thus, the image is composed by using the four-component image, as shown:

$$
\begin{array}{c|cc|cc|c}
\cdot & I & R & I & R & \cdot \\
\cdot & B & G & B & G & \cdot \\
\hline
\cdot & I & R & I & R & \cdot \\
\cdot & B & G & B & G & \cdot \\
\hline
\cdot & \cdot & \cdot & \cdot & \cdot & \cdot
\end{array}
$$

We call this model of transformation of the color image into the grayscale image the (2×2) model.

The new grayscale image of size 1496×2000, which is composed from the four-component "crane" image, is shown in Fig. 4.3. Many dark small squares in this image is due to low intensity of one of the channel in the image, which is the red color. This grayscale image uniquely represents the color "crane" image; all components of the color image can be extracted from its grayscale representation. We believe that such color-to-gray transformed images can be processed in different applications, for instance in enhancement and filtration, and these color images can be considered together with the original color images. It may be interesting to consider such images in the frequency domain as well, by using for instance the discrete Fourier, Hartley, Hadamard, cosine, and other unitary transforms [9]–[16].

4.1 2-D DFT of the Image

The two-dimensional discrete Fourier transform (2-D DFT) of the grayscale image $a_{n,m}$ at the frequency-point (p, s) is defined as [14]

$$
A_{p,s} = \sum_{n=0}^{N-1} \sum_{m=0}^{M-1} a_{n,m} W_M^{ms} W_N^{np} = \sum_{n=0}^{N-1} \left[\sum_{m=0}^{M-1} a_{n,m} W_M^{ms} \right] W_N^{np}, \tag{4.3}
$$

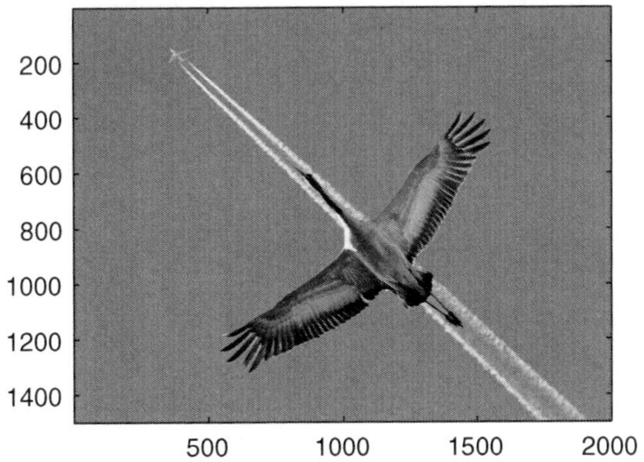

Figure 4.3 The grayscale image $a_{n,m}$ composed from the color "crane" image $f_{n,m}$.

where $p = 0 : (N - 1)$ and $s = 0 : (M - 1)$. The transform $\{A_{p,s}\}$ is called the $N \times M$-point 2-D DFT of the image $a_{n,m}$. Here, the basis exponential functions are $W_K = \exp(-2\pi i/K) = \cos(2\pi/K) - i\sin(2\pi/K)$, for $K = N$ and M. The 2-D DFT is a separable transform, meaning that the calculation of the transform can be accomplished by calculating the 1-D DFTs by rows first and then by columns, or by columns first and then by rows.

For images (non negative data), the maximum value of the 2-D DFT is at the frequency-point $(0, 0)$,

$$A_{0,0} = \sum_{n=0}^{N-1} \sum_{m=0}^{M-1} a_{n,m} = (\text{Area of the image}).$$

The transform is periodic by N and M in each directions, i.e.,

$$A_{p,s} = A_{p+N,s} = A_{p,s+M} = A_{p-N,s} = A_{p,s-M} = A_{p+N,s+M},$$

and it has symmetric corners when displaying the transform in the absolute scale, i.e.,

$$A_{N-p,M-s} = \overline{A_{p,s}}, \qquad A_{N-p,s} = \overline{A_{p,M-s}},$$

where $\bar{A}$ denotes the complex-conjugate operation. Therefore, the picture of the magnitude of the 2-D DFT looks symmetric and in many cases with the most information around the center when shifting to the middle,

$$A_{p+N/2,s+M/2} = \overline{A_{N/2-p,M/2-s}}, \qquad p, s = 0, 1, 2, 3, \cdots.$$

As an example, Fig. 4.4 shows the grayscale image of size 746×1000 in part (a), the 2-D DFT of the image in parts (b) and (c).

It is assumed that the numbers N and M are even. It should be noted that the shift of the transform to the middle frequency-point can be performing by

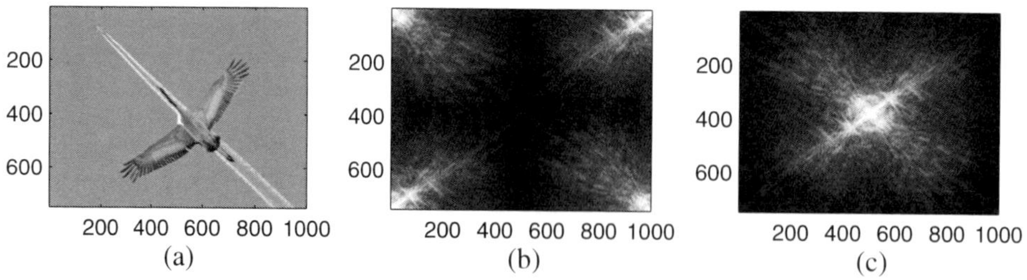

Figure 4.4 (a) The original image and the magnitude of the 2-D DFT of the image and (b) before and (c) after shifting to the middle.

changing the sign of every second value of the image, i.e., the following is valid for the 2-D DFT:

$$A_{p+N/2,s+M/2} = \sum_{n=0}^{N-1}\sum_{m=0}^{M-1} a_{n,m} W_M^{m(s+M/2)} W_N^{n(p+N/2)}$$
$$= \sum_{n=0}^{N-1}\sum_{m=0}^{M-1} a_{n,m} W_M^{ms} W_N^{np} W_M^{mM/2} W_N^{nN/2}$$

and since

$$W_M^{mM/2} W_N^{nN/2} = W_2^m W_2^n = (-1)^n(-1)^m = (-1)^{n+m},$$

we obtain

$$A_{p+N/2,s+M/2} = \sum_{n=0}^{N-1}\sum_{m=0}^{M-1} (-1)^{n+m} a_{n,m} W_M^{ms} W_N^{np}.$$

The 2-D DFT as a complex table can be represented by the magnitude $|A_{p,s}|$ and the phase

$$\vartheta(p,s) = \text{phase}[A(p,s)], \qquad p = 0:(N-1), s = 0:(M-1).$$

It is difficult to analyze the phase of the transform from its picture, as shown in Fig. 4.5(a) for the 2-D DFT of the same image in Fig. 4.4(a). This image was created by using the MATLAB® command "fft2," to calculate the 2-D DFT of the image and the command "phase," to calculate the phase (in radians) of the 2-D DFT. The phase is calculated with values in the interval $[\pi, \pi]$; therefore, for the above phase picture, the number π was added to make the phase data non negative and then the command "imagesc" was

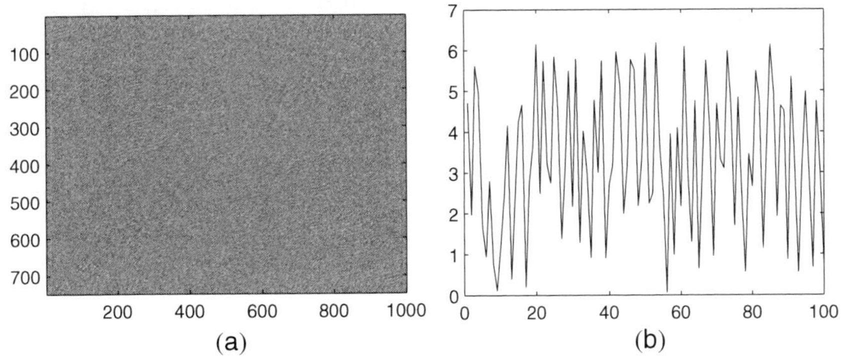

Figure 4.5 (a) The phase of the 2-D DFT of the image and (b) phase along one column.

used. In Fig. 4.5(b), the only first 100 values of the phase along the column number 100 are shown.

For comparison, we consider the spectrum of the 2-D DFT of the grayscale image composed from the color image in the (2×2) model. Figure 4.6 shows the 1498×2000-point 2-D DFT of the image in Fig. 4.3. The transform in absolute scale is shown in Fig. 4.6(a), and the transform shifted to the middle in Fig. 4.6(b).

4.2 The (3 × 2) Color-to-Gray Model

We consider another model of transferring colors to a grayscale image without the image component brightness, $i_{n,m}$. The mapping of the three primary colors into the image

$$(r_{n,m}, g_{n,m}, b_{n,m}) \to a_{n,m}$$

can be performed in different ways. As an example, we consider the grayscale image composition starting from the point $(0, 0)$ in the following way:

$r_{0,0}$	$g_{0,0}$	$b_{0,1}$	$r_{0,2}$	$g_{0,2}$	$b_{0,3}$	$\cdots$
$b_{0,0}$	$r_{0,1}$	$g_{0,1}$	$b_{0,2}$	$r_{0,3}$	$g_{0,3}$	$\cdots$
$r_{1,0}$	$g_{1,0}$	$b_{1,1}$	$r_{1,2}$	$g_{1,2}$	$b_{2,3}$	$\cdots$
$b_{1,0}$	$r_{1,1}$	$g_{1,1}$	$b_{1,2}$	$r_{2,3}$	$g_{2,3}$	$\cdots$
$\cdots$	$\cdots$	$\cdots$	$\cdots$	$\cdots$	$\cdots$	$\cdots$

In this mapping, each row of the three component color image is mapping into two rows in the new image $a_{n,m}$. This image has the size $2N \times 3(M/2)$. Figure 4.7 shows the grayscale image composed in this way from the color "crane" image.

Thus, the mapping of the colors into the grayscale image is performing as shown in Table 4.2.

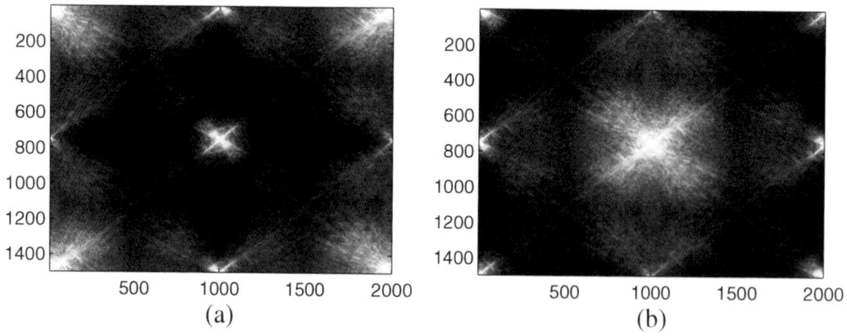

Figure 4.6 The magnitude of the 2-D DFT of the grayscale image (a) before and (b) after shifting to the middle.

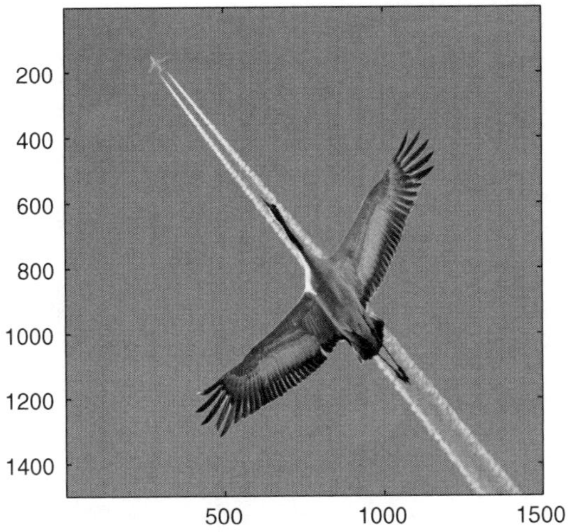

Figure 4.7 The color “crane” image after transforming into the (3×2) model.

Table 4.2 Mapping of the color image into the grayscale image.

.	.	.	.	.
.	$a_{2n,2m} : r_{n,m}$	$a_{2n,2m+1} : g_{n,m}$	$a_{2n,2m+2} : b_{n,m+1}$	.
.	$a_{2n+1,2m} : b_{n,m}$	$a_{2n+1,2m+1} : r_{n,m+1}$	$a_{2n+1,2m+2} : g_{n,m+1}$	.
.	$a_{2n+2,2m} : r_{n+1,m}$	$a_{2n+2,2m+1} : g_{n+1,m}$	$a_{2n+2,2m+2} : b_{n+1,m+1}$	.
.	$a_{2n+3,2m} : b_{n+1,m}$	$a_{2n+3,2m+1} : r_{n+1,m+1}$	$a_{2n+3,2m+2} : g_{n+1,m+1}$	.
.	.	.	.	.

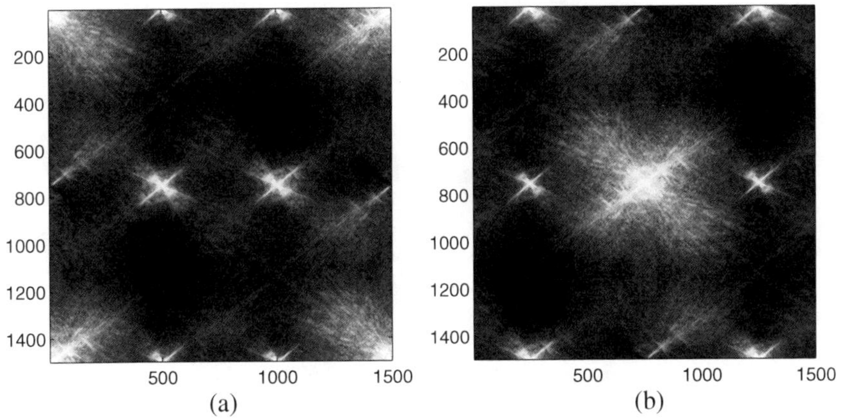

Figure 4.8 The magnitude of the 2-D DFT of the grayscale image $a_{n,m}$ (a) before and (b) after shifting to the middle frequency-point.

To see the imaging in the frequency domain, Fig. 4.8 shows the 1496×1500-point 2-D DFT of the image $a_{n,m}$ in part (a). In Fig. 4.8(b), the amplitude of the transform is shown after shifting to the middle frequency-point $(748, 750)$.

4.3 The (3 × 1) Color-to-Gray Model

We consider a simple model of transferring colors into one grayscale image by sorting the colors by columns. The mapping of the three primary colors into the grayscale image

$$(r_{n,m}, g_{n,m}, b_{n,m}) \to a_{n,m}$$

starts from point $(0, 0)$ and is performed in the following way:

$r_{0,0}$	$r_{0,1}$	$r_{0,2}$	$r_{0,3}$	$r_{0,4}$	$r_{0,5}$	...
$g_{0,0}$	$g_{0,1}$	$g_{0,2}$	$g_{0,3}$	$g_{0,4}$	$g_{0,5}$	...
$b_{0,0}$	$b_{0,1}$	$b_{0,2}$	$b_{0,3}$	$b_{0,4}$	$b_{0,5}$	...
$r_{1,0}$	$r_{1,1}$	$r_{1,2}$	$r_{1,3}$	$r_{1,4}$	$r_{1,5}$	...
$g_{1,0}$	$g_{1,1}$	$g_{1,2}$	$g_{1,3}$	$g_{1,4}$	$g_{1,5}$	...
$b_{1,0}$	$b_{1,1}$	$b_{1,2}$	$b_{1,3}$	$b_{1,4}$	$b_{1,5}$	...
...	...	...	...	...	...	...

In such mapping, the image rows with numbers $3n$, where $n = 0 : (N - 1)$, are composed by the rows in the red color component $r_{n,m}$. The rows with numbers $3n + 1$ and $3n + 2$ are composed by the rows of the green component $g_{n,m}$ and blue component $b_{n,m}$, respectively. The size of the composition is $3N \times M$.

As an example, Fig. 4.9 shows the grayscale image composed from the color "crane" image in this model.

The 2244×1000-point 2-D DFT of this image $a_{n,m}$ is shown in Fig. 4.10(a). The same spectrum but periodically shifted to the middle frequency-point $(1122, 500)$ is shown in Fig. 4.10(b).

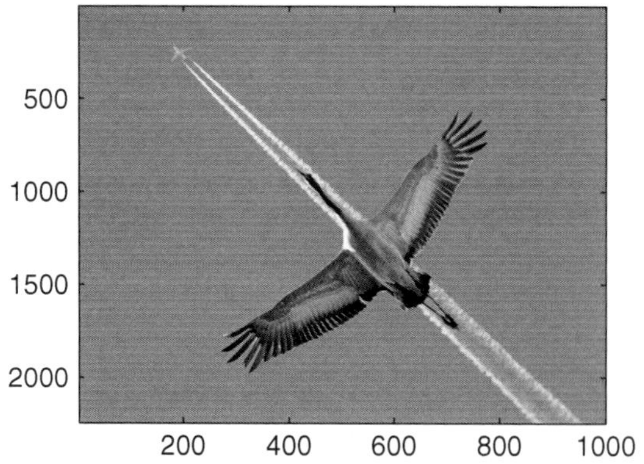

Figure 4.9 The color "crane" image transformed into the (3×1) model.

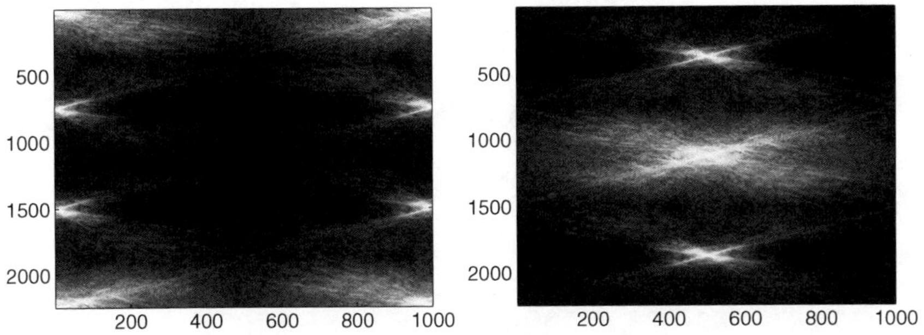

(a) (b)

Figure 4.10 The magnitude of the 2-D DFT of the grayscale image $a_{n,m}$ (a) before and (b) after shifting to the middle.

4.4 The (1 × 3) Color-to-Gray Model

This model is similar to the (3×1) model, only the colors are sorted in one grayscale image by rows, when starting from the point $(0, 0)$ in the following way:

$r_{0,0}$	$g_{0,0}$	$b_{0,0}$	$r_{0,1}$	$g_{0,1}$	$b_{0,1}$	$\cdots$
$r_{1,0}$	$g_{1,0}$	$b_{1,0}$	$r_{1,1}$	$g_{1,1}$	$b_{1,1}$	$\cdots$
$r_{2,0}$	$g_{2,0}$	$b_{2,0}$	$r_{2,1}$	$g_{2,1}$	$b_{2,1}$	$\cdots$
$\cdots$	$\cdots$	$\cdots$	$\cdots$	$\cdots$	$\cdots$	$\cdots$

In such mapping, the image columns with numbers $3n$, $n = 0 : (N-1)$, are composed from the columns of the red component image $r_{n,m}$. The columns with number $3n+1$ and $3n+2$ are composed from the rows in the green image $g_{n,m}$ and blue image $b_{n,m}$, respectively. The size of the composed grayscale image is $N \times 3M$.

As an example, Fig. 4.11 shows the grayscale image composed from the color "crane" image in this model.

The magnitude of the 748×3000-point 2-D DFT of this image $a_{n,m}$ is shown in Fig. 4.12(a). The same spectrum but periodically shifted to the middle frequency-point $(374, 1500)$ is shown in Fig. 4.12(b).

4.5 Model of Color-to-Octonion Image

The mapping of color images into octonion images is probably the most interesting case, when applying the octonion arithmetic to real imaging. Let us consider the RGB model with three color channels [1].

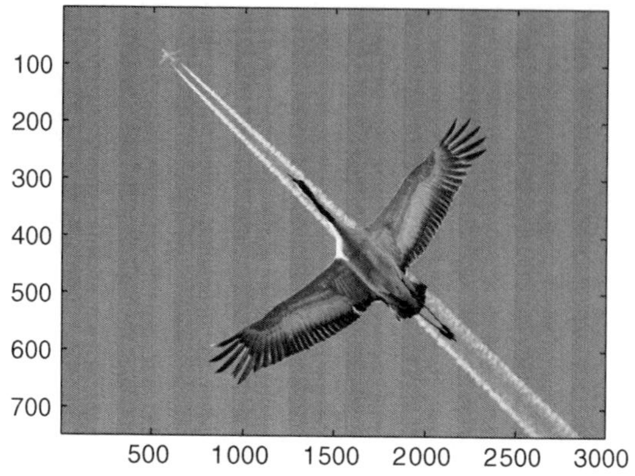

Figure 4.11 The color "crane" image transformed in the (1×3) model.

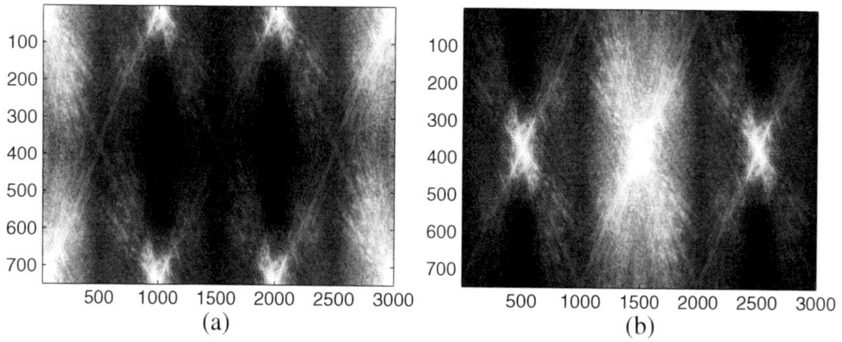

Figure 4.12 The magnitude of the 2-D DFT of the image $a_{n,m}$ (a) before and (b) after shifting to the middle.

4.5.1 Octonion image with RGB color model

In the RGB color model, we consider the color image $f_{n,m}$ with its red, green, and blue color components $f_{n,m} = (r_{n,m}, g_{n,m}, b_{n,m})$, where $n = 0 : (N-1)$, $m = 0 : (M-1)$, plus the grayscale image $i_{n,m}$ of the brightness, which is calculated by Eq. (4.1), $i_{n,m} = 0.3r_{n,m} + 0.59g_{n,m} + 0.11b_{n,m}$. Thus, we use the 4-D model of the image

$$\tilde{f}_{n,m} = (i_{n,m}, r_{n,m}, g_{n,m}, b_{n,m}).$$

Together with pixel (n, m), we consider the image at the next pixel (or neighbor), for instance $(n, m+1)$,

$$\tilde{f}_{n,m+1} = (i_{n,m+1}, r_{n,m+1}, g_{n,m+1}, b_{n,m+1}).$$

Thus, we unite each two columns of the image $f_{n,m}$ and compose the new image (assuming M is even)

$$o(f)_{n,m} = (\tilde{f}_{n,2m+1}, \tilde{f}_{n,2m+2}), \qquad n = 0:(N-1),\ m = 0:(M/2-1).$$

The image $\tilde{f}$ is four-dimensional and the new image is eight-dimensional. Therefore, we can transfer such a paired image $o(f)$ into the octonion space.

As an example, we consider the following mapping:

$$\begin{bmatrix} i_{n,2m+1} & r_{n,2m+1} \\ g_{n,2m+1} & b_{n,2m+1} \end{bmatrix} \cdot\times \begin{bmatrix} 1 & i \\ j & k \end{bmatrix} + \begin{bmatrix} i_{n,2m+2} & r_{n,2m+2} \\ g_{n,2m+2} & b_{n,2m+2} \end{bmatrix} \cdot\times \begin{bmatrix} E & I \\ J & K \end{bmatrix} \rightarrow o_{n_1,m_1}. \tag{4.4}$$

In other words, the octonion image is composed as

$$\begin{aligned} o_{n,m} &= \tilde{f}_{n,2m+1} + \tilde{f}_{n,2m+2}E \\ &= i_{n,2m+1} + ir_{n,2m+1} + jg_{n,2m+1} + kb_{n,2m+1} \\ &\quad + Ei_{n,2m+2} + Ir_{n,2m+2} + Jg_{n,2m+2} + Kb_{n,2m+2}, \end{aligned} \tag{4.5}$$

where $n = 0:(N-1)$ and $m = 0:(M/2-1)$.

It is important to note that, in this model, the color image is processed in two neighbor pixels as being united in one unit; the concepts of the color and distance are united into one octonion number. The octonion image is the pair of two quaternion images, one image is composed by 4-D image $\tilde{f}_{n,2m+1}$ and another one by shifted image $\tilde{f}_{n,2m+2}$.

As an example, Fig. 4.13 shows the original color "crane" image $f_{n,m}$ of size 748 × 1000 separated into the sub-image $f_{n,2m+1}$ in part (a) and the sub-image $f_{n,2m+1}$ in part (b) of size 374 × 500 each.

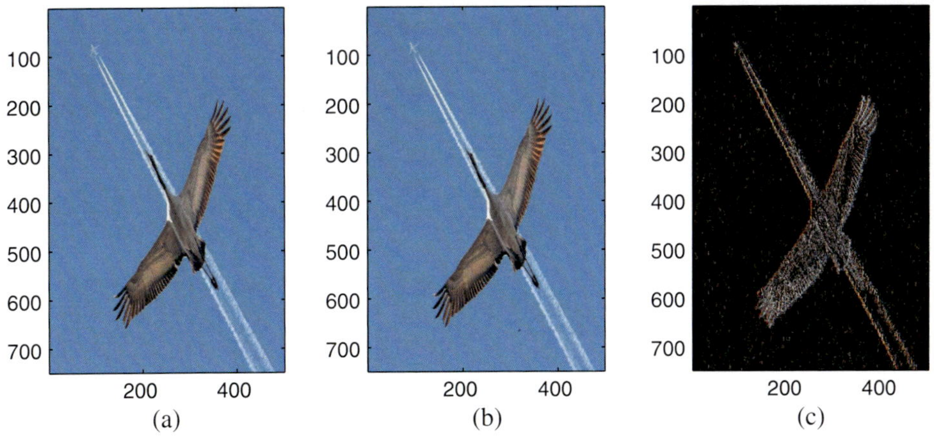

Figure 4.13 Two parts of images (a) $f_{n,2m+1}$ and (b) $f_{n,2m+2}$, and (c) the difference of these parts (which was scaled by the factor of 16).

In octonion space, one can also consider the image composed as

$$o_{n,m} = \tilde{f}_{n,2m+1} + [\tilde{f}_{n,2m+1} - \tilde{f}_{n,2m+2}]E, \tag{4.6}$$

where $n = 0 : (N-1)$ and $m = 0 : (M/2 - 1)$. In this model, the part of the image $\tilde{f}_{n,2m+1}$ is processed together with the difference image $[\tilde{f}_{n,2m+1} - \tilde{f}_{n,2m+2}]$. Such difference image is shown in Fig. 4.13(c). In this octonion image, some information of the gradient along the X-axis is also stored.

Two operations over the image can also be considered. For example, the octonion image can be calculated from the low-pass and high-pass components of the grayscale image [13, 14],

$$o_{n,m} = \frac{1}{2}\left[\tilde{f}_{n,2m+1} + \tilde{f}_{n,2m+2}\right] + \left[\tilde{f}_{n,2m+1} - \tilde{f}_{n,2m+2}\right]E. \tag{4.7}$$

Other models of the octonion image composed from the color image can also be considered by the equations similar to Eqs. (4.4) and (4.5).

4.5.2 Octonion image with RGBY color model

In the RGBY color model, we consider the color image $f_{n,m}$ with its red, green, blue, and yellow color components $f_{n,m} = (r_{n,m}, g_{n,m}, b_{n,m}, y_{n,m})$, where $n = 0 : (N-1)$, $m = 0 : (M-1)$.

Together with pixel (n, m), we consider the image in the next pixel, for instance $(n, m+1)$,

$$f_{n,m+1} = (r_{n,m+1}, g_{n,m+1}, b_{n,m+1}, y_{n,m+1}).$$

Thus, each two columns of the image $f_{n,m}$ are united to compose a new image (assuming M is even)

$$o(f)_{n,m} = (f_{n,2m+1}, f_{n,2m+2}), \qquad n = 0 : (N-1),\ m = 0 : (M/2 - 1).$$

We can transfer this image $o(f)$ with eight components into the octonion space, by using the following mapping:

$$\begin{bmatrix} r_{n,2m+1} & g_{n,2m+1} \\ b_{n,2m+1} & y_{n,2m+1} \end{bmatrix} \cdot\times \begin{bmatrix} 1 & i \\ j & k \end{bmatrix} + \begin{bmatrix} r_{n,2m+2} & g_{n,2m+2} \\ b_{n,2m+2} & y_{n,2m+2} \end{bmatrix} \cdot\times \begin{bmatrix} E & I \\ J & K \end{bmatrix} \to o_{n_1,m_1}. \tag{4.8}$$

In other words, the octonion image is composed as

$$\begin{aligned} o_{n,m} &= f_{n,2m+1} + f_{n,2m+2}E \\ &= r_{n,2m+1} + ig_{n,2m+1} + jb_{n,2m+1} + ky_{n,2m+1} \\ &\quad + Er_{n,2m+2} + Ig_{n,2m+2} + Jb_{n,2m+2} + Ky_{n,2m+2}, \end{aligned} \tag{4.9}$$

where $n = 0 : (N-1)$ and $m = 0 : (M/2 - 1)$.

4.5.3 Octonion image with CMYK color model

In the CMYK color model (see Section 3.3.4), the color image $f_{n,m}$ has four components

$$f_{n,m} = (c_{n,m}, m_{n,m}, y_{n,m}, k_{n,m}), \qquad n = 0 : (N-1),\ m = 0 : (M-1).$$

Together with the pixel (n, m), we consider the image at the next pixel $(n, m+1)$ similarly to the model described above for the RGBY model,

$$f_{n,m+1} = (c_{n,m+1}, m_{n,m+1}, y_{n,m+1}, k_{n,m+1}).$$

Therefore, we can compose the new eight-dimensional image

$$o(f)_{n,m} = (f_{n,2m+1}, f_{n,2m+2}), \qquad n = 0 : (N-1),\ m = 0 : (M/2-1),$$

and transfer it into the octonion space, by using the following mapping:

$$\begin{bmatrix} c_{n,2m+1} & m_{n,2m+1} \\ y_{n,2m+1} & k_{n,2m+1} \end{bmatrix} \cdot \times \begin{bmatrix} 1 & i \\ j & k \end{bmatrix} + \begin{bmatrix} c_{n,2m+2} & m_{n,2m+2} \\ y_{n,2m+2} & k_{n,2m+2} \end{bmatrix} \cdot \times \begin{bmatrix} E & I \\ J & K \end{bmatrix} \to o_{n_1,m_1}. \tag{4.10}$$

In other words, the octonion image is composed as

$$\begin{aligned} o_{n,m} &= \tilde{f}_{n,2m+1} + \tilde{f}_{n,2m+2}E \\ &= c_{n,2m+1} + im_{n,2m+1} + jy_{n,2m+1} + kk_{n,2m+1} \\ &\quad + Ec_{n,2m+2} + Im_{n,2m+2} + Jy_{n,2m+2} + Kk_{n,2m+2}, \end{aligned} \tag{4.11}$$

where $n = 0 : (N-1)$ and $m = 0 : (M/2-1)$. Thus, the color image in the CMYK model can be processed as such an image where at each pixel contains the information of the colors at two neighbor pixels.

4.5.4 Model with two color images

In octonion space, we can process two color images as one octonion image, by using the following model which can be used for different color models, such as the RBG, RGBY, CMY, CMYK, XYZ, and YCbCr models [1]–[6]. Here, we consider the RGB model since other models can be described similarly.

In the RGB color model, we consider the full 4-D representation of the color image $f_{n,m}$,

$$\tilde{f}_{n,m} = [(f_i)_{n,m}, (f_r)_{n,m}, (f_g)_{n,m}, (f_b)_{n,m}],$$

where $n = 0 : (N-1)$ and $m = 0 : (M-1)$. The intensity, red, green, and blue components of the image are denoted by $(f_i), (f_r), (f_g)$, and (f_b), respectively. Other image of the same size, which we call $v_{n,m}$, can similarly be considered

$$\tilde{v}_{n,m} = [(v_i)_{n,m}, (v_r)_{n,m}, (v_g)_{n,m}, (v_b)_{n,m}].$$

We can compose the new eight-dimensional image of the same size

$$o(f)_{n,m} = (\tilde{f}_{n,m}, \tilde{v}_{n,m}), \qquad n = 0 : (N-1),\ m = 0 : (M-1),$$

and transfer it into the octonion space, by using the following mapping:

$$\begin{bmatrix} (f_i)_{n,m} & (f_r)_{n,m} \\ (f_g)_{n,m} & (f_b)_{n,m} \end{bmatrix} \cdot \times \begin{bmatrix} 1 & i \\ j & k \end{bmatrix} + \begin{bmatrix} (v_i)_{n,m} & (v_r)_{n,m} \\ (v_g)_{n,m} & (v_b)_{n,m} \end{bmatrix} \cdot \times \begin{bmatrix} E & I \\ J & K \end{bmatrix} \to o_{n_1,m_1}. \tag{4.12}$$

The octonion image is composed as

$$\begin{aligned} o_{n,m} &= \tilde{f}_{n,m} + \tilde{v}_{n,m}E \\ &= (f_i)_{n,m} + i(f_r)_{n,m} + j(f_g)_{n,m} + k(f_b)_{n,m} \\ &\quad + E(v_i)_{n,m} + I(v_r)_{n,m} + J(v_g)_{n,m} + K(v_b)_{n,m}, \end{aligned} \tag{4.13}$$

where $n = 0 : (N-1)$ and $m = 0 : (M-1)$.

4.6 Color images and Quaternion Multiplication

In this section, we consider color images in the quaternion space and illustrate the operation of multiplication of images. Let $f_{n,m}$ be a color image of size $N \times M$, which is presented in the RGB color model as the three-component image with primary red, green, and blue colors,

$$f_{n,m} = [(f_r)_{n,m}, (f_g)_{n,m}, (f_b)_{n,m}], n = 0 : (N-1), m = 0 : (M-1).$$

We can extend this image to the 4-D image by adding the grayscale image calculated as the average of colors,

$$(f_a)_{n,m} = \frac{1}{3}[(f_r)_{n,m} + (f_g)_{n,m} + (f_b)_{n,m}].$$

The extended image is

$$\tilde{f}_{n,m} = [(f_a)_{n,m}, (f_r)_{n,m}, (f_g)_{n,m}, (f_b)_{n,m}].$$

Let $f_{n,m}$ and $y_{n,m}$ be two color images of the size $N \times M$ each. As an example, two such color images f and y are shown in Fig. 4.14(a,b), respectively. The size of both images is $N \times M = 524 \times 640$.

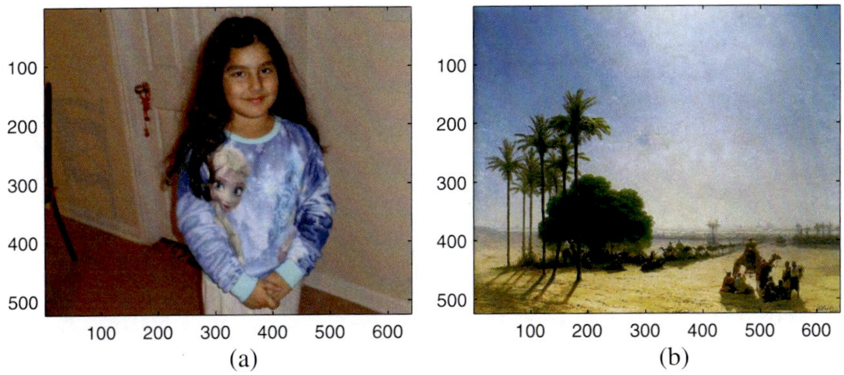

Figure 4.14 Color images: (a) the "girl Anoush" image, $f_{n,m}$, and (b) the "caravan" image, $y_{n,m}$, from Ivan Aivazovsky's paintings.

These two images can be mapped into quaternion space by composing the following quaternion images:

$$q(f)_{n,m} = (f_a)_{n,m} + i(f_r)_{n,m} + j(f_g)_{n,m} + k(f_b)_{n,m},$$
$$q(y)_{n,m} = (y_a)_{n,m} + i(y_r)_{n,m} + j(y_g)_{n,m} + k(y_b)_{n,m}.$$

We consider the operations of the inner and vector products and the multiplication of quaternions for images; these operations will be calculated point-wise. Thus, the multiplication of two quaternion images $q(f)$ and $q(y)$ is defined as the following quaternion data $q(fy)$:

$$q(fy)_{n,m} = q(f)_{n,m} \cdot q(y)_{n,m}, \quad n = 0:(N-1), \quad m = 0:(M-1).$$

Figure 4.15 shows the result of multiplication of two images of Fig. 4.14. In part (a), the real part of the multiplication is shown, and the imaginary part of the multiplication of these images as a color image in the RGB color space is shown in part (b). Since the components of multiplication of two quaternions may be positive and negative, both images are shown in the absolute scale.

4.6.1 Quaternion multiplication of images

To illustrate the multiplication of quaternion images step-by-step, we consider first the grayscale and color channels of the given images $f_{n,m}$ and $y_{n,m}$.

For the image $\tilde{f}_{n,m}$, Fig. 4.16 shows the grayscale component in part (a), and color channels for red, green, and blue components in parts (b), (c), and (d), respectively.

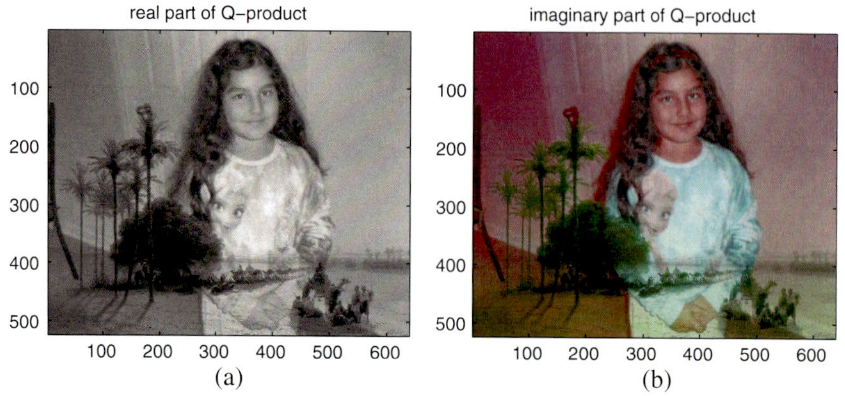

Figure 4.15 The quaternion multiplication, or *Q*-multiplication of images $q(f)$ and $q(y)$: (a) the real part and (b) the imaginary part.

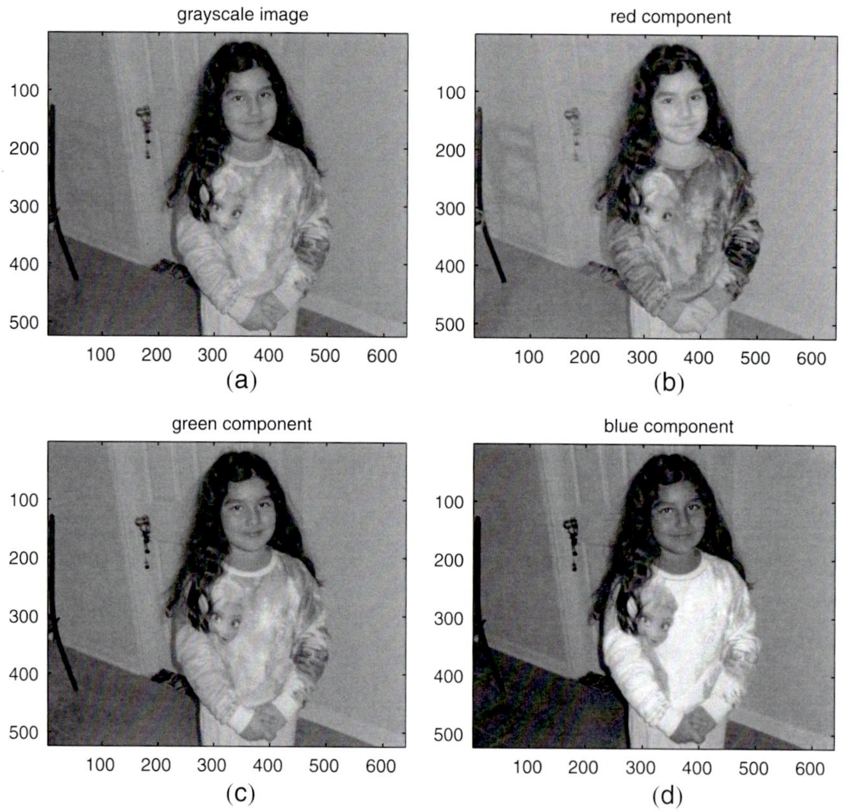

Figure 4.16 (a) The grayscale component, (b) the red, (c) the green, and (d) the blue components of the image $\tilde{f}_{n,m}$.

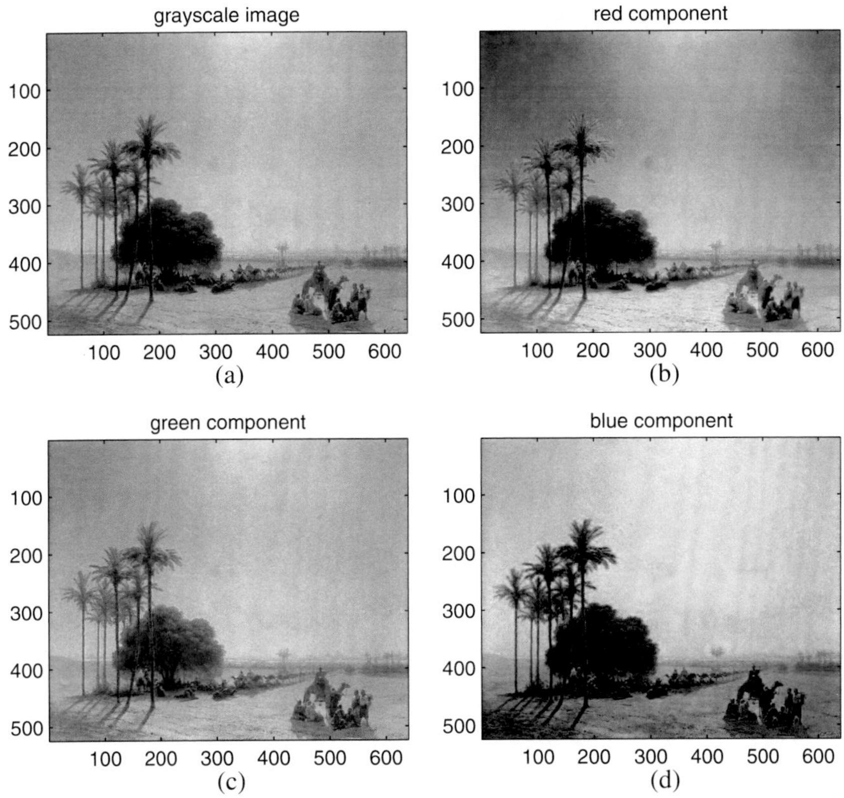

(a) (b) (c) (d)

Figure 4.17 (a) The grayscale component, (b) the red, (c) the green, and (d) the blue components of the image $\tilde{y}_{n,m}$.

Figure 4.17 shows the grayscale component of the image $\tilde{y}_{n,m}$ in part (a), and color channels for red, green, and blue components in parts (b), (c), and (d), respectively.

These two images can be mapped into the quaternion space by composing the following quaternion images:

$$\begin{aligned} q(f)_{n,m} &= (f_a)_{n,m} + q(f)'_{n,m}, \\ q(y)_{n,m} &= (y_a)_{n,m} + q(y)'_{n,m}. \end{aligned} \tag{4.14}$$

The imaginary components of these images are

$$\begin{aligned} q(f)'_{n,m} &= i(f_r)_{n,m} + j(f_g)_{n,m} + k(f_b)_{n,m}, \\ q(y)'_{n,m} &= i(y_r)_{n,m} + j(y_g)_{n,m} + k(y_b)_{n,m}. \end{aligned}$$

To show the modules of the imaginary parts of these images in the original scale $[0, 255]$, we consider the amplitude normalization by the factor of $1/\sqrt{3}$, as

$$\begin{aligned} |q(f)'_{n,m}| &= \sqrt{\frac{(f_r)^2_{n,m} + (f_g)^2_{n,m} + (f_b)^2_{n,m}}{3}}, \\ |q(y)'_{n,m}| &= \sqrt{\frac{(y_r)^2_{n,m} + (y_g)^2_{n,m} + (y_b)^2_{n,m}}{3}}. \end{aligned} \tag{4.15}$$

As an example, Fig. 4.18 shows the color image $f_{n,m}$ in part (a), and the imaginary part in the absolute scale, $|q(f)'_{n,m}|$, in part (b).

The module of the quaternion image can be calculated in a similar way with the factor of $1/\sqrt{4}$,

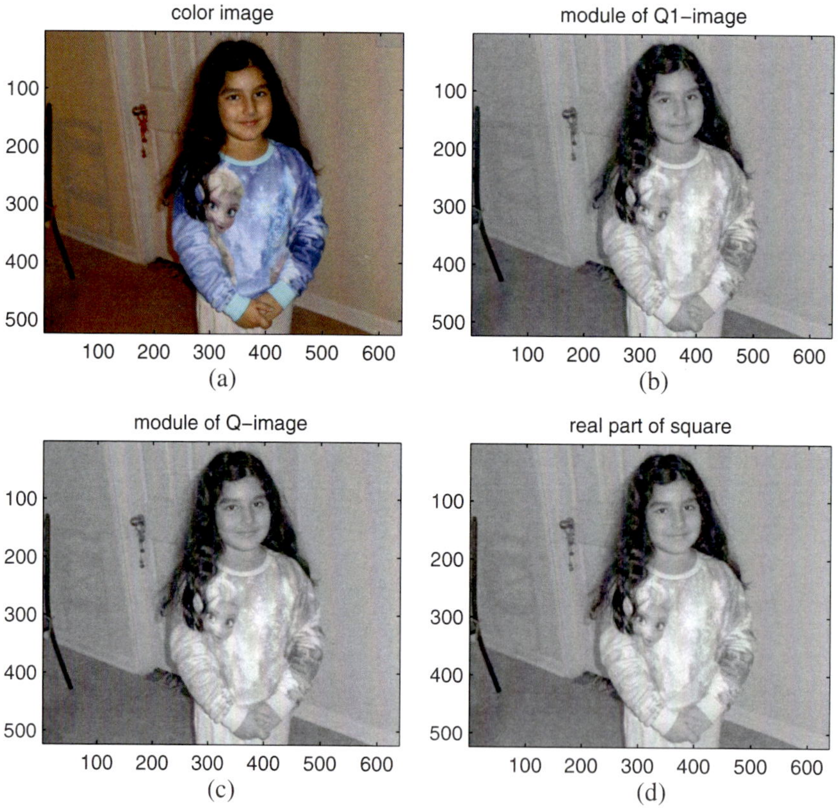

Figure 4.18 (a) The color "girl Anoush" image $f_{n,m}$, (b) the module $|q(f)'_{n,m}|$ of the imaginary part of the color-in-quaternion image, (c) the module $|q(f)_{n,m}|$ of the quaternion image, and (d) the real part of the square $q(f)^2_{n,m}$ of the quaternion image.

$$|q(f)_{n,m}| = \sqrt{\frac{(f_e)^2_{n,m} + (f_r)^2_{n,m} + (f_g)^2_{n,m} + (f_b)^2_{n,m}}{4}}, \tag{4.16}$$

which can also be written as

$$|q(f)_{n,m}| = \sqrt{\frac{1}{4}(f_e)^2_{n,m} + \frac{3}{4}|q(f)'_{n,m}|^2}.$$

Such module of the quaternion image "girl Anoush" is shown in part (c).

As follows from Eq. (1.11), the square of the quaternion number $q = a + q' = a + (ib + cd + kd)$ is calculated by

$$q^2 = [a^2 - (b^2 + c^2 + d^2)] + 2aq'.$$

The real part of this square is $a^2 - (b^2 + c^2 + d^2)$. For the above color images, we are considering the case when $a = (b + c + d)/3$. Therefore, the real part of the square of such quaternions can be shown in the positive scale after changing the sign, or calculating the real part by $(b^2 + c^2 + d^2) - a^2$. The square root is also can be used, to preserve the values of the real part inside the given interval of variables b, c, and d.

Thus, the real part of the square of the quaternion image $q(f)_{n,m}$ can be calculated as an image in the following form:

$$\Re\{q(f)^2_{n,m}\} = \sqrt{\frac{(f_r)^2_{n,m} + (f_g)^2_{n,m} + (f_b)^2_{n,m} - (f_e)^2_{n,m}}{2}}. \tag{4.17}$$

The image of the real part of the quaternion image $q(f)_{n,m}$ is shown in Fig. 4.18(d).

Figure 4.19 shows the images for the "caravan" image $q(y)_{n,m}$, which are calculated similarly to images given in Fig. 4.18.

As follows from the definition in Eq. (1.9), for given two quaternions $q_1 = a_1 + q'_1$ and $q_2 = a_2 + q'_2$, the multiplication $q = q_1 q_2$ is calculated

$$q = a_1 a_2 - (b_1 b_2 + c_1 c_2 + d_1 d_2) + [a_1 q'_2 + a_2 q'_1] + q'_1 \times q'_2. \tag{4.18}$$

Here, $(b_1 b_2 + c_1 c_2 + d_1 d_2)$ is the inner product (q'_1, q'_2).

When calculating point-wise the inner product of two quaternion images, we obtain the following data, which are denoted by (f, y):

$$(f, y)_{n,m} = (f_r)_{n,m}(y_r)_{n,m} + (f_g)_{n,m}(y_g)_{n,m} + (f_b)_{n,m}(y_b)_{n,m}.$$

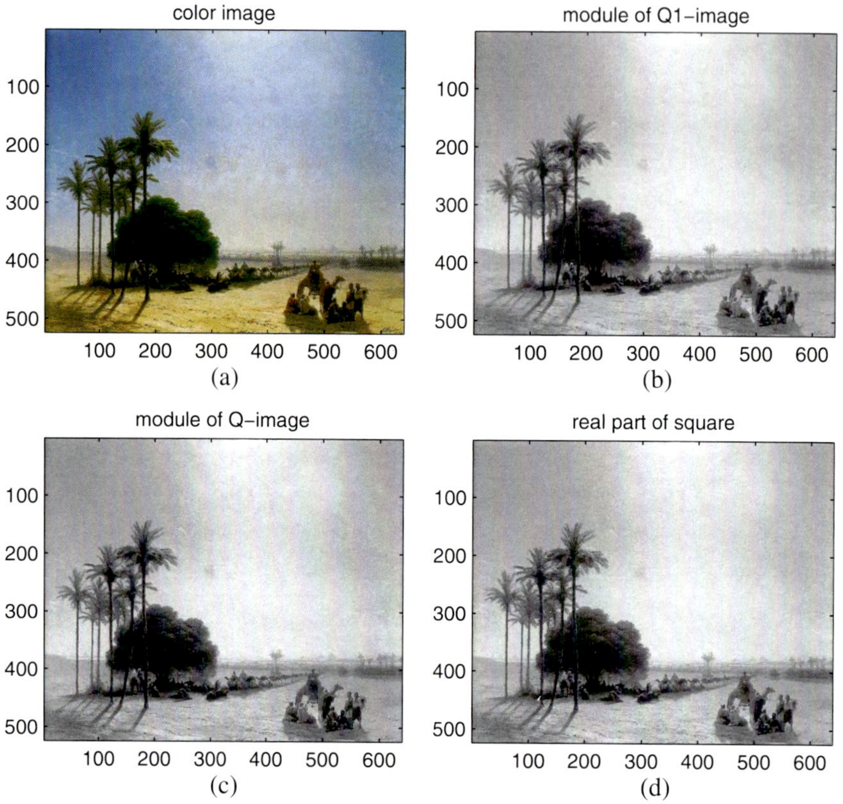

Figure 4.19 (a) The color image $y_{n,m}$, (b) the module $|q(y)'_{n,m}|$ of the imaginary part of the quaternion image $q(y)_{n,m}$, (c) the module $|q(y)_{n,m}|$ of the quaternion image, and (d) the real part of the square $q(y)^2_{n,m}$ of the quaternion image.

To calculate and display these data as an image, the square root and the normalizing coefficients can be used,

$$(f, y)_{n,m} = \sqrt{\frac{(f_r)_{n,m}(y_r)_{n,m} + (f_g)_{n,m}(y_g)_{n,m} + (f_b)_{n,m}(y_b)_{n,m}}{3}}.$$

The inner product in this presentation is shown in Fig. 4.20(a), for the quaternion "girl Anoush" image $q(f)_{n,m}$ and "caravan" image $q(y)_{n,m}$. The real part of the multiplication in (4.18) is

$$\Re\{q\} = a_1a_2 - (b_1b_2 + c_1c_2 + d_1d_2).$$

Therefore, the real part of the multiplication of two quaternion images $q(f)$ and $q(y)$ can be presented and visualized as the following image:

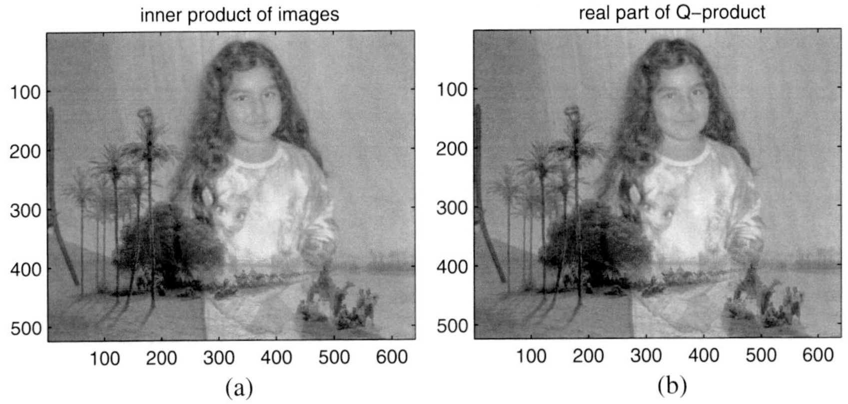

Figure 4.20 (a) The inner product of the color images $f_{n,m}$ and $y_{n,m}$, and (b) real part of the multiplication of images $q(f)_{n,m}$ and $q(y)_{n,m}$.

$$\Re\{q(fy)\}_{n,m} = \sqrt{\frac{|(f_e)_{n,m}(y_e)_{n,m} - (f_r)_{n,m}(y_r)_{n,m} - (f_g)_{n,m}(y_g)_{n,m} - (f_b)_{n,m}(y_b)_{n,m}|}{2}}.$$

The real part of the product of the images $q(f)$ and $q(y)$ in this presentation is shown in Fig. 4.20(b).

Now for these images, we apply the concept of the vector product of quaternions q_1 and q_2, which is

$$q_1 \times q_2 = \begin{vmatrix} i & j & k \\ b_1 & c_1 & d_1 \\ b_2 & c_2 & d_2 \end{vmatrix} = i(c_1 d_2 - c_2 d_1) - j(b_1 d_2 - b_2 d_1) + k(b_1 c_2 - b_2 c_1).$$

The i-, j-, and k-components of this product are $(c_1 d_2 - c_2 d_1)$, $(b_2 d_1 - b_1 d_2)$, and $(b_1 c_2 - b_2 c_1)$, respectively. Therefore, for the quaternion images $q(f)$ and $q(y)$, we can calculate the imaginary components by

$$\begin{aligned} ([f \times y]_i)_{n,m} &= (f_j)_{n,m}(y_k)_{n,m} - (f_k)_{n,m}(y_j)_{n,m}, \\ ([f \times y]_j)_{n,m} &= (f_k)_{n,m}(y_i)_{n,m} - (f_i)_{n,m}(y_k)_{n,m}, \\ ([f \times y]_k)_{n,m} &= (f_i)_{n,m}(y_j)_{n,m} - (f_j)_{n,m}(y_i)_{n,m}. \end{aligned}$$

Figure 4.21 shows these i-, j-, and k-components as grayscale images in the absolute and square root scale in parts (a), (b), and (c), respectively. These three components can be mapped to the RGB color model and therefore, the

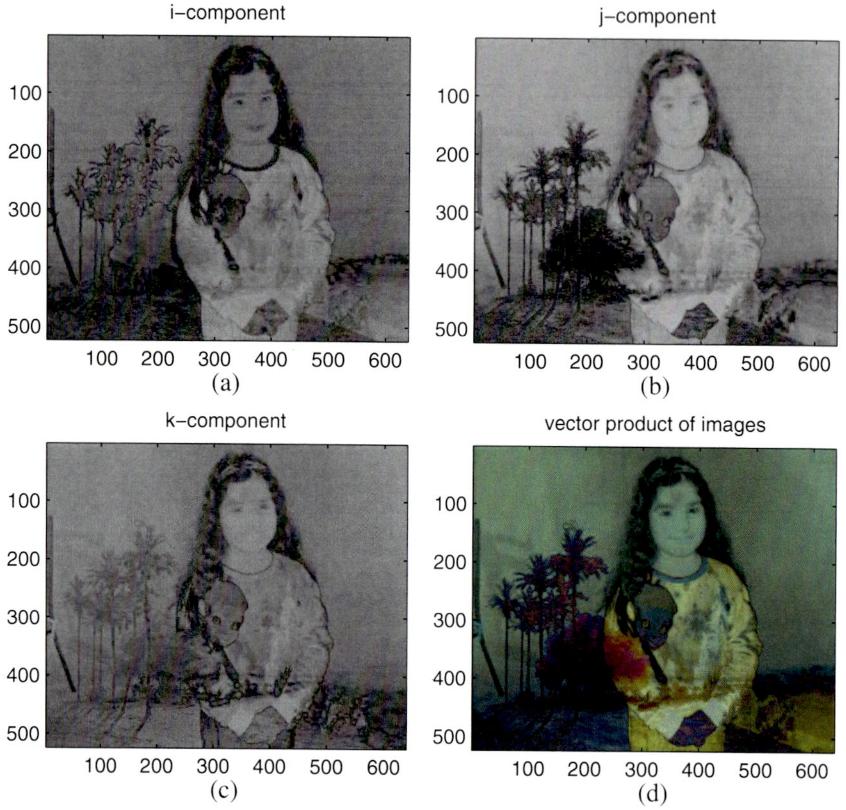

Figure 4.21 (a) The *i*-, (b) the *j*-, and (c) the *k*-components of the vector product of color "girl Anoush" and "caravan" images, and (d) the color image presenting this vector product.

vector product $q(f)' \times q(y)'$ can be illustrated as the color image shown in Fig. 4.21(d).

As follows from Eq. (4.18), the components of the imaginary part of the multiplication $q = q_1 q_2$ are calculated by

$$\begin{aligned}(q)_i &= (c_1 d_2 - c_2 d_1) + a_1 b_2 + a_2 b_1,\\ (q)_j &= (b_2 d_1 - b_1 d_2) + a_1 c_2 + a_2 c_1,\\ (q)_k &= (b_1 c_2 - b_2 c_1) + a_1 d_2 + a_2 d_1.\end{aligned} \tag{4.19}$$

To display these images, the components of the imaginary part of multiplication of images $q(y)$ and $q(y)$ are calculated as

$$q(fy)_i = \sqrt{\frac{|[(f \times y)_i]_{n,m} + (f_e)_{n,m}(y_i)_{n,m} + (y_e)_{n,m}(f_i)_{n,m}|}{2}},$$
$$q(fy)_j = \sqrt{\frac{|[(f \times y)_j]_{n,m} + (f_e)_{n,m}(y_j)_{n,m} + (y_e)_{n,m}(f_j)_{n,m}|}{2}}, \tag{4.20}$$
$$q(fy)_k = \sqrt{\frac{|[(f \times y)_k]_{n,m} + (f_e)_{n,m}(y_k)_{n,m} + (y_e)_{n,m}(f_k)_{n,m}|}{2}}.$$

Figure 4.22 shows the i-, j-, and k-components of the multiplication of quaternion images $q(f)$ and $q(y)$ in parts (a), (b), and (c), respectively. In the RGB color model, these three components present the color image shown in Fig. 4.22(d). As a result of the above calculations, we obtain the quaternion image presenting the multiplication of quaternion images $q(f)$ and $q(y)$, with the real and imaginary parts shown in the beginning of this section in Fig. 4.15.

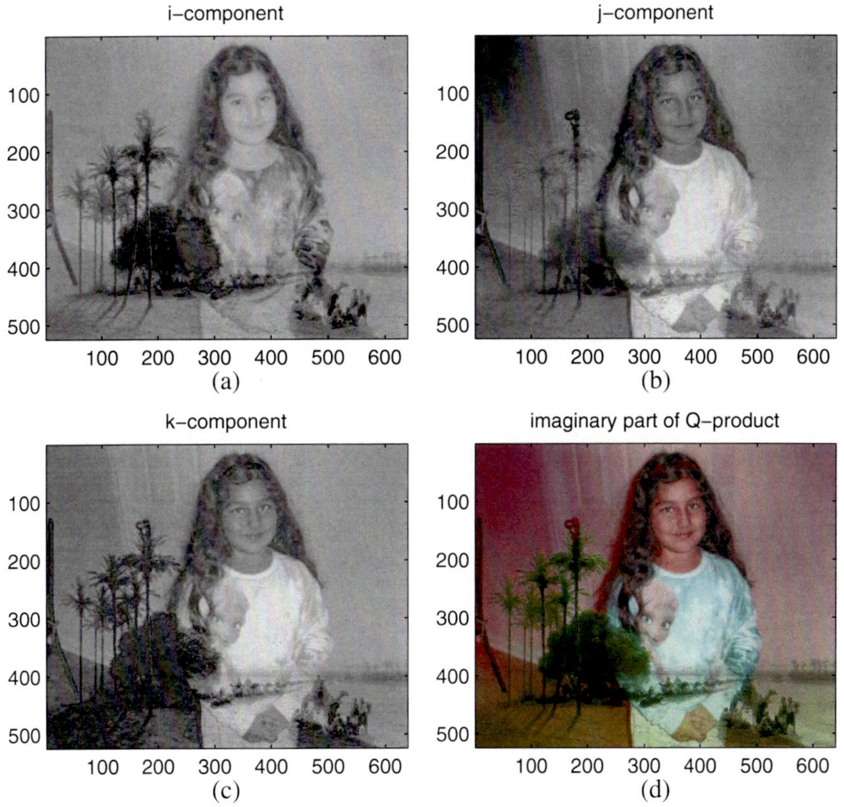

Figure 4.22 The (a) i-, (b) j-, and (c) k-components of the product of the color images $q(f)$ and $q(y)$, and (d) the color image presenting this product in the RGB space.

4.6.1.1 Script for the quaternion image multiplication

Below is the script of the MATLAB-based code "test_exampleQMI.m," which was used to calculate all shown above images in the example.

```
% -----------------------------------------------------------
% call: test_exampleQMI / Artyom Grigoryan, December 2015
% demo code for illustration of quaternion multiplication of
  two images.
   %I Read the images of size 524×640 each.
   clear all; close all;
   Y1=imread('image1.jpg','jpg'); [N M L]=size(Y1);
   Y2=imread('image2.bmp','bmp'); [N2 M2 L2]=size(Y2);
   Y=double(Y1);
   Qi=Y(:,:,1); Qj=Y(:,:,2); Qk=Y(:,:,3); Qe=round
   ((Qi+Qj+Qk)/3);
   Y=double(Y2);
   Yi=Y(:,:,1); Yj=Y(:,:,2); Yk=Y(:,:,3); Ye=round
   ((Yi+Yj+Yk)/3);
   % 1
   figure; colormap(gray(255));
   subplot(1,2,1); image(Y1); axis image; xlabel('(a)');
   subplot(1,2,2); image(Y2); axis image; xlabel('(b)');
   % 2
   figure; colormap(gray(255));
   subplot(2,2,1); image(Qe); axis image;
   title('grayscale image');
   subplot(2,2,2); image(Qi); axis image; title('red
   component');
   subplot(2,2,3); image(Qj); axis image; title('green
   component');
   subplot(2,2,4); image(Qk); axis image; title('blue
   component');
   % 3
   figure; colormap(gray(255));
   subplot(2,2,1); image(Ye); axis image;
   title('grayscale image');
   subplot(2,2,2); image(Yi); axis image; title('red
   component');
   subplot(2,2,3); image(Yj); axis image; title('green
   component');
   subplot(2,2,4); image(Yk); axis image; title('blue
   component');
```

```
% 4a
figure; colormap(gray(255));
subplot(2,2,1); image(Y1); axis image; title('color
image');
Q1mod=sqrt(Qi.*Qi + Qj.*Qj +Qk.*Qk)/sqrt(3);
subplot(2,2,2); image(Q1mod); axis image; title('modul
of Q1-image');
Qmod=sqrt(Qe.*Qe + Qi.*Qi + Qj.*Qj +Qk.*Qk)/sqrt(4);
subplot(2,2,3); image(Qmod); axis image; title('modul
of Q-image');
Qip=sqrt(-Qe.*Qe + Qi.*Qi + Qj.*Qj +Qk.*Qk)/sqrt(2);
subplot(2,2,4); image(Qip); axis image; title('real
part of square');
% 4b
figure; colormap(gray(255));
subplot(2,2,1); image(Y2); axis image; title('color
image');
Y1mod=sqrt(Yi.*Yi + Yj.*Yj +Yk.*Yk)/sqrt(3);
subplot(2,2,2); image(Y1mod); axis image; title('modul
of Q1-image');
Ymod=sqrt(Ye.*Ye + Yi.*Yi + Yj.*Yj +Yk.*Yk)/sqrt(4);
subplot(2,2,3); image(Ymod); axis image; title('modul
of Q-image');
Yip=sqrt(-Ye.*Ye + Yi.*Yi + Yj.*Yj +Yk.*Yk)/sqrt(2);
subplot(2,2,4); image(Yip); axis image; title('real
part of square');
% 5
QYip1=sqrt(Qi.*Yi + Qj.*Yj +Qk.*Yk)/sqrt(3);
figure; colormap(gray(255));
subplot(1,2,1); image(QYip1); axis image; title('inner
product of images');
QYip=sqrt(abs(-Qe.*Ye + Qi.*Yi + Qj.*Yj +Qk.*Yk))/
sqrt(2);
QYip=round(QYip);
subplot(1,2,2); image(QYip); axis image; title('real
part of Q-product');
% 6
Vi=sqrt(abs( Qj.*Yk - Qk.*Yj));
Vj=sqrt(abs( Qi.*Yk - Qk.*Yi));
Vk=sqrt(abs( Qi.*Yj - Qj.*Yi));
figure; colormap(gray(255));
subplot(2,2,1); image(Vi); axis image;
title('i-component');
```

```
    subplot(2,2,2); image(Vj); axis image;
    title('j-component');
    subplot(2,2,3); image(Vk); axis image;
    title('k-component');
    V=zeros(N2,M2,3);
    V(:,:,1)=Vi; V(:,:,2)=Vj; V(:,:,3)=Vk; V=uint8(V);
    subplot(2,2,4); image(V); axis image; title('vector
    product of images');
    % 7
    Si=(Qe.*Yi +Ye.*Qi)+(Qj.*Yk - Qk.*Yj); Si=
    sqrt(abs(Si/2));
    Sj=(Qe.*Yj +Ye.*Qj)-(Qi.*Yk - Qk.*Yi); Sj=
    sqrt(abs(Sj/2));
    Sk=(Qe.*Yk +Ye.*Qk)-(Qi.*Yj - Qj.*Yi); Sk=
    sqrt(abs(Sk/2));
    figure; colormap(gray(255));
    subplot(2,2,1); image(Si); axis image;
    title('i-component');
    subplot(2,2,2); image(Sj); axis image;
    title('j-component');
    subplot(2,2,3); image(Sk); axis image;
    title('k-component');
    S=zeros(N2,M2,3);
    S(:,:,1)=Si; S(:,:,2)=Sj; S(:,:,3)=Sk; S=uint8(S);
    subplot(2,2,4); image(S); axis image; title('imaginary
    part of Q-product');
    % 8
    figure; colormap(gray(255));
    subplot(2,2,3); image(QYip); axis image; title('real
    part of Q-product');
    subplot(2,2,4); image(S); axis image; title('imaginary
    part of Q-product');
% end of the script ==================================
```

References

[1] R. C. Gonzalez and R. E. Woods, *Digital Image Processing*, 2nd Edition, Prentice Hall, Upper Saddle River, New Jersey (2002).

[2] H. Palus, *Color Spaces*, Chapmann and Hall (1998).

[3] M. D. Fairchild, *Color Appearance Models*, Addison Wesley, Boston (1998).

[4] H. J. Trussell, E. Saber, and M. Vrhel, "Color image processing," *IEEE Signal Processing Magazine*, **22**(1), 14–22 (2005).

[5] Y. Murakumi et al., "Color conversion method for multi-primary display for spectral color reproduction," *Journal of Electronic Imaging*, **13**(4), 701–708 (2004).
[6] A. Tremeau, S. Tominaga, and K. N. Plataniotis, "Color in image and video processing: Most recent trends and future research directions," *EURASIP Journal on Image and Video Processing*, p. 26 (2008).
[7] P. Denis, P. Carre, and C. F. Maloigne, "Spatial and spectral quaternionic approaches for color images," *Computer Vision and Image understanding*, **107**, 74–87 (2007).
[8] A. M. Grigoryan and S. S. Agaian, "Retooling of color imaging in the quaternion algebra," *Applied Mathematics and Sciences: An International Journal (MathSJ)*, **1**(3), 23–39 (2014).
[9] L. Rabiner and B. Gold, *Theory and Application of Digital Signal Processing*, Prentice-Hall Inc., Upper Saddle River, New Jersey (1975).
[10] S. S. Agaian, "Advances and problems of the fast orthogonal transforms for signal-images processing applications (Part 1)," *Pattern Recognition, Classification, Forecasting. Yearbook, The Russian Academy of Sciences, Nauka*, Moscow, **3**, 146–215 (1990).
[11] J. S. Walker, *Fast Fourier Transforms*, CRC Press, Boca Raton, Florida (1991).
[12] A. M. Grigoryan and S. S. Agaian, "Split manageable efficient algorithm for Fourier and Hadamard transforms, *IEEE Trans. Signal Processing*, **48**(1), 172–183 (2000).
[13] A. M. Grigoryan, "2-D and 1-D multi-paired transforms: Frequency-time type wavelets," *IEEE Trans. on Signal Processing*, **49**(2), 344–353 (2001).
[14] A. M. Grigoryan and S. S. Agaian, *Multidimensional Discrete Unitary Transforms: Representation, Partitioning and Algorithms*, Marcel Dekker Inc., New York (2003).
[15] A. M. Grigoryan and M. M. Grigoryan, *Brief Notes in Advanced DSP: Fourier Analysis with MATLAB*, CRC Press, Taylor and Francis Group, Boca Raton, Florida (2009).
[16] S. S. Agaian, H. G. Sarukhanyan, K. O. Egiazarian, and J. Astola, *Hadamard Transforms*, SPIE Press, Bellingham, Washington (2011) [doi: 10.1117/3.890094].

Chapter 5
1-D Quaternion and Octonion Discrete Fourier Transforms

The quaternion discrete Fourier transform (QDFT) is an effective and well-suited tool for many color image processing applications, including the image analysis, filtration and enhancement, and watermarking [1]–[9]. In color models, such as RGB, RGBY, XYZ, CMY, and others [10], the QDFT processes all three or four color components simultaneously, it captures the inherent correlation between the components, it does not generate color artifacts or blending, and finally, it does not need an additional color restoration process.

The quaternion multiplication is not commutative. Therefore, the 1-D QDFT can be defined by the sum of signal components multiplied by the exponential kernel of the transform from the right or the left [12]. In the two- and multi-dimensional cases, if the DFT is separable, along each direction the quaternion exponential coefficients can be used in multiplication from the left or the right, therefore there is a number of definitions of the quaternion DFT [11]–[14]. We consider in this chapter, the 1-D quaternion DFTs, which are called the right-side and left-side ODFTs, and in the 2-D case, three 2-D DFTs are described; the left-, right-, and both-side QDFTs with fast algorithms [17, 18]. As known, in the tensor representation of images the 2-D DFT is reduced the calculation of the 2-D DFT to 1-D DFTs of the splitting-signals, or image-signal, which uniquely represent the image. This representation unites the concepts of the right-side and left-side 2-D DFTs, in a sense that both transforms have the same tensor algorithm. The both-side 2-D QDFT is defined to balance the multiplication from the image by the quaternion exponential coefficients left and right. In this sense this transform has a symmetry, but it is difficult to imagine that in the real system such multiplications could be accomplished from right (before getting the value of the image) and from the left (after getting the value of the image).

All fast methods of calculating the quaternion and octonion DFTs use the algorithms of fast Fourier transform (FFT) [23]–[26]. Therefore, we first will describe the concepts of the 1-D and 2-D DFTs and fast algorithms for calculating these transforms. In the 1-D case, the well-known Cooley-Tukey algorithm [19] and paired FFT [28] are described, and in the 2-D case, we consider the traditional row-column method and the effective tensor representation of the image, when the calculation of the 2-D DFT is split by the 1-D DFTs of image-signals [28]–[34].

5.1 Discrete Fourier Transform

Let $f = \{f_n\}$, $n = 0 : (N-1)$, be the discrete signal of length $N > 1$. The interval of integers $n = 0 : (N-1)$ is denoted by X_N. The N-point discrete Fourier transform (DFT) of the signal f_n is defined as

$$F_p = (\mathcal{F}_N \circ f)_p = \sum_{n=0}^{N-1} f_n W^{np}, \quad p = 0 : (N-1), \tag{5.1}$$

where $W = W_N = \exp(-2\pi i/N)$. The notation $\mathcal{F}_N$ is used for the discrete Fourier transformation (operation). If we denote by $[\mathcal{F}_N] = [W^{np}]_{n,p=0\,:\,(N-1)}$ the $N \times N$ matrix of the transformation, the DFT can be written in matrix form as $[F] = [\mathcal{F}_N][f]$, where $[F]$ and $[f]$ are the column-vectors composed by values of F_p and f_n, respectively. F_p is the value of the transform at frequency-point p, which corresponds to the frequency $\omega_p = (2\pi/N)p$.

The inverse N-point DFT is calculated by

$$f_n = (\mathcal{F}_N^{-1} \circ F)_n = \frac{1}{N} \sum_{p=0}^{N-1} F_p W^{-np}, \quad n = 0 : (N-1). \tag{5.2}$$

In matrix form, the inverse transform can be written as

$$[f] = [\mathcal{F}_N^{-1}][F] = \frac{1}{N}\overline{[\mathcal{F}_N]^T}[F] = \frac{1}{N}\overline{[\mathcal{F}_N]}^T[F].$$

Here, the operation A^T denotes the operation of transposition of a matrix A, and $\bar{A}$ denotes the operation of complex conjugate of A.

Example 5.1

Let $N = 8$ and let $\{F_p\}$ be the eight-point DFT

$$F_p = (\mathcal{F}_8 \circ f)_p = \sum_{n=0}^{7} f_n W^{np}, \quad p = 0 : 7, \tag{5.3}$$

where $W = W_8 = \exp(-2\pi i/8)$. It transforms the discrete signal $f = \{f_0, f_1, \ldots, f_7\}$ into the sequence $F = \{F_0, F_1, \ldots, F_7\}$. The 8-point DFT of the signal can be written in matrix form as

$$\begin{bmatrix} F_0 \\ F_1 \\ F_2 \\ F_3 \\ F_4 \\ F_5 \\ F_6 \\ F_7 \end{bmatrix} = [\mathcal{F}_8][f] = \begin{bmatrix} 1 & 1 & 1 & 1 & 1 & 1 & 1 & 1 \\ 1 & W^1 & -i & W^3 & -1 & W^5 & i & W^7 \\ 1 & -i & -1 & i & 1 & -i & -1 & i \\ 1 & W^3 & i & W^1 & -1 & W^7 & -i & W^5 \\ 1 & -1 & 1 & -1 & 1 & -1 & 1 & -1 \\ 1 & W^5 & -i & W^7 & -1 & W^1 & i & W^3 \\ 1 & i & -1 & -i & 1 & i & -1 & -i \\ 1 & W^7 & i & W^5 & -1 & W^3 & -i & W^1 \end{bmatrix} \begin{bmatrix} f_0 \\ f_1 \\ f_2 \\ f_3 \\ f_4 \\ f_5 \\ f_6 \\ f_7 \end{bmatrix}, \tag{5.4}$$

where the coefficients of rotation are

$$W^0 = 1,\ W^1 = 0.7071 - 0.7071i,\ W^2 = -i,\ W^3 = -0.7071 - 0.7071i,$$
$$W^4 = -1,\ W^5 = -0.7071 + 0.7071i,\ W^6 = i,\ W^7 = 0.7071 + 0.7071i.$$

Four coefficients W^1, W^3, W^5, and W^7 are not trivial for multiplication, i.e., $\neq \pm 1$ and $\neq \pm i$. All coefficients W^t, $t = 0:7$, are on the unit circle at the points where the circle is divided by eight equal sectors. The coefficients W^1 and W^7, W^2 and W^6, and W^3 and W^5 are complex conjugate to each other, i.e., $W^t = \overline{W^{8-t}}$, for $t = 1, 2, 3$.

The inverse eight-point DFT is calculated by

$$f_n = (\mathcal{F}_8^{-1} \circ F)_n = \frac{1}{8}\sum_{p=0}^{7} F_p W^{-np}, \quad n = 0:7,$$

and it can be written in matrix form as

$$\begin{bmatrix} f_0 \\ f_1 \\ f_2 \\ f_3 \\ f_4 \\ f_5 \\ f_6 \\ f_7 \end{bmatrix} = [\mathcal{F}_8]^{-1}[F] = \frac{1}{8}\begin{bmatrix} 1 & 1 & 1 & 1 & 1 & 1 & 1 & 1 \\ 1 & W^{-1} & i & W^{-3} & -1 & W^{-5} & -i & W^{-7} \\ 1 & i & -1 & -i & 1 & i & -1 & -i \\ 1 & W^{-3} & -i & W^{-1} & -1 & W^{-7} & i & W^{-5} \\ 1 & -1 & 1 & -1 & 1 & -1 & 1 & -1 \\ 1 & W^{-5} & i & W^{-7} & -1 & W^{-1} & -i & W^{-3} \\ 1 & -i & -1 & i & 1 & -i & -1 & i \\ 1 & W^{-7} & -i & W^{-5} & -1 & W^{-3} & i & W^{-1} \end{bmatrix} \begin{bmatrix} F_0 \\ F_1 \\ F_2 \\ F_3 \\ F_4 \\ F_5 \\ F_6 \\ F_7 \end{bmatrix}.$$

The N-point DFT is periodic, and N is called the order of the transform,

$$F_{N+p} = F_{kN+p} = F_{p-N} = F_{p-kN} = F_p,$$

for any value of p in the integer interval $[0, N-1]$ and integer k. Therefore, the transform is calculated only in the first period, i.e., for

$p = 0, 1, 2, \ldots, (N-1)$. The sum in Eq. (5.2) is also periodic with the same period, N, i.e., this equation defines the periodically extension of the original signal f_n,

$$f_{n+N} = f_{n+kN} = f_{n-N} = f_{n-kN} = f_n,$$

for integers $n = 0, 1, 2, \ldots, (N-1)$ and integers k.

5.1.1 Fast DFT, or FFT

Different methods of calculation exist for the DFT, and one of them is the decimation-in-time algorithm, which was introduced by Cooley-Tukey [19] and is widely used in applications. This algorithm is also called the radix-2 algorithm when the order of the transforms is a power of 2, i.e., $N = 2^r, r > 1$.

Radix-2 FFT is based on the following properties of the exponential function $W^t = W_N^t$:

$$W^{(2n+1)p} = W_N^{(2n+1)p} = W_N^{2np} W_N^p = W_{N/2}^{np} W_N^p = W_{N/2}^{np} W^p,$$

$$W_{N/2}^{n(p+N/2)} = W_{N/2}^{np}, \quad W_N^{p+N/2} = -W_N^p,$$

and $W^{(2n)p} = W_N^{(2n)p} = W_{N/2}^{np}$.

Therefore, by separating the signal components with even and odd numbers, we can split the sum in Eq. (5.1) into two sums as follows:

$$\begin{aligned} F_p &= \sum_{n=0}^{N/2-1} \left[f_{2n} W^{2np} + f_{2n+1} W^{(2n+1)p} \right] = \sum_{n=0}^{N/2-1} \left[f_{2n} W_{N/2}^{np} + f_{2n+1} W_{N/2}^{np} W^p \right] \\ &= \sum_{n=0}^{N/2-1} x_n W_{N/2}^{np} + \left[\sum_{n=0}^{N/2-1} y_n W_{N/2}^{np} \right] W^p, \end{aligned}$$

where $x_n = f_{2n}$ and $y_n = f_{2n+1}$, $n = 0 : (N/2 - 1)$. Here, two sums describe the $N/2$-point DFTs of the new signals x_n and y_n of length $N/2$ each, which we denote by X_p and Y_p, respectively.

Therefore, we can express the values of the N-point DFT of the signal f_n by the values of two $N/2$-point DFTs of signals x_n and y_n, as follows:

$$F_p = X_p + Y_p W^p, \tag{5.5}$$

$$F_{p+N/2} = X_p - Y_p W^p, \tag{5.6}$$

where $p = 0 : (N/2 - 1)$. Thus, the N-point is calculated by a combination of two $N/2$-point DFTs.

Example 5.2

The calculation of the DFT for $N = 8$ is given in Fig. 5.1. The 8-point DFT first is split by two 4-point DFTs and then four operations of "butterfly" are used to combine the results of these 4-point DFTs, as shown in the figure.

The algorithm of splitting the 8-point DFT by two 4-point DFTs can be written in matrix form as follows:

$$
[\mathcal{F}_8] = \begin{bmatrix}
1 & 0 & 0 & 0 & 1 & 0 & 0 & 0 \\
0 & 1 & 0 & 0 & 0 & W^1 & 0 & 0 \\
0 & 0 & 1 & 0 & 0 & 0 & W^2 & 0 \\
0 & 0 & 0 & 1 & 0 & 0 & 0 & W^3 \\
1 & 0 & 0 & 0 & -1 & 0 & 0 & 0 \\
0 & 1 & 0 & 0 & 0 & -W^1 & 0 & 0 \\
0 & 0 & 1 & 0 & 0 & 0 & -W^2 & 0 \\
0 & 0 & 0 & 1 & 0 & 0 & 0 & -W^3
\end{bmatrix}
$$

$$
\begin{bmatrix}
 & & & & 0 & 0 & 0 & 0 \\
 & & & & 0 & 0 & 0 & 0 \\
 & [\mathcal{F}_4] & & & 0 & 0 & 0 & 0 \\
 & & & & 0 & 0 & 0 & 0 \\
0 & 0 & 0 & 0 & & & & \\
0 & 0 & 0 & 0 & & [\mathcal{F}_4] & & \\
0 & 0 & 0 & 0 & & & & \\
0 & 0 & 0 & 0 & & & &
\end{bmatrix}
\begin{bmatrix}
1 & 0 & 0 & 0 & 0 & 0 & 0 & 0 \\
0 & 0 & 1 & 0 & 0 & 0 & 0 & 0 \\
0 & 0 & 0 & 0 & 1 & 0 & 0 & 0 \\
0 & 0 & 0 & 0 & 0 & 0 & 1 & 0 \\
0 & 1 & 0 & 0 & 0 & 0 & 0 & 0 \\
0 & 0 & 0 & 1 & 0 & 0 & 0 & 0 \\
0 & 0 & 0 & 0 & 0 & 1 & 0 & 0 \\
0 & 0 & 0 & 0 & 0 & 0 & 0 & 1
\end{bmatrix}.
$$

The first matrix (from the right) is the matrix of the permutation, $[P_8]$, of the signal f_n by the even- and odd-indexed components. The second matrix is the Kronecker sum of two matrices of the 4-point DFTs, and the last matrix, which we call $[4B_{2\times 2}]$, describes four operations of butterfly,

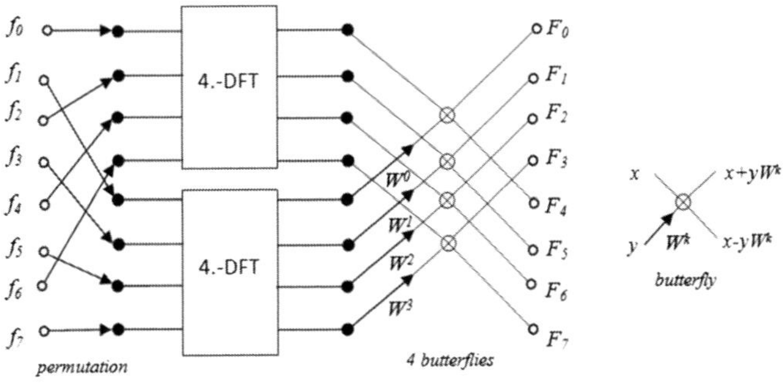

Figure 5.1 Signal flow-graph of the 8-point DFT splitting by two 4-point DFTs.

$$[\mathcal{F}_8] = [4B_{2\times 2}]([\mathcal{F}_4] \oplus [\mathcal{F}_4])[P_8].$$

Each of these 4-point DFTs can similarly be split by two 2-point DFTs. The matrix of the 4-point DFT can be written as

$$[\mathcal{F}_4] = \begin{bmatrix} 1 & 0 & 1 & 0 \\ 0 & 1 & 0 & W_4^1 \\ 1 & 0 & -1 & 0 \\ 0 & 1 & 0 & -W_4^1 \end{bmatrix} \begin{bmatrix} & [\mathcal{F}_2] & 0 & 0 \\ & & 0 & 0 \\ 0 & 0 & [\mathcal{F}_2] & \\ 0 & 0 & & \end{bmatrix} \begin{bmatrix} 1 & 0 & 0 & 0 \\ 0 & 0 & 1 & 0 \\ 0 & 1 & 0 & 0 \\ 0 & 0 & 0 & 1 \end{bmatrix},$$

where $W_4^1 = -i$. Denoting by $[P_4]$ the matrix of permutation and by $[2B_{2\times 2}]$ the matrix with two butterflies, we can write the matrix of the 4-point DFT as

$$[\mathcal{F}_4] = [2B_{2\times 2}]([\mathcal{F}_2] \oplus [\mathcal{F}_2])[P_4].$$

Thus, the 8-point DFT can be calculated by four 2-point DFTs, which represent themselves as simple butterflies $(x, y) \to (x + y, x - y)$ without a multiplication, i.e.,

$$[\mathcal{F}_2] = \begin{bmatrix} 1 & 1 \\ 1 & -1 \end{bmatrix}.$$

5.1.2 Inverse FFT, or inverse radix-2 FFT

It is clear that if we redirect the signal-flow graph in the decimation-in-time algorithm and change all butterflies by their inverse ones, we obtain the fast algorithm for the inverse N-point DFT.

The butterfly $B_{2\times 2}(W^k)$ and its inverse butterfly $B_{2\times 2}^{-1}(W^k)$ can be described by the following matrices, inverse to each other:

$$[B_{2\times 2}(W^k)] = \begin{bmatrix} 1 & W^k \\ 1 & -W^k \end{bmatrix},$$

$$[B_{2\times 2}^{-1}(W^k)] = [B_{2\times 2}(W^k)]^{-1} = \frac{1}{2}\begin{bmatrix} 1 & 1 \\ W^{-k} & -W^{-k} \end{bmatrix}.$$

Thus, the result of the inverse butterfly is

$$[B_{2\times 2}^{-1}(W^k)]\begin{bmatrix} x \\ y \end{bmatrix} = \frac{1}{2}\begin{bmatrix} 1 & 1 \\ W^{-k} & -W^{-k} \end{bmatrix}\begin{bmatrix} x \\ y \end{bmatrix} = \begin{bmatrix} \frac{x+y}{2} \\ \frac{x-y}{2} W^{-k} \end{bmatrix}.$$

Figure 5.2 illustrates the butterfly $B_{2\times 2}(W^k)$ and its inverse butterfly $B_{2\times 2}^{-1}(W^k)$.

Figure 5.3 shows the redirected signal flow-graph of the 8-point inverse DFT by two 4-point inverse DFTs. All four butterflies in the signal

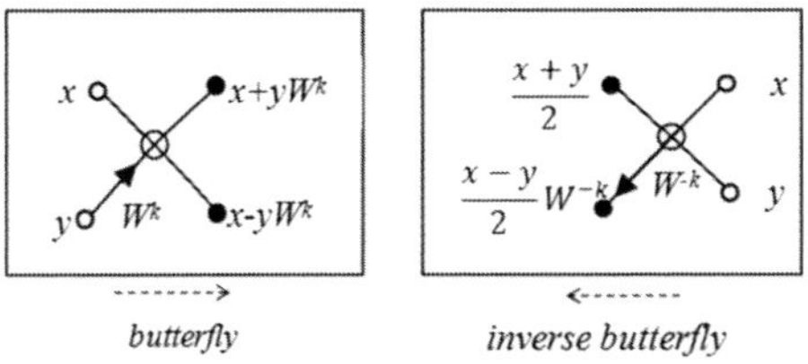

Figure 5.2 The direct and inverse butterflies 2×2.

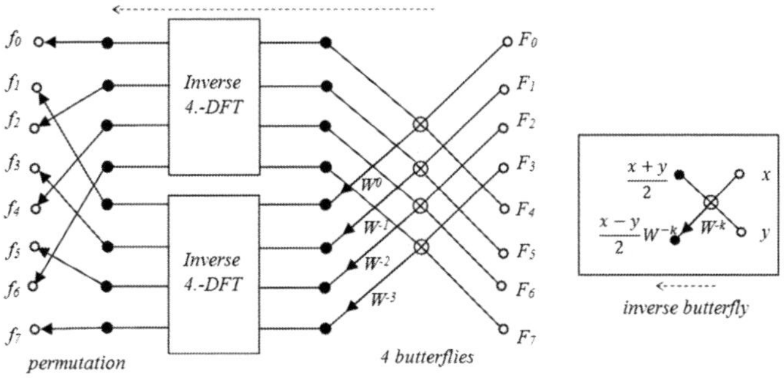

Figure 5.3 Signal flow-graph of the 8-point inverse DFT by two 4-point inverse DFTs.

flow-graph of the radix-2 algorithm were substituted by the inverse ones, and two 4-point DFTs by the inverse DFTs. The calculations are performed from the right to the left.

5.1.3 Decimation-in-frequency FFT

The N-point direct and inverse discrete Fourier transforms have similar formulas [Eqs. (5.1) and (5.2)], except the normalizing coefficient $1/N$. By changing F_p by f_p and the time point n by $-n$, we obtain the following inverse DFT:

$$\sum_{n=0}^{N-1} F_p W^{-np} \rightarrow \sum_{n=0}^{N-1} f_p W^{-np} \rightarrow \sum_{n=0}^{N-1} f_p W^{-(-n)p} = \sum_{n=0}^{N-1} f_p W^{np} = F_p.$$

It means that we can use the fast algorithm of the inverse DFT to calculate the direct DFT. For instance, in the above signal flow-graph of the 8-point DFT, which was obtained by redirecting the decimation-in-time algorithm, we can use the new butterflies, which are similar to the inverse butterflies $B_{2\times 2}^{-1}(W^k)$ but without the factor of $1/2$ and with W^k instead of W^{-k}, and then use two four-point DFTs instead of the inverse four-point DFTs. The signal flow-graph of the eight-point DFT obtained in such a way is shown in Fig. 5.4.

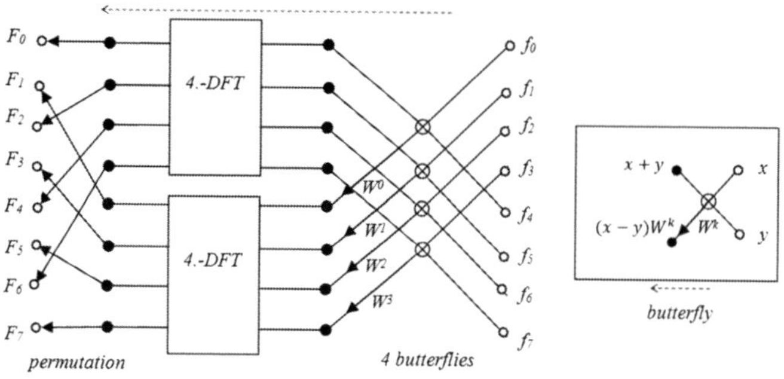

Figure 5.4 Signal flow-graph of the 8-point DFT by two 4-point DFTs.

For $N = 8$ case, the matrix $[\mathcal{F}_8]$ can be written as follows:

$$
\begin{bmatrix}
1&0&0&0&0&0&0&0\\
0&0&0&0&1&0&0&0\\
0&1&0&0&0&0&0&0\\
0&0&0&0&0&1&0&0\\
0&0&1&0&0&0&0&0\\
0&0&0&0&0&0&1&0\\
0&0&0&1&0&0&0&0\\
0&0&0&0&0&0&0&1
\end{bmatrix}
\begin{bmatrix}
 & & & & 0&0&0&0\\
 & & & & 0&0&0&0\\
 & [\mathcal{F}_4] & & & 0&0&0&0\\
 & & & & 0&0&0&0\\
0&0&0&0& & & & \\
0&0&0&0& & [\mathcal{F}_4] & & \\
0&0&0&0& & & & \\
0&0&0&0& & & &
\end{bmatrix}
\begin{bmatrix}
1&0&0&0&1&0&0&0\\
0&1&0&0&0&1&0&0\\
0&0&1&0&0&0&1&0\\
0&0&0&1&0&0&0&1\\
1&0&0&0&-1&0&0&0\\
0&W^1&0&0&0&-W^1&0&0\\
0&0&W^2&0&0&0&-W^2&0\\
0&0&0&W^3&0&0&0&-W^3
\end{bmatrix}.
$$

The similar calculation of the 4-point DFTs results in the following representation of the matrix of the transformation:

$$
[\mathcal{F}_4] =
\begin{bmatrix}
1&0&0&0\\
0&0&1&0\\
0&1&0&0\\
0&0&0&1
\end{bmatrix}
\begin{bmatrix}
 & & 0&0\\
 [\mathcal{F}_2] & & 0&0\\
0&0& [\mathcal{F}_2] & \\
0&0& &
\end{bmatrix}
\begin{bmatrix}
1&0&1&0\\
0&1&0&1\\
1&0&-1&0\\
0&W_4^1&0&-W_4^1
\end{bmatrix}.
$$

This method is known as the decimation-in-frequency algorithm of the FFT. In the first step of this algorithm, the splitting of the N-point DFT by two $N/2$-point DFTs performed by $N/2$ butterflies [which we call $B_{2\times 2}(W^n)$] as follows:

$$[B_{2\times 2}(W^n)]\begin{bmatrix} f_n \\ f_{n+N/2} \end{bmatrix} = \begin{bmatrix} 1 & 1 \\ W^n & -W^n \end{bmatrix}\begin{bmatrix} f_n \\ f_{n+N/2} \end{bmatrix} = \begin{bmatrix} f_n + f_{n+N/2} \\ (f_n - f_{n+N/2})W^n \end{bmatrix},$$

where $n = 0, 1, \ldots, (N/2-1)$. Therefore, the decimation-in-frequency algorithm is based on the following property:

$$\begin{aligned} F_{2p} &= \sum_{n=0}^{N/2-1} (f_n + f_{n+N/2}) W^p_{N/2}, \\ F_{2p+1} &= \sum_{n=0}^{N/2-1} [(f_n - f_{n+N/2}) W^n_N] W^p_{N/2}, \end{aligned} \tag{5.7}$$

when $p = 0, 1, \ldots, (N/2-1)$.

5.1.4 Paired FFT

The decimation-in-time and decimation-in-frequency algorithms reduce the calculation of the N-point DFT, when N is a power of two, to calculation of two $N/2$-point DFTs,

$$\mathcal{F}_N \to \{\mathcal{F}_{N/2}, \mathcal{F}_{N/2}\},$$

and then the obtained $N/2$-point DFTs are calculated by $N/4$-point DFTs in a similar way,

$$\{\mathcal{F}_{N/2}, \mathcal{F}_{N/2}\} \to \{\{\mathcal{F}_{N/4}, \mathcal{F}_{N/4}\}, \{\mathcal{F}_{N/4}, \mathcal{F}_{N/4}\}\},$$

and so on, until the calculation will be reduced to $N/2$ two-point DFTs.

In the paired FFT [27, 28], the splitting of the N-point DFT is not reduced to two $N/2$-point DFT, but to the $N/2$-,$N/4$-,$N/8$-, …,2-,1-,1-point DFTs,

$$\mathcal{F}_N \to \{\mathcal{F}_{N/2}, \mathcal{F}_{N/4}, \mathcal{F}_{N/8}, \mathcal{F}_{N/16}, \ldots, \mathcal{F}_4, \mathcal{F}_2, \mathcal{F}_1, \mathcal{F}_1\},$$

and then each of the obtained $N/2^k$-point DFT is split in a similar way. Here $k = 1, 2, \ldots, (\log_2 N - 2)$.

Example 5.3

The 8-point discrete Fourier transform can first be reduced to the set $\mathcal{F}_8 \to \{\mathcal{F}_4, \mathcal{F}_2, \mathcal{F}_1, \mathcal{F}_1\}$, with calculation of transform at the corresponding four subsets frequency-points $\{1, 3, 5, 7\}, \{2, 6\}, \{4\}$, and $\{0\}$. The 4-point

DFT can be calculated by the 2-,1-,1-point DFTs, $\mathcal{F}_4 \to \{\mathcal{F}_2, \mathcal{F}_1, \mathcal{F}_1\}$, with calculation of transform at the corresponding three subsets of frequency-points $\{1, 5\}, \{3\}$, and $\{7\}$. The signal-flow graph for computing the 8-point DFT is shown in Fig. 5.5. The splitting of the transform is performed by the discrete paired transforms (DPT), which for the $N = 8, 4$, and 2 have the following matrices:

$$[\chi'_8] = \begin{bmatrix} 1 & 0 & 0 & 0 & -1 & 0 & 0 & 0 \\ 0 & 1 & 0 & 0 & 0 & -1 & 0 & 0 \\ 0 & 0 & 1 & 0 & 0 & 0 & -1 & 0 \\ 0 & 0 & 0 & 1 & 0 & 0 & 0 & -1 \\ 1 & 0 & -1 & 0 & 1 & 0 & -1 & 0 \\ 0 & 1 & 0 & -1 & 0 & 1 & 0 & -1 \\ 1 & -1 & 1 & -1 & 1 & -1 & 1 & -1 \\ 1 & 1 & 1 & 1 & 1 & 1 & 1 & 1 \end{bmatrix},$$

and

$$[\chi'_4] = \begin{bmatrix} 1 & 0 & -1 & 0 \\ 0 & 1 & 0 & -1 \\ 1 & -1 & 1 & -1 \\ 1 & 1 & 1 & 1 \end{bmatrix}, \quad [\chi'_2] = \begin{bmatrix} 1 & -1 \\ 1 & 1 \end{bmatrix}.$$

Two components of the DFT, F_0 and F_4, are the outputs of the 8-point discrete paired transform (DPT). Four other components F_1, F_2, F_5, and F_6 are obtained form the 4- and 2-point DPTs. On the last stage of the algorithm, the components of the DFT, F_3 and F_7, are calculated by the 2-point DPT.

The components of the DFT are calculated in the order (from the top) $F_7, F_3, F_5, F_1, F_6, F_2, F_4$, and F_0. The bit-reversal transform

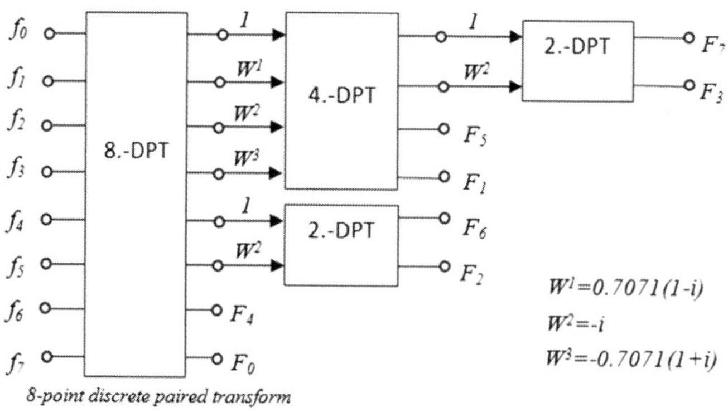

Figure 5.5 The signal-flow graph of the 8-point DFT.

$$(1,1,1) \to (1,1,1), (0,1,1) \to (1,1,0), (1,0,1) \to (1,0,1), (0,0,1) \to (1,0,0)$$
$$(1,1,0) \to (0,1,1), (0,1,0) \to (0,1,0), (1,0,0) \to (0,0,1), (0,0,0) \to (0,0,0),$$

reorder these outputs as $F_7, F_6, F_5, F_4, F_3, F_2, F_1$, and F_0.

The paired transform is the frequency-time representation of the signal f_n by the splitting-signals of lengths $N/2, N/4, N/8, \ldots, 2, 1, 1$,

$$\chi'_N : \{f_n\} \to \left\{\{f'_{2^k,2^k t}; t = 0, 1, 2, \ldots, N/2^k - 1\}; k = 0, 1, \ldots, r-1\}, f'_{0,0}\right\},$$

where the components of the signals are calculated by [28]

$$f'_{p,t} = \sum\{f_n;\, np = t \bmod N\} - \sum\{f_n;\, np = (t + N/2) \bmod N\},$$

for $t = 0, 1, 2, \ldots, (N/2 - 1)$. The N-point discrete paired transform is fast and requires only $(2N - 2)$ operations of addition and subtraction [30].

The paired FFT is based on the following property of the DFT:

$$F_{(2m+1)p \bmod N} = \sum_{t=0}^{N/2-1} f'_{p,t} W^{(2m+1)t} = \sum_{t=0}^{N/2-1} (f'_{p,t} W^t) W^{mt}_{N/2}, \tag{5.8}$$

when $m = 0 : (N/2 - 1)$.

The disjoint subsets of frequency-points $T'_0 = \{0\}$ and

$$T'_p = \{(2m+1)p \bmod N;\, m = 0 : N/(2p) - p\},$$

for $p = 2^k$, $k = 0, 1, \ldots, \log_2 N - 2$, cover the entire set of frequency-points $X_N = \{0, 1, 2, \ldots (N-1)\}$. Therefore, the matrix of the N-point DFT can be represented as

$$[\mathcal{F}_N] = \left(\bigoplus_{k=1}^{r} [\mathcal{F}_{N/2^k}] \oplus 1\right) [\bar{W}][\chi'_N], \tag{5.9}$$

where the diagonal matrix

$$[\bar{W}_N] = \bigoplus_{k=1}^{r} \operatorname{diag}\{1, W^1_{N/2^{k-1}}, W^2_{N/2^{k-1}}, \ldots, W^{N/2^k - 1}_{N/2^{k-1}}\} \oplus 1. \tag{5.10}$$

The number of operations of multiplication and addition required to compute the 2^r-point DFT equal

$$m_{2^r} = 2^{r-1}(r-3) + 2, \quad a_{2^r} = 2^r 6 - r^2 - 3r - 6, \quad r > 2. \tag{5.11}$$

The 16-point paired DFT uses $m_{16} = 10$ operations of complex multiplication.

Example 5.4

The matrix of the 8-point DFT in the paired algorithm can be written as

$$[\mathcal{F}_8] = ([\mathcal{F}_4] \oplus [\mathcal{F}_2] \oplus 1 \oplus 1)[\bar{W}_8][\chi'_8],$$

where the diagonal matrix is

$$\begin{aligned}[\bar{W}_8] &= \operatorname{diag}\{1, W_8^1, W_8^2, W_8^3, 1, W_4^1, 1, 1\} \\ &= \operatorname{diag}\{1, 0.7071(1-i), -i, -0.7071(1+i), 1, -i, 1, 1\}.\end{aligned}$$

Therefore, we can write the matrix $[\mathcal{F}_8]$ as

$$[\mathcal{F}_8] = \begin{bmatrix} & & & & 0 & 0 & 0 & 0 \\ & & & & 0 & 0 & 0 & 0 \\ & & [\mathcal{F}_4] & & 0 & 0 & 0 & 0 \\ & & & & 0 & 0 & 0 & 0 \\ 0 & 0 & 0 & 0 & 1 & -1 & 0 & 0 \\ 0 & 0 & 0 & 0 & 1 & 1 & 0 & 0 \\ 0 & 0 & 0 & 0 & 0 & 0 & 1 & 0 \\ 0 & 0 & 0 & 0 & 0 & 0 & 0 & 1 \end{bmatrix} \operatorname{diag} \begin{bmatrix} 1 \\ W^1 \\ -i \\ W^3 \\ 1 \\ -i \\ 1 \\ 1 \end{bmatrix}$$
$$\begin{bmatrix} 1 & 0 & 0 & 0 & -1 & 0 & 0 & 0 \\ 0 & 1 & 0 & 0 & 0 & -1 & 0 & 0 \\ 0 & 0 & 1 & 0 & 0 & 0 & -1 & 0 \\ 0 & 0 & 0 & 1 & 0 & 0 & 0 & -1 \\ 1 & 0 & -1 & 0 & 1 & 0 & -1 & 0 \\ 0 & 1 & 0 & -1 & 0 & 1 & 0 & -1 \\ 1 & -1 & 1 & -1 & 1 & -1 & 1 & 1 \\ 1 & 1 & 1 & 1 & 1 & 1 & 1 & 1 \end{bmatrix}.$$

The matrix $[\mathcal{F}_4]$ can similarly be decomposed by the 4-point DPT,

$$[\mathcal{F}_4] = \begin{bmatrix} 1 & -1 & 0 & 0 \\ 1 & 1 & 0 & 0 \\ 0 & 0 & 1 & 0 \\ 0 & 0 & 0 & 1 \end{bmatrix} \operatorname{diag} \begin{bmatrix} 1 \\ -i \\ 1 \\ 1 \end{bmatrix} \begin{bmatrix} 1 & 0 & -1 & 0 \\ 0 & 1 & 0 & -1 \\ 1 & -1 & 1 & -1 \\ 1 & 1 & 1 & 1 \end{bmatrix},$$

$$\left([\mathcal{F}_2] = \begin{bmatrix} 1 & -1 \\ 1 & 1 \end{bmatrix}\right).$$

Two multiplications by coefficients W and W^3 are required for calculating the 8-point DFT, i.e., $m_8 = 2$.

The paired transform describes the frequency-time representation of the signal and splits the mathematical structure of the N-point DFT by the DFTs of orders $N/2, N/4, \ldots, 2, 1$ (for more detail, see Refs. [30, 31]).

Figure 5.6 shows the signal of length 512 in part (a). The signal is the one period of the function

$$x(t) = 1 + 10\cos(2\omega_0 t + 1) + 4\sin(5\omega_0 t - 2) + \cos(10\omega_0 t), \quad t = 0:511,$$

where the fundamental frequency $\omega_0 = 2\pi/512$ (rad/s). The 512-point DPT of this signal is shown in part (b). The last component of the DPT is the sum of the signal and equals 512.

Another example with the signal of length 256 is shown in Fig. 5.7(a), along with the paired transform of the signal in part (b). The mean of the signal is zero, therefore the last component of the DPT is zero, and the maximum of the DPT at the point $N - 1 = 255$ and equals 174.4766. The first half of the

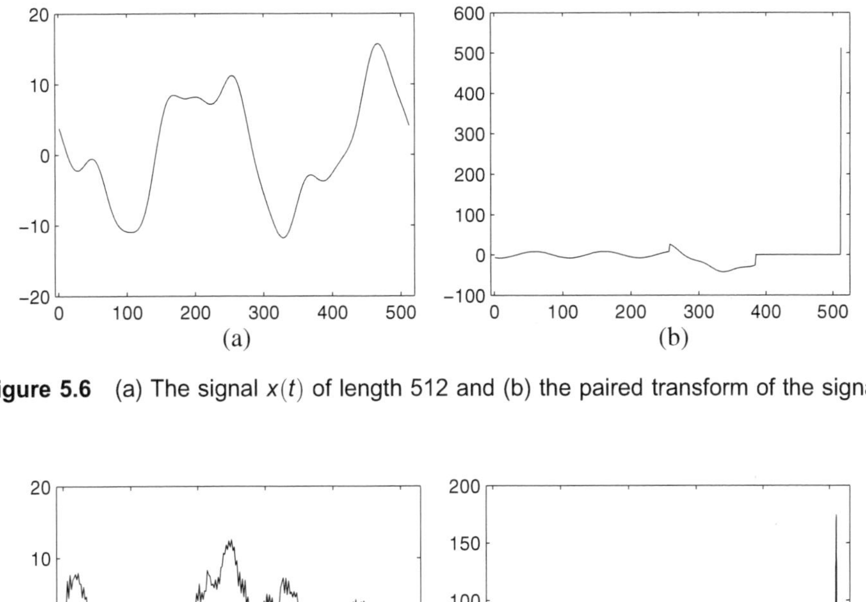

Figure 5.6 (a) The signal $x(t)$ of length 512 and (b) the paired transform of the signal.

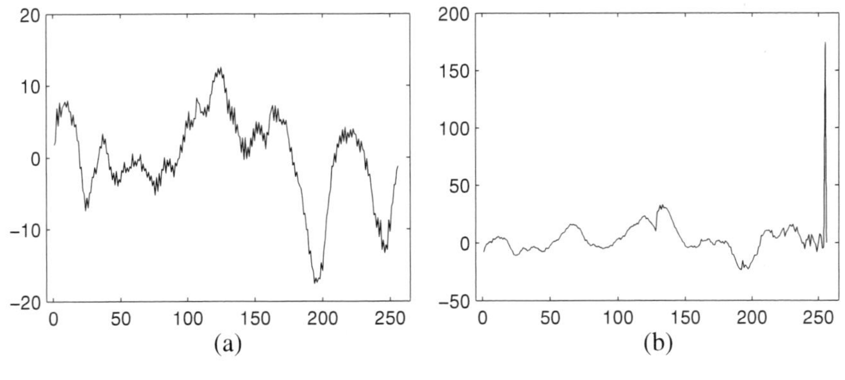

Figure 5.7 (a) The signal of length 256 and (b) the paired transform of the signal.

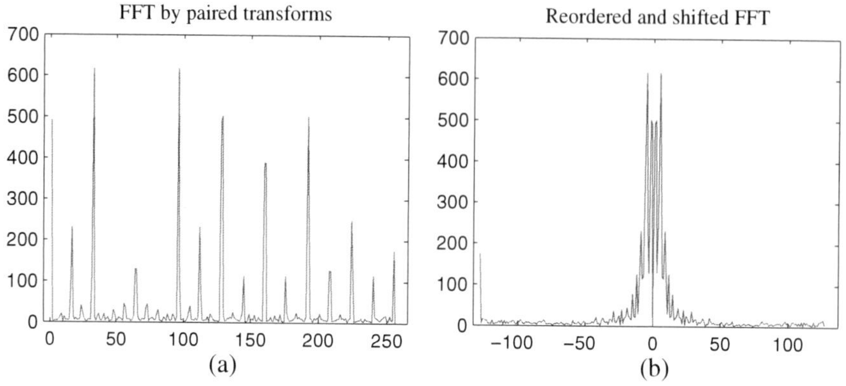

Figure 5.8 (a) The splitting of the 256-point DFT and (b) the permutation of the DFT.

transform is the splitting-signal $\{f'_{1,t}; t = 0 : 127\}$, the next 64 components together compose the splitting-signal $\{f'_{2,2t}; t = 0 : 63\}$, and so on.

The fast paired DFT requires a permutation after the calculation, which is known as the bit-reversal permutation. Figure 5.8 shows the result of calculation of the 256-point DFT of the signal of Fig. 5.7, before and after the permutation in the paired algorithm.

We include the simple demo code for calculating the 512-point paired FFT of the signal shown in Figure 5.6. The script "fft_art" is for the code calculating the paired FFT for $N = 2^r$, $r > 2$. The function "fastpaired.m" calculates the paired transform, and the permutation is performed by the function "fastpermut.m" as shown below.

```
  N=512;
  w0=2*pi/N; % 0.0123
  t=0:N-1;
  signal1=1+10*cos(2*w0*t+1) + 4*sin(5*w0*t-2)
  + 1*cos(10*w0*t);
  FTofsignal=fft_art(signal1,N);
  FTofsignal=fastpermut(FTofsignal);
%--------------------------------------------------
% 1-D fast direct Fourier transform 'fft_art(A,N)'
% Artyom M. Grigoryan, 1999
function B1=fft_art(A1,ND)
%1. N/2 Exponential coefficients for the diagonal matrices
[W]
  N2=ND/2;
  T=-1i*pi/N2; % or +1i*pi/N2;
  W=zeros(1,N2);  J=0;
  for I=0:N2-1
```

```
    W(I+1)=exp(J);  J=J+T;
  end
%2. Matrix decomposition by the paired transforms
(fastpaired.m)
  MD=log2(ND); NK1=ND; MD1=MD;
  B1=A1; IW=1;
  for I=1:MD
      LK=NK1/2;
      for J1=1:2*NK1:ND
          N_1=NK1+J1-1;
          B2=zeros(1,NK1); B2=B1(J1:N_1);
          B1(J1:N_1)=fastpaired(B2);
          LK1=LK; II=IW; JJ=J1;
          for I2=1:MD1-1
              J2=1;
              for J3=1:LK1
                  B1(JJ)=B1(JJ)*W(J2);
                  JJ=JJ+1; J2=J2+II;
              end
              LK1=LK1/2; II=II*2;
          end
      end
  NK1=LK; MD1=MD1-1; IW=IW*2;
  end
%-------------------------------------------------------
function output=fastpaired(f)
  N=length(f);
  n2=N/2; a=1; n22=n2; c=n22;
  F=f;
  while (n22 >= 1)
    for i=a:c
        i2=i+n22;
        f1=f(i); f2=f(i2);
        F(i)=f1-f2; F(i2)=f1+f2;
    end
    a=c+1; n22=n22/2; c=c+n22;
    f=F;
  end
  output=f;
%-------------------------------------------------------
function output=fastpermut(input)
  N=length(input);
  m=log2(N);
```

```
  output = input;
  for a = 1:N
    output(mod(binvec2dec(fliplr(dec2binvec(a - 1,m)))
    + 1,N) + 1) = input(a);
end
```

5.2 1-D Right-Side QDFT

In this section, we describe briefly the algorithm for calculating the one-dimensional quaternion discrete Fourier transform (1-D QDFT) [17]. Let $q_n = (a_n, b_n, c_n, d_n) = a_n + ib_n + jc_n + kd_n$ be the quaternion signal of length N. The right-side 1-D QDFT is defined as

$$Q_p = Q_1(p) + iQ_i(p) + jQ_j(p) + kQ_k(p) = \sum_{n=0}^{N-1} q_n W_\mu^{np}, \quad p = 0 : (N-1),$$

where μ is a pure unit quaternion number $\mu = m_1 i + m_2 j + m_3 k$, $\mu^2 = -1$. The kernel of the transformation equals

$$W_\mu = W_{N;\mu} = \exp(-\mu 2\pi/N) = \cos(2\pi/N) - \mu \sin(2\pi/N).$$

The direct calculations show that the multiplication of quaternion numbers $q_n = a_n + ib_n + jc_n + kd_n$ and $\exp(-\mu\phi)$ can be written as

$$\begin{aligned} q_n \exp(-\mu\phi) &= (a_n + ib_n + jc_n + kd_n)[\cos(\phi) - (m_1 i + m_2 j + m_3 k)\sin(\phi)] \\ &= a_n \cos(\phi) + (b_n m_1 + c_n m_2 + d_n m_3)\sin(\phi) \\ &\quad + i\,[\,b_n \cos(\phi) - (a_n m_1 - d_n m_2 + c_n m_3)\sin(\phi)\,] \\ &\quad + j\,[\,c_n \cos(\phi) - (d_n m_1 + a_n m_2 - b_n m_3)\sin(\phi)\,] \\ &\quad + k\,[\,d_n \cos(\phi) - (-c_n m_1 + b_n m_2 + a_n m_3)\sin(\phi)\,]. \end{aligned} \tag{5.12}$$

This equation is simple to implement in calculation of 1-D QDFT. Let φ_{np} be the angle $(2\pi/N)np$. For a real signal x_n of length N, we define the following cosine and sine transforms being the real and imaginary parts of the 1-D DFT of this signal:

$$C_x(p) = \sum_{n=0}^{N-1} x_n \cos(\varphi_{np}), \quad S_x(p) = \sum_{n=0}^{N-1} x_n \sin(\varphi_{np}), \tag{5.13}$$

where $p = 0 : (N-1)$.

We denote by $C(x)$ and $S(x)$ the vectors of coefficients $[C_x(0), C_x(1), C_x(2) \ldots, C_x(N-1)]$ and $[S_x(0), S_x(1), S_x(2) \ldots, S_x(N-1)]$, respectively.

The four vectors of coefficients $Q_1(p)$, $Q_i(p)$, $Q_j(p)$, and $Q_k(p)$ are also defined as Q_1, Q_i, Q_j, and Q_k. According to Eq. (5.12), the 1-D QDFT can be written as

$$\begin{aligned} Q_1 + iQ_i + jQ_j + kQ_k = {} & C(a) + m_1 S(b) + m_2 S(c) + m_3 S(d) \\ & + i[C(b) - m_1 S(a) + m_2 S(d) - m_3 S(c)] \\ & + j[C(c) - m_1 S(d) - m_2 S(a) + m_3 S(b)] \\ & + k[C(d) + m_1 S(c) - m_2 S(b) - m_3 S(a)] \end{aligned} \tag{5.14}$$

and

$$\begin{aligned} Q_1 + iQ_i + jQ_j + kQ_k = {} & C(a) + m_1 S(b) + m_2 S(c) + m_3 S(d) \\ & + i[-m_1 S(a) + C(b) - m_3 S(c) + m_2 S(d)] \\ & + j[-m_2 S(a) + m_3 S(b) + C(c) - m_1 S(d)] \\ & + k[-m_3 S(a) - m_2 S(b) + m_1 S(c) + C(d)]. \end{aligned} \tag{5.15}$$

Therefore, if we denote the N-point 1-D DFTs of the parts a_n, b_n, c_n, and d_n of the quaternion signal q_n by A_p, B_p, C_p, and D_p, respectively, we can calculate the 1-D QDFT as

$$Q_p = Q_1(p) + iQ_i(p) + jQ_j(p) + kQ_k(p), \quad p = 0 : (N-1),$$

with components calculated as

$$\begin{aligned} Q_1(p) &= \Re(A_p) + m_1 \Im(B_p) + m_2 \Im(C_p) + m_3 \Im(D_p), \\ Q_i(p) &= -m_1 \Im(A_p) + \Re(B_p) - m_3 \Im(C_p) + m_2 \Im(D_p), \\ Q_j(p) &= -m_2 \Im(A_p) + m_3 \Im(B_p) + \Re(C_p) - m_1 \Im(D_p), \\ Q_k(p) &= -m_3 \Im(A_p) - m_2 \Im(B_p) + m_1 \Im(C_p) + \Re(D_p). \end{aligned}$$

Here, $\Re(A)$ and $\Im(A)$ denote the real and imaginary parts of a complex number A, respectively. One can present these calculations in matrix form, by writing the above equalities as

$$\begin{aligned} Q_1(p) &= \Re(A_p) + m_1 \Im(B_p) + m_2 \Im(C_p) + m_3 \Im(D_p), \\ Q_i(p) &= \Re(B_p) - m_1 \Im(A_p) - m_3 \Im(C_p) + m_2 \Im(D_p), \\ Q_j(p) &= \Re(C_p) - m_2 \Im(A_p) + m_3 \Im(B_p) - m_1 \Im(D_p), \\ Q_k(p) &= \Re(D_p) - m_3 \Im(A_p) - m_2 \Im(B_p) + m_1 \Im(C_p), \end{aligned} \tag{5.16}$$

and therefore

$$\begin{bmatrix} Q_1(p) \\ Q_i(p) \\ Q_j(p) \\ Q_k(p) \end{bmatrix} = \begin{bmatrix} \Re(A_p) \\ \Re(B_p) \\ \Re(C_p) \\ \Re(D_p) \end{bmatrix} + \begin{bmatrix} 0 & m_1 & m_2 & m_3 \\ -m_1 & 0 & -m_3 & m_2 \\ -m_2 & m_3 & 0 & -m_1 \\ -m_3 & -m_2 & m_1 & 0 \end{bmatrix} \begin{bmatrix} \Im(A_p) \\ \Im(B_p) \\ \Im(C_p) \\ \Im(D_p) \end{bmatrix} \tag{5.17}$$

for each frequency-point $p = 0 : (N - 1)$. The 4×4 matrix in this equation, which we denote by $\mathbf{M} = \mathbf{M}_{r-s}$, is orthogonal and skew-symmetric, meaning that its transpose equals minus the matrix, $\mathbf{M}^T = -\mathbf{M}$, i.e., the following holds for the elements of the matrix: $m_{i_1,i_2} = -m_{i_2,i_1}$ for all $i_1, i_2 = 0, 1, 2, 3$.

If the real part is zero, $a_n = 0$, and $q_n = 0 + ib_n + jc_n + kd_n$, the number of operations of multiplication and addition can be estimated as $m_{QFT}(N) = 3m_{FT}(N) + 9N$ and $a_{QFT}(N) = 3a_{FT}(N) + 9N$, respectively. Here, m_{FT} and a_{FT} are the number of operations of multiplication and addition in the N-point DFT of the real signal. Thus, per one sample, the 1-D QDFT of this signal uses nine more operations of multiplication and addition, than the 1-D DFTs of three signals b_n, c_n, and d_n, when calculating separately.

As an example, Fig. 5.9 shows separately the three components of the "color" signal $q_n = 0 + ib_n + jc_n + kd_n$. The components of the imaginary parts b_n, c_n, and d_n are shown in parts (a), (b), and (c), respectively. Each signal is of length $N = 512$.

The 1-D Fourier transforms of the signals parts b_n, c_n, and d_n are shown in parts (a), (b), and (c) in Fig. 5.10, respectively. Thus, the vector signal $\{(b_n, c_n, d_n); n = 0 : 511\}$ is represented by three spectral characteristics in the frequency domain. The same signal in quaternion algebra, $q_n = 0 + ib_n + jc_n + kd_n$ is represented by four spectral characteristics in the frequency domain in the cost of $9N$ operations of multiplication and addition. Figure 5.11 shows four components of the N-point 1-D QDFT in parts (a),

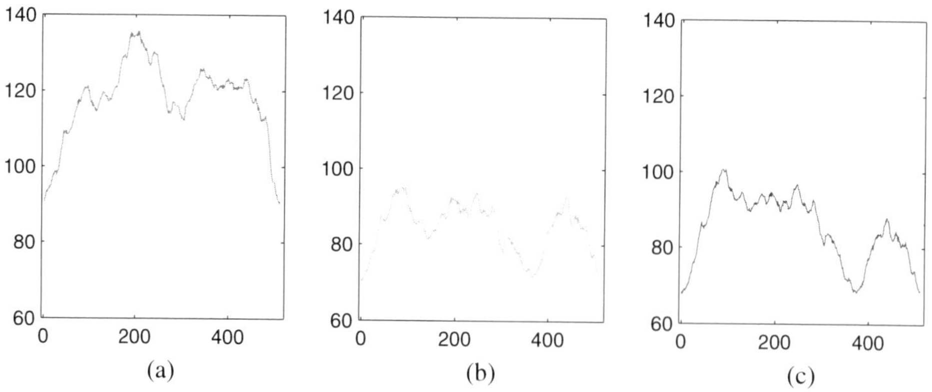

Figure 5.9 Imaginary components of the quaternion signal: (a) b_n, (b) c_n, and (c) d_n.

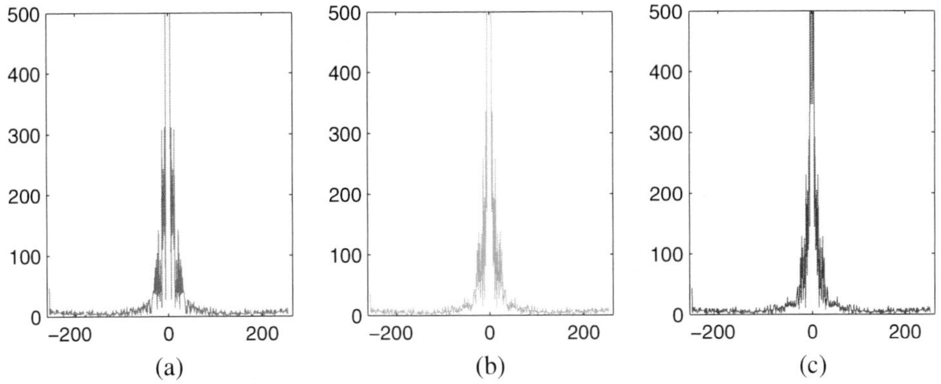

Figure 5.10 Magnitudes of the 512-point DFTs of the signals (a) b_n, (b) c_n, and (c) d_n.

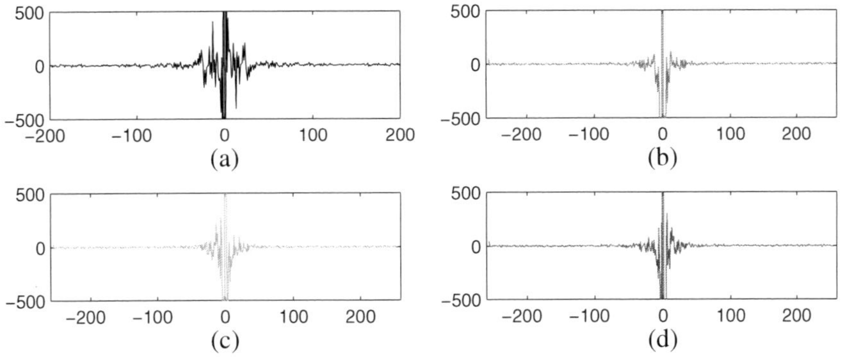

Figure 5.11 Four components of the 512-point QDFT of the quaternion signal; (a) $Q_1(p)$, (b) $Q_i(p)$, (c) $Q_j(p)$, and (d) $Q_k(p)$. (All transforms are shifted to the center.)

(b), (c), and (d). These four characteristics are real, and the three characteristics in Eq. 10 are complex.

As follows from Eq. (5.16), in the general case of the quaternion signal f_n, the number of operations of multiplication and addition can be estimated as $m_{QFT}(N) = 4m_{FT}(N) + 12N$ and $a_{QFT}(N) = 4a_{FT}(N) + 12N$, respectively. When $N = 2^r$, $r > 2$, the fast paired transform algorithm for the N-point DFT uses the following numbers of real multiplications and additions:

$$
\begin{aligned}
4 \times \mu_{FT}(N) &= 4 \times [2^{r-1}(r-3)+2],\\
2 \times \alpha_{FT}(N) &= 2 \times [(2^r 6 - r^2 - 3r - 6) + \mu_{FT}(N)].
\end{aligned}
\tag{5.18}
$$

For the fast N-point discrete paired FFT transform, the estimation for multiplications is $\mu_{FT}(N) = 2^{r-1}(r-3)+2$ and $\alpha_{FT}(N) = (2^r 6 - r^2 - 3r - 6)$ is the number of additions, as shown in Eq. (5.11). Here, we assume that the complex multiplication is performed with two additions and four multiplications. The calculation of two N-point DFTs over the real signals can be reduced to one N-point DFT of a complex signal [31]. Therefore, we consider that

$$2m_{FT}(N) = 4 \times \mu_{FT}(N) \quad \text{and} \quad 2a_{FT}(N) = 2 \times \alpha_{FT}(N).$$

The number of operations for calculating the N-point DFT of the real signal is twice less than the number of operations for the transform of complex signal. The number of operations for the 1-D QDFT can be estimated as

$$\begin{aligned} m_{QFT}(N) &= 8\mu_{FT}(N) + 12N = 4Nr + 16, \\ a_{QFT}(N) &= 4\alpha_{FT}(N) + 12N = 2N(r+15) - 4(r^2 + 3r + 4). \end{aligned} \tag{5.19}$$

5.2.1 Special cases

Consider the case when $\mu = (i + j + k)/\sqrt{3}$. When considering the RGB color model [10] with all colors (r, g, b) in the unit cube, one can say that the vector $\mu = (1, 1, 1)/\sqrt{3}$ is directed to the grayscale line [2]. Equation 5.17 can be written as

$$Q_p = Q_1(p) + iQ_i(p) + jQ_j(p) + kQ_k(p),$$

$$\begin{bmatrix} Q_1(p) \\ Q_i(p) \\ Q_j(p) \\ Q_k(p) \end{bmatrix} = \begin{bmatrix} \Re(A_p) \\ \Re(B_p) \\ \Re(C_p) \\ \Re(D_p) \end{bmatrix} + \frac{1}{\sqrt{3}} \begin{bmatrix} 0 & 1 & 1 & 1 \\ -1 & 0 & -1 & 1 \\ -1 & 1 & 0 & -1 \\ -1 & -1 & 1 & 0 \end{bmatrix} \begin{bmatrix} \Im(A_p) \\ \Im(B_p) \\ \Im(C_p) \\ \Im(D_p) \end{bmatrix}, \tag{5.20}$$

where $p = 0 : (N-1)$. Therefore, $8N$ operations of real multiplication can be saved for $m_{QFT}(N)$ in Eq. (5.18).

Now, we consider the pure unit quaternion $\mu = (0, 1, 1)/\sqrt{2}$ directed to the cyan color line. In the RGB model, the color $(0, 1, 1)$ is cyan. Equation 5.17 can be reduced to the following:

$$\begin{bmatrix} Q_1(p) \\ Q_i(p) \\ Q_j(p) \\ Q_k(p) \end{bmatrix} = \begin{bmatrix} \Re(A_p) \\ \Re(B_p) \\ \Re(C_p) \\ \Re(D_p) \end{bmatrix} + \frac{1}{\sqrt{2}} \begin{bmatrix} 0 & 0 & 1 & 1 \\ 0 & 0 & -1 & 1 \\ -1 & 1 & 0 & 0 \\ -1 & -1 & 0 & 0 \end{bmatrix} \begin{bmatrix} \Im(A_p) \\ \Im(B_p) \\ \Im(C_p) \\ \Im(D_p) \end{bmatrix}, \tag{5.21}$$

where $p = 0 : (N-1)$. In comparison to the first case, $4N$ more additions will be saved. Therefore for the 1-D QDFT, the number of operations of multiplication and addition can be calculated by

$$\begin{aligned} m_{QFT}(N) &= 8\mu_{FT}(N) + 4N = 4N(r-2) + 16, \\ a_{QFT}(N) &= 4\alpha_{FT}(N) + 8N = 2N(r+13) - 4(r^2 + 3r + 4). \end{aligned} \tag{5.22}$$

In the RGB model, the color $(1, 0, 1)$ is magenta and if the pure unit quaternion is directed to the magenta color line, i.e., $\mu = (1, 0, 1)/\sqrt{2}$, Eq. (5.17) can be reduced to the following:

$$\begin{bmatrix} Q_1(p) \\ Q_i(p) \\ Q_j(p) \\ Q_k(p) \end{bmatrix} = \begin{bmatrix} \Re(A_p) \\ \Re(B_p) \\ \Re(C_p) \\ \Re(D_p) \end{bmatrix} + \frac{1}{\sqrt{2}} \begin{bmatrix} 0 & 1 & 0 & 1 \\ -1 & 0 & -1 & 0 \\ 0 & 1 & 0 & -1 \\ -1 & 0 & 1 & 0 \end{bmatrix} \begin{bmatrix} \Im(A_p) \\ \Im(B_p) \\ \Im(C_p) \\ \Im(D_p) \end{bmatrix}, \tag{5.23}$$

where $p = 0 : (N-1)$. The number of operations of multiplication and addition are calculated as in Eq. (5.22).

In the RGB model, the color $(1, 1, 0)$ is yellow. If the pure unit quaternion is directed to the yellow color line, i.e., $\mu = (1, 1, 0)/\sqrt{2}$, Eq. (5.17) can be reduced to the following:

$$\begin{bmatrix} Q_1(p) \\ Q_i(p) \\ Q_j(p) \\ Q_k(p) \end{bmatrix} = \begin{bmatrix} \Re(A_p) \\ \Re(B_p) \\ \Re(C_p) \\ \Re(D_p) \end{bmatrix} + \frac{1}{\sqrt{2}} \begin{bmatrix} 0 & 1 & 1 & 0 \\ -1 & 0 & 0 & 1 \\ -1 & 0 & 0 & -1 \\ 0 & -1 & 1 & 0 \end{bmatrix} \begin{bmatrix} \Im(A_p) \\ \Im(B_p) \\ \Im(C_p) \\ \Im(D_p) \end{bmatrix}. \tag{5.24}$$

5.2.2 Codes for the 1-D right-side QDFTs

In this section, the scripts of the programs for calculating the N-point 1-D QDFT are given. The $N = 8$ case is considered with the quaternion signal q_n written as the following 4×8 matrix whose four rows are components a_n, b_n, c_n, and d_n, $n = 0 : 7$,

$$\mathbf{q} = \begin{bmatrix} 1 & -1 & 2 & 1 & 5 & -1 & 2 & 1 \\ 1 & 2 & 4 & 1 & -2 & 3 & -1 & 2 \\ 0 & 2 & 1 & 2 & 1 & -1 & 1 & 2 \\ 1 & -1 & 2 & 1 & 2 & 0 & 1 & -2 \end{bmatrix}.$$

The pure unit quaternion number $\mu = (i + j + k)/\sqrt{3}$ is used for the 1-D QDFT, which equals

$$\mathbf{Q}_{1,1,1} = \begin{bmatrix} 10 & -8.6888 & 2.5774 & -1.7606 & 10 & -6.2394 & 1.4226 & 0.6888 \\ 10 & 2.8309 & -8.0415 & 3.9856 & -6 & 2.0144 & 0.0415 & 3.1691 \\ 8 & 0.4449 & -4.4641 & 0.8211 & -2 & -7.0637 & 2.4641 & 1.7977 \\ 4 & -2.9829 & 0.5774 & -3.0996 & 8 & 6.7564 & -0.5774 & -4.6739 \end{bmatrix}.$$

The values of the real part $Q_1(p)$, $p = 0:7$, are written in the first row of this matrix. The components $Q_i(p)$, $Q_j(p)$, and $Q_k(p)$ of the imaginary part of the 1-D QDFT are in the rows number 2, 3 and 4, respectively.

The following scripts of programs are presented:

"test_1Dqfft.m" – the main code to calculate the 8-point quaternion DFT of the given quaternion signal;

"q1dfft.m" – function to calculate the quaternion 1-D DFT;

"iq1dfft.m" – function to calculate the inverse quaternion 1-D DFT.

The direct and inverse transforms are calculated by two complex 1-D DFTs, which are calculated by the MATLAB® functions "fft.m" and "ifft.m" for the direct DFT and inverse DFT, respectively.

```
% ==========================================================
% call: test_1Dqfft.m / Artyom Grigoryan, May 10, 2014
%-----------------------------------------------------------
  % The quaternion signal composition
  N=8;
  fq=zeros(4,N);
  fq=[1 -1  2  1   5  -1    2   1
        1  2  4  1 -2    3   -1   2
        0  2  1  2   1  -1    1   2
        1 -1  2  1   2    0    1  -2];
  qmu=[1,1,1]/sqrt(3);          % pure unit quaternion
  FQ=q1dfft(fq,qmu,N);          % the N-point QDFT
  fqi=iq1dfft(FQ,qmu,N);        % the N-point inverse QDFT
%-----------------------------------------------------------
% call: q1dfft.m
% Calculate the 1-D QDFT ("FQ") of signal q(n)
% qmu is the pure unit such that (qmu)^2= -1
% Two complex 1-D DFTs are used.
  function FQ=q1dfft(fq,qmu,N)
    fa=fq(1,:);fb=fq(2,:);fc=fq(3,:);fd=fq(4,:);
    m1=qmu(1); m2=qmu(2); m3=qmu(3);
%   Calculate two complex 1-D DFTs:
    % one is instead of A=fft(fa) and B=fft(fb)
    x=fa+1i*fb; X=fft(x);
    Y=conj([X(1),X(N:-1:2)]);
    A=(X+Y)/2;    Ca=real(A);    Sa=imag(A);
    B=(X-Y)/2i;   Cb=real(B);    Sb=imag(B);
    % another is instead of C=fft(fc) and D=fft(fd)
    x=fc+1i*fd;   X=fft(x);
    Y=conj([X(1),X(N:-1:2)]);
```

```
    C= (X+Y)/2;    Cc=real(C);    Sc=imag(C);
    D= (X-Y)/2i;   Cd=real(D);    Sd=imag(D);
%   1-D QDFT Q= (Q1,Qi,Qj,Qk) by Equation 5.16
    Q1=zeros(1,N); Qi=Q1; Qj=Q1; Qk=Q1;
    Q1=Ca + m1*Sb + m2*Sc + m3*Sd;   % **
    Qi=Cb - m1*Sa + m2*Sd - m3*Sc;   % **
    Qj=Cc - m1*Sd - m2*Sa + m3*Sb;   % **
    Qk=Cd + m1*Sc - m2*Sb - m3*Sa;   % **
    FQ=zeros(4,N);
    FQ(1,:)=Q1;FQ(2,:)=Qi;FQ(3,:)=Qj;FQ(4,:)=Qk;
%-----------------------------------------------------
```

```
% call: iq1ddft.m
% To calculate the inverse 1-D QDFT with two 1-D DFTs
% qmu is the unit such that (qmu)^2= -1
  function fqi=iq1dfft(FQ,qmu,N)
    Q1=zeros(1,N); Qi=Q1; Qj=Q1; Qk=Q1;
    Q1=FQ(1,:); Qi=FQ(2,:); Qj=FQ(3,:); Qk=FQ(4,:);
    m1=qmu(1); m2=qmu(2); m3=qmu(3);
% Calculate two complex 1-D DFTs:
  % one instead of A1=ifft(Q1) and B1=ifft(Qi):
    A1=zeros(1,N); B1=A1; C1=A1; D1=A1;
    x=Q1+1i*Qi; X=ifft(x);
    Y=conj([X(1),X(N:-1:2)]);
    A1= (X+Y)/2; C1a=real(A1); S1a=imag(A1);
    B1= (X-Y)/2i; C1b=real(B1); S1b=imag(B1);
  % and another instead of C1=ifft(Qj) and D1=ifft(Qk)
    x=Qj+1i*Qk; X=ifft(x);
    Y=conj([X(1),X(N:-1:2)]);
    C1= (X+Y)/2; C1c=real(C1); S1c=imag(C1);
    D1= (X-Y)/2i; C1d=real(D1); S1d=imag(D1);
% Calculation of the quaternion signal (q1,qi,qj,qk)
    q1 =zeros(1,N); qi=q1; qj=q1; qk=q1;
    q1=C1a + m1*S1b + m2*S1c + m3*S1d;   % **
    qi=C1b - m1*S1a + m2*S1d - m3*S1c;   % **
    qj=C1c - m1*S1d - m2*S1a + m3*S1b;   % **
    qk=C1d + m1*S1c - m2*S1b - m3*S1a;   % **
    fqi=zeros(4,N);
    fqi(1,:)=q1; fqi(2,:)=qi; fqi(3,:)=qj; fqi(4,:)=qk;
%-----------------------------------------------------
```

For comparison, we also consider the results of this program for the eight-point QDFT of the same signal **q**, when the pure unit quaternion number is $\mu = (i+j)/\sqrt{2}$,

$$\mathbf{Q}_{1,1,0} = \begin{bmatrix} 10 & -8.0355 & 2.7071 & -0.9645 & 10 & -7.0355 & 1.2929 & 0.0355 \\ 10 & 1.2929 & -6.8284 & 2.7071 & -6 & 3.2929 & -1.1716 & 4.7071 \\ 8 & 2.8284 & -3.8284 & -2.8284 & -2 & -3.4142 & 1.8284 & -0.5858 \\ 4 & -2.7929 & 3.5355 & -4.2071 & 8 & 7.8640 & -3.5355 & -4.8640 \end{bmatrix}.$$

When the pure unit quaternion number is $\mu = (j + k)/\sqrt{2}$,

$$\mathbf{Q}_{0,1,1} = \begin{bmatrix} 10 & -7.2071 & 4.1213 & -5.7929 & 10 & -2.2071 & -0.1213 & -0.7929 \\ 10 & 2.7929 & -6.1213 & 4.2071 & -6 & 1.7929 & -1.8787 & 3.2071 \\ 8 & -1.4142 & -5.2426 & 1.4142 & -2 & -7.6569 & 3.2426 & 3.6569 \\ 4 & -1.2929 & -1.4142 & -2.7071 & 8 & 6.3640 & 1.4142 & -6.3640 \end{bmatrix}.$$

For the case when the pure unit quaternion number is $\mu = (i + k)/\sqrt{2}$,

$$\mathbf{Q}_{1,0,1} = \begin{bmatrix} 10 & -8.2426 & 0.5858 & 0.2426 & 10 & -8.2426 & 3.4142 & 0.2426 \\ 10 & 4.5000 & -8.9497 & 4.5000 & -6 & 1.5000 & 0.9497 & 1.5000 \\ 8 & 0.2929 & -2.4142 & 1.7071 & -2 & -7.9497 & 0.4142 & 1.9497 \\ 4 & -5.3284 & -0.7071 & 0.3284 & 8 & 3.3284 & 0.7071 & -2.3284 \end{bmatrix}.$$

When the input signal is pure quaternion, i.e., $q_n = ib_n + jc_n + kd_n$, the calculation of the 1-D QDFT can be simplified, since the DFT of the real part is not calculated, it is zero. The transform in Eq. (5.16) can be written as

$$Q_p = Q_1(p) + iQ_i(p) + jQ_j(p) + kQ_k(p), \quad p = 0 : (N-1),$$

with components calculated as

$$\begin{aligned} Q_1(p) &= m_1\Im(B_p) + m_2\Im(C_p) + m_3\Im(D_p), \\ Q_i(p) &= \Re(B_p) - m_3\Im(C_p) + m_2\Im(D_p), \\ Q_j(p) &= m_3\Im(B_p) + \Re(C_p) - m_1\Im(D_p), \\ Q_k(p) &= -m_2\Im(B_p) + m_1\Im(C_p) + \Re(D_p). \end{aligned} \tag{5.25}$$

Therefore, the 1-D QDFT can be calculated by one real and one complex N-point DFTs. The number of operations for the 1-D QDFT can be estimated as

$$\begin{aligned} m_{QFT}(N) &< 8\mu_{FT}(N) + 9N = N(4r-3) + 16, \\ a_{QFT}(N) &< 4\alpha_{FT}(N) + 9N = 2N(r+12) - 4(r^2 + 3r + 4). \end{aligned}$$

$3N$ operations of multiplication and $3N$ additions can be saved, when comparing with the estimations given in Eq. (5.19).

The script "qreal1dfft.m" of the 1-D QDFT for a pure quaternion signal is shown below.

```
% ============================================================
% call: qreal1dfft.m / Artyom Grigoryan, May 10, 2014
%------------------------------------------------------------
% To calculate the quaternion 1-D DFT of the pure signal fq(n)
% qmu is the unit such that (qmu)^2= -1
% One real and one complex 1-D DFTs are used
function FQ=qreal1dfft(fq,qmu,N)
  fb=fq(1,:); fc=fq(2,:); fd=fq(3,:);
  m1=qmu(1); m2=qmu(2); m3=qmu(3);
% calculate one real 1-D DFT
  B=fft(fb); Cb=real(B); Sb=imag(B);
% Calculate one complex 1-D DFT
% instead of C=fft(fc); D=fft(fd);
  x=fc+1i*fd; X=fft(x);
  Y=conj([X(1),X(N:-1:2)]);
  C=(X+Y)/2;   Cc=real(C);   Sc=imag(C);
  D=(X-Y)/2i;  Cd=real(D);   Sd=imag(D);
% 1-D QDFT F=(Q1,Qi,Qj,Qk) by Eq. 5.25
  Q1=zeros(1,N); Qi=Q1; Qj=Q1; Qk=Q1;
  Q1=m1*Sb + m2*Sc + m3*Sd;
  Qi=Cb + m2*Sd - m3*Sc;
  Qj=Cc - m1*Sd + m3*Sb;
  Qk=Cd + m1*Sc - m2*Sb;
  FQ=zeros(4,N);
  FQ(1,:)=Q1; FQ(2,:)=Qi; FQ(3,:)=Qj; FQ(4,:)=Qk;
```

5.3 1-D Left-Side QDFT

Let $q_n = (a_n, b_n, c_n, d_n) = a_n + ib_n + jc_n + kd_n$ be the quaternion signal of length N. The left-side 1-D quaternion DFT (ls-QDFT) is defined as [18]

$$Q_p = Q_1(p) + iQ_i(p) + jQ_j(p) + kQ_k(p) = \sum_{n=0}^{N-1} W_\mu^{np} q_n,$$

where $p = 0 : (N-1)$ and μ is a pure unit quaternion $\mu = im_1 + jm_2 + km_3$, $\mu^2 = -1$. The kernel of this transform is

$$W_\mu = W_{N;\mu} = \exp(-\mu 2\pi/N) = \cos(2\pi/N) - \mu \sin(2\pi/N).$$

The inverse left-side 1-D QDFT is defined as

$$q_n = \frac{1}{N}\sum_{p=0}^{N-1} W_{\mu}^{-np} Q_p, \quad n = 0 : (N-1).$$

Given an angle ϕ, the multiplication of two quaternion numbers $\exp(-\mu\phi)$ and $q = a + ib + jc + kd$ can be written as

$$\begin{aligned}
\exp(-\mu\phi)\cdot q &= [\cos(\phi) - [im_1 + jm_2 + km_3]\sin(\phi)]\cdot(a + ib + jc + kd) \\
&= a\cos(\phi) + (bm_1 + cm_2 + dm_3)\sin(\phi) \\
&\quad + i[b\cos(\phi) - (am_1 + dm_2 - cm_3)\sin(\phi)] \\
&\quad + j[c\cos(\phi) - (-dm_1 + am_2 + bm_3)\sin(\phi)] \\
&\quad + k[d\cos(\phi) - (cm_1 - bm_2 + am_3)\sin(\phi)].
\end{aligned} \tag{5.26}$$

Let φ_{np} be the angle $(2\pi/N)np$. Considering the traditional N-point DFT of a signal x_n,

$$X_p = \sum_{n=0}^{N-1} W^{np} x_n, \quad p = 0 : (N-1),$$

where the kernel $W = W_{N;i} = \exp(-i2\pi/N) = \cos(2\pi/N) - i\sin(2\pi/N)$, we define the real and imaginary parts of this 1-D DFT by

$$\begin{aligned}
C_x(p) &= \Re(X_p) = \sum_{n=0}^{N-1} x_n \cos(\varphi_{np}), \\
S_x(p) &= -\Im(X_p) = \sum_{n=0}^{N-1} x_n \sin(\varphi_{np}),
\end{aligned} \tag{5.27}$$

where $p = 0 : (N-1)$. Let $C(x)$ be the vector of coefficients $[C_x(0), C_x(1), C_x(2), \ldots, C_x(N-1)]$ and let $S(x)$ be the vector of coefficients $[S_x(0), S_x(1), S_x(2), \ldots, S_x(N-1)]$. The four vectors composed by the coefficients $Q_1(p)$, $Q_i(p)$, $Q_j(p)$, and $Q_k(p)$ are denoted by Q_1, Q_i, Q_j, and Q_k, respectively. It directly follows from (5.26) that the 1-D left-side QDFT can be written as

$$\begin{aligned}
Q_1 + iQ_i + jQ_j + kQ_k &= C(a) + m_1 S(b) + m_2 S(c) + m_3 S(d) \\
&\quad + i[C(b) - m_1 S(a) - m_2 S(d) + m_3 S(c)] \\
&\quad + j[C(c) + m_1 S(d) - m_2 S(a) - m_3 S(b)] \\
&\quad + k[C(d) - m_1 S(c) + m_2 S(b) - m_3 S(a)],
\end{aligned} \tag{5.28}$$

or

$$\begin{aligned}Q_1 + iQ_i + jQ_j + kQ_k = C(a) &+ m_1S(b) + m_2S(c) + m_3S(d)\\ &+ i[-m_1S(a) + C(b) + m_3S(c) - m_2S(d)]\\ &+ j[-m_2S(a) - m_3S(b) + C(c) + m_1S(d)]\\ &+ k[-m_3S(a) + m_2S(b) - m_1S(c) + C(d)].\end{aligned} \tag{5.29}$$

Denoting by A_p, B_p, C_p, and D_p the N-point 1-D DFTs of the parts a_n, b_n, c_n, and d_n, respectively, we obtain the following algorithm of calculation of the 1-D left-side QDFT:

$$\begin{aligned}Q_1(p) &= \Re(A_p) + m_1\Im(B_p) + m_2\Im(C_p) + m_3\Im(D_p),\\ Q_i(p) &= -m_1\Im(A_p) + \Re(B_p) + m_3\Im(C_p) - m_2\Im(D_p),\\ Q_j(p) &= -m_2\Im(A_p) - m_3\Im(B_p) + \Re(C_p) + m_1\Im(D_p),\\ Q_k(p) &= -m_3\Im(A_p) + m_2\Im(B_p) - m_1\Im(C_p) + \Re(D_p),\end{aligned} \tag{5.30}$$

where $p = 0 : (N-1)$. No more than $12N$ operations of real multiplications plus 4 times $m_{\mathrm{DFT}}(N)$ are used for $m_{QFT}(N)$. Thus, the number of operations of multiplication and addition equal $m_{QFT}(N) = 4m_{FT}(N) + 12N$ and $a_{QFT}(N) = 4a_{FT}(N) + 12N$. The numbers of multiplication and addition for the left-side are the same as for the 1-D right-side QDFT.

In the special case when $\mu = (i + j + k)/\sqrt{3}$, Eq. (5.30) can be written as

$$F_p = Q_1(p) + iQ_i(p) + jQ_j(p) + kQ_k(p),$$

$$\begin{aligned}Q_1(p) &= \Re(A_p) + [\Im(B_p) + \Im(C_p) + \Im(D_p)]/\sqrt{3},\\ Q_i(p) &= \Re(B_p) - [\Im(A_p) - \Im(C_p) + \Im(D_p)]/\sqrt{3},\\ Q_j(p) &= \Re(C_p) - [\Im(A_p) + \Im(B_p) - \Im(D_p)]/\sqrt{3},\\ Q_k(p) &= \Re(D_p) - [\Im(A_p) - \Im(B_p) + \Im(C_p)]/\sqrt{3},\end{aligned} \tag{5.31}$$

where $p = 0 : (N-1)$. The number of operations of multiplication equals $m_{QFT}(N) = 4m_{FT}(N) + 4N$, which means the saving of $8N$ operations of real multiplication.

The system of equalities in Eq. (5.30) can be written in matrix form as

$$\begin{bmatrix}Q_1(p)\\ Q_i(p)\\ Q_j(p)\\ Q_k(p)\end{bmatrix} = \begin{bmatrix}\Re(A_p)\\ \Re(B_p)\\ \Re(C_p)\\ \Re(D_p)\end{bmatrix} + \begin{bmatrix}0 & m_1 & m_2 & m_3\\ -m_1 & 0 & m_3 & -m_2\\ -m_2 & -m_3 & 0 & m_1\\ -m_3 & m_2 & -m_1 & 0\end{bmatrix}\begin{bmatrix}\Im(A_p)\\ \Im(B_p)\\ \Im(C_p)\\ \Im(D_p)\end{bmatrix} \tag{5.32}$$

for each frequency-point $p = 0 : (N - 1)$. The 4×4 matrix in this equation, which we denote by $\mathbf{M} = \mathbf{M}_{l-s}$ is skew-symmetric. This matrix differs from the skew-symmetric matrix for the right-side 1-D QDFT. One can note that the sum of these two matrices

$$\mathbf{M}_{r-s} + \mathbf{M}_{l-s} = 2\begin{bmatrix} 0 & m_1 & m_2 & m_3 \\ -m_1 & 0 & 0 & 0 \\ -m_2 & 0 & 0 & 0 \\ -m_3 & 0 & 0 & 0 \end{bmatrix}. \tag{5.33}$$

The $\mathbf{M}_{r-s}$ can be obtained from $\mathbf{M}_{l-s}$ and vice versa. That means that the 1-D QDFT left-side and right-side can be transformed to each other.

The codes for the left-side direct and inverse 1-D QDFTs can be written similarly to the above given functions "q1dfft.m" and "iq1dfft.m" for the right-side QDFT. To save the space in this text, we only mention the difference between these functions and functions "q1lsdfft.m" and "iq1lsdfft.m" written for the left-side direct and inverse 1-D QDFTs, respectively. To write script for "q1lsdfft.m" from "q1dfft.m" we only need to change four lines marked from right as %** by the following lines:

```
%-------------------------------------------------------------
  Q1 = Ca + m1*Sb + m2*Sc + m3*Sd;
  Qi = Cb - m1*Sa - m2*Sd + m3*Sc;
  Qj = Cc + m1*Sd - m2*Sa - m3*Sb;
  Qk = Cd - m1*Sc + m2*Sb - m3*Sa;
%-------------------------------------------------------------
```

To write script for "iq1lsdfft.m" from "iq1dfft.m" we need change only four lines marked from right as %** by the following lines:

```
%-------------------------------------------------------------
  q1 = C1a + m1*S1b + m2*S1c + m3*S1d;
  qi = C1b - m1*S1a - m2*S1d + m3*S1c;
  qj = C1c + m1*S1d - m2*S1a - m3*S1b;
  qk = C1d - m1*S1c + m2*S1b - m3*S1a;
%-------------------------------------------------------------
```

To compare the left-side and right-side QDFTs, we consider the above quaternion signal of length eight,

$$\mathbf{q} = \begin{bmatrix} 1 & -1 & 2 & 1 & 5 & -1 & 2 & 1 \\ 1 & 2 & 4 & 1 & -2 & 3 & -1 & 2 \\ 0 & 2 & 1 & 2 & 1 & -1 & 1 & 2 \\ 1 & -1 & 2 & 1 & 2 & 0 & 1 & -2 \end{bmatrix}$$

and the pure unit quaternion number $\mu = (i + j + k)/\sqrt{3}$. The left-side 1-D QDFT of this signal is

$$\mathbf{Q}_{1,1,1;l-s} = \begin{bmatrix} 10 & -8.6888 & 2.5774 & -1.7606 & 10 & -6.2394 & 1.4226 & 0.6888 \\ 10 & 3.1691 & -4.5774 & 2.0144 & -6 & 3.9856 & -3.4226 & 2.8309 \\ 8 & 1.7977 & -2.1547 & -7.0637 & -2 & 0.8211 & 0.1547 & 0.4449 \\ 4 & -4.6739 & -5.1962 & 6.7564 & 8 & -3.0996 & 5.1962 & -2.9829 \end{bmatrix}.$$

The difference between the right-side and left-side 1-D QDFTs is

$$\Delta\mathbf{Q}_{1,1,1} = \begin{bmatrix} 0 & 0 & 0 & 0 & 0 & 0 & 0 & 0 \\ 0 & -0.3382 & -3.4641 & 1.9712 & 0 & -1.9712 & 3.4641 & 0.3382 \\ 0 & -1.3528 & -2.3094 & 7.8848 & 0 & -7.8848 & 2.3094 & 1.3528 \\ 0 & 1.6910 & 5.7735 & -9.8560 & 0 & 9.8560 & -5.7735 & -1.6910 \end{bmatrix}.$$

In this matrix, two 3×3 non-zero sub-matrices are the mirror transforms of each other with respect to the 4th column up to the sign minus.

5.4 Octonion Discrete Fourier Transform

The 1-D octonion discrete Fourier transform (ODFT) is a new concept that allows for processing up to eight signals as one octonion 8-D signal in the spectral domain. This concept generalizes the traditional complex and quaternion 1-D DFTs and can be effectively used in signal and image processing. The basic functions for the ODFT are the octonion exponential functions, which are defined for octonions o as

$$e^{o} = \exp(o) = 1 + \sum_{n=1}^{\infty} \frac{o^n}{n!}.$$

The octonion $o = a + ib + jc + dk + EA + IB + JC + KD$ can also be denoted as the vector $o = (a, b, c, d, A, B, C, D)$.

We consider the exponential function for a pure unit octonion

$$\lambda = \lambda' = i\lambda_i + j\lambda_j + k\lambda_k + E\lambda_E + I\lambda_I + J\lambda_J + K\lambda_K \quad (\lambda^2 = -1).$$

Given a real number, or angle ϑ, the exponential function $\exp(\lambda\vartheta)$ is

$$e^{\lambda\vartheta} = \exp(\lambda\vartheta) = \cos(\vartheta) + \lambda\sin(\vartheta) \tag{5.34}$$

and the conjugate function $\exp(-\lambda\vartheta) = \cos(\vartheta) - \lambda\sin(\vartheta)$ is inverse to $\exp(\lambda\vartheta)$, i.e., $\exp(\lambda\vartheta)\exp(-\lambda\vartheta) = 1$. It is important to mention that for any angles ϑ_1 and ϑ_2, we octonion exponent has the following property:

$$\exp(\lambda\vartheta_1)\exp(\lambda\vartheta_2) = \exp[\lambda(\vartheta_1 + \vartheta_2)].$$

Therefore, we have $(e^{\lambda\vartheta})^n = e^{n\lambda\vartheta}$, for any integer n.

We also consider the following properties of the multiplication of quaternion numbers by octonion units E, I, J, and K, which will be used for calculating the octonion ODFTs.

1. For any quaternion $q = a + ib + jc + kd$, the following products are equal:

$$Eq = \bar{q}E, \quad Iq = \bar{q}I, \quad Jq = \bar{q}J, \quad Kq = \bar{q}K. \tag{5.35}$$

Indeed for instance, when considering the unit E, the following calculations hold:

$$\begin{aligned} Eq &= E(a + ib + jc + kd) = Ea + Eib + Ejc + Ekd \\ &= aE - iEb - jEc - kEd = (a - ib - jc - kd)E = \bar{q}E. \end{aligned}$$

2. For any quaternions $q_1 = a_1 + ib_1 + jc_1 + kd_1$ and $q_2 = a_2 + ib_2 + jc_2 + kd_2$, the following products are equal:

$$\begin{aligned} (Eq_1)q_2 = (\bar{q}_1\bar{q}_2)E, \quad (Iq_1)q_2 = (\bar{q}_1\bar{q}_2)I, \\ (Jq_1)q_2 = (\bar{q}_1\bar{q}_2)J, \quad (Kq_1)q_2 = (\bar{q}_1\bar{q}_2)K. \end{aligned} \tag{5.36}$$

For instance, when considering the unit E, it directly follows from the 1st property that

$$(Eq_1)q_2 = (\bar{q}_1E)q_2 = \bar{q}_1(Eq_2) = \bar{q}_1(\bar{q}_2E) = (\bar{q}_1\bar{q}_2)E.$$

Example 5.5

For the quaternion $q_1 = 2 - i + 3j + 4k$, the product Eq_1 is the octonion

$$o = Eq_1 = E(2 - i + 3j + 4k) = 2E + I - 3J - 4K.$$

If we multiply this product by the quaternion $q_2 = 4 - 2i + j + 5k$, we obtain

$$(Eq_1)q_2 = (2E + I - 3J - 4K)(4 - 2i + j + 5k) = -17E + 19I - 17J - 21K.$$

Its is not difficult to verify that the same octonion can be calculated by

$$(\bar{q}_1\bar{q}_2)E = (2 + i - 3j - 4k)(4 + 2i - j - 5k)E = -17E + 19I - 17J - 21K.$$

Below is the script of the code for calculating the above products in the example. The function "mult2octs.m" is used for multiplication of octonions.

```
%------------------------------------------------------------
% call: test_obyqmult.m / Artyom Grigoryan, March 28 2016
% to verify that the multiplication Eq1q2=conj(q1)conj(q2)E
  E=[0 0 0 0 1 0 0 0];           % the unit E
  o1=[2 -1 3 4 0 0 0 0];         % the 1st quaternion q1
  o2=[4 -2 1 5 0 0 0 0];         % the 2nd quaternion q2
  Eo1=mult2octs(E,o1);           %   0  0   0   0   2  1  -3  -4
  Eo12=mult2octs(Eo1,o2);        %   0  0   0   0 -17 19 -17 -21
  o1c=[o1(1) -o1(2:8)];          %   2  1  -3  -4   0  0   0   0
  o2c=[o2(1) -o2(2:8)];          %   4  2  -1  -5   0  0   0   0
  o12c=mult2octs(o1c,o2c);       % -17 19 -17 -21   0  0   0   0
  o12E=mult2octs(o12c,E);        %   0  0   0   0 -17 19 -17 -21
%------------------------------------------------------------
```

Another important property of multiplication of quaternions with the unit E, I, J, and K should also be mentioned. We describe it for the unit E.

3. For any quaternions q_1 and q_2, the following equalities hold:

$$(Eq_1)q_2 = (\overline{q_2 q_1})E = E(q_2 q_1) \tag{5.37}$$

and therefore,

$$(q_1 E)q_2 = (E\bar{q}_1)q_2 = E(q_2\bar{q}_1) = \overline{(q_2\bar{q}_1)}E = (q_1\bar{q}_2)E. \tag{5.38}$$

Example 5.6

For the quaternions $q_1 = 2 - i + 3j + 4k$ and $q_2 = 4 - 2i + j + 5k$, the following calculations hold:

$$\begin{aligned} Eq_1 &= E(2 - i + 3j + 4k) = 2E + I - 3J - 4K, \\ (Eq_1)q_2 &= (2E + I - 3J - 4K)(4 - 2i + j + 5k) = -17E + 19I - 17J - 21K, \end{aligned}$$

and

$$\begin{aligned} q_2 q_1 &= (4 - 2i + j + 5k)(2 - i + 3j + 4k) = -17 - 19i + 17j + 21k, \\ \overline{q_2 q_1} &= -17 + 19i - 17j - 21k, \\ (\overline{q_2 q_1})E &= (-17 + 19i - 17j - 21k)E = -17E + 19I - 17J - 21K, \\ E(q_2 q_1) &= E(-17 - 19i + 17j + 21k) = -17E + 19I - 17J - 21K. \end{aligned}$$

To verify the result in Eq. (5.38), we consider the calculations

$$
\begin{aligned}
(q_1E)q_2 &= [(2 - i + 3j + 4k)E](4 - 2i + j + 5k)\\
&= (2E - I + 3J + 4K)(4 - 2i + j + 5k) = 33E - 11I + 13J + K,\\
q_1\bar{q}_2 &= (2 - i + 3j + 4k)(4 + 2i - j - 5k) = 33 - 11i + 13j + k,\\
(q_1\bar{q}_2)E &= (33 - 11i + 13j + k)E = 33E - 11I + 13J + K.
\end{aligned}
$$

For the unit numbers I, J, and K, the similar properties hold.

$3'$. For octonion unit numbers $U = E, I, J$, and K and any quaternions q_1 and q_2, the following equalities hold:

$$(Uq_1)q_2 = (\overline{q_2q_1})U = U(q_2q_1) \tag{5.39}$$

and therefore,

$$(q_1U)q_2 = (q_1\bar{q}_2)U. \tag{5.40}$$

5.4.1 1-D right-side ODFT

The 1-D octonion discrete Fourier transform (ODFT) can be defined by the sum of signal components multiplied by the exponential kernel of the transform from the right or the left. Two definitions of the 1-D ODFT can be considered.

Let $o_n = (a_n, b_n, c_n, d_n, A_n, B_n, C_n, D_n)$ be the octonion signal of length N, i.e.,

$$o_n = a_n + ib_n + jc_n + kd_n + EA_n + IB_n + JC_n + KD_n,$$

where $n = 0 : (N - 1)$. Given a pure unit octonion

$$\lambda = i\lambda_i + j\lambda_j + k\lambda_k + E\lambda_E + I\lambda_I + J\lambda_J + K\lambda_K,$$

we consider the octonion exponential function

$$W_\lambda^t = W_{\lambda;N}^t = \cos(2\pi/N) - \lambda\sin(2\pi/N),$$

for integers $t = 0, \pm 1, \pm 2, \ldots$.

The 1-D N-point right-side ODFT (rs-ODFT) is defined as

$$O_p = O_p^{(rs)} = \sum_{n=0}^{N-1} o_n W_\lambda^{np}, \quad p = 0 : (N - 1). \tag{5.41}$$

The inverse transform is

$$o_n = \frac{1}{N}\sum_{p=0}^{N-1} O_p W_\lambda^{-np}, \quad n = 0 : (N - 1). \tag{5.42}$$

The definition of the rs-ODFT is not unique because of the octonion pure unit λ. The selection of such a parameter is beyond our discussion in this book. Among the pure unit octonions, we consider the numbers $\lambda = \mu$ in the quaternion subspace, i.e, $\mu = i\mu_i + j\mu_j + k\mu_k$, such that $\mu^2 = -1$. In this case, the 1-D N-point rs-ODFT is defined as

$$O_p = O_p^{(rs)} = \sum_{n=0}^{N-1} o_n W_\mu^{np}, \quad p = 0 : (N-1). \tag{5.43}$$

The inverse 1-D N-point rs-ODFT is calculated by

$$o_n = \frac{1}{N}\sum_{p=0}^{N-1} O_p W_\mu^{-np}, \quad n = 0 : (N-1). \tag{5.44}$$

Indeed, let us consider the first and second quaternion signals of the octonion signal

$$o_n = q_n + f_n E, \quad n = 0 : (N-1).$$

The operation of multiplication of quaternion numbers is associative. Therefore, for the first signal, the following calculations hold:

$$\sum_{p=0}^{N-1}\left(\sum_{k=0}^{N-1} q_k W_\mu^{kp}\right) W_\mu^{-np} = \sum_{p=0}^{N-1}\sum_{k=0}^{N-1} q_k W_\mu^{(k-n)p} = \sum_{k=0}^{N-1} q_k \sum_{p=0}^{N-1} W_\mu^{(k-n)p}$$

and we can show that

$$\sum_{p=0}^{N-1} W_\mu^{(k-n)p} = \begin{cases} N, & \text{if } k = n, \\ 0, & \text{if } k \neq n. \end{cases}$$

In the $k \neq n$ case, we consider that, for any integer $t = k - n \neq 0 \bmod N$,

$$\sum_{p=0}^{N-1} W_\mu^{tp} \cdot (1 - W_\mu^t) = [1 + W_\mu^t + W_\mu^{2t} + \ldots + W_\mu^{(N-1)t}](1 - W_\mu^t)$$
$$= 1 - W_\mu^{Nt} = 1 - W_{\mu;N}^{Nt} = 0.$$

Therefore,

$$\sum_{k=0}^{N-1} q_k \sum_{p=0}^{N-1} W_\mu^{(k-n)p} = \sum_{k=n} q_k \sum_{p=0}^{N-1} W_\mu^{(k-n)p} = q_n N.$$

Due to property 3,

$$[(f_k E)W_\mu^{kp}]W_\mu^{-np} = [(f_k W_\mu^{-kp})E]W_\mu^{-np}$$
$$= [(f_k W_\mu^{-kp})W_\mu^{np}]E = [f_k W_\mu^{(n-k)p}]E.$$

Therefore, for the second quaternion signal, the following calculations hold:

$$\sum_{p=0}^{N-1}\left[\sum_{k=0}^{N-1}(f_k E)W_\mu^{kp}\right]W_\mu^{-np} = \left[\sum_{k=0}^{N-1} f_k \sum_{p=0}^{N-1} W_\mu^{(n-k)p}\right]E = (f_n N)E.$$

As a result of above calculations, we obtain the formula for the inverse 1-D ODFT,

$$\sum_{p=0}^{N-1} O_p W_\mu^{-np} = \sum_{p=0}^{N-1}\left[\sum_{k=0}^{N-1}(q_k + f_k E)W_\mu^{kp}\right]W_\mu^{-np} = (q_n + f_n E)N = o_n N,$$

where q_n and f_n are quaternion signals.

The N-point rs-ODFT can be calculated by using fast algorithms of the 1-D right-side quaternion DFT (rs-QDFT). Equation 5.43 can be written as

$$O_p = \sum_{n=0}^{N-1}(q_n + f_n E)W_\mu^{np} = \sum_{n=0}^{N-1} q_n W_\mu^{np} + \sum_{n=0}^{N-1}(f_n E)W_\mu^{np} \tag{5.45}$$

when $p = 0 : (N-1)$. The product $(f_n E)W_\mu^{np}$ equals $(f_n W_\mu^{-np})E$, therefore

$$O_p = \sum_{n=0}^{N-1} q_n W_\mu^{np} + \left(\sum_{n=0}^{N-1} f_n W_\mu^{-np}\right)E. \tag{5.46}$$

Let Q_p be the N-point rs-QDFTs of the signal q_n and let F_p the N-point rs-QDFTs of f_n. Since $W_\mu^{-np} = W_\mu^{n(N-p)}$, we obtain

$$O_p = Q_p + F_{N-p}E, \quad p = 0 : (N-1). \tag{5.47}$$

This also can be opened as

$$O_p = (Q_e)_p + i(Q_i)_p + j(Q_j)_p + k(Q_k)_p$$
$$+ [(F_e)_{N-p} + i(F_i)_{N-p} + j(F_j)_{N-p} + k(F_k)_{N-p}]E,$$

and therefore,

$$\begin{aligned}O_p = & (Q_e)_p + i(Q_i)_p + j(Q_j)_p + k(Q_k)_p \\ & + E(F_e)_{N-p} + I(F_i)_p + J(F_j)_{N-p} + K(F_k)_{N-p}.\end{aligned}$$

Thus, the N-point rs-ODFT of the octonion signal o_n can be reduced to calculation of two N-point right-side QDFTs of the quaternion signals q_n and f_n.

The principle of conservation of energy holds for the 1-D rs-ODFT,

$$E(o) = \sum_{n=0}^{N-1} |o_n|^2 = E(O) = \frac{1}{N}\sum_{p=0}^{N-1} |O_p|^2, \tag{5.48}$$

where

$$\begin{aligned}|o_n|^2 &= (o_e)_n^2 + (o_i)_n^2 + (o_j)_n^2 + (o_k)_n^2 + (o_E)_n^2 + (o_I)_n^2 + (o_J)_n^2 + (o_K)_n^2, \\ |O_p|^2 &= (O_e)_p^2 + (O_i)_p^2 + (O_j)_p^2 + (O_k)_p^2 + (O_E)_p^2 + (O_I)_p^2 + (O_J)_p^2 + (O_K)_p^2.\end{aligned}$$

Here, for the quaternion units $u = e, i, j$, and k, we denote $(O_u)_p = (Q_u)_p$, and for the corresponding octonion units $U = E, I, J$, and K, we denote $(O_U)_p = (F_u)_{N-p}$, when $p = 1 : (N-1)$, and $(O_U)_0 = (F_u)_0$.

Example 5.7

Consider the $N = 8$ case with the following octonion signal:

$$\mathbf{o} = \begin{bmatrix}\mathbf{o}_e \\ \mathbf{o}_i \\ \mathbf{o}_i \\ \mathbf{o}_j \\ \mathbf{o}_E \\ \mathbf{o}_I \\ \mathbf{o}_J \\ \mathbf{o}_K\end{bmatrix} = \begin{bmatrix} -1 & 2 & -1 & 2 & 2 & 0 & 1 & 1 \\ 2 & 0 & -3 & 3 & -1 & -3 & 1 & -2 \\ 0 & 0 & -2 & 0 & -2 & -2 & 1 & -1 \\ 2 & -2 & -2 & 0 & -1 & -1 & 0 & 1 \\ -2 & 2 & -2 & -1 & -2 & 2 & 0 & 2 \\ -1 & 1 & -2 & 2 & -2 & -3 & -1 & -3 \\ -2 & -1 & 0 & -1 & 3 & -3 & 1 & 3 \\ -2 & 0 & -3 & -2 & 3 & -2 & -2 & 2 \end{bmatrix}.$$

The real part of the octonion signal o_n is written in the first row of the matrix, and the seven-component imaginary part of the signal are written in the remaining seven rows.

With the pure unit quaternion $\mu = (i + j + k)/\sqrt{3}$, the 8-point rs-ODFT of this signal is

$$\mathbf{O} = \begin{bmatrix} \mathbf{O}_e \\ \mathbf{O}_i \\ \mathbf{O}_i \\ \mathbf{O}_j \\ \mathbf{O}_E \\ \mathbf{O}_I \\ \mathbf{O}_J \\ \mathbf{O}_K \end{bmatrix}$$

$$= \begin{bmatrix} 6 & -0.7710 & 6.1962 & -12.5775 & -4 & 5.1633 & -4.1962 & -3.8148 \\ -3 & 3.1197 & 4.1547 & 9.4123 & 1 & -0.5838 & 1.8453 & 0.0519 \\ -6 & -0.1506 & -1.5774 & -1.5648 & 0 & 4.1506 & -0.4226 & 5.5648 \\ -3 & 4.5339 & 0.6906 & 7.9980 & 1 & -1.9980 & 5.3094 & 1.4661 \\ -1 & 2.4305 & -7.1962 & 1.6520 & -11 & -5.8946 & 3.1962 & 1.8121 \\ -9 & 2.6723 & 0.5774 & 1.7772 & -3 & 1.6371 & -0.5774 & -2.0866 \\ 0 & 6.1128 & -1.1547 & -4.6819 & 4 & -13.8034 & 1.1547 & -7.6275 \\ -6 & -2.8686 & 1.3812 & -13.6633 & -2 & -4.8220 & 10.6188 & 1.3539 \end{bmatrix}.$$

We also can verify the principle of conservation of energy for this 1-D rs-ODFT,

$$E(o) = \sum_{n=0}^{7} |o_n|^2 = 212, \quad E(O) = \frac{1}{8}\sum_{p=0}^{7} |O_p|^2 = \frac{1}{8}1696 = 212.$$

When using the pure unit quaternion $\mu = (i + k)/\sqrt{2}$, the 8-point rs-ODFT of this signals is

$$\mathbf{O} = \begin{bmatrix} \mathbf{O}_e \\ \mathbf{O}_i \\ \mathbf{O}_j \\ \mathbf{O}_k \\ \mathbf{O}_E \\ \mathbf{O}_I \\ \mathbf{O}_J \\ \mathbf{O}_K \end{bmatrix}$$

$$= \begin{bmatrix} 6 & -1.0503 & 6.6569 & -10.9497 & -4 & 3.5355 & -4.6569 & -3.5355 \\ -3 & 1.0503 & 1.5858 & 10.9497 & 1 & -2.1213 & 4.4142 & 2.1213 \\ -6 & -0.8787 & -1.0000 & -5.1213 & 0 & 7.7071 & -1.0000 & 6.2929 \\ -3 & 3.7071 & 3.0000 & 2.2929 & 1 & 3.7071 & 3.0000 & 2.2929 \\ -1 & 4.2071 & -4.1213 & 2.7929 & -11 & -7.0355 & 0.1213 & 0.0355 \\ -9 & 4.9142 & 2.1213 & 2.0858 & -3 & 1.3284 & -2.1213 & -4.3284 \\ 0 & 4.7426 & 0.7071 & -3.7426 & 4 & -14.7426 & -0.7071 & -6.2574 \\ -6 & 0.4497 & -0.3640 & -9.4497 & -2 & -9.0355 & 12.3640 & -1.9645 \end{bmatrix}.$$

Below is the script of the demo code "test_rs-odft.m" for calculating the N-point rs-ODFTs for the above example. Two functions "o1dfft.m" and "oi1dfft.m" are written, to calculate the direct and inverse 1-D rs-QDFT, respectively.

```
%--------------------------------------------------------------
% test_1Dofft.m / Artyom Grigoryan, March 30, 2016
% The fast 1-D N-point ODFT (with mu equal a pure unit
  quaternion)
%--------------------------------------------------------------
 N=8;
 oct_signal=zeros(8,N);
 oct_signal=[-1  2 -1  2  2  0  1  1
              2  0 -3  3 -1 -3  1 -2
              0  0 -2  0 -2 -2  1 -1
              2 -2 -2  0 -1 -1  0  1
             -2  2 -2 -1 -2  2  0  2
             -1  1 -2  2 -2 -3 -1 -3
             -2 -1  0 -1  3 -3  1  3
             -2  0 -3 -2  3 -2 -2  2 ];
%1. Calculate the direct N-point rs-ODFT of the octonion
   signal
   qmu=[1,1,1]/sqrt(3); % or qmu=[1,0,1]/sqrt(2);
   FT_ofsignal=o1dfft(oct_signal,qmu,N);
%2. Calculate the inverse N-point rs-ODFT to verify the
   result
   osignal=oi1dfft(FT_ofsignal,qmu,N);
%3. To check the Parseval's equality: Energy preservation
   Energy_ofsignal=0;
   for im=1:8
       for n=1:N
           Energy_ofsignal=Energy_ofsignal+oct_signal
           (im,n)^2;
       end
   end
   Energy_ofODFT=0;
   for im=1:8
       for n=1:N
           Energy_ofODFT=Energy_ofODFT+FT_ofsignal
           (im,n)^2;
       end
   end
 R=Energy_ofODFT/Energy_ofsignal % R is N (or 8)
%--------------------------------------------------------------
```

```
% call: oldfft.m / Art Grigoryan, April 2, 2016
% calculate the N-point octonion right-side DFT
%       oct_signal = q1_part + (q2_part) E
%-----------------------------------------------------------
function FT_ofsignal=oldfft(oct_signal,qmu,N)
  q1_part=oct_signal(1:4,:);
  q2_part=oct_signal(5:8,:);
  FQ1 = q1dfft(q1_part,qmu,N); % 1-D rs-QDFT of q1_part
  FQ2 = q1dfft(q2_part,qmu,N); % 1-D rs-QDFT of q2_part
  % Composition of the N-point ODFT from two 1-D QDFTs
  FQ2n =zeros(4,N);
  FQ2n(:,1)=FQ2(:,1);
  FQ2n(:,2:N)=FQ2(:,N:-1:2);
  FT_ofsignal=zeros(8,N);
  FT_ofsignal(1:4,:)=FQ1; FT_ofsignal(5:8,:)=FQ2n;
%-----------------------------------------------------------
% call: oldfft.m / Art Grigoryan, April 2, 2016
% calculate the inverse N-point octonion right-side DFT
function osignal=oildfft(FT_ofsignal,qmu,N)
  FQ1=FT_ofsignal(1:4,:); FQ2=FT_ofsignal(5:8,:);
  q1i=iq1dfft(FQ1,qmu,N); % inverse rs-QDFT of the 1st part
  q2i=iq1ddft(FQ2,qmu,N); % inverse rs-QDFT of the 2nd part
  % signal composition from two inverse QDFTs
  q2in =zeros(4,N);
  q2in(:,1)=q2i(:,1);
  q2in(:,2:N)=q2i(:,N:-1:2);
  osignal=zeros(8,N);
  osignal(1:4,:)=q1i; osignal(5:8,:)=q2in;
%-----------------------------------------------------------
```

5.4.2 1-D left-side ODFT

The concept of the 1-D left-side octonion discrete Fourier transform (ls-ODFT) is described similarly to the rs-ODFT. Let o_n be the octonion signal of length N, i.e.,

$$o_n = a_n + ib_n + jc_n + kd_n + EA_n + IB_n + JC_n + KD_n,$$

where $n = 0 : (N-1)$. The N-point ls-ODFT of the signal is defined as

$$O_p^{(ls)} = O_p = \sum_{n=0}^{N-1} W_\lambda^{np} o_n, \quad p = 0 : (N-1). \tag{5.49}$$

Here, λ is a pure unit octonion, $\lambda^2 = -1$. The inverse transform is

$$o_n = \frac{1}{N}\sum_{p=0}^{N-1} W_\lambda^{-np} O_p, \quad n = 0 : (N-1). \tag{5.50}$$

The proof of this formula can be obtained by calculations similar to ones that were given above for the inverse right-side ODFT in Eq. (5.42). Considering the number λ to be a pure quaternion $\mu = i\mu_i + j\mu_j + k\mu_k$, $\mu^2 = -1$, we define the 1-D N-point ls-ODFT as

$$O_p = \sum_{n=0}^{N-1} W_\mu^{np} o_n, \quad p = 0 : (N-1), \tag{5.51}$$

where the quaternion exponential function

$$W_\mu^{np} = W_{\mu;N}^{np} = \cos(2\pi np/N) - \mu \sin(2\pi np/N).$$

This transform can be calculated by using fast algorithms of the 1-D left-side quaternion DFT (ls-QDFT) and rs-QDFT. The octonion signal is the double quaternion signal

$$o_n = q_n + f_n E, \quad n = 0 : (N-1),$$

where q_n and f_n are quaternion signals. Equation 5.49 can be written as

$$O_p = \sum_{n=0}^{N-1} W_\mu^{np}(q_n + f_n E) = \sum_{n=0}^{N-1} W_\mu^{np} q_n + \sum_{n=0}^{N-1} W_\mu^{np}(f_n E) \tag{5.52}$$

when $p = 0 : (N-1)$.

It should be noted that for many quaternions q_1 and q_2, $q_2(q_1E) \neq (q_2q_1)E$. For instance, when $q_1 = 2 - i + 3j + 4k$ and $q_2 = 4 - 2i + j + 5k$, we obtain

$$\begin{aligned}(2 - i + 3j + 4k)[(4 - 2i + j + 5k)E] &= (2 - i + 3j + 4k)(4E - 2I + J + 5K)\\ &= -17E + 3I + 11J + 31K,\end{aligned}$$

and

$$\begin{aligned}[(2 - i + 3j + 4k)(4 - 2i + j + 5k)]E &= (-17 - 19i + 17j + 21k)E\\ &= -17E - 19I + 17J + 21K.\end{aligned}$$

To obtain the algorithm for the ls-ODFT, we consider the multiplication $q = q_1(q_2E)$ for two quaternions $q_1 = a_1 + ib_1 + jc_1 + kd_1$ and $q_2 = a_2 + ib_2 + jc_2 + kd_2$. The following calculations hold:

$$\begin{aligned} q_1(q_2E) &= q_1(a_2E + b_2I + c_2J + d_2K) \\ &= (a_1 + ib_1 + jc_1 + kd_1)(a_2E + b_2I + c_2J + d_2K) \\ &= (a_1a_2 - b_1b_2 - c_1c_2 - d_1d_2)E \\ &\quad + (a_1b_2 + b_1a_2 - c_1d_2 + d_1c_2)I \\ &\quad + (a_1c_2 + c_1a_2 + b_1d_2 - d_1b_2)J \\ &\quad + (a_1d_2 + d_1a_2 - b_1c_2 + c_1b_2)K. \end{aligned}$$

We can write the calculation of the E, I, J, and K components of this multiplication in matrix form,

$$\begin{bmatrix} q_E \\ q_I \\ q_J \\ q_K \end{bmatrix} = \begin{bmatrix} a_2 & -b_2 & -c_2 & -d_2 \\ b_2 & a_2 & -d_2 & c_2 \\ c_2 & d_2 & a_2 & -b_2 \\ d_2 & -c_2 & b_2 & a_2 \end{bmatrix} \begin{bmatrix} a_1 \\ b_1 \\ c_1 \\ d_1 \end{bmatrix}. \tag{5.53}$$

As directly follows form the matrix **A** of the multiplication of two quaternions in Eq. (1.24), the 4×4 matrix in the above equation corresponds to the multiplication q_2q_1. Since by the definition, $iE = I, jE = J$, and $kE = K$, we obtain that

$$q_1(q_2E) = (q_2q_1)E. \tag{5.54}$$

For instance, when $q_1 = 2 - i + 3j + 4k$ and $q_2 = 4 - 2i + j + 5k$, we obtain

$$(2 - i + 3j + 4k)(4E - 2I + J + 5K) = -17E - 19I + 17J + 21K$$

and

$$\begin{aligned} (4 - 2i + j + 5k)(2 - i + 3j + 4k) &= -17 - 19i + 17j + 21k, \\ (-17 - 19i + 17j + 21k)E &= -17E - 19I + 17J + 21K. \end{aligned}$$

Thus, each addend in the second sum in Eq. (5.52) can be written as

$$W_\mu^{np}(f_nE) = (f_nW_\mu^{np})E,$$

and therefore,

$$\sum_{n=0}^{N-1} W_\mu^{np}(f_nE) = \sum_{n=0}^{N-1}(f_nW_\mu^{np})E = \left(\sum_{n=0}^{N-1} f_nW_\mu^{np}\right)E.$$

Let Q_p be the N-point ls-QDFT of the signal q_n and let F_p be the rs-QDFT of f_n. The ls-ODFT in Eq. (5.52) can therefore be written as

$$O_p = Q_p + F_p E = Q_p^{(ls)} + F_p^{(rs)} E, \quad p = 0 : (N-1), \tag{5.55}$$

which also can be opened as

$$O_p = (Q_e)_p + i(Q_i)_p + j(Q_j)_p + k(Q_k)_p + [(F_e)_p + i(F_i)_p + j(F_j)_p + k(F_k)_p]E.$$

Thus,

$$O_p = (Q_e)_p + i(Q_i)_p + j(Q_j)_p + k(Q_k)_p + E(F_e)_p + I(F_i)_p + J(F_j)_p + K(F_k)_p.$$

Example 5.8

Consider the $N = 8$ case with the octonion signal of Example 5.7:

$$o = o_e + o_i + o_j + o_k + o_E + o_I + o_J + o_K.$$

When the pure unit quaternion is $\mu = (i + j + k)/\sqrt{3}$, the 8-point ls-ODFT of this signal equals

$$\mathbf{O} = \begin{bmatrix} \mathbf{O}_e \\ \mathbf{O}_i \\ \mathbf{O}_i \\ \mathbf{O}_j \\ \mathbf{O}_E \\ \mathbf{O}_I \\ \mathbf{O}_J \\ \mathbf{O}_K \end{bmatrix}$$

$$= \begin{bmatrix} 6 & -0.7710 & 6.1962 & -12.5775 & -4 & 5.1633 & -4.1962 & -3.8148 \\ -3 & 0.1919 & 0.6906 & 4.1751 & 1 & 4.6534 & 5.3094 & 2.9796 \\ -6 & 5.7049 & -1.5774 & 8.9095 & 0 & -6.3237 & -0.4226 & -0.2907 \\ -3 & 1.6062 & 4.1547 & 2.7609 & 1 & 3.2391 & 1.8453 & 4.3938 \\ -1 & 1.8121 & 3.1962 & -5.8946 & -11 & 1.6520 & -7.1962 & 2.4305 \\ -9 & -2.0866 & -0.5774 & 1.6371 & -3 & 1.7772 & 0.5774 & 2.6723 \\ 0 & -7.6275 & 1.1547 & -13.8034 & 4 & -4.6819 & -1.1547 & 6.1128 \\ -6 & 1.3539 & 10.6188 & -4.8220 & -2 & -13.6633 & 1.3812 & -2.8686 \end{bmatrix}.$$

With the pure unit quaternion $\mu = (i + k)/\sqrt{2}$, the 8-point ls-ODFT of this signal is

$$
\mathbf{O} = \begin{bmatrix} \mathbf{O}_e \\ \mathbf{O}_i \\ \mathbf{O}_i \\ \mathbf{O}_j \\ \mathbf{O}_E \\ \mathbf{O}_I \\ \mathbf{O}_J \\ \mathbf{O}_K \end{bmatrix}
$$

$$
= \begin{bmatrix}
6 & -1.0503 & 6.6569 & -10.9497 & -4 & 3.5355 & -4.6569 & -3.5355 \\
-3 & 1.0503 & 1.5858 & 10.9497 & 1 & -2.1213 & 4.4142 & 2.1213 \\
-6 & -0.8787 & -1.0000 & -5.1213 & 0 & 7.7071 & -1.0000 & 6.2929 \\
-3 & 3.7071 & 3.0000 & 2.2929 & 1 & 3.7071 & 3.0000 & 2.2929 \\
-1 & 4.2071 & -4.1213 & 2.7929 & -11 & -7.0355 & 0.1213 & 0.0355 \\
-9 & 4.9142 & 2.1213 & 2.0858 & -3 & 1.3284 & -2.1213 & -4.3284 \\
0 & 4.7426 & 0.7071 & -3.7426 & 4 & -14.7426 & -0.7071 & -6.2574 \\
-6 & 0.4497 & -0.3640 & -9.4497 & -2 & -9.0355 & 12.3640 & -1.9645
\end{bmatrix}.
$$

Below are two scripts for calculating the direct and inverse N-point 1-D ls-ODFTs.

```
%-------------------------------------------------------------
% call: o1dlsfft.m / Artyom Grigoryan, April 4, 2016
% The fast 1-D N-point ls-ODFT of the octonion signal
%          oct_signal = q1_part + (q2_part) E
function FT_ofsignal=o1dlsfft(oct_signal,qmu,N)
  q1_part=oct_signal(1:4,:);
  q2_part=oct_signal(5:8,:);
  FQ1 = q1lsdfft(q1_part,qmu,N); % 1-D ls-QDFT of q1_part
  FQ2 = q1dfft(q2_part,qmu,N);    % 1-D rs-QDFT of q2_part
  % Composition of the N-point ls-ODFT from these 1-D QDFTs
  FT_ofsignal=zeros(8,N);
  FT_ofsignal(1:4,:)=FQ1; FT_ofsignal(5:8,:)=FQ2;
%-------------------------------------------------------------
% call: oi1dlsfft.m / Artyom Grigoryan, April 4, 2016
% The fast inverse 1-D N-point ls-ODFT
function osignal=oi1dlsfft(FT_ofsignal,qmu,N)
  FQ1=FT_ofsignal(1:4,:); FQ2=FT_ofsignal(5:8,:);
  q1i=iq1lsdfft(FQ1,qmu,N); % inverse ls-QDFT of the 1st part
  q2i=iq1ddft(FQ2,qmu,N); % inverse rs-QDFT of the 2nd part
```

```
    % signal composition from two inverse QDFTs
    osignal=zeros(8,N);
    osignal(1:4,:)=q1i; osignal(5:8,:)=q2i;
%-----------------------------------------------------------
```

The principle of conservation of energy holds for the 1-D ls-ODFT,

$$E(o) = \sum_{n=0}^{N-1} |o_n|^2 = E(O) = \frac{1}{N}\sum_{p=0}^{N-1} |O_p|^2, \tag{5.56}$$

where

$$|o_n|^2 = (q_e)_n^2 + (q_i)_n^2 + (q_j)_n^2 + (q_k)_n^2 + (f_e)_n^2 + (f_i)_n^2 + (f_j)_n^2 + (f_k)_n^2,$$
$$|O_p|^2 = (Q_e)_p^2 + (Q_i)_p^2 + (Q_j)_p^2 + (Q_k)_p^2 + (F_e)_p^2 + (F_i)_p^2 + (F_j)_p^2 + (F_k)_p^2.$$

References

[1] Z. Yang and S.-I. Kamata, "Hypercomplex polar Fourier analysis for color image," *Proc. International Conference on Image Processing (ICIP)*, 2117–2120 (2011).

[2] A. M. Grigoryan and S. S. Agaian, "Retolling of color imaging in the quaternion algebra," *Applied Mathematics and Sciences: An International Journal (MathSJ)*, **1**(3), 23–39 (2014).

[3] B. C. Ujang, C. C. Took, and D. P. Mandic, "On quaternion analyticity: Enabling quaternion-valued nonlinear adaptive filtering," *ICASSP, IEEE International Conference on Acoustics, Speech and Signal Processing - Proceedings* (2012).

[4] A. Greenblatt, C. Mosquera-Lopez, and S. S. Agaian, "Quaternion neural networks applied to prostate cancer Gleason grading," in Proceedings *Proc. Man and Cybernetics, SMC 2013. IEEE International Conference*, Manchester, 1144–1149 (2013).

[5] A. M. Grigoryan and S. S. Agaian, "Alpha-rooting method of color image enhancement by discrete quaternion Fourier transform," [9019–3], *Proc. SPIE*, **9019**, 901904 (2014) [doi: 10.1117/12.2040596].

[6] B. Chen, G. Coatrieux, G. Chen, J.-L. Coatrieux, and H. S. Full, "4-D Quaternion discrete Fourier transform based watermarking for color images," *Digital Signal Processing*, **28**, 106–119 (2014).

[7] A. M. Grigoryan, J. Jenkinson, and S. S. Agaian, "Quaternion Fourier transform based alpha-rooting method for color image measurement and enhancement," *SIGPRO-D-14-01083R1, Signal Processing*, **109**, 269–289 (2015) [doi: 10.1016/j.sigpro.2014.11.019]

[8] S. Qi, J. Ma, H. Li, S. Zhang, and J. Tian, "Infrared small target enhancement via phase spectrum of quaternion Fourier transform," *Infrared Physics & Technology*, **62**, 50–58 (2014).

[9] A. M. Grigoryan and S. S. Agaian, "Color enhancement and correction for camera cell phone medical images using quaternion tools," Ch. 4, 77–117. In book: Electronic Imaging Applications in Mobile Healthcare, (J. Tang, S. S. Agaian, and J. Tan, Eds.,) *SPIE Press*, Bellingham, Washington (2016) [doi: 10.1117/3.2204748.ch4].

[10] R. C. Gonzalez and R. E. Woods, *Digital Image Processing*, 2nd Edition, Prentice Hall, Upper Saddle River, New Jersey (2002).

[11] S. J. Sangwine, "Fourier transforms of colour images using quaternion, or hypercomplex, numbers," *Electronics Letters*, **32**(21), 1979–1980 (1996).

[12] S. J. Sangwine, "The discrete quaternion-Fourier transform," *Proc. IPA97*, Conference Publication, **443**, 790–793 (1997).

[13] E. Hitzer, "Quaternion Fourier transform on quaternion fields and generalizations," *Advances in Applied Clifford Algebras*, **17**(3), 497–517 (2007).

[14] M. H. Yeh, "Relationships among various 2-D quaternion Fourier transforms," *IEEE Signal Processing Letters*, **15**, 669–672 (2008).

[15] S. J. Sangwine, T. A. Ell, J. M. Blackledge, and M. J. Turner, "The discrete Fourier transform of a color image," in Proc. *Image Processing II Mathematical Methods, Algorithms and Applications*, 430–441 (2000).

[16] S. J. Sangwine and T. A. Ell, "Hypercomplex Fourier transforms of color images," in Proc. *IEEE Intl. Conf. Image Process.*, **1**, 137–140 (2001).

[17] A. M. Grigoryan and S. S. Agaian, "Tensor transform-based quaternion Fourier transform algorithm," *Information Sciences*, **320**, 62–74 (2015).

[18] A. M. Grigoryan and S. S. Agaian, "2-D Left-side quaternion discrete Fourier transform fast algorithms," *Proc. IS &T International Symposium, 2016 Electronic Imaging: Algorithms and Systems XIV*, San Francisco, California (2016).

[19] J. W. Cooley and J. W. Tukey, "An algorithm the machine computation of complex Fourier series," *Math. Comput*, **9**(2), 297–301 (1965).

[20] D. F. Elliot and K. R. Rao, *Fast transforms – Algorithms, Analyzes, and Applications*, Academic, New York (1982).

[21] L. Rabiner and R. W. Schaefer, *Speech Signal Processing*, Prentice Hall, Englewood Cliffs, New Jersey (1983).

[22] R. E. Blahut, *Fast Algorithms for Digital Signal Processing*, Addison-Wesley, Reading, Mass (1985).

[23] J. S. Walker, *Fast Fourier Transforms*, Florida: CRC Press (1991).

[24] S. Agaian and D. Gevorkian, "Synthesis of a class of orthogonal transforms. Parallel SIMD-algorithms and specialized processors," *Pattern Recognition and Image Analysis*, **2**, 394–408 (1992).

[25] C. Hsiao, Y. Chen, and C. Lee "A generalized mixed-radix algorithm for memory-based FFT processors," *IEEE Trans. Circuits Syst. II, Exp. Briefs*, **57**(1), 26–30 (2010).

[26] Y. Zhou, W. Cao, L. Liu, S. Agaian, and C. L. P. Chen, "Fast Fourier transform using matrix decomposition," *Information Sciences*, **291**, 172–183 (2015).
[27] A. M. Grigoryan, "New algorithms for calculating discrete Fourier transforms," *Journal Vichislitelnoi Matematiki i Matematicheskoi Fiziki*, AS USSR, Moscow, **26**(9), 1407–1412 (1986). (Available in http://fasttransforms.com/Art-USSR-papers/Art_JVMATofASUSSR1986.pdf)
[28] A. M. Grigoryan, "2-D and 1-D multi-paired transforms: Frequency-time type wavelets," *IEEE Trans. on Signal Processing*, **49**(2), 344–353 (2001).
[29] A. M. Grigoryan and S. S. Agaian, "Split manageable efficient algorithm for Fourier and Hadamard transforms, *IEEE Trans. Signal Processing*, **48**(1), 172–183 (2000).
[30] A. M. Grigoryan and S. S. Agaian, *Multidimensional discrete unitary transforms: Representation, partitioning and algorithms*, Marcel Dekker Inc., New York (2003).
[31] A. M. Grigoryan and M. M. Grigoryan, *Brief notes in advanced DSP: Fourier analysis with MATLAB*, CRC Press, Taylor and Francis Group, Boca Raton, New York (2009).
[32] N. Ranganadh, A. M. Grigoryan, and B. D. Tushara, "Implementation of FFT by using MATLAB: SIMULINK on Xilinx Virtex-4 FPGAs: Performance of a paired transform based FFT," *International Journal of Advanced Computer Research*, **3**(2), 72–78 (2013).
[33] N. Ranganadh and M. Guravaiah, "Performance case study of Grigoryan FFT over Cooley-Tukey FFT using TMS DSP processors," *International Journal of Advanced Computer Research*, **2**(4), 452–457 (2012).
[34] R. Narayanam, P. Parimal, and A. M. Grigoryan, "Performances of Texas Instruments DSP and Xilinx FPGAs for Cooley-Tukey and Grigoryan FFT algorithms," *SOURCE Journal of Engineering & Technology*, **1**(2), p. 83 (2011).

Chapter 6
2-D Quaternion and Octonion Discrete Fourier Transforms

There were many attempts to generalize the known concept of the 2-D DFT, including the 2-D quaternion discrete Fourier transform tailored to color images [2]–[12]. The two-side quaternion discrete Fourier transformation (QDFT) was introduced in [1] for the analysis of 2-D linear time-invariant partial-differential systems. The classical fast algorithms are based on representation of the QDFT by a combinations of a few classical DFT transforms. This allows us to obtain QDFT fast numerical implementation with the standard FFT algorithms [13]–[15]. In this chapter, we first describe the concept of the 2-D DFT and then the two-side, right, and left-side QDFTs with MATLAB®-based scripts. We start with the effective tensor transform-based algorithm that reduces the 2-D DFT transform to calculation of the separate 1-D DFTs [22]–[25]. Then, we describe the 2-D QDFTs, which include the tensor algorithm for calculating the 2-D QDFT [26, 27] and 2-D octonion DFT (ODFT).

6.1 Tensor Representation of the Image

The $N \times N$-point DFT of the discrete image $f_{n,m}$ is calculated by

$$F_{p,s} = (\mathcal{F}_{N,N} \circ f)_{p,s} = \sum_{n=0}^{N-1} \sum_{m=0}^{N-1} f_{n,m} W^{np+ms}, \tag{6.1}$$

where $W = \exp(-2\pi j/N)$. The frequency-points (p, s) are from the set of integers $X_{N,N} = \{(p, s); p, s = 0 : (N - 1)\}$. For simplicity of calculations, we consider the image of the size $N \times N$ and the cases of the most interest, when $N = 2^r$, $r > 1$, and $N > 2$ is prime.

The main idea of the tensor representation of the image can be described in the following way [16, 22]. The set (or lattice) of frequency-point $X_{N,N}$ can be covered by a family of sets $(T_k)_{k=1:l}$, where $(N + 1) \leq l \leq 2N$, in a way that the 2-D Fourier transform of the 2-D image $f_{n,m}$ at the set T_k becomes an image of the 1-D N-point Fourier transform, $\mathcal{F}_N$, of an 1-D signal, $f^{(k)}$, of length N. When $N > 2$ is prime, the number $l = (N+1)$, and when $N = 2^r$, the number of such subsets $L = N + N/2$. This statement means the following:

(a) l 1-D transformations $\mathcal{F}_N$ describe a splitting of the 2-D DFT $\mathcal{F}_{N,N}$,

$$\mathcal{F}[f] \leftrightarrow \left\{\mathcal{F}_N[f^{(1)}], \mathcal{F}_N[f^{(2)}], \ldots, \mathcal{F}_N[f^{(l)}]\right\}. \tag{6.2}$$

(b) The set of splitting-signals $f^{(k)}$, $k = 1 : l$, defines completely the image

$$\{f^{(1)}, f^{(2)}, \ldots, f^{(l)}\} \leftrightarrow f.$$

Given an integer triplet (p, s, t), where $p, s, t \in [0, N-1]$, we consider the following set of points in the lattice $X_{N,N}$:

$$V_{p,s,t} = \{(n,m);\ n, m = 0 : (N-1),\ \overline{np + ms} = t\}, \tag{6.3}$$

where $\bar{t} = t \bmod N$, and characteristic function of this set

$$\chi_{p,s,t}(n,m) = \begin{cases} 1, & \text{if } (n,m) \in V_{p,s,t}, \\ 0, & \text{otherwise.} \end{cases} \tag{6.4}$$

$V_{p,s,t}$, if it is not empty, is the set of points (n, m) along a maximum of $(p+s)$ parallel straight lines at an angle of $\psi = \arctan(p/s)$ to the horizontal x-axis. The equations for the lines are

$$\left.\begin{array}{l} xp + ys = t \\ xp + ys = t + N \\ \ldots \quad\quad \cdot \quad \ldots \\ xp + ys = t + (p+s-1)N \end{array}\right\} \tag{6.5}$$

in the square $[0, N] \times [0, N]$. We define this set of parallel lines by $\mathcal{L}_{p,s,t}$. As an example, Fig. 6.1 shows the elements of the set $V_{1,2,2}$ on the discrete lattice $X_{8,8}$, which lie on three lines. All points of the set $V_{p,s,t}$ lie on the parallel lines passing through the knots of the discrete net $X_{N,N}$ traced on the square. It means that all '1's in the mask of the 2-D function $\chi_{p,s,t}(n,m)$ are located on parallel lines.

We now describe the tensor representation of the image for the Fourier transform (Grigoryan, 1984, [16]). The sets T are considered to be the groups

$$T_{p,s} = \{(\overline{kp}, \overline{ks});\quad k = 0 : (N-1)\},\quad (T_{0,0} = \{(0,0)\}). \tag{6.6}$$

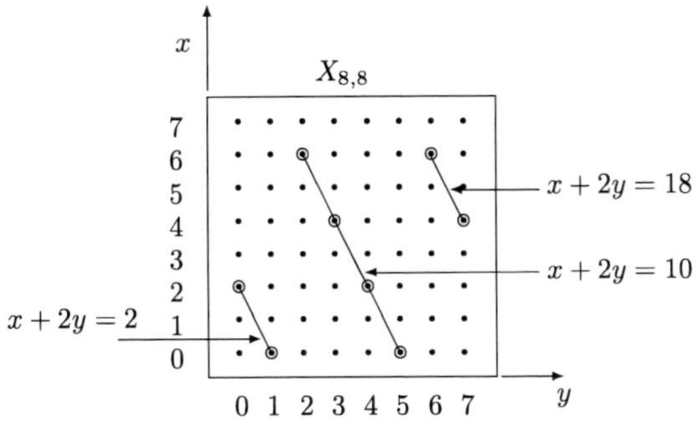

Figure 6.1 Eight points (n, m) of the set $V_{1,2,2}$ lie on the three straight lines.

The points of the set $T_{p,s}$ lie on a few parallel lines at the angle of $\theta = \arctan(s/p)$ to the horizontal axis x. This direction is perpendicular to the parallel straight lines on which the points of the set $V_{p,s,t}$ are located, for all $t = 0 : (N-1)$, i.e., $\theta = \psi + \pi/2$. The same lattice $X_{N,N}$ is considered in the image plane and the frequency domain.

For instance, when $N = 3$, the frequency-points of the sets $T_{1,1}$, $T_{1,0}$, $T_{1,2}$, and $T_{0,1}$ cover the lattice $X_{3,3}$,

$$\begin{aligned} T_{1,0} &= \{(0,0),(1,0),(2,0)\}, \\ T_{1,1} &= \{(0,0),(1,1),(2,2)\}, \\ T_{1,2} &= \{(0,0),(1,2),(2,1)\}, \\ T_{0,1} &= \{(0,0),(0,1),(0,2)\}. \end{aligned}$$

For the $N = 4$, we can consider the following covering of the lattice $X_{4,4}$:

$$\begin{aligned} T_{1,0} &= \{(0,0),(1,0),(2,0),(3,0)\}, \\ T_{1,1} &= \{(0,0),(1,1),(2,2),(3,3)\}, \\ T_{1,2} &= \{(0,0),(1,2),(2,0),(3,2)\}, \\ T_{1,3} &= \{(0,0),(1,3),(2,2),(3,1)\}, \\ T_{0,1} &= \{(0,0),(0,1),(0,2),(0,3)\}, \\ T_{2,1} &= \{(0,0),(2,1),(0,2),(2,3)\}. \end{aligned}$$

Given a frequency-point (p, s), we can compose the 1-D signal, which we call *the splitting-signal*, or *the image-signal*, by the characteristic functions as follows:

$$f_{p,s,t} = \chi_{p,s,t} \circ f = \sum_{n=0}^{N-1} \sum_{m=0}^{N-1} \chi_{p,s,t}(n,m) f_{n,m} = \sum_{(n,m)\in V_{p,s,t}} f_{n,m} \tag{6.7}$$

when $t = 0 : (N-1)$. For the example shown in Fig. 6.1, two points of the set $V_{1,2,2}$ are on the line $x + 2y = 2$, four points are on the line $x + 2y = 10$, and two points are on the line $x + 2y = 18$. Therefore,

$$f_{1,2,2} = (f_{0,1} + f_{2,0}) + (f_{0,5} + f_{2,4} + f_{4,3} + f_{6,2}) + (f_{4,7} + f_{6,6}).$$

The splitting-signal generated by the frequency-point (p, s) is

$$f_{T_{p,s}} = \{f_{p,s,0}, f_{p,s,1}, \ldots, f_{p,s,N-1}\}. \tag{6.8}$$

The following important relation holds in the frequency domain:

$$F_{\overline{kp},\overline{ks}} = (\mathcal{F}_N[f_{T_{p,s}}])_k = \sum_{t=0}^{N-1} f_{p,s,t} W^{kt}, \quad k = 0 : (N-1). \tag{6.9}$$

Thus, the 2-D DFT of the image $f_{n,m}$ at frequency-points of the group $T_{p,s}$ is the N-point DFT of the splitting-signal $f_{T_{p,s}}$.

Example 6.1

Consider the $N = 5$ case and the image

$$[f] = \begin{bmatrix} \underline{1} & 2 & 1 & 2 & 1 \\ 2 & 1 & 2 & 2 & 3 \\ 5 & 2 & 1 & 2 & 1 \\ 4 & 1 & 2 & 3 & 2 \\ 3 & 2 & 3 & 2 & 1 \end{bmatrix}, \qquad \sum_{n=0}^{4} \sum_{m=0}^{4} f_{n,m} = 51.$$

For the frequency-point $(p, s) = (1, 2)$, all values of t in the equations $t = (np + ms) \bmod N$ can be written in the form of the following matrix:

$$\|t = (n \cdot 1 + m \cdot 2) \bmod 4\|_{n,m=0:4} = \begin{bmatrix} 0 & 2 & 4 & 1 & 3 \\ 1 & 3 & 0 & 2 & 4 \\ 2 & 4 & 1 & 3 & 0 \\ 3 & 0 & 2 & 4 & 1 \\ 4 & 1 & 3 & 0 & 2 \end{bmatrix}.$$

Therefore, the splitting-signal $f_{T_{1,2}}$ of the image is calculated as follows:

$$f_{T_{1,2}} = \begin{cases} f_{1,2,0} = f_{0,0} + f_{1,2} + f_{2,4} + f_{3,1} + f_{4,3} = 1+2+1+1+2 = 7, \\ f_{1,2,1} = f_{0,3} + f_{1,0} + f_{2,2} + f_{3,4} + f_{4,1} = 2+2+1+2+2 = 9, \\ f_{1,2,2} = f_{0,1} + f_{1,3} + f_{2,0} + f_{3,2} + f_{4,4} = 2+2+5+2+1 = 12, \\ f_{1,2,3} = f_{0,4} + f_{1,1} + f_{2,3} + f_{3,0} + f_{4,2} = 1+1+2+4+3 = 11, \\ f_{1,2,4} = f_{0,2} + f_{1,4} + f_{2,1} + f_{3,3} + f_{4,0} = 1+3+2+3+3 = 12. \end{cases}$$

One can note that

$$f_{1,2,0} + f_{1,2,1} + f_{1,2,2} + f_{1,2,3} + f_{1,2,4} = 7 + 9 + 12 + 11 + 12 = 51.$$

For the generator $(p, s) = (1, 1)$, the corresponding matrix for $t = (np + ms) \bmod N$ equals

$$\|t = (n \cdot 1 + m \cdot 1) \bmod 4\|_{n,m=0:3} = \begin{bmatrix} 0 & 1 & 2 & 3 & 4 \\ 1 & 2 & 3 & 4 & 0 \\ 2 & 3 & 4 & 0 & 1 \\ 3 & 4 & 0 & 1 & 2 \\ 4 & 0 & 1 & 2 & 3 \end{bmatrix},$$

and the splitting-signal $f_{T_{1,1}}$ of the image f is calculated by

$$f_{T_{1,1}} = \begin{cases} f_{1,1,0} = f_{0,0} + f_{1,4} + f_{2,3} + f_{3,2} + f_{4,1} = 1+3+2+2+2 = 10, \\ f_{1,1,1} = f_{0,1} + f_{1,0} + f_{2,4} + f_{3,3} + f_{4,2} = 2+2+1+3+3 = 11, \\ f_{1,1,2} = f_{0,2} + f_{1,1} + f_{2,0} + f_{3,4} + f_{4,3} = 1+1+5+2+2 = 11, \\ f_{1,1,3} = f_{0,3} + f_{1,2} + f_{2,1} + f_{3,0} + f_{4,4} = 2+2+2+4+1 = 11, \\ f_{1,1,4} = f_{0,4} + f_{1,3} + f_{2,2} + f_{3,1} + f_{4,0} = 1+2+1+1+3 = 8. \end{cases}$$

The sum of the components of this signal also equals 51,

$$f_{1,1,0} + f_{1,1,1} + f_{1,1,2} + f_{1,1,3} + f_{1,1,4} = 10 + 11 + 11 + 11 + 8 = 51.$$

In general, for any frequency-point (p, s),

$$\sum_{t=0}^{N-1} f_{p,s,t} = \sum_{n=0}^{N-1} \sum_{m=0}^{N-1} f_{n,m}.$$

The splitting-signals for the generators $(p, s) = (1, 0), (1, 3), (1, 4)$, and (0, 1) are calculated similarly,

$$f_{T_{1,0}} = \{7, 10, 11, 12, 11\}, \quad f_{T_{1,3}} = \{10, 7, 13, 12, 9\},$$
$$f_{T_{1,4}} = \{7, 9, 14, 10, 11\}, \quad f_{T_{0,1}} = \{15, 8, 9, 11, 8\}.$$

The sets $T_{p,s}$ with generators

$$J_{5,5} = \{(1,0),(1,1),(1,2),(1,3),(1,4),(0,1)\}$$

cover all 25 frequency-points of the lattice $X_{5,5}$. These six splitting-signals compose the tensor transform of the image. Writing these signals along the rows of the 6×5 matrix, we obtain the following tensor transform:

$$[f] \rightarrow T[f] = \begin{bmatrix} 7 & 10 & 11 & 12 & 11 \\ 10 & 11 & 11 & 11 & 8 \\ 7 & 9 & 12 & 11 & 12 \\ 10 & 7 & 13 & 12 & 9 \\ 7 & 9 & 14 & 10 & 11 \\ 15 & 8 & 9 & 11 & 8 \end{bmatrix}.$$

This matrix contains more than 25 coefficients, but the tensor transform is not redundant since the sum of coefficients of this matrix along each row equals 51. Therefore, this matrix can be reduced to the incomplete matrix with 25 coefficients as

$$T[f] = \begin{bmatrix} 7 & 10 & 11 & 12 & 11 \\ 10 & 11 & 11 & 11 & \cdot \\ 7 & 9 & 12 & 11 & \cdot \\ 10 & 7 & 13 & 12 & \cdot \\ 7 & 9 & 14 & 10 & \cdot \\ 15 & 8 & 9 & 11 & \cdot \end{bmatrix}. \tag{6.10}$$

The redundancy of the tensor transform has been removed; the transform has the same number of points as the image.

Below is the script of the MATLAB-based code "fft2tt_Nisprime.m" for the tensor 2-D DFT of the image of size $N \times N$ when N is a prime number. The splitting-signals are calculated by the function "f_pst.m."

```
% ==================================================
% Call: fft2tt_Nisprime.m  / Art Grigoryan 03/9/2016
% Calculate the NxN-point DFT by (N+1) N-point DFTs,
% for the case when N is a prime number >2.
% --------------------------------------------------
% Tree image can be found in the library of the SIPI, USC,
% by address  http://sipi.usc.edu/database/
% --------------------------------------------------
  % Open file and read the binary image:
  fid=fopen('tree.img','rb');
  X=fread(fid,[256,256]); fclose(fid); clear fid; X=X';
  N=257;  % extend this image to the size of 257x257
  X(:,257)=X(:,256); X(257,:)=X(256,:);
```

```
% The set of (N+1) generators JN:
N1=N+1;
JN=ones(2,N1);
JN(1,1:N)=0:N-1; JN(2,N1)=0;
% JN(2,1:N)=0:N-1; JN(1,N1)=0;  % also can be used
FT=zeros(N,N);  % for the 2-D DFT of the image
t=0:N-1;
for k=1:N1
      s=JN(1,k); p=JN(2,k);
      image_signal=zeros(1,N);
      image_signal=f_pst(X,p,s);
      TDFFT_signal=fft(image_signal);
      T_ps=zeros(N,2);
      T_ps(:,1)=mod(p*t,N)+1;
      T_ps(:,2)=mod(s*t,N)+1;
      for k=1:N
          p1=T_ps(k,1); s1=T_ps(k,2);
          FT(p1,s1)=TDFFT_signal(k);
      end
end
% ============================================================
function image_signal=f_pst(X,p,s)
  N=size(X,1);
  image_signal=zeros(1,N);
  t=0;
  for n=1:N
      for m=1:N
          if (t>=N) t=mod(t,N); end
          image_signal(t+1)=image_signal(t+1)+X(n,m);
          t=t+s;
      end
      t=t+p;
  end
% ------------------------------------------------------------
```

The calculation of the 2-D DFT by the tensor transform when N is a power of two can be accomplished by a similar program, by changing only the set of generators as shown below in the script "fft2tt_Npower2.m" written for the tree image of size 256 × 256. The 2-D DFT is calculated by 384 256-point DFTs, instead of 512 256-point DFTs in the traditional column-row method.

```
% ------------------------------------------------------------
% Call: fft2tt_Npower2.m   / Art Grigoryan 03/9/2016
% Calculate the NxN-point DFT by 3N/2 N-point DFTs,
```

```
% for the case when N>2 is a power of 2.
% ------------------------------------------------------------
  fid=fopen('tree.img','rb');
  X=fread(fid,[256,256]); fclose(fid); clear fid; X=X';
  N=256;
  % Set of 3N/2 generators JN
  N1=N+N/2;
  JN=ones(2,N1);
  JN(1,1:N)=0:N-1; JN(2,N+1:N1)=0:2:N-2;
  % 2-D DFT by the tensor transform
  FT=zeros(N,N);
  t=0:N-1;
  for k=1:N1
      s=JN(1,k); p=JN(2,k);
      image_signal=zeros(1,N);
      image_signal=f_pst(X,p,s);
      TDFFT_signal=fft(image_signal);
      T_ps=zeros(N,2);
      T_ps(:,1)=mod(p*t,N)+1;
      T_ps(:,2)=mod(s*t,N)+1;
      for k=1:N
          p1=T_ps(k,1); s1=T_ps(k,2);
          FT(p1,s1)=TDFFT_signal(k);
      end
  end
% ------------------------------------------------------------
```

Figure 6.2 shows the "tree" image of size 256 × 256 in part (a) and the splitting-signal $f_{T_{5,1}}$ of length 256 in part (b). The 1-D DFT over this splitting-signal is shown in Fig. 6.2(c). The transform is shifted to the center and plotted in the absolute scale. The 2-D DFT of the image is shown in absolute scale in Fig. 6.2(d), and all 256 frequency-points of the subset $T_{5,1}$, at which the 2-D DFT of the image equals the 1-D DFT of the splitting-signal, are highlighted along five parallel lines.

We also consider another example. Figure 6.3 shows the grayscale image of size 512 × 512 in part (a), and the splitting-signal $\{f_{4,1,t};\ t = 0 : 511\}$ in part (b), which is generated by the frequency-point $(p, s) = (4, 1)$. The 512-point 1-D DFT of this signal in absolute scale is shown in part (c). The 1-D DFT equals the part of the 2-D DFT of the image at 512 frequency-points of the group $T_{4,1}$. The location of these 512 frequency-points is highlighted in the picture of the 2-D DFT in part (d).

In 3-D space, one can identify the opposite sides of boundaries of the square $[0, N]^2$ and consider it as a torus and the 2-D lattice $X_{N,N}$ as a net traced on the torus. Then, for a given (p, s), the straight lines of $\mathcal{L}_{p,s,t}$ in

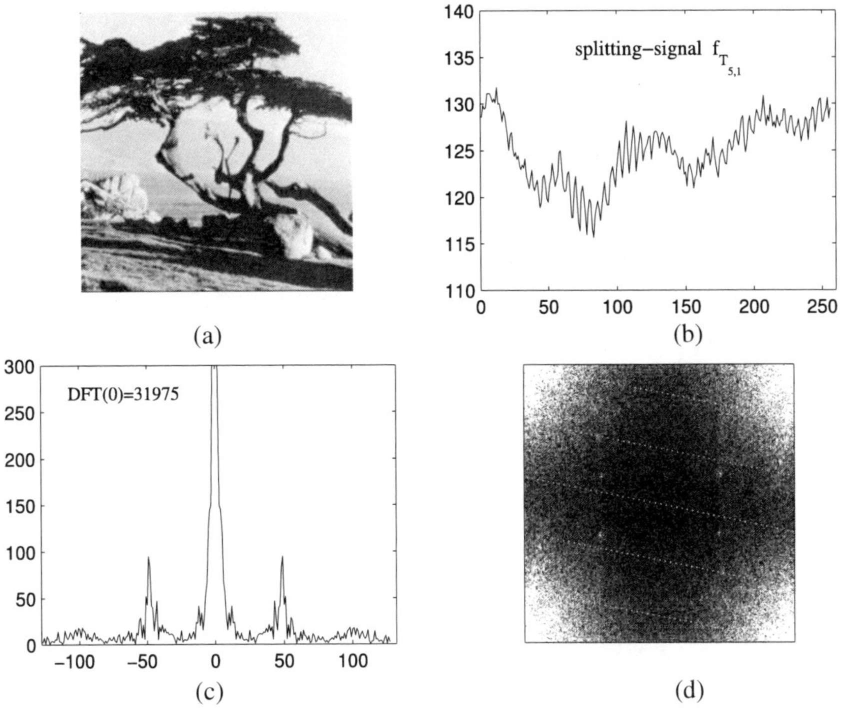

Figure 6.2 (a) The "tree" image, (b) the splitting-signal $f_{T_{5,1}}$, (c) the 1-D DFT of the splitting-signal, and (d) the 2-D DFT of the image with points of the set $T_{5,1}$. (The DFTs are shown in the absolute scale.)

Eq. (6.5) compose one discrete spiral $S_{p,s,t}$ on the torus, because of the equality $\overline{xp+ys} = \overline{t+kN} = t$ for all integers k. As an example for the $N = 32$ case, Fig. 6.4 shows the locus of two spirals $S_{1,1,3}$ and $S_{1,1,7}$ on the net, for the frequency-point $(p, s) = (1, 1)$. The values of the components $f_{1,1,3}$ and $f_{1,1,7}$ of the splitting-signal $f_{T_{1,1}} = \{f_{1,1,t};\ t = 0:31\}$ are the sums of the image along the spirals $S_{1,1,3}$ and $S_{1,1,7}$, respectively.

In the general case $N > 1$, the splitting-signal $\{f_{p,s,t};\ t = 0:(N-1)\}$ determines the 2-D DFT of the image $f_{n,m}$ at frequency-points of the group $T_{p,s}$. If we also consider the lattice $X_{N,N}$ of all frequency-points as a net on the torus, the group $T_{p,s}$ will be situated on one spiral, which we denote by $S_{p,s}$. If we imagine spirals $S_{p,s,t}$, $t = 0:(N-1)$, and the spiral $S_{p,s}$ on the same torus, the spiral $S_{p,s}$ will pass through the initial point (0, 0) of the net and make the angle $\pi/2$ with the spirals $S_{p,s,t}$.

As an example for the $N = 256$ case, Fig. 6.5 illustrates the locus of spiral $S_{5,1}$ on the 3-D net with the knots of the grid 256×256 in part (a). The net $X_{256,256}$ on the torus is not shown. The frequency-points of the spiral $S_{5,1}$ compose five circuits around the torus. This spiral corresponds to five parallel

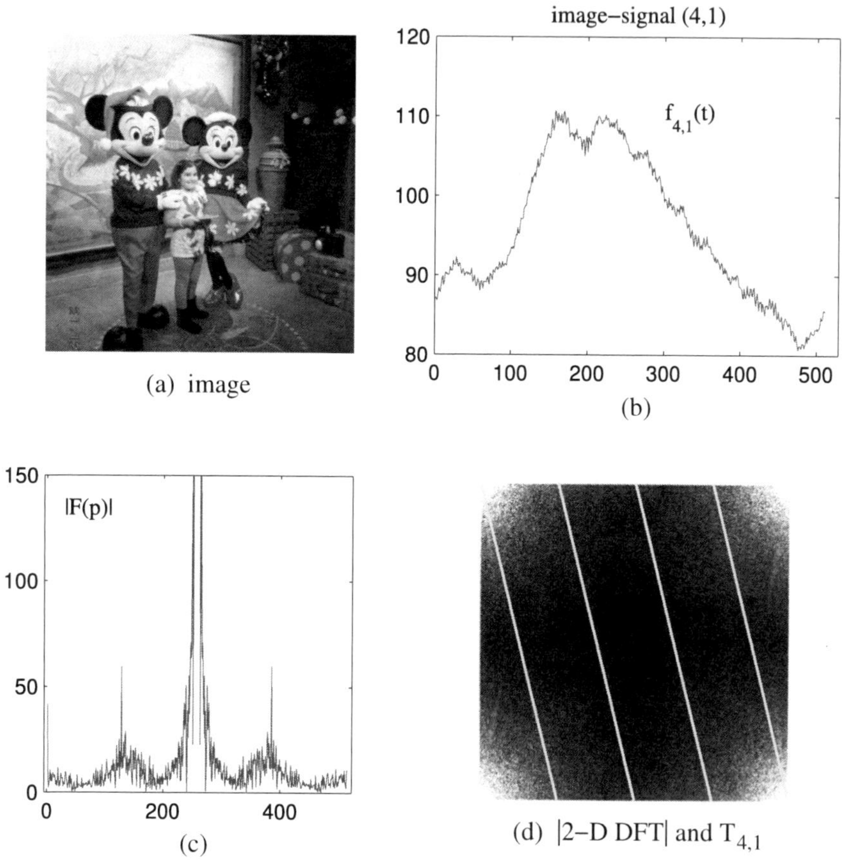

Figure 6.3 (a) The "Miki-Anoush-Mini" image, (b) splitting-signal generated by $(p, s) = (4, 1)$, (c) magnitude of the shifted to the middle 1-D DFT of the signal, and (d) the 2-D DFT of the image with the frequency-points of the set $T_{4,1}$.

lines shown in Fig. 6.2(d). At points of this spiral, the 2-D 256 × 256-point DFT of the image $f_{n,m}$ is calculated by the 256-point DFT of the splitting-signal $f_{T_{5,1}}$ of length 256. Three other spirals $S_{p,s}$ are also shown in Fig. 6.5. The spiral $S_{127,1}$ is shown in Fig. 6.5(b), the spiral $S_{11,1}$ in Fig. 6.5(c), and the spiral $S_{125,1}$ in Fig. 6.5(d).

The set $J_{N,N}$ of generators (p, s) of the splitting-signals can be taken as

$$J_{N,N} = \{(1,0), (1,1), \ldots, (1, N-1), (0,1), (2,1), (4,1), \ldots, (N-2, 1)\} \tag{6.11}$$

for the $N = 2^r$ case. The set $J_{N,N}$ contains $3N/2$ generators, and this number is the minimum of the cyclic groups covering the whole lattice $X_{N,N}$. Such set of generators is used in the code "fft2tt_Npower2.m" for the $N = 256$ case.

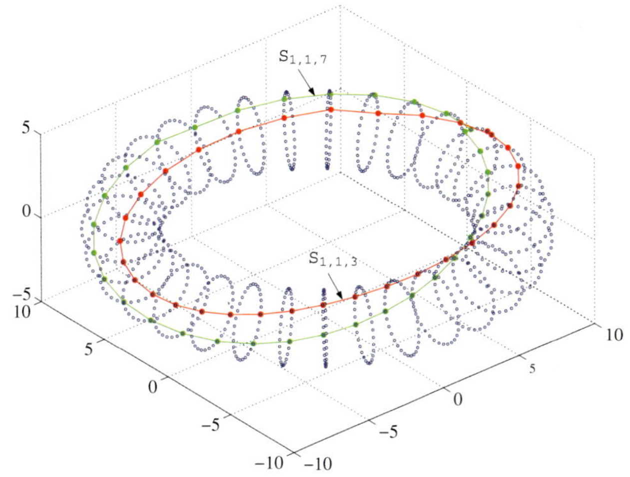

Figure 6.4 The square lattice $X_{32,32}$ on the grid 32×32 in 3-D space with locus of two spirals $S_{1,1,3}$ and $S_{1,1,7}$.

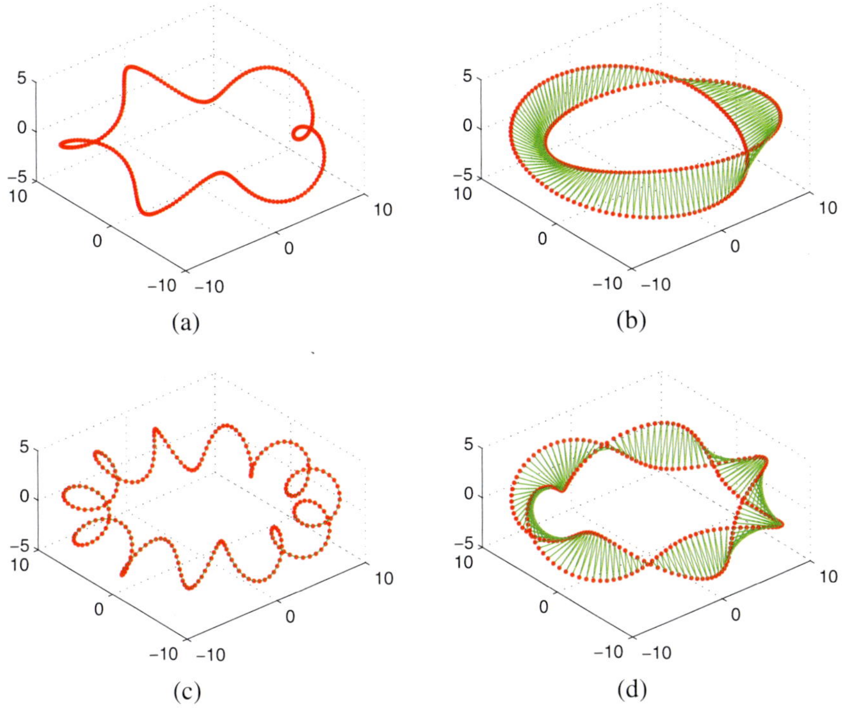

Figure 6.5 Four spirals (a) $S_{5,1}$, (b) $S_{127,1}$, (c) $S_{11,1}$, and (d) $S_{125,1}$ in the grid 256×256 on the torus.

When N is prime, this set can be taken as

$$J_{N,N} = \{(1,0), (1,1), (1,2), \ldots, (1, N-1), (0,1)\}, \tag{6.12}$$

with $(N+1)$ generators. In the given above script "fft2tt_Nisprime.m," this set of generator is used for the $N = 257$ case. One can use other sets, for instance the set $J_{N,N} = \{(0,1),(1,1),(2,1),\ldots,(N-1,1),(1,0)\}$. Important to mention that the minimum number of generators of such sets is $(N+1)$.

Thus, unlike the conventional representation of the image as the 2-D matrix, or the set of rows or columns, in the tensor representation, the image is considered as a set of 1-D signals that can be processed separately. These signals describe uniquely the original image and at the same time split the mathematical structure of the 2-D DFT.

6.1.1 Tensor transform and direction images

The splitting-signals of the tensor representation of the image define the direction components of the image [24],[32]–[38]. The image can be composed by the sum of direction images that correspond to the splitting signals. In the case when N is prime, the sum of such direction images equals the original image. When N is a power of 2, the tensor representation is redundant and that requires a factorization of direction images before their summing.

Given a frequency-point $(p,s) \in J_{N,N}$, the splitting signal

$$f_{T_{p,s}} = \{f_{p,s,0}, f_{p,s,1}, f_{p,s,2}, \ldots, f_{p,s,N-1}\}$$

defines the following direction image:

$$d_{n,m} = (d_{p,s})_{n,m} = \frac{1}{N} f_{p,s,(np+ms)\,\mathrm{mod}\,N}, \quad n,m = 0:(N-1). \tag{6.13}$$

In this image, N values of the splitting-signal $f_{p,s,t}$ are located along the parallel lines $(np+ms)\,\mathrm{mod}\,N$, which pass through the knots of the lattice. As an example, Fig. 6.6 shows the image of size 128×128 in part (a), along with the splitting-signal of the image, which is generated by the frequency-point

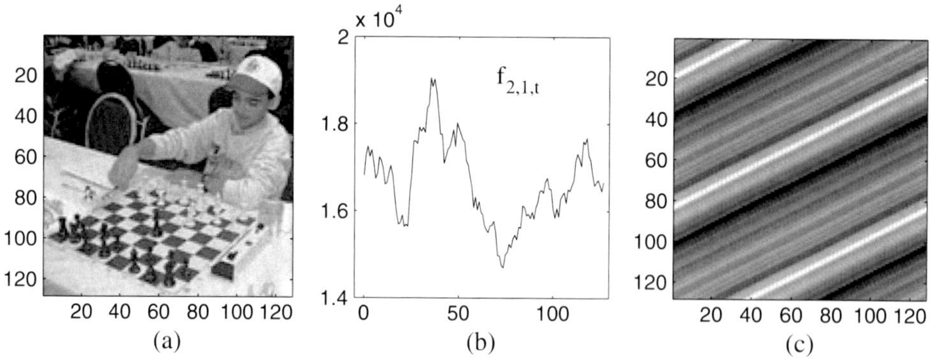

Figure 6.6 (a) "Sarkis" image of the size 128×128, (b) the splitting-signal $\{f_{2,1,t}\, t = 0:127\}$ of the image, and (c) the corresponding direction image.

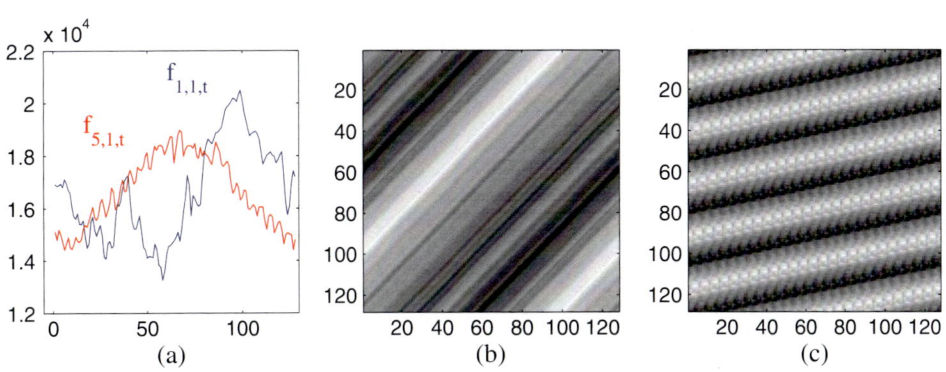

Figure 6.7 (a) The splitting-signals $\{f_{1,1,t}\}$ and $\{f_{5,1,t}\}$ of the "Sarkis" image and two direction images generated by (b) $(p,s) = (1,1)$ and (c) $(p,s) = (5,1)$.

(2, 1) in part (b). The direction image that corresponds to this splitting-signal is shown in Fig. 6.6(c). Figure 6.7 shows two splitting-signals of the image, which are generated by the frequency-points (1, 1) and (5, 1) in part (a). The direction images calculated by these splitting-signals are shown in Fig. 6.7(b,c).

The image $f_{n,m}$ is the sum or linear combination of direction images.

Case $N > 2$ is a prime. The set of $(N+1)$ splitting-signals of length N each defines $(N+1)$ different direction components of the image and the following image decomposition holds:

$$\begin{aligned} f_{n,m} &= \sum_{p=0}^{N-1} (d_{p,1})_{n,m} + (d_{1,0})_{n,m} - E[f] \\ &= \frac{1}{N}\left[\sum_{p=0}^{N-1} f_{p,1,(np+m) \bmod N} + f_{1,0,n}\right] - E[f], \end{aligned} \tag{6.14}$$

where $n, m = 0 : (N-1)$. Here, $E[f]$ is the mean of the splitting-signals

$$E[f] = \frac{1}{N}\sum_{t=0}^{N-1} f_{1,0,t} = \frac{1}{N}\sum_{t=0}^{N-1} f_{p,1,t} = \frac{1}{N}\sum_{n=0}^{N-1}\sum_{m=0}^{N-1} f_{n,m}.$$

Thus, the decomposition of the image by its direction components, or the inverse tensor transform of the image does not required complex operations.

The constant $E[f]$ presents in the inverse formula of the tensor transform because of "the redundancy" of the tensor transform in the frequency-point (0, 0). For the given set of generators in Eq. (6.12), all groups $T_{p,s}$ intersect at one points (0, 0). To remove this constant from the formula, we could consider the image after centering,

$$f_{n,m} \to f_{n,m} - \frac{1}{N^2}\sum_{n=0}^{N-1}\sum_{m=0}^{N-1} f_{n,m}.$$

The tensor transform of the image, as the set of $(N+1)$ splitting-signals of length N each, can be written into the $(N+1)\times N$ matrix. Figure 6.8 shows the image in part (a), along with the tensor transform in part (b). This matrix contains (N^2+N) coefficients, which is greater than the number of points in the image, N^2. This redundancy can be removed, considering the fact that the sum of components of all splitting signals is the same, $NE[f]$. This property was used in Example 6.1. Therefore, we can take one full splitting-signal, for instance $f_{1,0,t}$, $t=0:(N-1)$, and keep remaining N splitting-signals without the last components $f_{p,1,N-1}$, which can be calculated by

$$f_{p,1,N-1} = \sum_{t=0}^{N-1} f_{1,0,t} - \sum_{t=0}^{N-2} f_{p,1,t} = NE[f] - \sum_{t=0}^{N-2} f_{p,1,t}.$$

Total number of required components of the tensor transform therefore equals $N + N(N-1) = N^2$.

Case $N = 2^r > 2$. The set of generators $J_{N,N}$ defined in Eq. (6.11) contains $3N/2$ generators. Therefore, the tensor transform of the image is a family of $3N/2$ splitting-signals of length N each. The total number of components is $N^2 + N^2/2$. We can separate the set of generators by two parts. The first part is with the generators $\{(p,1);\, p=0:(N-1)\}$ and the second part is for the splitting-signals with the remaining generators $\{(1,2s);\, s=0:(N/2-1)\}$.

As an example, Fig. 6.9 shows the image of size 256×256 in part (a), along with the tensor transform in part (b). The set of all splitting-signals is divided by two parts, or matrices. The first part is with the generators $\{(p,1);\, p=0:255\}$ and the second part with the generators

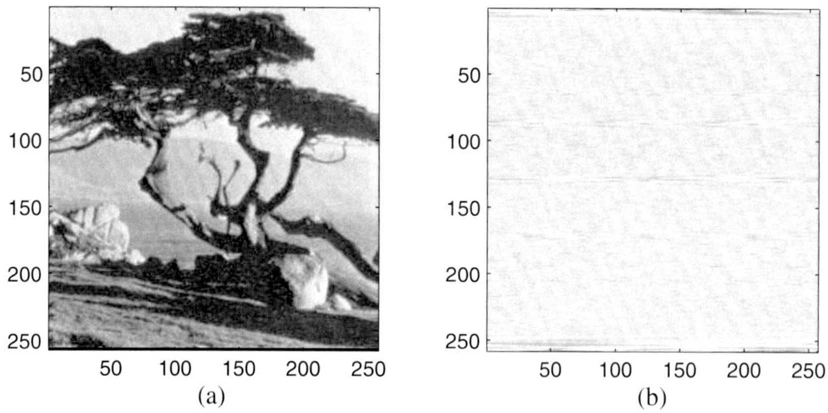

Figure 6.8 (a) Image of size 257×257 and (b) the tensor transform of the image.

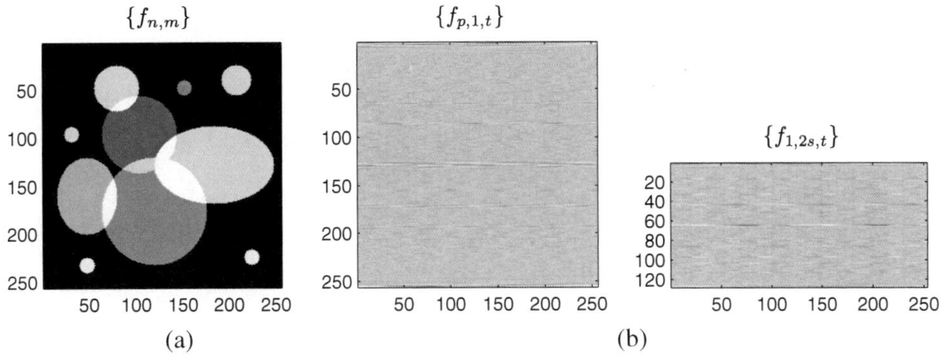

Figure 6.9 (a) The image and (b) two parts of the tensor transform of the image.

$\{(1,2s);\ s=0:127\}$ of the set $J_{256,256}$. The splitting-signals are written along the rows in these two matrices.

The tensor representation is redundant (the total number of components $\{f_{p,s,t}\}$ is greater than N^2) but it is unique. The image can be defined through the inverse 2-D DFT of the 2-D DFT calculated by Eq. (6.9), or can be defined directly from the tensor transform. According to the principle of superposition [22, 34, 35] the image $f_{n,m}$ can be composed from splitting-signals as follows:

$$f_{n,m}=\frac{1}{2N}\sum_{k=0}^{r-1}\frac{1}{2^k}\sum_{(p,s)\in J_{N/2^k,N/2^k}}\left[f_{2^kp,2^ks,t}-f_{2^kp,2^ks,t+N/2}\right]+\frac{1}{N^2}f_{0,0,0}, \qquad (6.15)$$

where to simplify the notations we denote

$$t=(n2^kp+m2^ks) \bmod N.$$

All components $f_{2^kp,2^ks,t}$ in this equation can be calculated directly by Eq. (6.7), or by using the following recursive formula:

$$f_{2^kp,2^ks,2^kt_1}=f_{2^{k-1}p,2^{k-1}s,2^{k-1}t_1}+f_{2^{k-1}p,2^{k-1}s,2^{k-1}t_1+N/2}, \qquad (6.16)$$

where $k=1:(r-1)$, and p or s equals 1, and $t_1=0:(N/2^{k+1}-1)$. In other words, these components can be calculated from $3N/2$ splitting-signals $f_{T_{p,s}}$ generated by the frequencies $(p,s)\in J_{N,N}$.

It should be mentioned that the inverse formula in (6.15) can also be used for other sets of $3N/2$ generators (p,s), with the condition that the covering of the lattice $N\times N$ by groups $T_{p,s}$ is irreducible. For instance, we can take the set of generators $\{(2p,1);\ p=0:(N/2-1)\}\cup\{(1,s);\ s=0:(N-1)\}$, or change the generator (1, 3) by (3, 1) in this set.

6.2 Mapping of 2-D DFTs in Quaternion Algebra

Analyzing the fast algorithms for 2-D QDFTs, it is not difficult to notice that fast calculations of such transforms are reduced to calculations of the classic DFTs. The calculation of the 2-D QDFTs, for instance right-side 2-D QDFT, includes the calculation of the discrete Fourier transforms of all color channels of the image. Therefore, we can embed a few 2-D DFTs of real images into the quaternion or octonion spaces.

For the color image in the RGB color space $f_{n,m} = (r_{n,m}, g_{n,m}, b_{n,m})$, we can define the transforms of all three channels

$$R_{p,s} = \sum_{n=0}^{N-1}\sum_{m=0}^{M-1} r_{n,m} W_M^{ms} W_N^{np}, \quad G_{p,s} = \sum_{n=0}^{N-1}\sum_{m=0}^{M-1} g_{n,m} W_M^{ms} W_N^{np},$$

$$B_{p,s} = \sum_{n=0}^{N-1}\sum_{m=0}^{M-1} b_{n,m} W_M^{ms} W_N^{np},$$

when $p = 0 : (N-1)$, $s = 0 : (M-1)$. Here, we use the complex exponential kernels (with $i^2 = -1$),

$$W_N = \exp(-i2\pi/N) = \cos(2\pi/N) - i\sin(2\pi/N),$$

and $W_N = \exp(-j2\pi/N)$.

These 2-D DFTs can be considered in the quaternion algebra as

$$\begin{aligned} Q_{p,s} &= (\Re\{R_{p,s}\} + i\Im\{R_{p,s}\}) + (\Re\{G_{p,s}\} + j\Im\{G_{p,s}\}) + (\Re\{B_{p,s}\} + k\Im\{B_{p,s}\}) \\ &= (\Re\{R_{p,s}\} + \Re\{G_{p,s}\} + \Re\{B_{p,s}\}) + i\Im\{R_{p,s}\} + j\Im\{G_{p,s}\} + k\Im\{B_{p,s}\}. \end{aligned}$$

$\Re$ and $\Im$ denotes the real and imaginary parts, respectively. Thus, by using the 2-D DFTs of color channels, one can compose a quaternion array of size $N \times M$. The real part of the transform is the sum of real parts of channels

$$\Re\{Q_{p,s}\} = \Re\{R_{p,s}\} + \Re\{G_{p,s}\} + \Re\{B_{p,s}\},$$

and three imaginary parts are imaginary parts of the channel transforms. However, this quaternion transform can not be used for reconstruction of original three channels of the color image.

We also may use representation of these three 2-D DFTs in the quaternion algebra in the following unique way:

$$\begin{aligned} S_{p,s} &= i\Re\{R_{p,s}\} + j\Re\{G_{p,s}\} + k\Re\{B_{p,s}\}, \\ V_{p,s} &= i\Im\{R_{p,s}\} + j\Im\{G_{p,s}\} + k\Im\{B_{p,s}\}. \end{aligned} \tag{6.17}$$

Thus, three complex 2-D DFTs of color components of the image in the RGB color model are represented by two pure quaternion 2-D data in the frequency domain. This representation is unique, because each component of the 2-D DFT of a primary color can be extracted from these quaternion data. For instance,

$$\Re\{G_{p,s}\} = (S_{p,s})_j \text{ and } \Im\{B_{p,s}\} = (V_{p,s})_k.$$

By mapping the Fourier transforms of the primary colors of the image into two such quaternion data, $S_{p,s}$ and $V_{p,s}$, we can process these 2-D data in the quaternion frequency space and then to obtain the new Fourier transforms of the primary colors. Such color image processing can be referred to as the color image processing with the traditional 2-D DFT.

The octonion space can also be used, especially when the image is the quaternion $f_{n,m} = (a_{n,m}, r_{n,m}, g_{n,m}, b_{n,m})$ with a non zero real part $a_{n,m}$. In this case, the 2-D DFT of this part

$$A_{p,s} = \sum_{n=0}^{N-1}\sum_{m=0}^{M-1} a_{n,m} W_M^{ms} W_N^{np}, \quad p,s = 0:(N-1),(M-1),$$

should be considered, too. Then, the four 2-D DFTs in octonion space can be represented in the following unique way:

$$\begin{aligned} O_{p,s} = \Re\{A_{p,s}\} + i\Re\{R_{p,s}\} + j\Re\{G_{p,s}\} + k\Re\{B_{p,s}\} + \\ + E\Im\{R_{p,s}\} + I\Im\{R_{p,s}\} + J\Im\{G_{p,s}\} + K\Im\{B_{p,s}\}. \end{aligned} \tag{6.18}$$

The idea of computing the Fourier transform of a color image has only recently been realized [2, 5]. As the generalization of the traditional Fourier transform, the quaternion Fourier transform was first defined by Ell to process quaternion signals [1]. Later, some practical works related to the discrete quaternion Fourier transform and its application in color image processing are presented in [4] and [8]. Hypercomplex Fourier transforms based on quaternions have been proposed by several groups. So far, however, there has not been a coherent explanation of what the spectral domain coefficients produced by a hypercomplex Fourier transform represent. The 2-D QDFT was applied to color images by filling in the vacant real part of the quaternion image with zeros, which is in fact not effective or not the best choice [10].

6.3 2-D Two-Side QDFT

The quaternion multiplication is not commutative and the definition of the 2-D QDFT is not unique. Different QDFT can be used in image processing, including the following transform: the two-side 2-D QDFT, the right-side and 2-D left-side QDFTs. Here, we describe the 2-D two-side QDFT. The quaternion (or color-in-quaternion) image $q_{n,m}$ is considered of the size $N \times M$ and the transform is defined as

$$Q_{p,s} = \sum_{n=0}^{N-1} \sum_{m=0}^{M-1} W_j^{np} q_{n,m} W_k^{ms} = \sum_{n=0}^{N-1} W_j^{np} \left(\sum_{m=0}^{M-1} q_{n,m} W_k^{ms} \right), \qquad (6.19)$$

$$p = 0 : (N-1), \ s = 0 : (M-1).$$

The basis functions are the exponential coefficients

$$W_j = \cos(2\pi/N) - j\sin(2\pi/N), \text{ and } W_k = \cos(2\pi/M) - k\sin(2\pi/M).$$

Figure 6.10 shows the color image "tree" of size 256×256 in part (a), the magnitude of the 256×256-point two-side 2-D QDFT of this image in part (b), and the difference of magnitudes of this transform and the 256×256-point 2-D DFT in part (c).

The magnitude of the 2-D DFT is defined as the square root of the sum

$$|DFT(R)|^2 + |DFT(G)|^2 + |DFT(B)|^2$$

at each frequency-point (p, s), where $p, s = 0 : 255$.

Thus, the concept of the 2-D DFT transform was generalized a hypercomplex, or quaternion operator, however only for pure unit quaternions μ such that $\mu^2 = -1$, like j and k. There are an infinite number of square roots μ of -1 that are located on the unit sphere in three imaginary subspace of the four-dimensional quaternion space. The operators i, j, and k are the special cases of μ, and the classic complex Fourier transform is the special $\mu = i$ case, when the function to be transformed is complex, the result $Q_{p,s}$ also being complex.

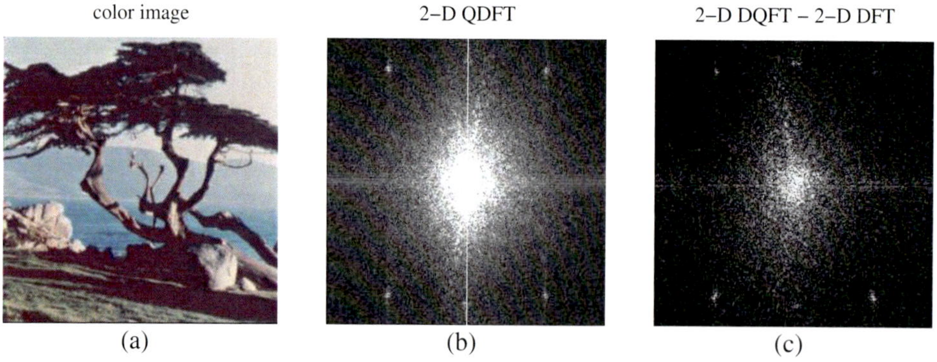

Figure 6.10 (a) The color image, the magnitudes of (b) the 2-D two-side QDFT of the quaternion image, and (d) difference of the 2-D QDFT and 2-D DFT.

The choice of μ is arbitrary, but not without consequence we determine a direction in color space and an obvious choice for color images is the direction corresponding to the luminance axis. In the RGB color space [31], this is the gray line and would be with $\mu = (i + j + k)/\sqrt{3}$. The term inverse transform is used to denote the transform from frequency domain to the spatial domain.

The inverse 2-D two-side QDFT is calculated by the similar formula

$$q_{n,m} = \frac{1}{NM}\sum_{n=0}^{N-1}\sum_{m=0}^{M-1} W_j^{-np} Q_{p,s} W_k^{-ms} = \frac{1}{NM}\sum_{p=0}^{N-1} W_j^{-np}\left(\sum_{s=0}^{M-1} Q_{p,s} W_k^{-ms}\right) \tag{6.20}$$

when $n = 0 : (N-1)$ and $m = 0 : (M-1)$.

Each multiplication $Q = W_j^{np} q_{n,m} W_k^{ms}$ is calculated by

$$\begin{aligned} Q &= Q_r + iQ_i + jQ_j + kQ_k \\ &= [\cos(\varphi) - j\sin(\varphi)][a + ib + jc + kd][\cos(\psi) - k\sin(\psi)] \\ &= [a\cos(\varphi)\cos(\psi) - b\sin(\varphi)\sin(\psi) + c\sin(\varphi)\cos(\psi) + d\cos(\varphi)\sin(\psi)] \\ &\quad + i[a\sin(\varphi)\sin(\psi) + b\cos(\varphi)\cos(\psi) - c\cos(\varphi)\sin(\psi) - d\sin(\varphi)\cos(\psi)] \\ &\quad + j[-a\sin(\varphi)\cos(\psi) + b\cos(\varphi)\sin(\psi) + c\cos(\varphi)\cos(\psi) - d\sin(\varphi)\sin(\psi)] \\ &\quad + k[-a\cos(\varphi)\sin(\psi) + b\sin(\varphi)\cos(\psi) + c\sin(\varphi)\sin(\psi) + d\cos(\varphi)\cos(\psi)], \end{aligned} \tag{6.21}$$

where $\varphi = (2\pi/N)np$ and $\psi = (2\pi/M)ms$. The quaternion number $a + ib + jc + kd$ stands for $f_{n,m}$. In matrix form, this multiplication is written as

$$Q = \begin{pmatrix} Q_r \\ Q_i \\ Q_j \\ Q_k \end{pmatrix} = R_{4,4}(\varphi,\psi)\begin{pmatrix} a \\ b \\ c \\ d \end{pmatrix}, \tag{6.22}$$

where the matrix $R_{4,4}(\varphi,\psi)$ is

$$R_{4,4}(\varphi,\psi) = \begin{bmatrix} \cos(\varphi)\cos(\psi) & \sin(\varphi)\sin(\psi) & \sin(\varphi)\cos(\psi) & \cos(\varphi)\sin(\psi) \\ \sin(\varphi)\sin(\psi) & \cos(\varphi)\cos(\psi) & -\cos(\varphi)\sin(\psi) & -\sin(\varphi)\cos(\psi) \\ -\sin(\varphi)\cos(\psi) & \cos(\varphi)\sin(\psi) & \cos(\varphi)\cos(\psi) & -\sin(\varphi)\sin(\psi) \\ -\cos(\varphi)\sin(\psi) & \sin(\varphi)\cos(\psi) & -\sin(\varphi)\sin(\psi) & \cos(\varphi)\cos(\psi) \end{bmatrix}. \tag{6.23}$$

The determinant of this matrix equals 1 and the inverse matrix equals its transpose, $R_{4,4}^{-1}(\varphi,\psi) = R_{4,4}^{T}(\varphi,\psi)$. The inverse matrix can also be obtained by the point-wise multiplication of the matrix $R_{4,4}(\varphi, \psi)$ by the binary matrix as

$$R_{4,4}^{-1}(\varphi,\psi) = R_{4,4}(\varphi,\psi).\times \begin{pmatrix} 1 & 1 & -1 & -1 \\ 1 & 1 & -1 & -1 \\ -1 & -1 & 1 & 1 \\ -1 & -1 & 1 & 1 \end{pmatrix},$$

and it changes the signs of only eight components of the matrix $R_{4,4}(\varphi, \psi)$.

Below is the script of the code "direct_dQDFT.m" for calculating the 2-D QDFT.

```
% ------------------------------------------------------------
% call: direct_dQDFT.m
% calculation of the two-side 2-D DQDFT NxM
%   X - is the image NxMx4
%   QDFT - is the 2-D QDFT
% ------------------------------------------------------
function QDFT=direct_dQDFT(X)
  [M, N, L]=size(X);
  pi2=2*pi; % pi2=-2*pi;  % for the inverse 2-D DQDFT
  QDFT=zeros(M,N,4);
  a1=pi2/N; a2=pi2/M; a2u=0;
  for u=0:M-1
      u1=u+1;  a1v=0;
      for v=0:N-1
          Q1=zeros(1,4); % for 2-D QDFT
          v1=v+1; a=0;
          for m=0:M-1
              c2=cos(a); s2=sin(a);
              m1=m+1; b=0;
              for n=0:N-1
                  c1=cos(b);    s1=sin(b);
                  c2c1=c2*c1; s2s1=s2*s1;
                  s2c1=s2*c1; c2s1=c2*s1;
                  % ----- Matrix R(a,b) ------
                  R= [ c2c1 s2s1  s2c1  c2s1
                        s2s1 c2c1 -c2s1 -s2c1
                       -s2c1 c2s1  c2c1 -s2s1
                       -c2s1 s2c1 -s2s1  c2c1];
                  % ----------------------
                  n1=n+1;
                  q(1:4)=X(m1,n1,1:4); Q=R*q'; Q1=Q1+Q';
                  b=b+a1v;
              end
```

```
            a=a+a2u;
        end
        QDFT(u1,v1,1:4)=Q1; a1v=a1v+a1;
    end
    a2u=a2u+a2;
end
% QDFT=QDFT/N/M; % for the inverse 2-D DQDFT
% ---------------------------------------------------
```

For the inverse 2-D QDFT, we use the similar code "inverse_dQDFT.m," wherein the normalization of the transform by the factor of NM is considered, as well as the change of the sign in the angles. For that, we change only the command "pi2=-2*pi;." We can also use the additional command for the transpose matrix "R=R′" after calculation of the matrix R in the script. Below is the script of the test-code for calculating the direct and inverse 2-D QDFTs of an image randomly generated from integer numbers in the interval $[0, 5]$.

```
% ------------------------------------------
% call: test_1.m
  N=32; M=N;
  X=round(5*rand(N,M,4));    % real part, R,G,B
  X(:,:,1)=zeros(N,M);       % real part is 0
  % ----------------------------------------
  QDFT=direct_dQDFT(X);
  Y=inverse_dQDFT(QDFT);
  Y1=round(Y);
  err1=sum(sum(sum(X-Y1)));
% ------------------------------------------
```

6.3.1 Column-row algorithm of the two-side QDFT

To perform the calculation of the 2-D QDFT by the column-row method, we consider the multiplications of type $q = q_{n,m}W_k^{ms}$, which are performed as

$$\begin{aligned} q = q_e + iq_i + jq_j + kq_k &= [a + ib + jc + kd][\cos(\psi) - k\sin(\psi)] \\ &= [a\cos(\psi) + d\sin(\psi)] + i[b\cos(\psi) - c\sin(\psi)] \\ &\quad + j[b\sin(\psi) + c\cos(\psi)] + k[-a\sin(\psi) + d\cos(\psi)]. \end{aligned} \tag{6.24}$$

In matrix form, this multiplication is defined by the matrix

$$R_{4,4}(\psi) = \begin{bmatrix} \cos(\psi) & 0 & 0 & \sin(\psi) \\ 0 & \cos(\psi) & -\sin(\psi) & 0 \\ 0 & \sin(\psi) & \cos(\psi) & 0 \\ -\sin(\psi) & 0 & 0 & \cos(\psi) \end{bmatrix}. \tag{6.25}$$

Thus, in 4-D space, this matrix describes two rotations. The first is the rotation of the vector $(a,d)'$ by the angle $-\psi$ in the 2-D plane "1-k" of quaternions $C_k = \{q = a + kd\}$). The second is the rotation of the vector $(b,c)'$ by the angle ψ in the 2-D plane "i-j" of quaternions $C_k = \{q = ib + jd\}$.

On the second stage of the column-row method, after the calculations by rows, the multiplications of type $Q = W_j^{np} q$ are performed as

$$\begin{aligned} Q &= Q_e + iQ_i + jQ_j + kQ_k \\ &= [\cos(\varphi) - k\sin(\varphi)](q_e + iq_i + jq_j + kq_k) \\ &= [q_e\cos(\varphi) + q_j\sin(\varphi)] + i[q_i\cos(\varphi) - q_k\sin(\varphi)] \\ &\quad + j[-q_e\sin(\varphi) + q_j\cos(\varphi)] + k[q_i\sin(\varphi) + q_k\cos(\varphi)]. \end{aligned} \tag{6.26}$$

In matrix form, this multiplication is defined by the matrix

$$R_{4,4}(\varphi) = \begin{bmatrix} \cos(\varphi) & 0 & \sin(\varphi) & 0 \\ 0 & \cos(\varphi) & 0 & -\sin(\varphi) \\ -\sin(\varphi) & 0 & \cos(\varphi) & 0 \\ 0 & \sin(\varphi) & 0 & \cos(\varphi) \end{bmatrix}. \tag{6.27}$$

In 4-D space, this matrix describes two rotations. The first is the rotation of the vector $(q_r, q_j)'$ by the angle $-\varphi$ in the 2-D plane "1-j." The second is the rotation of the vector $(q_i, q_k)'$ by the angle φ in the 2-D plane "i-k." It should be noted that

$$R_{4,4}(\varphi, \psi) = R_{4,4}(\psi)R_{4,4}(\varphi) = R_{4,4}(\varphi)R_{4,4}(\psi).$$

The script of the code "rowcol_dQDFT.m" for calculating the 2-D QDFT by the column-row method is given below. For the inverse 2-D QDFT, we use the similar code "rowcol_idQDFT.m," wherein the normalization of the transform by the factor of NM is considered and the change of the sign in the angles, by changing only the command "pi2=-2*pi;."

```
% ---------------------------------------------------
% call: rowcol_dQDFT.m
% row-column calculation of the two-side 2-D QDFT of image
%  X - is image NxMx4 as the quaternion-image
%  QDFT2 - is the 2-D QDFT
% ---------------------------------------------------
function QDFT2=rowcol_dQDFT(X)
 [M,N,L] = size(X);
% for the inverse 2-D QDFT, use pi2=-2*pi;
```

```
pi2=2*pi; a1=pi2/N; a2=pi2/M;
% I Processing by rows
QDFT1=zeros(M,N,4);
for m=0:M-1
    m1=m+1; a1v=0;
    for v=0:N-1
       Q1=zeros(1,4);
       v1=v+1; b=0;
       for n=0:N-1
           c1=cos(b); s1=sin(b);
           % ------ Matrix R(b) ------
           R= [c1  0    0  s1
                0  c1 -s1   0
                0  s1  c1   0
              -s1   0   0  c1];
         % ---------------------
         n1=n+1;
         q(1:4)=X(m1,n1,1:4); Q=R*q'; Q1=Q1+Q';
         b=b+a1v;
       end
       QDFT1(m1,v1,1:4)=Q1; a1v=a1v+a1;
    end
end
% II Processing by columns
QDFT2=zeros(M,N,4);
for v=0:N-1
    v1=v+1; a2u=0;
    for u=0:N-1
      Q1=zeros(1,4);
      u1=u+1; a=0;
      for m=0:N-1
           c2=cos(a); s2=sin(a);
           % ----- Matrix R(a) -----
           R= [  c2    0   s2      0
                  0   c2    0    -s2
                -s2    0   c2      0
                  0   s2    0     c2];
           % -------------------
        m1=m+1;
        q(1:4)=QDFT1(m1,v1,1:4); Q=R*q'; Q1=Q1+Q';
        a=a+a2u;
      end
```

```
        QDFT2(u1,v1,1:4)=Q1; a2u=a2u+a2;
      end
  end
  % QDFT2=QDFT2/N/M; % for the inverse 2-D QDFT
% ----------------------------------------------------
```

Below is the script of the test-code, for calculating the direct and inverse 2-D QDFTs of the image, which is randomly generated from integer numbers in the interval [0, 5]. One can see that, for images of size larger than the size of 128×128, the calculation of both direct and inverse 2-D QDFT takes minutes.

```
% call: test_2.m
% Test --------------------------------------
N=128; M=N;
X=round(5*rand(N,M,4)); % real part, R,G,B
X(:,:,1)=zeros(N,M);     % real part is 0
% -------------------------------------------
QDFT=rowcol_dQDFT(X);
Y=rowcol_idQDFT(QDFT);
Y1=round(Y);
err1=sum(sum(sum(X-Y1)));
fprintf(' Error of calculation is %f \n',err1);
% ---- TIME for Direct and Inverse QDFTs -----
%      Time =  0.19(s) for [N,M] = [16,16]
%      Time =  1.48(s) for [N,M] = [32,32]
%      Time = 11.03(s) for [N,M] = [64,64]
%      Time = 40.76(s) for [N,M] = [100,100]
%      Time =  1.43(m) for [N,M] = [128,128]
%      Time = 11.42(m) for [N,M] = [256,256]
% -------------------------------------------
```

6.3.2 Fast algorithms for the 2-D QDFT

For fast calculation of the 2-D QDFT, we first consider the real two-side QDFT

$$G_{p,s} = \sum_{n=0}^{N-1} \sum_{m=0}^{M-1} W_j^{np} g_{n,m} W_k^{ms} = \sum_{n=0}^{N-1} W_j^{np} \left(\sum_{m=0}^{M-1} g_{n,m} W_k^{ms} \right), \quad (6.28)$$
$$p = 0 : (N-1), s = 0 : (M-1),$$

where $g_{n,m}$ is a real 2-D sequence or image. It directly follows that

$$
\begin{aligned}
G_{p,s} &= \sum_{n=0}^{N-1}\sum_{m=0}^{M-1}[\cos(np\varphi) - j\sin(np\varphi)]g_{n,m}[\cos(ms\psi) - k\sin(ms\psi)] \\
&= \sum_{n=0}^{N-1}\sum_{m=0}^{M-1} g_{n,m}\cos(np\varphi)\cos(ms\psi) - j\sum_{n=0}^{N-1}\sum_{m=0}^{M-1} g_{n,m}\sin(np\varphi)\cos(ms\psi) \\
&\quad - k\sum_{n=0}^{N-1}\sum_{m=0}^{M-1} g_{n,m}\cos(np\varphi)\sin(ms\psi) + (jk)\sum_{n=0}^{N-1}\sum_{m=0}^{M-1} g_{n,m}\sin(np\varphi)\sin(ms\psi) \\
&= (G_e)_{p,s} + j(G_j)_{p,s} + k(G_k)_{p,s} + (jk)(G_{jk})_{p,s},
\end{aligned}
\tag{6.29}
$$

where $\varphi = 2\pi/N$ and $\psi = 2\pi/M$.

For fast calculation of $G_{p,s}$, we use the fast algorithms of the DFT in the following way. First we apply the 1-D M-point DFTs over the each row of the real sequence $g_{n,m}$, which results in the following 2-D sequence:

$$
\begin{aligned}
S_{n,s} &= \sum_{m=0}^{M-1} g_{n,m}\exp(-ism\psi), \; s = 0 : (M-1), \\
\Re(S_{n,s}) + i\,\Im(S_{n,s}) &= \sum_{m=0}^{M-1} g_{n,m}\cos(ms\psi) - i\sum_{m=0}^{M-1} g_{n,m}\sin(ms\psi).
\end{aligned}
\tag{6.30}
$$

Thus

$$
\sum_{m=0}^{M-1} g_{n,m}\cos(ms\psi) = \Re(S_{n,s}), \quad \sum_{m=0}^{M-1} g_{n,m}\sin(ms\psi) = -\Im(S_{n,s}).
$$

Now, we use similar calculations of the N-point DFTs over the columns of the obtained sequences $\Re(S_{n,s})$ and $\Im(S_{n,s})$,

$$
\begin{aligned}
C_{p,s} &= \sum_{n=0}^{N-1} \Re(S_{n,s})\exp(-inp\varphi), \; p = 0 : (N-1), \\
\Re(C_{p,s}) + i\,\Im(C_{p,s}) &= \sum_{n=0}^{N-1} \Re(S_{n,s})\cos(np\varphi) - i\sum_{n=0}^{N-1} \Re(S_{n,s})\sin(np\varphi),
\end{aligned}
\tag{6.31}
$$

and

$$
\begin{aligned}
D_{p,s} &= \sum_{n=0}^{N-1} \Im(S_{n,s}) \exp(-inp\varphi), \ p = 0 : (N-1), \\
\Re(D_{p,s}) + i\,\Im(D_{p,s}) &= \sum_{n=0}^{N-1} \Im(S_{n,s}) \cos(np\varphi) - i \sum_{n=0}^{N-1} \Im(S_{n,s}) \sin(np\varphi).
\end{aligned} \tag{6.32}
$$

It follows directly from the above equations that four components of the 2-D QDFT of the real sequence $g_{n,m}$ can be calculated by

$$
\begin{aligned}
(G_e)_{p,s} &= \sum_{n=0}^{N-1}\sum_{m=0}^{M-1} g_{n,m} \cos(np\varphi)\cos(ms\psi) = \Re\,(C_{p,s}), \\
(G_j)_{p,s} &= -\sum_{n=0}^{N-1}\sum_{m=0}^{M-1} g_{n,m} \cos(np\varphi)\sin(ms\psi) = \Im\,(C_{p,s}), \\
(G_k)_{p,s} &= -\sum_{n=0}^{N-1}\sum_{m=0}^{M-1} g_{n,m} \sin(np\varphi)\cos(ms\psi) = \Re\,(D_{p,s}), \\
(G_{jk})_{p,s} &= \sum_{n=0}^{N-1}\sum_{m=0}^{M-1} g_{n,m} \sin(np\varphi)\sin(ms\psi) = \Im\,(D_{p,s}).
\end{aligned} \tag{6.33}
$$

Therefore, the fast calculation of the 2-D QDFT can be described by the following algorithm with four steps.

This algorithm uses N M-point DFTs and $2M$ N-point DFTs. $3N$ N-point DFTs are used in the $N = M$ case.

Below is the script of the code for this algorithm.

```
% call: merQDFT_image1.m
  function [CC CS SC SS] =merQDFT_image1(X)
    A=fft(X.'); C=real(A); S=imag(A);
    A=fft(C.'); CC=real(A); CS=imag(A);
    A=fft(S.'); SC=real(A); SS=imag(A);
```

Algorithm 1. Main steps of the 2-D QDFT

1. Input the image (or 2-D real signal) **g** of a size $N \times M$.
2. Calculate the M-point DFTs over all rows of the image **g**. This 2-D complex sequence is denoted by **S**.
3. Calculate the N-point DFTs over all columns of the real part of the sequence **S**. This 2-D complex sequence is denoted by **C**.
4. Calculate the N-point DFTs over all columns of the imaginary part of the sequence **S**. This 2-D complex sequence is denoted by **D**.

$\Re(\mathbf{C})$, $\Re(\mathbf{C})$, $\Re(\mathbf{D})$, $\Re(\mathbf{D})$ are four output components of the 2-D QDFT.

It follows directly from Eqs. (6.31) and (6.32) that the 2-D DFT of the image $g_{n,m}$ is

$$\begin{aligned}\sum_{n=0}^{N-1}\sum_{m=0}^{M-1} g_{n,m} W_i^{ms} W_i^{np} &= [\Re(C_{p,s}) - \Re(D_{p,s})] - i[\mathrm{imag}(C_{p,s}) + \Re(D_{p,s})] \\ &= [(G_e)_{p,s} - (G_{jk})_{p,s}] - i[(G_j)_{p,s} + (G_k)_{p,s}],\end{aligned}$$

when $p = 0 : (N-1)$ and $s = 0 : (M-1)$.

Now, we consider the two-side QDFT for the 2-D image $q_{n,m}$ in quaternion space,

$$\begin{aligned}Q_{p,s} &= \sum_{n=0}^{N-1}\sum_{m=0}^{M-1} W_j^{np} q_{n,m} W_k^{ms} \\ &= \sum_{n=0}^{N-1}\sum_{m=0}^{M-1} [\cos(np\varphi) - j\sin(np\varphi)] q_{n,m} [\cos(ms\psi) - k\sin(ms\psi)].\end{aligned} \tag{6.34}$$

To simplify the further calculations, we consider the $N = M$ case and represent the 1-D DFT by two matrices with the coefficients $C = C_{n,s} = \cos(np\varphi)$ and $S = S_{n,s} = \sin(np\varphi)$. The 2-D image $q_{n,m}$ is considered as

$$q_{n,m} = [E_{n,m}, I_{n,m}, J_{n,m}, K_{n,m}] = eE_{n,m} + (jk)I_{n,m} + jJ_{n,m} + kK_{n,m},$$

when $n = 0 : (N-1)$ and $m = 0 : (M-1)$. We also use the short notation for this image, $q = eE + iI + jJ + kK$. Note that $E, I, J,$ and K are image components not octonion units.

Above Eq. (6.34) can be written in the following symbolic form:

$$Q = (Ce - jS)q(Ce - kS) = C(eq)C - S(jq)C - C(qk)S + S(jqk)S. \tag{6.35}$$

Note that

$$\begin{aligned}&eqe = eq = q = eE + iI + jJ + kK, \\ &jqe = jq = jeE + jiI + jjJ + jkK = jE - kI - eJ + iK, \\ &eqk = qk = ekE + ikI + jkJ + kkK = kE - jI + iJ - eK, \\ &jqk = jkE + jikI + jjkJ + jkkK = iE + eI - kJ - jK.\end{aligned} \tag{6.36}$$

Therefore, in Eq. (6.35) we obtain

$$\begin{aligned}Q &= C[eE + iI + jJ + kK]C - S[jE - kI - eJ + iK]C- \\ &\quad - C[kE - jI + iJ - eK]S + S[iE + eI - kJ - jK]S \\ &= e(CEC + SJC + CKS + SIS) + i(CIC - SKC - CJS + SES) \\ &\quad + j(CJC - SEC + CIS - SKS) + k(CKC + SIC - CES - SJS).\end{aligned} \tag{6.37}$$

The $N \times M$-point QDFT can thus be calculating as follows:

$$\begin{aligned}(Q_e) &= CEC + SJC + CKS + SIS, \\ (Q_i) &= CIC - SKC - CJS + SES, \\ (Q_j) &= CJC - SEC + CIS - SKS, \\ (Q_k) &= CKC + SIC - CES - SJS.\end{aligned} \tag{6.38}$$

The symmetric Hadamard matrix [39] lies in this equation, i.e.,

$$H_4 = \begin{bmatrix} +1 & +1 & +1 & +1 \\ +1 & -1 & -1 & +1 \\ +1 & -1 & +1 & -1 \\ +1 & +1 & -1 & -1 \end{bmatrix}$$

The script of the code for calculating the $N \times M$-point two-side QDFT is given.

```
.........................................
% call: merQDFT_image4.m
function [E1,I1,J1,K1]=merQDFT_image4(E,I,J,K)
   [CEC CES SEC SES]=merQDFT_image1(E);
   [CIC CIS SIC SIS]=merQDFT_image1(I);
   [CJC CJS SJC SJS]=merQDFT_image1(J);
   [CKC CKS SKC SKS]=merQDFT_image1(K);
   E1=  CEC + SIS + SJC + CKS;
   I1=  SES + CIC - CJS - SKC;
   J1= -SEC + CIS + CJC - SKS;
   K1= -CES + SIC - SJS + CKC;
.........................................
```

The two-side 2-D QDFT can be calculated by using the column-row algorithm over each image component in the quaternion field, and the fast Fourier transform can be used in the calculation [13]–[15]. In the $N = M$ case, this calculation requires $4(3N)$ real valued N-point DFTs. When N is a power of two, the number of operations of multiplication in the real valued N-point

DFT equals $[N/2(\log_2 N-3)+2]/2$ [23]. The total number of such operations for the 2-D QDFT equals $3N^2(\log_2 N-3)+12N$. For a pure quaternion image $q_{n,m}$ the number of these operations is reduced to $9/4N^2(\log_2 N-3)+9N$.

6.4 2-D Right-Side QDFT

The known concepts of the 2-D right-side and left-side QDFTs [2, 4, 8, 26, 27] can be described similarly. In this section, we consider the 2-D right-side QDFT. In the RGB model, the discrete color image $q_{n,m}$ is considered to be of size $N \times M$ pure quaternion table, or image.

Let μ be the pure unit quaternion. The 2-D right-side QDFT as a generalized quaternion Fourier transform is defined as follows:

$$Q_{p,s}=\sum_{n=0}^{N-1}\sum_{m=0}^{N-1} q_{n,m}W_{\mu}^{np+ms},\quad p,s=0:(N-1). \tag{6.39}$$

The inverse 2-D right-side QDFT is calculated by

$$q_{n,m}=\frac{1}{N^2}\sum_{p=0}^{N-1}\sum_{s=0}^{N-1} Q_{p,s}W_{\mu}^{-(np+ms)},\quad n,m=0:(N-1).$$

As an example, Fig. 6.11 shows the color image of size 362×500 in part (a), together with the real and imaginary parts of the 2-D QDFT

$$Q_{p,s}=\Re(Q_{p,s})+\Im(Q_{p,s})=(Q_{p,s})_e+[i(Q_{p,s})_i+j(Q_{p,s})_j+k(Q_{p,s})_k]$$

in the absolute scale in parts (b) and (c), respectively. The magnitude of the 2-D QDFT is given in Fig. 6.11(d). One can see that the real part of the 2-D QDFT is not a dominant component in the quaternion spectrum of the color image.

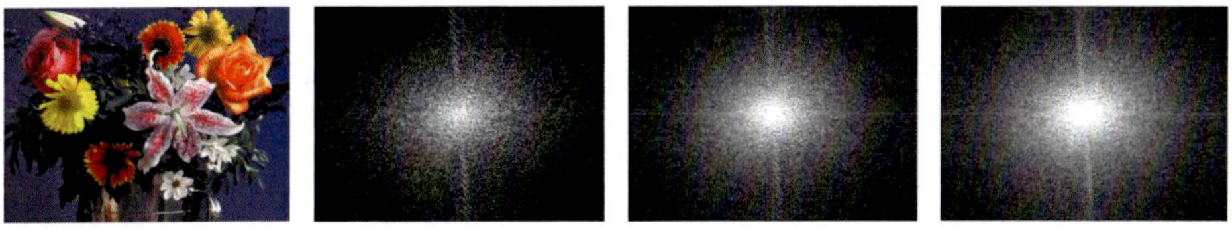

(a) (b) (c) (d)

Figure 6.11 (a) The color image and the magnitudes of (b) real component, $|(Q_{p,s})_e|$, (c) imaginary components, $\sqrt{(Q_{p,s})_i^2+(Q_{p,s})_j^2+(Q_{p,s})_k^2}$, and (d) 2-D QDFT of the color image. (All components of the 2-D QDFT were cyclically shifted to the center.)

The kernel of the 2-D right-side QDFT is defined by the exponential function $W_\mu = W_{\mu;N} = \exp(-2\pi\mu/N)$. Here, the equality $W_\mu^{np} = (W_\mu)^{np} = \exp(-\mu 2\pi np/N)$ holds. For instance, for the $np = 2$ case, the following calculations are valid:

$$\begin{aligned}(W_\mu)^2 &= [\cos(2\pi/N) - \mu\sin(2\pi/N)]^2 \\ &= \cos^2(2\pi/N) + \mu^2\sin^2(2\pi/N) - 2\mu\cos(2\pi/N)\sin(2\pi/N) \\ &= \cos(2\cdot 2\pi/N) - \mu\sin(2\cdot 2\pi/N) = \exp(-\mu 2\cdot 2\pi/N) = W_\mu^2,\end{aligned}$$

since $\mu^2 = -1$. Then, for $np = 3$,

$$\begin{aligned}(W_\mu)^3 &= (W_\mu)^2(W_\mu) = [\cos(4\pi/N) - \mu\sin(4\pi/N)][\cos(2\pi/N) - \mu\sin(2\pi/N)] \\ &= \cos(4\pi/N)\cos(2\pi/N) - \sin(4\pi/N)\sin(2\pi/N) - \\ &\quad - \mu[\cos(4\pi/N)\sin(2\pi/N) + \sin(4\pi/N)\cos(2\pi/N)] \\ &= \cos(3\cdot 2\pi/N) - \mu\sin(3\cdot 2\pi/N) = \exp(-\mu 3\cdot 2\pi/N) = W_\mu^3.\end{aligned}$$

Similarly, other cases of integer np can be considered.

The definition of the 2-D right-side QDFT has been originated from the traditional 2-D DFT over the grayscale image $f_{n,m}$, which can be written as

$$F_{p,s} = \sum_{n=0}^{N-1}\sum_{m=0}^{N-1} f_{n,m} W^{np+ms}, \quad p,s = 0:(N-1),$$

where $W = W_N = \exp(-2\pi i/N)$.

6.4.1 2-D QDFT with column-row algorithm

For a pure unit quaternion μ and any integers t and s, the following property holds for the exponential quaternions: $W_\mu^{t+s} = W_\mu^t W_\mu^s = W_\mu^s W_\mu^t$. Therefore, the 2-D right-side QDFT is a separable transformation and for calculation we can use the formula

$$Q_{p,s} = \sum_{n=0}^{N-1}\left[\sum_{m=0}^{N-1} q_{n,m} W_\mu^{ms}\right] W_\mu^{np}, \quad p,s = 0:(N-1).$$

The calculation of the 2-D QDFT can thus be performed by columns and then rows, or vice versa, which requires $2N$ N-point QDFTs. Each of the 1-D QDFTs requires two N-point DFTs. Therefore, the column-row method uses $4N$ N-point DFTs and the multiplications in number

$$m_{QDFT}(N,N) = 4N[2N\log_2 N + 8] = 8N^2\log_2 N + 32N. \tag{6.40}$$

Here, we mention that the classic 2-D DFT of the complex image can be calculated by using the operations of multiplication in number [23, 25]

$$\begin{aligned} m_{DFT}(N,N) &= 4[2Nm_{DFT}(N)] = 8Nm_{DFT}(N) \\ &= 8N[N/2(\log_2 N - 3) + 2] = 4N^2(\log_2 N - 3) + 16N, \end{aligned}$$

if we use the fast paired transform-based 1-D DFT, which requires operations of multiplication in number $m_{DFT}(N) = N/2(\log_2 N - 3) + 2$ [22]. The number of multiplications in the $N \times N$-point 2-D QDFT is almost twice the multiplications in the $N \times N$-point 2-D DFT.

Below is the script of the code "qrs2dfft.m" for calculating the direct 2-D right-side QDFT.

```
% ----------------------------------------------------------
% call: qrs2dfft.m  /  Art Grigoryan, July 29 2014
% To calculate the right-side quaternion 2-D QDFT
% of the quaternion image of size NxMx3 or NxMx4.
% qmu is the pure quaternion unit, (qmu)^2 = -1.
% ----------------------------------------------------------
function FQ2D=qrs2dfft(image_in,qmu,N,M)
  L=size(image_in,3);
  m1=qmu(1); m2=qmu(2); m3=qmu(3);
  FQ2byrow=zeros(N,M,4);
  for n=1:N
      fq=zeros(4,M);
      if L==3
          for k=2:4
              fq(k,:)=image_in(n,:,k-1);
          end
      elseif L==4
          for k=1:4
              fq(k,:)=image_in(n,:,k);
          end
      end
      FQn=q1dfft(fq,qmu,M);
      for k=1:4
          FQ2byrow(n,:,k)=FQn(k,:);
      end
  end
  FQ2D=zeros(N,M,4);
  for m=1:M
      fq=zeros(4,N);
      for k=1:4
```

```
            fq(k,:)=FQ2byrow(:,m,k);
        end
        FQn=q1dfft(fq,qmu,N);
        for k=1:4
            FQ2D(:,m,k)=FQn(k,:);
        end
    end
% ----------------------------------------------------
```

To calculate the $N \times M$-point real 2-D QDFT, the following scripts can be used.

```
% ----------------------------------------------------
% call: realqrs2dfft.m
function FQ2D=realqrs2dfft(image_in,qmu,N,M)
    m1=qmu(1); m2=qmu(2); m3=qmu(3);
    FQ2byrow=zeros(N,M,4);
    for n=1:N
        fq=zeros(3,M);
        for k=1:3
            fq(k,:)=image_in(n,:,k);
        end
        FQn=qreal1dfft(fq,qmu,N);
        for k=1:4
            FQ2byrow(n,:,k)=FQn(k,:);
        end
    end
    FQ2D=zeros(N,M,4);
    for m=1:M
        fq=zeros(4,M);
        for k=1:4
            fq(k,:)=FQ2byrow(:,m,k);
        end
        FQn=q1dfft(fq,qmu,N);
        for k=1:4
            FQ2D(:,m,k)=FQn(k,:);
        end
    end
% ----------------------------------------------------
```

6.5 Method of Symplectic Decomposition

We consider the known method of calculation of the 2-D QDFT

$$Q_{p,s} = \sum_{n=0}^{N-1}\sum_{m=0}^{N-1} q_{n,m} W_{\mu}^{np+ms}, \quad p,s = 0:(N-1), \tag{6.41}$$

by using the symplectic decomposition of the color image [40, 41].

Let μ_2 be a pure unit quaternion, which is perpendicular to $\mu_1 = \mu$, and let $\mu_3 = \mu_2\mu_1$. The quaternion image $q_{n,m} = a_{n,m} + b_{n,m}i + c_{n,m}j + d_{n,m}k$ can be written as

$$\begin{aligned} q_{n,m} &= a_{n,m} + b'_{n,m}\mu_1 + c'_{n,m}\mu_2 + d'_{n,m}\mu_3 \\ &= (a_{n,m} + b'_{n,m}\mu_1) + \mu_2(c'_{n,m} + d'_{n,m}\mu_1), \end{aligned} \tag{6.42}$$

where $b'_{n,m}$, $c'_{n,m}$, and $d'_{n,m}$ are 2-D real sequences. The image $q_{n,m}$ is presented by two quaternion images

$$q^{(1)}_{n,m} = a_{n,m} + b'_{n,m}\mu_1 \text{ and } q^{(2)}_{n,m} = c'_{n,m} + d'_{n,m}\mu_1 . \tag{6.43}$$

For each point (n, m), the transformation

$$(b_{n,m}, c_{n,m}, d_{n,m}) \to (b'_{n,m}, c'_{n,m}, d'_{n,m})$$

is described by a 3×3 matrix and requires nine real multiplications. Indeed, if $\mu_1 = \mu_{1,1}i + \mu_{1,2}j + \mu_{1,3}k$, $\mu_2 = \mu_{2,1}i + \mu_{2,2}j + \mu_{2,3}k$, and $\mu_3 = \mu_{3,1}i + \mu_{3,2}j + \mu_{3,3}k$, then it is not difficult to obtain the following matrix equation for the above transformation:

$$\begin{bmatrix} b'_{n,m} \\ c'_{n,m} \\ d'_{n,m} \end{bmatrix} = \begin{bmatrix} \mu_{1,1} & \mu_{1,2} & \mu_{1,3} \\ \mu_{2,1} & \mu_{2,2} & \mu_{2,3} \\ \mu_{3,1} & \mu_{3,2} & \mu_{3,3} \end{bmatrix} \begin{bmatrix} b_{n,m} \\ c_{n,m} \\ d_{n,m} \end{bmatrix}. \tag{6.44}$$

The determinant of this matrix, which we denote by D, is ± 1 and the inverse matrix is its transpose, i.e., $D^{-1} = D^T$.

The number μ_2 can be calculated as $\mu_2 = (\mu_1 p - p\mu_1)/|\mu_1 p - p\mu_1)|$, where p is any pure quaternion that is not parallel to μ_1. Indeed, according to property **P8** in Section 1.1.5, for any two pure quaternions q_1 and q_2,

$$\begin{aligned} q_1q_2 - q_2q_1 &= 2\begin{vmatrix} i & j & k \\ b_1 & c_1 & d_1 \\ b_2 & c_2 & d_2 \end{vmatrix} \\ &= 2i(c_1d_2 - d_1c_2) - 2j(b_1d_2 - d_1b_2) + 2k(b_1c_2 - c_1b_2). \end{aligned}$$

The pure quaternion $q_1q_2 - q_2q_1$ is perpendicular to q_1,

$$(q_1q_2 - q_2q_1, q_1) = 2(c_1d_2 - d_1c_2)b_1 - 2(b_1d_2 - d_1b_2)c_1 + 2(b_1c_2 - c_1b_2)d_1 = 0.$$

Therefore, the pure unit quaternion μ_2 is perpendicular to μ_1.

Example 6.2

The representation of the quaternions in form of Eq. (6.42) is a generalization of the original definition of the quaternions as the dual complex numbers. Let the pure unit quaternion be $\mu_1 = (0,1,0,0) = i$ and $\mu_2 = (0,0,1,0) = j$. Then, the third pure quaternion is $\mu_3 = ji = -k = (0,0,0,-1)$, and any quaternion

$$q = a + bi + cj + dk = (a + bi) + j(c - di) = (a + b\mu_1) + \mu_2(c - d\mu_1).$$

The transformation $(b,c,d) \to (b',c',d') = (b,c,-d)$ is described by the diagonal 3×3 matrix

$$D = \begin{bmatrix} 1 & 0 & 0 \\ 0 & 1 & 0 \\ 0 & 0 & -1 \end{bmatrix}, \quad \det(D) = -1, \ D^{-1} = D^T = D.$$

Example 6.3

Let the pure unit quaternion be

$$\mu_1 = \frac{(0,1,1,1)}{\sqrt{3}} = \frac{(i + j + k)}{\sqrt{3}} = 0.57735027(i + j + k).$$

We consider the pure quaternion $p = (0,1,1,0) = i + j$. The unit quaternion μ_2 is calculated as

$$\mu_2 = \frac{\mu_1 p - p\mu_1}{|\mu_1 p - p\mu_1|} = \frac{(i + j + k)(i + j) - (i + j)(i + j + k)}{\sqrt{4 + 4}} = \frac{j2 - i2}{\sqrt{8}}$$
$$= \frac{1}{\sqrt{2}}(j - i) = 0.70710678(j - i).$$

One can verify that $|\mu_2|^2 = 1$ and $\mu_2^2 = -1$. The next unit quaternion is

$$\mu_3 = \mu_2\mu_1 = \frac{(j - i)}{\sqrt{2}}\frac{(i + j + k)}{\sqrt{3}} = \frac{i + j - 2k}{\sqrt{6}} = 0.40824829(i + j - 2k).$$

Therefore, the matrix D of the transformation $(b_{n,m}, c_{n,m}, d_{n,m}) \to (b'_{n,m}, c'_{n,m}, d'_{n,m})$ is

$$D = \begin{bmatrix} \mu_{1,1} & \mu_{1,2} & \mu_{1,3} \\ \mu_{2,1} & \mu_{2,2} & \mu_{2,3} \\ \mu_{3,1} & \mu_{3,2} & \mu_{3,3} \end{bmatrix}$$

$$= \begin{bmatrix} 0.57735027 & 0.57735027 & 0.57735027 \\ -0.70710678 & 0.70710678 & 0 \\ 0.40824829 & 0.40824829 & -0.81649658 \end{bmatrix}, \quad \det(D) = 1.$$

This matrix can also be written as

$$D = \begin{bmatrix} \frac{1}{\sqrt{3}} & 0 & 0 \\ 0 & \frac{1}{\sqrt{2}} & 0 \\ 0 & 0 & \frac{1}{\sqrt{6}} \end{bmatrix} \begin{bmatrix} 1 & 1 & 1 \\ -1 & 1 & 0 \\ 1 & 1 & -2 \end{bmatrix}.$$

Now, we consider the quaternion $q = 1 + 4i + 2j + 7k$ and its imaginary part as the row-vector (4, 2, 7) and the multiplication

$$(4, 2, 7)D^T = (7.50555350, \ -1.41421356, \ -3.26598632).$$

It not difficult to verify that

$$q = 1 + 7.50555350\mu_1 - 1.41421356\mu_2 - 3.26598632\mu_3.$$

All numbers are given with precision of eight digits after the point.

The 2-D QDFT of the image $q_{n,m}$ can be calculated by two 2-D QDFTs $Q^{(1)}_{p,s}$ and $Q^{(2)}_{p,s}$ of the images $q^{(1)}_{n,m}$ and $q^{(2)}_{n,m}$, respectively,

$$Q_{p,s} = Q^{(1)}_{p,s} + \mu_2 Q^{(2)}_{p,s} = \sum_{n=0}^{N-1}\sum_{m=0}^{N-1} q^{(1)}_{n,m} W^{np+ms}_{\mu_1} + \mu_2 \left[\sum_{n=0}^{N-1}\sum_{m=0}^{N-1} q^{(2)}_{n,m} W^{np+ms}_{\mu_1}\right].$$

These two 2-D QDFTs can be calculated from the 2-D DFTs of the complex images

$$f^{(1)}_{n,m} = a_{n,m} + ib'_{n,m} \text{ and } f^{(2)}_{n,m} = c'_{n,m} + id'_{n,m}.$$

Indeed, the following calculations hold:

$$
\begin{aligned}
Q^{(1)}_{p,s} &= \sum_{n=0}^{N-1}\sum_{m=0}^{N-1}(a_{n,m} + \mu_1 b'_{n,m})[\cos_N(np+ms) - \mu_1 \sin_N(np+ms)] \\
&= \sum_{n=0}^{N-1}\sum_{m=0}^{N-1}[a_{n,m}\cos_N(np+ms) + b'_{n,m}\sin_N(np+ms)] \\
&\quad + \mu_1 \sum_{n=0}^{N-1}\sum_{m=0}^{N-1}[b'_{n,m}\cos_N(np+ms) - a_{n,m}\sin_N(np+ms)],
\end{aligned}
$$

where $\cos_N(t) = \cos(2\pi t/N)$ and $\sin_N(t) = \sin(2\pi t/N)$. The first sum equals the real part of the 2-D DFT, $F^{(1)}_{p,s}$, of the complex image $f^{(1)}_{n,m}$,

$$
\begin{aligned}
F^{(1)}_{p,s} &= \sum_{n=0}^{N-1}\sum_{m=0}^{N-1}(a_{n,m} + ib'_{n,m})[\cos_N(np+ms) - i\sin_N(np+ms)] \\
&= \sum_{n=0}^{N-1}\sum_{m=0}^{N-1}[a_{n,m}\cos_N(np+ms) + b'_{n,m}\sin_N(np+ms)] \\
&\quad + i \sum_{n=0}^{N-1}\sum_{m=0}^{N-1}[b'_{n,m}\cos_N(np+ms) - a_{n,m}\sin_N(np+ms)].
\end{aligned}
$$

The second sum equals the imaginary part of this 2-D DFT, and therefore

$$
Q^{(1)}_{p,s} = \Re F^{(1)}_{p,s} + \mu_1 \left[\Im F^{(1)}_{p,s}\right].
$$

The 2-D QDFT of the quaternion image $q^{(2)}_{n,m}$ can similarly be calculated from the 2-D DFT, $F^{(2)}_{p,s}$, of the complex image $f^{(2)}_{n,m}$,

$$
Q^{(2)}_{p,s} = \Re F^{(2)}_{p,s} + \mu_1 \left[\Im F^{(2)}_{p,s}\right],
$$

where the 2-D DFT of the complex image $f^{(2)}_{n,m}$ is

$$
\begin{aligned}
F^{(2)}_{p,s} &= \sum_{n=0}^{N-1}\sum_{m=0}^{N-1}(c_{n,m} + id'_{n,m})[\cos_N(np+ms) - i\sin_N(np+ms)] \\
&= \sum_{n=0}^{N-1}\sum_{m=0}^{N-1}[c_{n,m}\cos_N(np+ms) + d'_{n,m}\sin_N(np+ms)] \\
&\quad + i \sum_{n=0}^{N-1}\sum_{m=0}^{N-1}[d'_{n,m}\cos_N(np+ms) - c_{n,m}\sin_N(np+ms)].
\end{aligned}
$$

Thus, we can write the following:

$$\begin{aligned} Q_{p,s} &= Q^{(1)}_{p,s} + \mu_2 Q^{(2)}_{p,s} \\ &= \Re F^{(1)}_{p,s} + \mu_1\left[\Im F^{(1)}_{p,s}\right] + \mu_2\left[\Re F^{(2)}_{p,s}\right] + \mu_3\left[\Im F^{(2)}_{p,s}\right]. \end{aligned} \tag{6.45}$$

Calculation of each component $Q_{p,s}$ in the new $(1, \mu_1, \mu_2, \mu_3)$ system requires nine real multiplications, if μ_1 and μ_2 are not among the imaginary units i, j, or k. The image decomposition with the transformation in Eq. (6.42) requires $9N^2$ real multiplications and $6N^2$ additions. Thus, the total number of real multiplications required to calculate the 2-D QDFT by Eq. (6.45) can be estimated as

$$m_{QFT}(N,N) = (9N^2 + 2m_{FT}(N,N)) + 9N^2 = 18N^2 + 2m_{FT}(N,N). \tag{6.46}$$

Here, we denote by $m_{FT}(N,N)$ the number of real multiplications in the complex $N \times N$-point 2-D DFT. If estimate $m_{FT}(N,N)$ by the column-row method and fast 1-D paired DFT with $m_{FT}(N,N) = 4N^2(\log_2 N - 3) + 16N$ real multiplications, we obtain the following estimation:

$$\begin{aligned} m_{QFT}(N,N) &= 18N^2 + 2[4N^2(\log_2 N - 3) + 16N] \\ &= 8N^2(\log_2 N - 3/2) + 32N. \end{aligned} \tag{6.47}$$

Comparing this estimation with the number of multiplications obtained in Eq. (6.40), we see that the application of the symplectic decomposition for the 2-D right-side QDFT uses $6N^2$ real multiplications less than the simple column-row method. It should be noted from Eq. (6.42) that the reconstruction of the original quaternion image $q_{n,m}$ from two images $q^{(1)}_{n,m}$ and $q^{(2)}_{n,m}$ requires $9N^2$ real multiplications.

The case of $\mu_1 = i$ and $\mu_2 = j$ considered in Example 6.2, we have the following:

$$\begin{aligned} Q_{p,s} &= Q^{(1)}_{p,s} + jQ^{(2)}_{p,s} \\ &= \Re F^{(1)}_{p,s} + i\left[\Im F^{(1)}_{p,s}\right] + j\left[\Re F^{(2)}_{p,s}\right] - k\left[\Im F^{(2)}_{p,s}\right], \end{aligned} \tag{6.48}$$

and the number of multiplications is reduced to $m_{QFT}(N,N) = 2m_{FT}(N,N))$, i.e.,

$$m_{QFT}(N,N) = 2[4N^2(\log_2 N - 3) + 16N] = 8N^2(\log_2 N - 3) + 32N.$$

6.6 Tensor Representation of the 2-D Right-Side QDFT

In this section, the concept of tensor representation is described for color-in-quaternion and quaternion images. The kernel of the 2-D right-side QDFT, which is defined in Eq. (6.41), is a periodic function,

$$W_{\mu}^{t+N} = \cos[2\pi(t+N)/N] - \mu \sin[2\pi(t+N)/N]$$
$$= \cos(2\pi t/N) - \mu \sin(2\pi t/N) = W_{\mu}^{t}$$

when $t = 0 : (N-1)$. Therefore, for calculating the 2-D right-side QDFT, we can apply the concept of the tensor representation, which was originally proposed for the grayscale images, and reduce the calculation and processing of this transform to separate processing 1-D signals, which we call *the quaternion splitting-signals*.

Similar to the grayscale images, the tensor representation of the discrete quaternion image $q_{n,m}$ is the 2-D frequency and 1-D time representation, when the image is described by a set of 1-D splitting-signals each of length N,

$$\chi : \{q_{n,m}\} \to \{q_{T_{p,s}} = \{q_{p,s,t}; t = 0 : (N-1)\}\}_{(p,s)\in J_{N,N}}.$$

The components $q_{p,s,t}$ are periodic by t, i.e., $q_{p,s,t+N} = q_{p,s,t}$. The tensor representation is unique, and the image can be defined through the 2-D DFT calculated by Eq. (6.9), or directly from the tensor transform, as shown in Section 6.1. In the $N = 2^r$ case, $3N/2$ splitting-signals compose the tensor representation of the image, the calculation of the 2-D DFT is reduced to $3N/2$ one-dimensional N-point DFT, instead of $2N$ such transformation in the traditional row-column method. It should be noted that the total number of components of $3N/2$ splitting-signals equals $N^2 + N^2/2$, which exceeds the number N^2 of points in the image. On the other side, many sets

$$T_{p,s} = \{(kp \bmod N, ks \bmod N); k = 0, 1, \ldots, (N-1)\},$$

with generators $(p,s) \in J_{N,N}$ have intersections at frequency-points. The tensor transform is therefore redundant. This redundancy can be removed and the 2-D DFT can be calculated in a more effective way, when using the modification of the tensor representation, which is called the paired transform [22, 24]. The tensor representation is very effective in the case of most interest when N is a prime, since the number of required sets $T_{p,s}$ is $(N+1)$ and the sets intersect only at the point (0, 0).

Now, we generalize the concept of the tensor representation for the quaternion images. We consider the full quaternion image

$$q_{n,m} = a_{n,m} + (r_{n,m}i + g_{n,m}j + b_{n,m}k)$$

in the general case, without the constrain that $a_{n,m}, r_{n,m}, g_{n,m}, b_{n,m} \geq 0$. When the real part $a_{n,m}$ is zero and $r_{n,m}, g_{n,m}, b_{n,m} \geq 0$, the image is referred to as a color image in the RGB or XYZ color space. For a given generator (p, s) from the set $J_{N,N}$ in Eq. (6.11), we define the following N subsets in the Cartesian lattice:

$$V_{p,s,t} = \{(n, m) \in X;\ np + ms = t \bmod N\},\ t = 0 : (N-1).$$

N pixel on each of these sets $V_{p,s,t}$ are located on parallel lines.

The quaternion splitting-signal $q_{T_{p,s}} = \{q_{p,s,t};\ t = 0 : (N-1)\}$ is defined as

$$\begin{aligned} q_{p,s,t} &= \sum_{(n,m)\in V_{p,s,t}} q_{n,m} = a_{p,s,t} + (r_{p,s,t})i + (g_{p,s,t})j + (b_{p,s,t})k \\ &= \left[\sum_{(n,m)\in V_{p,s,t}} a_{n,m}\right] + \left[\sum_{(n,m)\in V_{p,s,t}} r_{n,m}\right] i \\ &\quad + \left[\sum_{(n,m)\in V_{p,s,t}} g_{n,m}\right] j + \left[\sum_{(n,m)\in V_{p,s,t}} b_{n,m}\right] k. \end{aligned} \tag{6.49}$$

The components of this splitting-signal are calculated by

$$\begin{aligned} a_{p,s,t} &= \sum_{(n,m)\in X} \{a_{n,m};\ np + ms = t \bmod N\} \\ r_{p,s,t} &= \sum_{(n,m)\in X} \{r_{n,m};\ np + ms = t \bmod N\} \\ g_{p,s,t} &= \sum_{(n,m)\in X} \{g_{n,m};\ np + ms = t \bmod N\} \\ b_{p,s,t} &= \sum_{(n,m)\in X} \{b_{n,m};\ np + ms = t \bmod N\}. \end{aligned}$$

The imaginary components of the splitting-signals are the splitting-signals of the red, green, and blue channels of the color image, if $a_{n,m} \equiv 0$. The 2-D right-side QDFT at N frequency-points of the cyclic group $T_{p,s}$ is defined by the right-side 1-D QDFT of the quaternion splitting-signal,

$$Q_{kp \bmod N, ks \bmod N} = \sum_{t=0}^{N-1} q_{p,s,t} W_{\mu}^{kt},\ k = 0 : (N-1). \tag{6.50}$$

The union of the family of disjoint subsets $V_{p,s,t}$, $t = 0 : (N-1)$, cover the Cartesian grid and, therefore, the following calculations hold:

$$\begin{aligned}\sum_{t=0}^{N-1} q_{p,s,t} W_{\mu}^{kt} &= \sum_{t=0}^{N-1} \left[\sum_{np+ms=t \bmod N} a_{n,m} + (r_{n,m})i + (g_{n,m})j + (b_{n,m})k \right] W_{\mu}^{kt} \\ &= \sum_{t=0}^{N-1} \left[\sum_{np+ms=t \bmod N} a_{n,m} + (r_{n,m})i + (g_{n,m})j + (b_{n,m})k \right] W_{\mu}^{k(np+ms)} \\ &= \sum_{n=0}^{N-1}\sum_{m=0}^{N-1} [a_{n,m} + (r_{n,m})i + (g_{n,m})j + (b_{n,m})k] W_{\mu}^{n(kp)+m(ks)} \\ &= \sum_{n=0}^{N-1}\sum_{m=0}^{N-1} q_{n,m} W_{\mu}^{n(kp)+m(ks)} = Q_{kp \bmod N, ks \bmod N}.\end{aligned}$$

Thus,

$$\sum_{t=0}^{N-1} q_{p,s,t} W_{\mu}^{kt} = \sum_{n=0}^{N-1}\sum_{m=0}^{N-1} q_{n,m} W_{\mu}^{n(kp)+m(ks)} = Q_{kp \bmod N, ks \bmod N}.$$

In the $N = 2^r$ case when $r > 1$, the color image is described by $3N/2$ quaternion splitting-signals and 2-D QDFT of the image is split by the $3N/2$ 1-D QDFT of these signals. As an example, Fig. 6.12 shows the color image of size 512×512 in part (a), and the color splitting-signals of the red, green, and blue channels, which are calculated for the frequency-points $(p, s) = (1, 2)$ and (2, 1) in parts (b) and (c), respectively. The splitting-signals have the length equal to 512. This is the case of the pure quaternion image, with the real part equal to zero. Therefore, the splitting-signals for such an image are called *the color splitting-signals.* The set of $(512 + 256) = 768$ color splitting-signals of length $N = 512$ each is the tensor transform of the color-in-quaternion image, which describes the image and carries its spectral information, i.e., the 2-D QDFT.

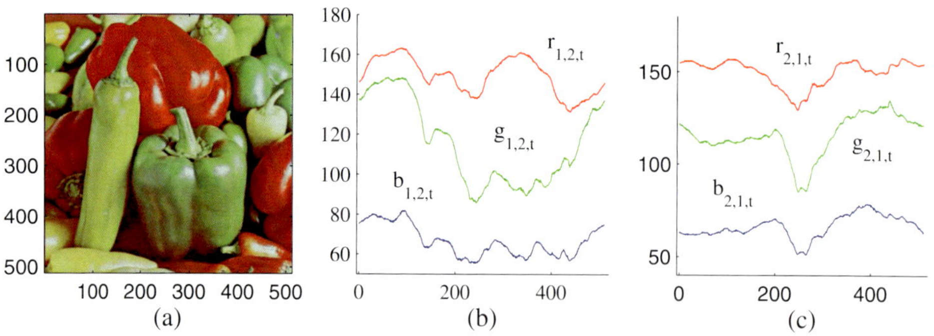

Figure 6.12 (a) The RGB color image and three components of the color splitting-signal for (b) $(p, s) = (1, 2)$ and (c) $(p, s) = (2, 1)$.

The signal $q_{1,2,t} = (r_{1,2,t})i + (g_{1,2,t})j + (b_{1,2,t})k$ is called the color splitting-signal in quaternion space, which is generated by the frequency-point (1, 2), or the frequency $(2\pi/N, 0\pi/N)$. Figure 6.13 shows the 512-point 1-D QDFT of this splitting-signal in absolute scale [and with the normalized factor of $1/(512)^2$] in part (a). The 2-D QDFT of the color image in the absolute scale is shown in Fig. 6.13(b). In Fig. 6.13(c), this spectrum is shifted to the center. The 1-D QDFT of the color splitting-signal coincides with the 2-D QDFT of the image at frequency-points of the set $T_{1,2}$, which are shown in Fig. 6.13(b,c).

We also consider the $(p, s) = (2, 1)$ case. Figure 6.14 shows the 512-point 1-D QDFT of the color splitting-signal $q_{2,1,t}$ in part (a) in absolute scale [and with the normalized factor of $1/(512)^2$]. The amplitude spectra of the 2-D QDFT of the color image is shown in Fig. 6.14(b,c). The 1-D QDFT of the color splitting-signal coincides with the 2-D QDFT of the image at frequency-

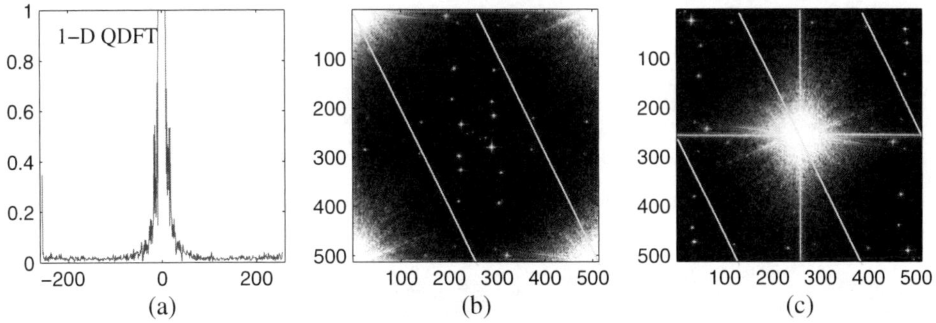

Figure 6.13 (a) The 512-point right-side QDFT of the color splitting-signal $q_{1,2,t}$, (b) the amplitude spectrum of the 2-D color-in-quaternion image and the marked location of frequency-points of the set $T_{1,2}$, and (c) the same spectrum with $T_{1,2}$, after cyclically shifting to the center.

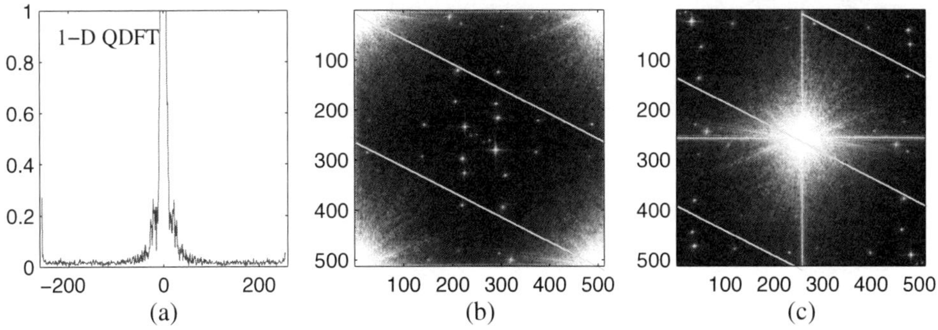

Figure 6.14 (a) The amplitude spectrum of the color splitting-signal $q_{2,1,t}$, (b) the amplitude spectrum of the 2-D color-in-quaternion image and the marked location of frequency-points of the set $T_{2,1}$, and (c) the same spectrum with $T_{2,1}$, after cyclically shifting to the center.

points of the set $T_{2,1}$, which also are shown in Fig. 6.14(b,c). All quaternion DFTs were calculated for the unit pure quaternion $\mu = (i + j + k)/\sqrt{3}$.

6.6.1 Program for tensor transform-based 2-D QDFT

Below is the script of the program for calculating the $N \times N$-point 2-D QDFT for the case, when $N \geq 3$ is a prime:

"test_2dQDFTbyTT.m" – the main code to calculate the 2-D right-side QDFT;

"fcol_pst.m" – to calculate the color splitting-signal $\{f_{p,s,t};\ t = 0 : N - 1\}$;

"qreal1dfft.m" – to calculate the quaternion 1-D DFT of the quaternion splitting-signal with zero real part;

"t_ps.m" – to calculate the cyclic group $T_{p,s}$ of order N.

```
% ==========================================================
% call: test_2dQDFTbyTT.m
% The tensor NxN-point 2-D QDFT (N is prime)
% Artyom Grigoryan, June 14, 2014
% ----------------------------------------------------------
% Read the color image 256x256 and extend it to size 257x257
  Y=imread('couple_color.tiff','tiff');
  image_in=double(Y); N=257;
  image_in(:,N,:)=Y(:,N-1,:); image_in(N,1:N-1,:)=
  Y(N-1,:,:);
% Set of generators: (p,1),p=0:N-1, (1,0)
  number_ofgenerators=N+1;
  Jps=ones(2,number_ofgenerators);
  Jps(1,1:N)=0:N-1; Jps(2,number_ofgenerators)=0;
% Select unit pure quaternion, for instance mu=[1,1,1]/
  sqrt(3)
  qmu=[1,1,1]; qmu=qmu/norm(qmu);
% (N+1) 1-D QDFTs of splitting-signals
  Q2dDFT=zeros(N,N,4);
  for k=1:number_ofgenerators
      p=Jps(1,k); s=Jps(2,k);
      % 3-D quaternion splitting-signal
      col_ssignal=fcol_pst(image_in,N,p,s);
      % calculate the real 1-D QDFT
      FQ=qreal1dfft(col_ssignal,qmu,N);
      Q1=FQ(1,:); Qi=FQ(2,:); Qj=FQ(3,:); Qk=FQ(4,:);
      % Place 1-D QDFT into the 2-D QDFT
      [Tp,Ts]=t_ps(p,s,N);
      for n=1:N
          p1=Tp(n); s1=Ts(n);
```

```
        Q2dDFT(p1,s1,1)=Q1(n); Q2dDFT(p1,s1,2)=Qi(n);
        Q2dDFT(p1,s1,3)=Qj(n); Q2dDFT(p1,s1,4)=Qk(n);
      end
  end
% -------------------------------------------------------------
% Call: fcol_pst.m
% Calculate the color splitting-signal f_p,s,t
  function B=fcol_pst(A,N,p,s)
    B=zeros(3,N);
    ks=0;
    for n=1:N
      t=ks;
      for m=1:N
        if t>=N t=t-N; end
        t1=t+1; t=t+p;
        B(1,t1)=B(1,t1) + A(m,n,1);
        B(2,t1)=B(2,t1) + A(m,n,2);
        B(3,t1)=B(3,t1) + A(m,n,3);
      end
      ks=ks+s; if ks>=N ks=ks-N; end
  end
% -----------------------------------------------
% Calculate the cyclic group T(p,s)
  function [TP,TS]=t_ps(p,s,N)
        TP=zeros(1,N); TS=zeros(1,N);
        kp=0; ks=0;
        for n=1:N
          TP(n)=kp+1; TS(n)=ks+1;
          kp=kp+p; if kp>=N kp=kp-N; end
          ks=ks+s; if ks>=N ks=ks-N; end
        end
% -----------------------------------------------
```

The scripts of codes given above can be used for calculating the $N \times N$-point QDFT for any integer $N > 2$, and can be modified for the $N \times M$-point QDFT when $M \neq N$. For each of these cases, the corresponding set of generators "Jps" should be calculated (methods of calculating such sets can be found in Ref. [24]). It should be mentioned that unlike other algorithms the tensor transform-based 2-D QDFT reduces the calculation to parallel 1-D QDFTs of splitting-signals.

In the $N = 2^r$ case when $r > 2$, the color image is described by $3N/2$ quaternion splitting-signals and the 2-D QDFT of the image is split by the $3N/2$ 1-D QDFTs of these signals. Therefore, the total number of real multiplications required for the 2-D QDFT with this splitting equals

Table 6.1 Number of DFTs and real multiplications for calculating the $N \times N$-point 2-D right-side QDFT, when $N = 2^r$, $r > 2$.

2-D QDFT	C-R	SD	TT
# 1-D DFT	$4N$		$3N$
# 2-D DFT		2	
# mult.s	$8N^2r + 32N$	$8N^2r - 6N^2 + 32N$	$6N^2r + 24N$

$$\begin{aligned} m_{QFT}(N,N) &= 3N/2 m_{QFT}(N) = 3N/2[4N\log_2 N + 16] \\ &= 6N^2\log_2 N + 24N. \end{aligned}$$

Table 6.1 shows the number of real multiplications used in calculation of the 2-D QDFT, by using the column-row (C-R), symplectic decomposition (SD), and tensor transform (TT) methods. The required number of 1-D DFTs is also given for both C-R and TT methods. For the SD method, we only mention the required two complex 2-D DFTs that can be calculated by using different algorithms.

6.7 Tensor Representation of Grayscale and Color Images

We consider the quaternion image, which is composed from both grayscale and RGB color images. Figure 6.15 shows the quaternion image in parts (a) and (b). The real part of the image is in Fig. 6.15(a) and the imaginary part of the quaternion image is the color image in the RGB format and shown in Fig. 6.15(b). The volume of this quaternion image is $512 \times 512 \times 4$. The quaternion splitting-signal that is calculated for the frequency-point $(p,s) = (1,5)$ and composed from four components is shown in part Fig. 6.15(c).In this example, the grayscale image is the average of the color channels, therefore, the real component $a_{1,5,t}$, (colored in black) of the

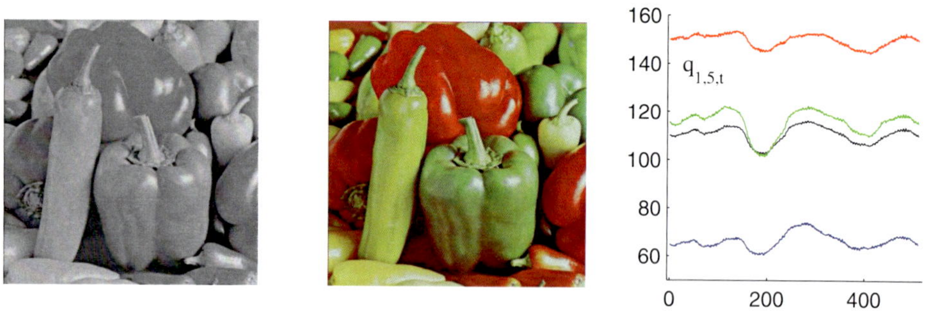

(a) (b) (c)

Figure 6.15 (a) The grayscale image, (b) the RGB color image, and (c) four-component quaternion splitting-signal with $(p,s) = (1,5)$.

quaternion splitting-signal $q_{1,5,t}$ is the average of color components of the imaginary part of $q_{1,5,t}$,

$$a_{1,5,t} = (r_{1,5,t} + g_{1,5,t} + b_{1,5,t})/3, \quad t = 0 : 511.$$

Figure 6.16 shows the 512-point 1-D rs-QDFT of the quaternion splitting-signal $q_{1,5,t}$ in part (a) in the absolute scale [and with the normalized factor of $1/(512)^2$]. The amplitude spectra of the 2-D rs-QDFT of the quaternion image is shown in Fig. 6.16(b,c). The 1-D rs-QDFT of the quaternion splitting-signal coincides with the 2-D rs-QDFT of the quaternion image at 512 frequency-points of the set $T_{1,5}$, which are also shown in Fig. 6.16(b,c). The 1-D and 2-D rs-QDFTs were calculated for the pure unit quaternion $\mu = (i + k)/\sqrt{2}$.

6.8 Direction Components of Quaternion Images

The tensor representation of quaternion images and color images allows calculation of the 2-D QDFT by 1-D QDFTs of the splitting-signals. The tensor representation allows for image composition by the finite number of direction images, or direction image components. In color imaging, the color image can be composed by the direction color images, i.e., such images that along each line in the given direction, all pixels of the image have the same color. Each of such direction images is defined by the corresponding splitting-signal, number of such signals is defined by the set $J_{N,M}$ of generators (p, s). These generators are the selected frequency-points. Thus, the tensor transform, which is the frequency-time representation of images (quaternion, color, or grayscale images), contains the information of direction components in images. The processing of images by their splitting-signals means the image processing by directions.

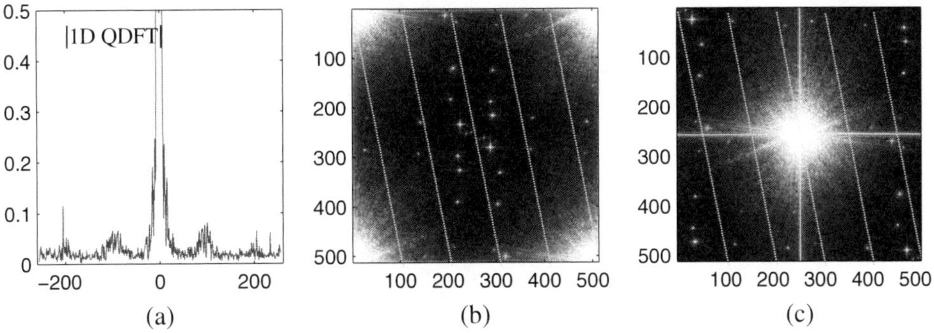

Figure 6.16 (a) The amplitude spectrum of the quaternion splitting-signal $q_{1,5,t}$, (b) the amplitude spectrum of the 2-D quaternion image with the location of frequency-points of the set $T_{1,5}$, and (c) the same spectrum with $T_{1,5}$, after cyclically shifting to the middle.

In the case when N is prime, the number of quaternion splitting-signals equals $(N+1)$, because the set of generators can be taken as

$$J_{N,N} = \{(1,s); s = 0 : (N-1)\} \cup \{(0,1)\}.$$

The number of the N-point 1-D QDFTs that are required to calculate the $N \times N$-point 2-D QDFT equals $(N+1)$. If we consider the above "peppers" image of size 512×512 inside the matrix 521×521, the number of splitting-signals in the tensor representation for this new image will have only 522 splitting-signals of length 521 each. For comparison, the original image of size 512×512 is represented by 768 splitting-signals of length 512 each.

It is important to note that by using the tensor representation, the quaternion image can be directly reconstructed from its splitting-signals. To show this fact, we consider the concept of the quaternion direction image that is defined by the corresponding splitting-signal. Similar to the grayscale images, the direction quaternion image is defined from the quaternion splitting-signal $q_{p,s,t}$ by

$$d_{n,m} = d_{n,m;p,s} = \frac{1}{N} q_{p,s,(np+ms) \bmod N} \quad n,m = 0 : (N-1). \tag{6.51}$$

By using the set of the direction images generated by frequency-points $(p,s) \in J_{N,N}$, we can compose the original quaternion image from its direction components. The tensor representation of the quaternion image is very effective, since all $(N+1)$ required groups $T_{p,s}$ intersect only at the point (0, 0). The following statement holds.

Principle of Superposition 1: The discrete quaternion image $q_{n,m}$ of size $N \times N$, where $N > 2$ is a prime, can be calculated from $(N+1)$ quaternion directional images as follows:

$$\begin{aligned} q_{n,m} &= \sum_{(p,s) \in J_{N,N}} d_{n,m;p,s} - E[q_{T_{0,1}}] \\ &= \frac{1}{N}\left[\sum_{s=0}^{N-1} q_{1,s,(n+ms) \bmod N} + q_{0,1,m}\right] - E[q_{T_{0,1}}], \\ & n,m = 0 : (N-1). \end{aligned} \tag{6.52}$$

Here, $E[q_{T_{0,1}}]$ is the mean of the splitting-signal,

$$E[q_{T_{0,1}}] = \frac{1}{N}\sum_{t=0}^{N-1} q_{0,1,t},$$

which can also be calculated by using other splitting-signals, since

$$\sum_{t=0}^{N-1} q_{0,1,t} = \sum_{t=0}^{N-1} q_{1,s,t} = \sum_{n=0}^{N-1}\sum_{m=0}^{N-1} q_{n,m}, \quad s = 0 : (N-1).$$

Thus, $E[q_{T_{0,1}}] = NE[q]$. If the quaternion image $q_{n,m}$ has non-zero mean, it can be centered, by changing the image by $q_{n,m} \to q_{n,m} - E[q]$. Then, $E[q_{T_{0,1}}] = NE[q] = 0$ and the image is the sum of direction components

$$q_{n,m} = \sum_{s=0}^{N-1} d_{n,m;1,s} + d_{n,m;1,1} = \frac{1}{N}\left[\sum_{s=0}^{N-1} q_{1,s,(np+m) \bmod N} + q_{0,1,m}\right].$$

As an example, we consider a pure quaternion image $q_{n,m} = i(q_{n,m})_i + j(q_{n,m})_j + k(q_{n,m})_k$, which can be referred to as the color image if all components $(q_{n,m})_i$, $(q_{n,m})_j$, and $(q_{n,m})_k$ are non negative. Figure 6.17 shows the color image of size 509 × 509 in part (a), and the direction images generated by 11 different frequency-points (p, s) in parts (b)

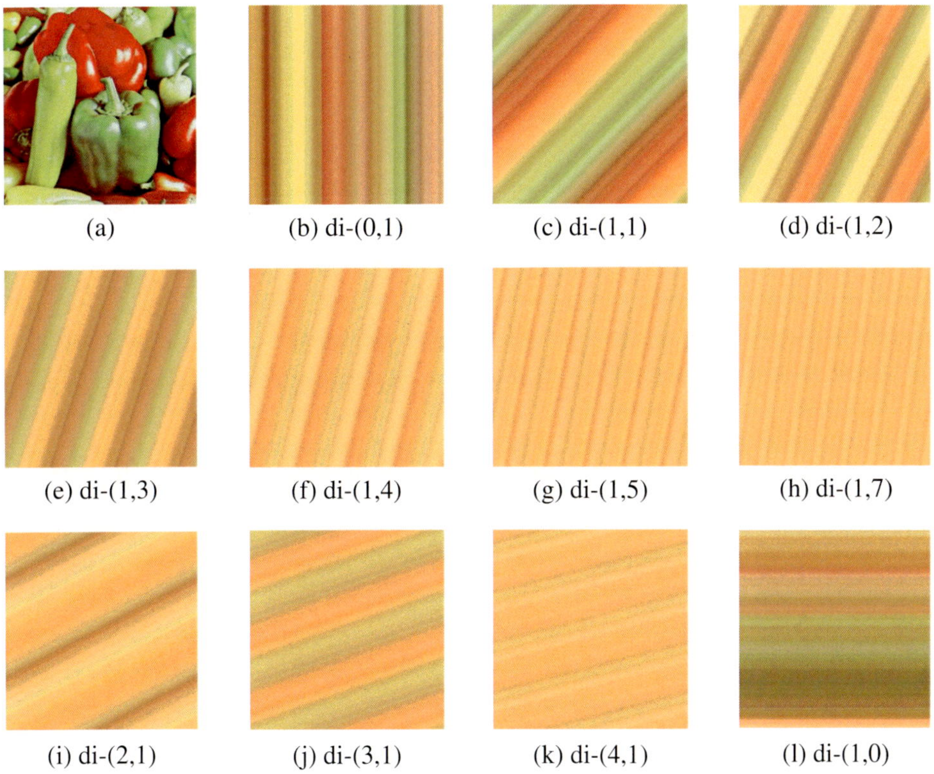

Figure 6.17 (a) The color image and its direction components generated by the frequency-points (b) (0, 1), (c) (1, 1), (d) (1, 2), (e) (1, 3), (f) (1, 4), (g) (1, 5), (h) (1, 7), (i) (2, 1), (j) (3, 1), (k) (4, 1), and (l) (1, 0).

through (l). All these color direction images were normalized to the intensity 255. The number $N = 509$ is prime and the sum of 510 such direction images $\{d_{n,m;p,s};\ (p,s) \in J_{509,509}\}$ defines the color image.

In image composition by Eq. (6.52), it is assumed that the set of generators $J_{N,N}$ is $\{(1,s);\ s = 0 : (N-1)\} \cup (0,1)$. If we take the transpose set of generators $\{(p,1);\ p = 0 : (N-1)\} \cup (1,0)$, the quaternion image can be written as the following sum of $(N+1)$ direction components:

$$q_{n,m} = \sum_{p=0}^{N-1} d_{n,m;p,1} + d_{n,m;1,0} = \frac{1}{N}\left[\sum_{p=0}^{N-1} q_{p,1,(np+m) \bmod N} + q_{1,0,n}\right].$$

Among the quaternion or color direction images, we can select "the main" direction components, meaning the direction images with high energy that contribute to the image composition in Eq. (6.52). For the above "peppers" image, Fig. 6.18 shows the graph of the energy for all 510 color direction images. The energy of the quaternion direction image (or splitting-signal) generated by the frequency-point (p,s) is calculated (in the square-root scale) by

$$E(p,s) = \sqrt{\frac{1}{N}\sum_{t=0}^{N-1} |q_{p,s,t}|^2},\ (p,s) \in J_{N,N}.$$

One can see from the graph that the (1, 1)-direction image has the highest energy, $E(1,1) = 1.0318 \cdot 10^5$. Below are the next 11 frequency-points (p,s) for which the direction images have high energies together with their energies ($\times 10^5$)

$(0,1)$	$(2,1)$	$(127,1)$	$(170,1)$	$(255,1)$	$(382,1)$	$(504,1)$	$(505,1)$	$(507,1)$	$(508,1)$	$(1,0)$
1.0316	1.0230	1.0230	1.0243	1.0274	1.0221	1.0221	1.0221	1.0249	1.0285	1.0288.

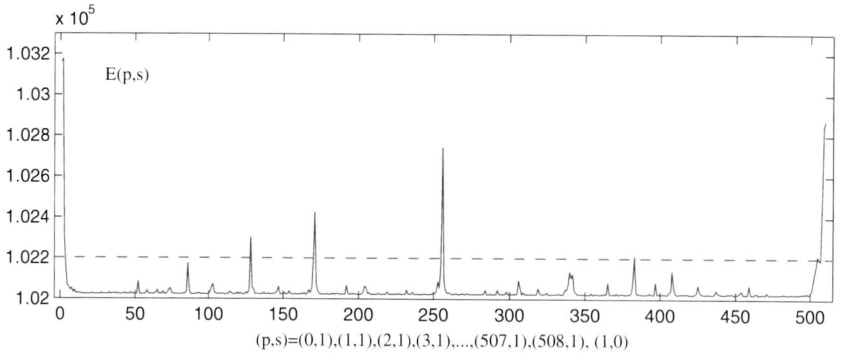

Figure 6.18 The energy graph of 510 color splitting-signals of the 509 × 509 "peppers" image.

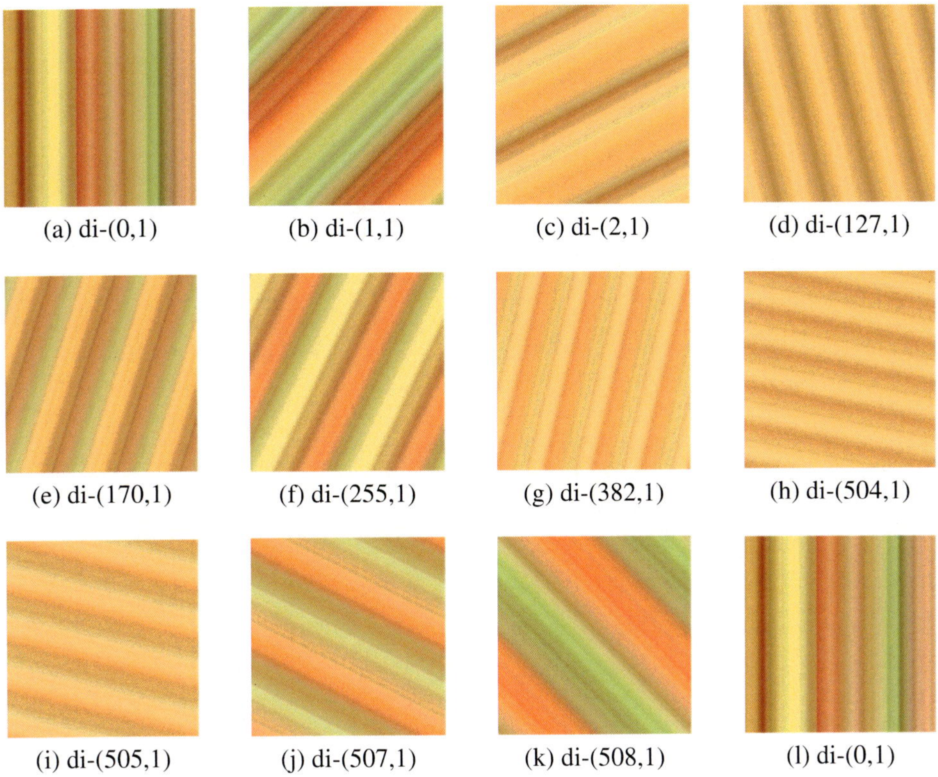

Figure 6.19 (a)-(l) The twelve direction components with the highest energies (greater than $1.02195 \cdot 10^5$) for the “peppers” image.

Figure 6.19 shows the direction images generated by these twelve frequency-points.

The number of real multiplications required for calculating the 2-D QDFT by $(N+1)$ quaternion N-point DFTs can be estimated as

$$m_{QDFT}(N,N) = (N+1)m_{QDFT}(N) = 2(N+1)m_{DFT}(N). \tag{6.53}$$

This estimation is twice than the traditional $N \times N$-point DFT requires, i.e., $m_{QDFT}(N,N) \approx 2m_{DFT}(N,N)$. The 2-D QDFT with symplectic decomposition requires $18N^2 + 2m_{DFT}(N,N)$ multiplications, i.e., $18N^2$ more multiplications, than the method of tensor representation.

In the $N = 2^r$ case, when the integer $r > 1$, the tensor representation of the quaternion image exceeds 1.5 times the size of the image. $3N/2$ splitting-signals are used with the total of $3N/2N = N^2 + N^2/2$ components. The representation is redundant in a sense that there are many frequency-points at which the values of the 2-D QDFT of the image can be obtained from 1-D QDFTs of different splitting-signals. To remove such a redundancy, the modified tensor representation can be used. This representation is called the 2-D paired representation or paired transform [22],[34]–[36]. The components of the paired representation are calculated from the tensor representation by

$$q'_{p,s,t} = q_{p,s,t} - q_{p,s,t+N/2}, \quad t = 0 : (N/2 - 1).$$

Thus, the splitting signal $\{q_{p,s,0}, q_{p,s,1}, q_{p,s,2}, \ldots, q_{p,s,N-1}\}$ of length N is folding into the new signal $\{q'_{p,s,0}, q'_{p,s,1}, \ldots, q'_{p,s,N/2-1}\}$ of length $N/2$. It should be noted that, if g.c.d.$(p,s) = 2^k$, $k > 0$, this signals is composed only by the components $q'_{p,s,t}$ with $t = 2^k t_1$, where $t_1 = 0, 1, \ldots, N/2^{k+1} - 1$, since all other components are zero.

Each paired splitting-signal carries the spectral information of the 2-D QDFT of the quaternion image in the subset of frequency-points

$$T'_{p,s} = \{([2k+1]p \bmod N, [2k+1]s \bmod N); k = 0, 1, \ldots, (N/2 - 1)\}.$$

Indeed, as shown in Section 1.4, $e^{\mu\vartheta_1}e^{\mu\vartheta_2} = e^{\mu(\vartheta_1+\vartheta_2)}$ for any two angles ϑ_1 and ϑ_2. Therefore, the following holds:

$$Q_{(2k+1)p \bmod N, (2k+1)s \bmod N} = \sum_{t=0}^{N/2-1} [q'_{p,s,t} W^t_{N/2,\mu}] W^{kt}_{\mu}, \quad k = 0 : (N/2 - 1). \tag{6.54}$$

It is clear that, by folding $3N/2$ splitting-signals of the tensor representation into $3N/2$ such new signals, we will have only $(3N/2)N/2 = 3/4N^2$ components instead of required N^2 for any unitary transform of the image $N \times N$. In other words, the set of generators $J_{N,N}$ should be extended and new splitting-signals in this new representation, which we call the paired representation. The extended set of generators of the paired splitting-signals can be defined as (see for more details [23]–[25])

$$J'_{N,N} = J_{N,N} \cup 2J_{N/2,N/2} \cup 4J_{N/4,N/4} \cup \cdots \cup N/2J_{2,2} \cup \{(0,0)\}.$$

The number of generators in this set is $(3N - 2)$. The paired splitting-signals generated by (p,s) of the subset $J_{N,N}$ has the length $N/2$ each, and when generated by (p,s) of the subset $J_{N/2,N/2}$ has the length $N/4$ each, and so on.

Introducing the quaternion direction images for this set of generators by

$$d'_{n,m;p,s} = q'_{p,s,(np+ms) \bmod N} = q_{p,s,(np+ms) \bmod N} - q_{p,s,(np+ms+N/2) \bmod N},$$

the quaternion image $q_{n,m}$ can be calculated by the direction components, or splitting-signals generated by $(p,s) \in J'_{N,N}$.

Principle of Superposition 2: The discrete quaternion image of size $N \times N$, where $N = 2^r, r > 1$, can be calculated from its $(3N-2)$ quaternion direction images as

$$q_{n,m} = \sum_{(p,s)\in J'_{N,N}} d'_{n,m;p,s}$$
$$= \frac{1}{2N}\sum_{k=0}^{r-1}\frac{1}{2^k}\sum_{(p,s)\in 2^k J_{N/2^k,N/2^k}} q'_{p,s,(np+ms)\bmod N} + E[q]. \tag{6.55}$$

The last addendum in the formula is the mean value of the image,

$$E[q] = \frac{1}{N^2}\sum_{n=0}^{N-1}\sum_{m=0}^{N-1} q_{n,m} = \frac{1}{N^2} q'_{0,0,0}.$$

6.9 2-D Left-Side QDFT and Tensor Transform

In this section, we consider the 2-D left-side QDFT (ls-QDFT). The color image $f_{n,m}$ is considered to be of size $N \times M$, which is transformed form the RGB color model [31] to the quaternion subspace, namely the subspace of pure quaternions. Let N_0 be the g.c.d.(N, M) and $N = N_0N_1$, $M = N_0M_1$, and $K = M_1N = N_1M$, where integers $N_1, M_1 \geq 1$. The 2-D ls-QDFT of the complex-in-quaternion or full quaternion image $q_{n,m}$ is defined as

$$Q_{p,s} = \sum_{n=0}^{N-1} W^{np}_{N;\mu}\left[\sum_{m=0}^{M-1} W^{ms}_{M;\mu} q_{n,m}\right], \tag{6.56}$$

where $p = 0 : (N-1)$ and $s = 0 : (M-1)$. This transform is based on the N-point ls-QDFT, which is defined on the quaternion signal q_n of length N as

$$Q_p = \sum_{n=0}^{N-1} W^{np}_{\mu} q_n,$$

where $p = 0 : (N-1)$ and μ is a pure unit quaternion $\mu = im_1 + jm_2 + km_3$, $\mu^2 = -1$. The kernel of this transform is $W_\mu = W_{N;\mu} = \exp(-\mu 2\pi/N) = \cos(2\pi/N) - \mu\sin(2\pi/N)$. The 2-D ls-QDFT can also be written as [4]

$$Q_{p,s} = \sum_{n=0}^{N-1}\sum_{m=0}^{M-1} W^{M_1np+N_1ms}_{K;\mu} q_{n,m}, \tag{6.57}$$

where $p = 0 : (N - 1)$ and $s = 0 : (M - 1)$. Here, for a given quaternion unit number μ, the basis exponential function is

$$W^t_{K;\mu} = \exp(-2\pi\mu t/K) = \cos(2\pi t/K) - \mu\sin(2\pi t/K),\ t = 0 : (K - 1).$$

The inverse 2-D ls-QDFT is calculated by

$$q_{n,m} = \frac{1}{NM}\sum_{p=0}^{N-1}\sum_{s=0}^{M-1} W_{K;\mu}^{-(M_1 np + N_1 ms)} Q_{p,s}, \tag{6.58}$$

where $n = 0 : (N - 1)$ and $m = 0 : (M - 1)$.

With the direct calculation by the column-row method, the separable 2-D ls-QDFT requires N M-point ls-QDFTs and M N-point ls-QDFTs. The 1-D ls-QDFT requires two 1-D DFTs. Therefore, the row-column method uses $2N$ M-point DFTs and $2M$ N-point DFTs.

6.9.1 Tensor representation of the 2-D left-side QDFT

The discrete left-side QDFT can be applied for any quaternion images, but we consider images with the zero real parts, for instance, the color images that can be transformed to quaternion space from the RGB model. For the simplicity of calculation, we consider the $N = M = K$ case, i.e., when the $N \times N$-point ls-QDFT of the color-in-quaternion image $q_{n,m}$ is

$$Q_{p,s} = \sum_{n=0}^{N-1}\sum_{m=0}^{N-1} W_\mu^{np+ms} q_{n,m},\ p, s = 0 : (N - 1). \tag{6.59}$$

It is important to note that the kernel of this transform is periodic,

$$\begin{aligned} W_\mu^{t+N} &= \cos[2\pi(t + N)/N] - \mu\sin[2\pi(t + N)/N] \\ &= \cos(2\pi t/N) - \mu\sin(2\pi t/N) = W_\mu^t \end{aligned}$$

when $t = 0 : (N - 1)$. Therefore, for the 2-D ls-QDFT, we can apply the concept of the tensor representation and reduce the calculation of this transform to parallel processing the 1-D signals, which we call *the color splitting-signals,* or *the quaternion splitting-signals.*

Figure 6.20 shows the color image of size 1223×1223 in part (a). The number 1223 is prime. In the absolute scale, the 2-D ls-QDFT of the color image is shown in Fig. 6.20(b). This transform is shifted to the center.

We consider the color image $q_{n,m} = (r_{n,m}i + g_{n,m}j + b_{n,m}k)$ with zero real part in the quaternion algebra. The tensor representation of the discrete image is the 2-D frequency and 1-D time representation when the image is described by a set of 1-D splitting-signals of length N each,

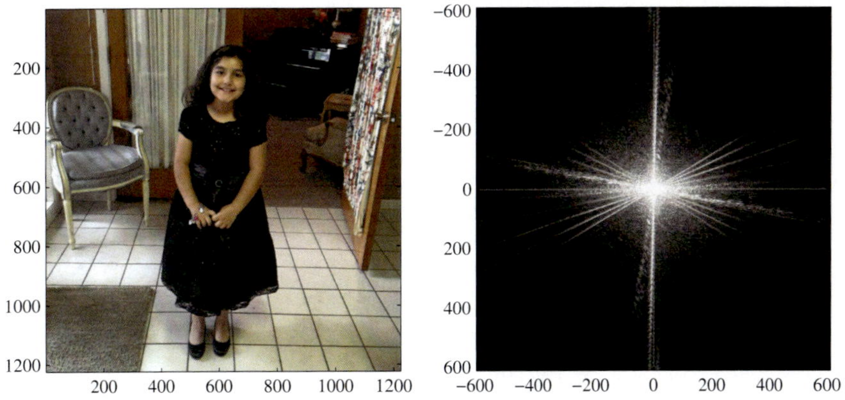

(a) (b)

Figure 6.20 (a) The color "girl Anoush" image and (b) the ls-QDFT of the image in the absolute scale and shifted cyclically to the center.

$$\chi : \{q_{n,m}\} \to \{q_{T_{p,s}} = \{q_{p,s,t}; t = 0 : (N-1)\}\}_{(p,s)\in J_{N,N}}.$$

This transform is also named as the discrete Radon transform in many publications (see Ref. [35]). Given (p, s), the components of the splitting-signal $q_{T_{p,s}}$ are the following sums (or ray-sums) of the image $q_{n,m}$ along parallel lines on the Cartesian lattice $X = X_{N,N} = \{(n,m);\ n,m = 0, 1, ..., (N-1)\}$:

$$q_{p,s,t} = \sum_{(n,m)\in X} \{q_{n,m};\ np + ms = t \bmod N\}.$$

This quaternion splitting-signal $q_{T_{p,s}} = \{q_{p,s,t};\ t = 0 : (N-1)\}$ is calculated by

$$\begin{aligned} q_{p,s,t} &= i(r_{p,s,t}) + j(g_{p,s,t}) + k(b_{p,s,t}) \\ &= \sum_{(n,m)\in V_{p,s,t}} f_{n,m} = i\left[\sum_{(n,m)\in V_{p,s,t}} r_{n,m}\right] \\ &\quad + j\left[\sum_{(n,m)\in V_{p,s,t}} g_{n,m}\right] + k\left[\sum_{(n,m)\in V_{p,s,t}} b_{n,m}\right]. \end{aligned} \tag{6.60}$$

Here, for a given generator (p, s) from the set $J_{N,N}$, we define the following N subsets in the Cartesian lattice:

$$V_{p,s,t} = \{(n,m) \in X;\ np + ms = t \bmod N\},\ \ t = 0 : (N-1).$$

The imaginary components of the splitting-signals are the splitting-signals of the red, green, and blue channels of the color image.

The frequency-point (p, s) is called the generator of the splitting-signal. The components $q_{p,s,t}$ are periodic by t, i.e., $q_{p,s,t+N} = q_{p,s,t}$. We denote this splitting-signal by the set $T_{p,s} = \{(kp \bmod N, ks \bmod N);\ k = 0 : (N-1)\}$, since the signal carries the information about the 2-D ls-QDFT at N frequency-points of this set,

$$Q_{kp \bmod N, ks \bmod N} = \sum_{t=0}^{N-1} W_{\mu}^{kt} q_{p,s,t},\ k = 0 : (N-1). \tag{6.61}$$

Indeed, the union of the family of disjoint subsets $V_{p,s,t}$, $t = 0 : (N-1)$, is the Cartesian grid and, therefore, the following calculations hold:

$$\begin{aligned}
&\sum_{t=0}^{N-1} W_{\mu}^{kt} q_{p,s,t} \\
&= \sum_{t=0}^{N-1} W_{\mu}^{kt} \left[\sum_{np+ms=t \bmod N} (r_{n,m})i + (g_{n,m})j + (b_{n,m})k \right] \\
&= \sum_{t=0}^{N-1} W_{\mu}^{k(np+ms)} \left[\sum_{np+ms=t \bmod N} (r_{n,m})i + (g_{n,m})j + (b_{n,m})k \right] \\
&= \sum_{n=0}^{N-1} \sum_{m=0}^{N-1} W_{\mu}^{n(kp)+m(ks)} [(r_{n,m})i + (g_{n,m})j + (b_{n,m})k] \\
&= \sum_{n=0}^{N-1} \sum_{m=0}^{N-1} W_{\mu}^{n(kp)+m(ks)} q_{n,m} = Q_{kp \bmod N, ks \bmod N}.
\end{aligned}$$

The set $J_{N,N}$ of frequency-points (p, s), or generators, of the splitting-signals is selected in a way that covers the Cartesian lattice $X_{N,N}$ with a minimum number of subsets $T_{p,s}$. For instance, when $N = 2^r$, $r > 1$, the set $J_{N,N}$ contains $3N/2$ generators and can be defined as

$$J_{N,N} = \{(1, s); s = 0 : (N-1)\} \cup \{(2p, 1); p = 0 : (N/2 - 1)\}.$$

The tensor representation is unique, and the image can be defined through the 2-D QDFT calculated by Eq. (6.61), or directly from the tensor transform. Since $3N/2$ splitting-signals compose the tensor representation, the calculation of the 2-D ls-QDFT is reduced to $3N/2$ 1-D N-point ls-QDFTs, instead of $2N$ such transformation in the traditional row-column method.

It should be noted that the total number of components of $3N/2$ splitting-signals equals $N^2 + N^2/2$, which exceeds the number N^2 of points in the image. On the other side, many sets $T_{p,s}$, with generators $(p, s) \in J_{N,N}$ have intersections at frequency-points. In the $N = 2^r$ case when $r > 1$, the color image is described by $3N/2$ quaternion splitting-signals and 2-D ls-QDFT of the image is split by the $3N/2$ 1-D ls-QDFTs of these signals. The tensor transform is therefore redundant. This redundancy can be removed and the 2-D ls-QDFT can be calculated in a more effective way, by using the modification of the tensor representation, which is called the paired transform [27], also shown above for the 2-D rs-QDFT.

The tensor representation is very effective in the case of most interest when $N > 1$ is a prime, since the number of required sets $T_{p,s}$ is $(N + 1)$ and the sets intersect only at the point (0, 0). Now, we implement the concept of the tensor representation for the quaternion images. In this case, the number of such signals and 1-D ls-QDFT equals $(N + 1)$ and the set of generators can be taken as

$$J_{N,N} = \{(1, s); s = 0 : (N - 1)\} \cup \{(0, 1)\}.$$

We consider the representation of the "girl Anoush" at the frequency-point $(p, s) = (1, 4)$. The components of the quaternion splitting-signals of the red, green, and blue channels, which are generated by this frequency-point (1, 4) are shown in Fig. 6.21 in parts (a), (b), and (c), respectively. This is the $N = 1223$ case. The set of $N + 1 = 1224$ such triplet splitting-signals of length N each describe the color-in-quaternion image as well as its 2-D ls-QDFT. The splitting-signal

$$q_{1,4,t} = (r_{1,4,t})i + (g_{1,4,t})j + (b_{1,4,t})k,\ \ t = 0 : (N - 1),$$

is referred to as the color splitting-signal in quaternion space. The N-point ls-QDFT of this splitting-signal is shown in Fig. 6.22(a) in the absolute scale (and with the scale factor of $1/N^2$) only for the real part of the transform, and the i-component of the imaginary part in Fig. 6.22(b).

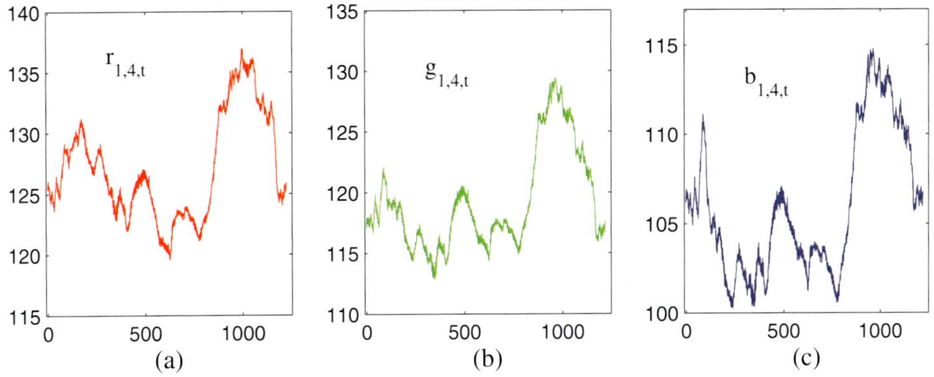

Figure 6.21 The (a) red, (b) green, and (c) blue components of the splitting-signal $q_{1,4,t}$.

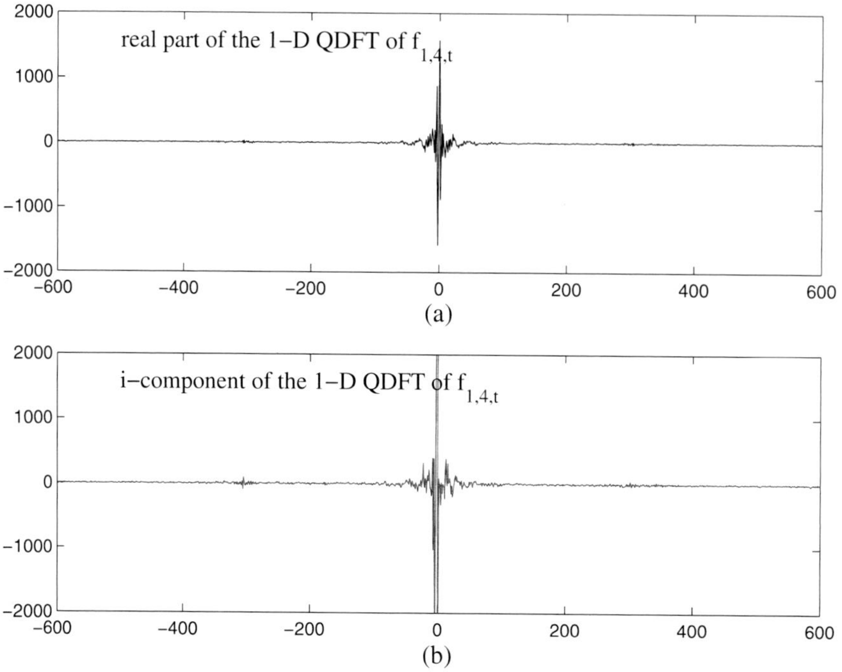

Figure 6.22 The (a) real and (b) i-component of the 1223-point ls-QDFT of the quaternion splitting-signal $q_{1,4,t}$.

In the absolute scale, the 1-D ls-QDFT of the splitting-signal is shown in Fig. 6.23(a) after shifting to the center. The 1-D ls-QDFT of this splitting-signal coincides with the 2-D ls-QDFT of the image at all frequency-points of the set $T_{1,4}$ that are shown in Fig. 6.23(b).

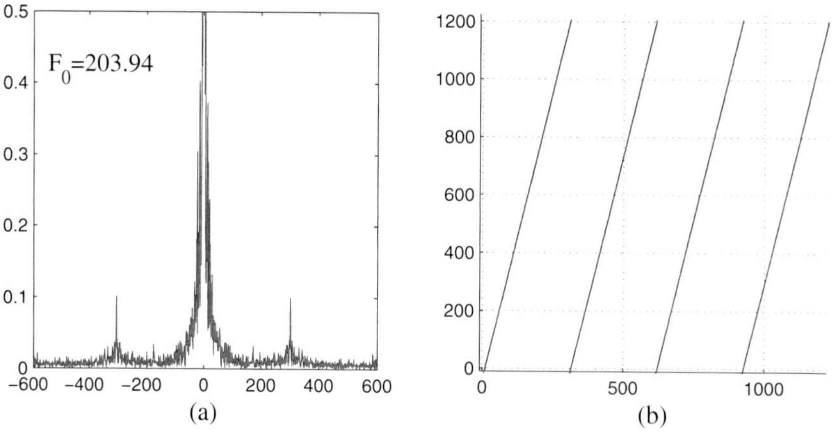

Figure 6.23 (a) The amplitude spectrum of the color splitting-signal $q_{1,4,t}$ and (b) the location of frequency-points of the set $T_{3,1}$ on the Cartesian grid 1223 × 1223.

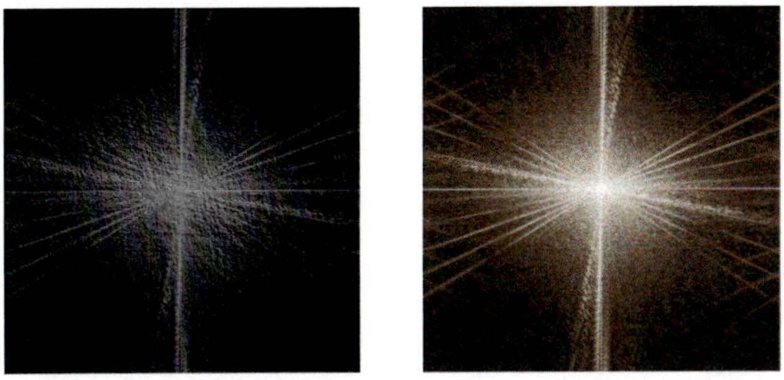

Figure 6.24 (a) The real part and (b) the imaginary part of the 2-D ls-QDFT of the color-in-quaternion "girl Anoush" image in the absolute scale and cyclically shifted to the center.

Figure 6.24 illustrates the 2-D ls-QDFT. The real part of this quaternion transform is shown in Fig. 6.24(a) in the absolute scale, and the imaginary part as three-component data is shown in the RGB color model in Fig. 6.24 (b). All 1-D and 2-D quaternion ls-DFTs were calculated for the unit number $\mu = (i + 2j + k)/\sqrt{6}$.

For $N > 2$ prime, the number of N-point 1-D QDFTs required to calculate the $N \times N$-point 2-D ls-QDFT equals $N + 1$. It should be noted that the tensor transform is of size $N(N+1) = N^2 + N > N^2$, but it is not redundant since we can represent the image by one full splitting-signal, let say $\{q_{1,0,t};\ t = 0 : (N-1)\}$, and N other signals as $\{q_{p,1,t};\ t = 0 : (N-2)\}$. The sum of components of the splitting-signal is the area (S) of the image, i.e.,

$$\sum_{t=0}^{N-1} q_{p,1,t} = S = \sum_{n=0}^{N-1} \sum_{m=0}^{N-1} q_{n,m},$$

for all $p = 0 : (N-1)$. Therefore,

$$q_{p,1,N-1} = S - \sum_{t=0}^{N-2} q_{p,1,t}.$$

Each N-point ls-QDFT can be calculated by two complex N-point DFTs. Here, we note for comparison that the $N \times N$-point 2-D ls-QDFT in the traditional row-column wise algorithm requires $2N$ N-point ls-QDFTs, or $4N$ N-point complex 1-D DFTs. In addition, it is important to mention that, in the tensor transform-based algorithm, the calculation of the 2-D ls-QDFT can be accomplished in parallel, i.e., by separated 1-D ls-QDFTs of the splitting-signals.

The representation of the color and quaternion images by the 1-D splitting-signals allows us not only calculate the 2-D ls-QDFT, but to reconstruct the image directly from its splitting-signals, or the direction images, which are defined by the splitting-signals. Similar to the grayscale images, the direction image from the signal $q_{p,s,t}$ is calculated by

$$d_{n,m} = d^{n,m;p,s} = \frac{1}{N} q_{p,s,(np+ms) \bmod N} \tag{6.62}$$

where $n, m = 0 : (N-1)$.

Figure 6.25 shows the color image in part (a), and the direction images generated by three frequency-points $(p, s) = (1, 1)$, (1, 2), and (1, 4) in parts (b), (c), and (d), respectively. These direction images were amplified by factors of 1.5, 1.5, and 1.85, respectively. For this example when $N = 1223$, the color image 1223×1223 is the sum of such 1224 direction images $\{d_{n,m;p,s};\ (p, s) \in J_{1223,1223}\}$, which represents the inverse tensor transform.

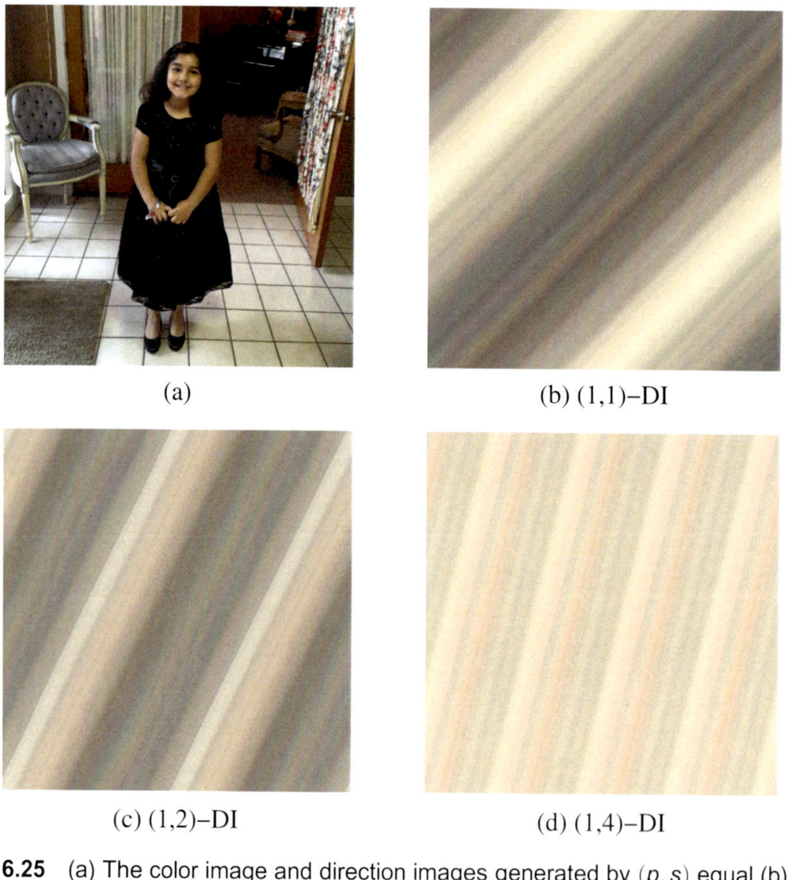

(a) (b) (1,1)–DI

(c) (1,2)–DI (d) (1,4)–DI

Figure 6.25 (a) The color image and direction images generated by (p, s) equal (b) (1, 1), (c) (1, 2), and (d) (1, 4).

In general case when N is prime, all $(N+1)$ cyclic-groups $T_{p,s}$ intersect only at points (0, 0). The following statement holds:

Principle of Superposition 3: The quaternion image $q_{n,m}$ of size $N \times N$, where $N > 2$ is a prime, can be calculated as the sum of $(N+1)$ directional images:

$$q_{n,m} = \sum_{(p,s)\in J_{N,N}} d_{n,m;p,s} - E[f_{T_{1,0}}]$$
$$= \frac{1}{N}\left[\sum_{p=0}^{N-1} q_{p,1,(np+m) \bmod N} + q_{1,0,n}\right] - E[q_{T_{1,0}}], \tag{6.63}$$

where $n, m = 0 : (N-1)$.

One can see that this statement and the statement in Eq. (6.52) are the same for both right-side and left side 2-D QDFTs. The two different concepts of the 2-D QDFT have the same tensor representation. In other words, this representation unites these 2-D quaternion transforms. The algorithm of calculation of these two transforms is the same. The quaternion image is described by the same set of quaternion splitting-signals

$$\{q_{p,s,0}, q_{p,s,1}, q_{p,s,2}, \ldots, q_{p,s,N-1}\}, \quad (p,s) \in J_{N,N},$$

and each of this signal carries the information of the 2-D right-side and the ls-QDFTs at the same set of frequency-points,

$$Q^{(rs)}_{kp \bmod N,\, ks \bmod N} = \sum_{t=0}^{N-1} q_{p,s,t} W^{kt}_{\mu},$$
$$Q^{(ls)}_{kp \bmod N,\, ks \bmod N} = \sum_{t=0}^{N-1} W^{kt}_{\mu} q_{p,s,t},$$

where $k = 0 : (N-1)$.

For images of size $N \times N$, where $N = 2^r$, $r > 1$, the principle of superposition of the direction components of the image is the same as described above principle of superposition 2 for the 2-D rs-QDFT. It should also be noted that the concept of the tensor representation of quaternion image was applied for calculating the 2-D right- and left-side QDFTs. The property of splitting the 2-D DFT by the tensor transform, which was originated from the grayscale images, remains for these quaternion transforms. For the two-side 2-D QDFT, the tensor representation cannot be used, and probably there is no need in considering 2-D QDFTs, which differ from the 2-D right- and left-side QDFTs.

6.10 2-D QDFT on the Hexagonal Lattice

In this section, we present a concept of the quaternion discrete Fourier transform on the two-dimensional hexagonal lattice, which is called the two-dimensional hexagonal quaternion discrete Fourier transform (2-D HQDFT). The concept of the right-side 2-D HQDFT is described and the left-side 2-D HQDFT is similarly considered. To calculate the transform, the image on the hexagonal lattice is described in the corresponding tensor representation when the image is presented by a set of 1-D signals, or splitting-signals, which can be separately processed in the frequency domain. The 2-D HQDFT can be calculated by a set of 1-D quaternion discrete Fourier transforms (QDFT) of the splitting-signals. The most efficient lattices for image sampling are not rectangular, the hexagonal lattices are important for many problems in image processing [42]–[48]. It is known that the vision system relates best to the regular hexagonal tessellation, which has a lower number of neighbors than the rectangular lattice (see Fig. 6.26). The representation of the image in the hexagonal lattice requires 13.4% less samples, than the rectangular lattice does. The color image is considered as a triplet of separate 2-D grayscale images and each of these images represents red, green, or blue component of the color. Color images can be transformed into a subspace of quaternion numbers and processed there.

Between two different spatial and frequency spaces exists an intermediate space, the so-called frequency-and-time space, or representation of the image in the form of 1-D signals, which are generated by a specific set of frequencies. In color tensor representation on the hexagonal lattice, three components of the image in the RGB space are described by one dimensional signal in quaternion algebra. The tensor representation is effective, since it allows us to process the color image by 1-D quaternion signals that can be processed separately. This representation also splits the algebraic structure of the 2-D HQDFT, since each of the quaternion signals defines the 2-D QDFT in the corresponding subset of frequency-points. The tensor transform-based algorithm of the 2-D QDFT is effective and simple to apply and design,

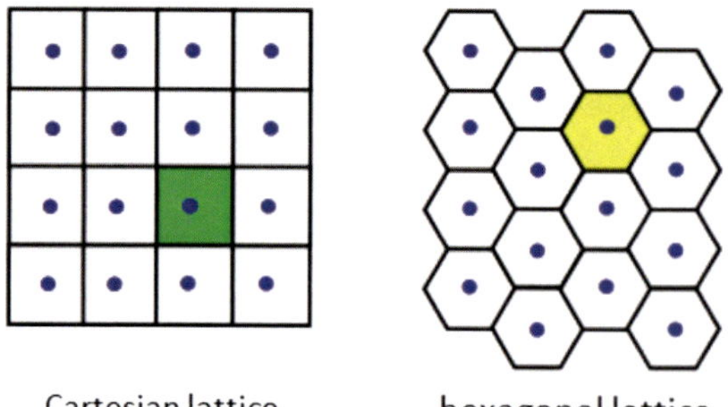

Figure 6.26 The Cartesian and hexagonal lattices.

which makes it very practical in color image processing in the frequency domain.

The 2-D Fourier transforms on the hexagonal lattice are complex, because of the hexagonal lattice, not separable, and kernels of the transforms have more complex forms than in the traditional 2-D DFT defined on the Cartesian lattice [49]–[55]. The method of tensor representation was applied to the 2-D hexagonal discrete Fourier transform (HDFT), as shown in [56, 57, 22]. When the transform orders are equal a power of two, the redundant tensor representation can be modified to the unitary paired transform which is also performed without multiplications. The set of 1-D signals can be processed separately, whether the application is for computing the discrete Fourier spectrum or for linear filtering [58, 24].

6.10.1 2-D Hexagonal DFT

We consider the case when both 2-D image $f = f_{n,m}$ and the transform are arranged on the following hexagonal lattice of size $2N \times N$:

$$X_{2N,N} = \{(p + [s], s);\ p = 0 : (2N - 1),\ \ s = 0 : (N - 1)\}, \tag{6.64}$$

where $[s] = [1 - (-1)^s]/4$, for all integer $s = 0 : (N - 1)$. For even s, $[s] = 0$, and $[s] = 0.5$ for odd s. Each second row of knots in the lattice $X_{2N,N}$ is shifted by 0.5.

The 2-D HDFT of the image $f_{n,m}$ is defined as

$$F_{p+[s],s} = \sum_{n=0}^{N-1}\sum_{m=0}^{N-1} f_{n+[m],m}\varphi_{p+[s],s}(n + [m], m), \quad n, m = 0 : (N - 1), \tag{6.65}$$

where the kernel of the transform is composed by the exponential functions

$$\varphi_{p+[s],s}(n + [m], m) = W^{(n+[m])(p+[s])+ms},$$

$W = W_N$ and $[m] = [1 - (-1)^m]/4$ when $m = 0 : (N - 1)$. The basic functions are not separable, the 2-D HDFT is not separable transform and the traditional row-column method cannot directly be applied for fast computing the 2-D HDFT. It should be noted that the definition of the 2-D HDFT is given in Eq. (6.65) for $n, m = 0 : (N - 1)$, but the lattice of size $2N \times N$, as defined in Eq. (6.64), is the fundamental period of the 2-D DHFT not the hexagonal lattice $X_{N,N}$ of size $N \times N$. Indeed, $F_{(p+2N)+[s+N],s+N} = F_{p+[s],s}$ for any integer p and s. Therefore, we define the 2-D HDFT as

$$F_{p+[s],s} = \sum_{n=0}^{2N-1}\sum_{m=0}^{N-1} f_{n+[m],m}\varphi_{p+[s],s}(n + [m], m), \quad (p + [s], s) \in X_{2N,N}, \tag{6.66}$$

and consider $\{f_{n+[m],m}\}$ as the image extended to or defined on the hexagonal lattice $X_{2N,N}$.

6.10.2 2-D Quaternion HDFT

On the hexagonal lattice $X_{2N,N}$, we consider the quaternion "image", i.e., the set of four images in the quaternion space,

$$q_{n+[m],m} = f_{n+[m],m} + ir_{n+[m],m} + jg_{n+[m],m} + kb_{n+[m],m},$$

where $f_{n+[m],m}$, $r_{n+[m],m}$, $g_{n+[m],m}$, and $b_{n+[m],m}$ are real images on the lattice, which are not necessary to be related to the gray and red, green, and blue color components of the image in the RGB color model.

We define the 2-D quaternion HDFT as the following right-side transform

$$Q_{p+[s],s} = \sum_{n=0}^{2N-1} \sum_{m=0}^{N-1} q_{n+[m],m} \varphi_{p+[s],s;\mu}(n+[m],m), \quad (p+[s],s) \in X_{2N,N}. \quad (6.67)$$

Here μ is a pure unit quaternion number and the basic functions of the transform are

$$\varphi_{p+[s],s;\mu}(n+[m],m) = W_{\mu}^{(n+[m])(p+[s])+ms},$$
$$(n+[m],m),\ (p+[s],s) \in X_{2N,N}.$$

The concept of the tensor representation can be applied for images on the hexagonal lattice, by constructing a covering of $X_{2N,N}$ by subsets similar to $T_{p,s}$ for the Cartesian lattice. We define the following subsets of frequency-point on the hexagonal lattice $X_{2N,N}$,

$$T_{p+[s],s} = \{(\overline{k(p+[s])}, \overline{ks}), k = 0 : (2N-1)\}, \quad (6.68)$$

where the operation $\bar{l}$ denotes the $l \bmod (2N)$ or $l \bmod (N)$, depending on the dimension,

$$(\overline{k(p+[s])}, \overline{ks}) = (k(p+[s]) \bmod 2N, ks \bmod N).$$

Example 6.4

For the $N = 7$ case, we consider subsets $T_{p+[s],s}$ of the lattice $X_{14,7}$ for the frequency-points $(p,s) = (1,0), (1,1), (1,2), (1,3)$, and $(2,1)$,

$$T_{1+[0],0} = T_{1,0} = \{(0,0),(1,0),(2,0),(3,0),(4,0),\ldots,(12,0),(13,0)\},$$

$$T_{1+[1],1} = T_{1.5,1} = \left\{\begin{array}{l}(0,0),(1.5,1),(3,2),(4.5,3),(6,4),(7.5,5),(9,6),\\(10.5,0),(12,1),(13.5,2),(1,3),(2.5,4),(4,5),(5.5,6)\\(7,0),(8.5,1),(10,2),(11.5,3),(13,4),(0.5,5),(2,6),\\(3.5,0),(5,1),(6.5,2),(8,3),(9.5,4),(11,5),(12.5,6)\end{array}\right\},$$

$$T_{1+[2],2} = T_{1,2} = \left\{\begin{array}{l}(0,0),(1,2),(2,4),(3,6),(4,1),(5,3),(6,5),\\(7,0),(8,2),(9,4),(10,6),(11,1),(12,3),(14,5)\end{array}\right\},$$

$$T_{1+[3],3} = T_{1.5,3} = \left\{\begin{array}{l}(0,0),(1.5,3),(3,6),(4.5,2),(6,5),(7.5,1),(9,4),\\(10.5,0),(12,3),(13.5,6),(1,2),(2.5,5),(4,1),(5.5,4)\\(0,5),(1.5,1),(3,4),(4.5,0),(6,3),(7.5,6),(9,2),\\(10.5,5),(12,1),(13.5,4),(1,0),(2.5,3),(4,6),(5.5,2)\end{array}\right\},$$

$$T_{2+[1],1} = T_{2.5,1} = \left\{\begin{array}{l}(0,0),(2.5,1),(5,2),(7.5,3),(10,4),(12.5,5),(1,6),\\(3.5,0),(6,1),(8.5,2),(11,3),(13.5,4),(2,5),(4.5,6)\\(7,0),(9.5,1),(12,2),(0.5,3),(3,4),(5.5,5),(8,6),\\(10.5,0),(13,1),(1.5,2),(4,3),(6.5,4),(9,5),(11.5,6)\end{array}\right\}.$$

Here, two subsets contain 14 points each, when $s = 0$ and 2, i.e., s is even. The remaining subsets are of 28 points, since s is odd. For instance, when $p = 2$ and $s = 1$, the first solution of the equation $((p + [s])k, k) = (2.5k, k) = (14, 7) = (0, 0)$ by modulo (14, 5) is for $k = 28$, not $k = 14$.

Given frequency-point $(p + [s], s) \in X_{2N,N}$ and integer $4t \in [0, 4N - 1]$, we consider the subsets of knots on the parallel lines passing through the hexagonal lattice $X_{2N,N}$ in the image plane,

$$V_t = V_{p+[s],s,t} = \{(n + [m], m); (n + [m])(p + [s]) + ms = t \bmod N\}, \quad (6.69)$$

and the following components of the splitting-signal generated by the frequency-point:

$$q_{p+[s],s,t} = \sum_{V_{p+[s],s,t}} q_{n+[m],m}, \quad 4t \in [0, 4N - 1]. \quad (6.70)$$

When both p and s are odd, like (1, 1), the equation $(n + [m])(p + [s]) + ms = t$ has the form $(n + [m])(p + 0.5) + ms = t$ and for m odd, it can be written as $t = (n + 0.5)(p + 0.5]) + ms = np + ms + (n + p)/2 + 1/4$. Thus, t runs with the step $\Delta = 1/4$. When p is even and s is odd, the step is $\Delta = 1/2$, and when p and s even, the step is 1. We can compose the quaternion splitting-signal

$$q_{T_{p+[s],s}} = \{q_{p+[s],s,t}; t = 0 : \Delta : (N - \Delta)\}$$

of length N, $2N$ or $4N$, depending on the frequency-point $(p + [s], s)$ and values of N. The index t varies from 0 through N with step $\Delta = 1/4$ or $1/2$. This 4-D signal is calculated component-wise as

$$\begin{aligned} q_{p+[s],s,t} &= f_{p+[s],s,t} + i(r_{p+[s],s,t}) + j(g_{p+[s],s,t}) + k(b_{p+[s],s,t}) \\ &= \sum_{(n,m)\in V_t} f_{n,m} + i\left[\sum_{(n,m)\in V_t} r_{n,m}\right] + j\left[\sum_{(n,m)\in V_t} g_{n,m}\right] + k\left[\sum_{(n,m)\in V_t} b_{n,m}\right]. \end{aligned} \tag{6.71}$$

The imaginary components of the quaternion splitting-signals are the splitting-signals of the red, green, and blue channels of the color image. The components $q_{p+[s],s,t}$ are periodic by t, i.e., $q_{p+[s],s,t+2N} = f_{p+[s],s,t}$.

The 1-D DFT of this splitting-signals equals 2-D HDFT of the image at the frequency-points of the subset $T_{p+[s],s}$. Indeed, the following holds:

$$Q_{\overline{k(p+[s]),ks}} = \sum_{4t=0}^{N-1} q_{p+[s],s,t} W_{\mu}^{kt}, \tag{6.72}$$

where k is an integer. This statement follows from the following calculations:

$$\begin{aligned} &\sum_{4t=0}^{4N-1} q_{p+[s],s,t} W_{\mu}^{kt} \\ &= \sum_{4t=0}^{4N-1} \left[\sum_{(n+[m])(p+[s])+ms=t \bmod 2N} f_{n,m} + (r_{n,m})i + (g_{n,m})j + (b_{n,m})k \right] W_{\mu}^{kt} \\ &= \sum_{4t=0}^{4N-1} \left[\sum_{(n+[m])(p+[s])+ms=t \bmod 2N} f_{n,m} + (r_{n,m})i + (g_{n,m})j \right. \\ &\quad \left. + (b_{n,m})k \right] W_{\mu}^{k\{(n+[m])(p+[s])+ms\}} \\ &= \sum_{4t=0}^{4N-1} \left[\sum_{(n+[m])(p+[s])+ms=t \bmod 2N} f_{n,m} + (r_{n,m})i + (g_{n,m})j \right. \\ &\quad \left. + (b_{n,m})k \right] W_{\mu}^{(n+[m])k(p+[s])+m(ks)} \\ &= \sum_{n=0}^{N-1} \sum_{m=0}^{N-1} q_{n,m} W_{\mu}^{(n+[m])k(p+[s])+m(ks)} = Q_{k(p+[s]) \bmod 2N, ks \bmod N}. \end{aligned}$$

If we construct an irreducible covering of $X_{2N,N}$, which is composed of the subsets Eq. (6.68),

$$\sigma = \left(T_{p+[s],s}\right)_{(p+[s],s)\in J} \tag{6.73}$$

for a certain subset of frequency-points $J \subseteq X_{2N,N}$, then the 2-D HQDFT will be split by the 1-D QDFTs of the splitting-signals.

In the $N = 2^r$ case, the number of elements of the subset $T_{p_1+[s],s}$ shows that the 2-D HQDFT is split by the 1-D QDFT, which has the following order M:

$$M = \begin{cases} 2N, & \text{if } p \text{ is odd, } s \text{ is even} \\ 4N, & \text{if } s \text{ is odd} \\ N/2^n, & \text{if } p \text{ and } s \text{ are even, } g.c.d.(p,s) = 2^n. \end{cases} \tag{6.74}$$

For instance, when $N = 4$, we can consider the following set of generators:

$$J = \{(1,0), (0+[1],1), (1+[1],1), (2,2), (0,2), (4,2)\}.$$

The subsets T with these generators are

$$\begin{aligned}
T_{1+[0],0} &= T_{1,0} = \{(0,0),(1,0),(2,0),(3,0),(4,0),(5,0),(6,0),(7,0)\},\\
T_{0+[1],1} &= T_{0.5,1} = \left\{\begin{matrix}(0,0),(0.5,1),(1,2),(1.5,3),(2,0),(2.5,1),(3,2),(3.5,3)\\(4,0),(4.5,1),(5,2),(5.5,3),(6,0),(6.5,1),(7,2),(7.5,3)\end{matrix}\right\},\\
T_{1+[1],1} &= T_{1.5,1} = \left\{\begin{matrix}(0,0),(1.5,1),(3,2),(4.5,3)(6,0),(7.5,1),(1,2),(2.5,3)\\(4,0),(5.5,1),(7,2),(0.5,3)(2,0),(3.5,1),(5,2),(6.5,3)\end{matrix}\right\},\\
T_{2+[2],2} &= T_{2,2} = \{(0,0),(2,2),(4,0),(6,2)\},\\
T_{0+[2],2} &= T_{0,2} = \{(0,0),(0,2)\},\\
T_{4+[2],2} &= T_{4,2} = \{(0,0),(4,2)\}.
\end{aligned}$$

The subset $T_{1,0}$ has eight elements and, therefore, the 8-point QDFT is used for the quaternion splitting-signal $\{q_{1,0,t};\ t = 0, 0.5, 1, 1.5, \ldots, 3.5\}$. The next two subsets $T_{0.5,1}$ and $T_{1.5,1}$ have 16 elements each. Therefore, the 16-point QDFT is used for each of the corresponding quaternion splitting-signals to fill the 2-D 8×4-point HQDFT at frequency-points of these subsets. At frequency-points of subsets $T_{2,2}$, $T_{0,2}$, and $T_{4,2}$, the 8×4-point HQDFT is calculated by the 4, 2, and 2-point QDFTs, respectively.

One can notice that the subsets have intersections at many frequency-points $(p+[s],s)$ with $s = 0$ and 2. The redundancy of the tensor algorithm at these intersection can be removed by the method of paired transforms, which is described in Ref. [56]. For the $N = 8$ case, such set is $J = \{(p+[1],1);\ p = 0:3\} \cup \{(1,0),(1,4),(2,0),(6,2)\} \cup \{(4p,2);\ p = 0:3\}$. For any integer number $N > 2$, the set of generators for splitting the $2N \times N$-point 2-D HQDFT can also be found, by writing a simple code for that. The set of generators is not unique, but for effective calculation of the transform such a set should be taken with a minimum number of subsets T and minimum number of intersections between the subsets. Such sets of generators for different N are described in detail in Refs. [24, 36].

Thus, the 2-D rs-HQDFT can be described in the tensor representation in quaternion algebra, wherein the color image can be transformed from different color models, such as the RGB and XYZ models. The color and quaternion images on the hexagonal lattice can uniquely be presented by a set of quaternion splitting-signals which allow to calculate the 2-D rs-HQDFT by a minimum number of 1-D rs-QDFTs.

6.11 2-D Two-Side Octonion DFT

The octonion discrete Fourier transform (ODFT) is a new concept which allowing for processing a few grayscale or color images, as well as images with their local information as one octonion 8-D image in the spectral domain. This concept generalizes the traditional complex and quaternion 2-D DFTs and can be effectively used for parallel processing up to eight of grayscale images or two color images. The basic functions for the ODFT are the octonion exponential functions. We first consider the two-side 2-D ODFT and, then, the 2-D right- and left-side ODFTs of the octonion image of size $N \times M$. The octonion numbers are dual quaternions, i.e., any octonion o is represented by $q_1 + q_2 E$, where q_1 and q_2 are quaternions. The 2-D octonion DFTs can be calculated by 1-D and 2-D QDFTs.

Given two pure unit octonions λ_1 and λ_2, which we consider equal to pure unit quaternions μ_1 and μ_2, the two-side OQFT of the octonion image $o_{n,m}$ can be defined as

$$O_{p,s} = \sum_{n=0}^{N-1}\sum_{m=0}^{M-1} W_{\mu_1}^{np}\left(o_{n,m} W_{\mu_2}^{ms}\right) = \sum_{n=0}^{N-1} W_{\mu_1}^{np}\left(\sum_{m=0}^{M-1} o_{n,m} W_{\mu_2}^{ms}\right), \tag{6.75}$$

where $p = 0:(N-1)$ and $s = 0:(M-1)$. The basis functions are defined by the exponential coefficients

$$W_{\mu_1} = W_{\mu_1;N} = \cos(2\pi/N) - \mu_1 \sin(2\pi/N),$$
$$W_{\mu_2} = W_{\mu_2;M} = \cos(2\pi/M) - \mu_2 \sin(2\pi/M).$$

We denote the kernel of the transform with such functions by (W_{μ_1}, W_{μ_2}). To calculate the two-side ODFT, the row-column method can be used. Along each row (when n is fixed) the M-point rs-ODFT is calculated, and then along each column (when m is fixed) of the obtained data, the N-point ls-ODFT is calculated.

One can also apply the 2-D QDFTs to calculate the two-side ODFT. To show that, we consider the first and second quaternion components of the image

$$o_{n,m} = q_{n,m} + f_{n,m}E$$

and write

$$\begin{aligned} O_{p,s} &= \sum_{n=0}^{N-1} W_{\mu_1}^{np}\left[\sum_{m=0}^{M-1}(q_{n,m} + f_{n,m}E)W_{\mu_2}^{ms}\right] \\ &= \sum_{n=0}^{N-1} W_{\mu_1}^{np}\left(\sum_{m=0}^{M-1} q_{n,m}W_{\mu_2}^{ms}\right) + \sum_{n=0}^{N-1} W_{\mu_1}^{np}\left[\sum_{m=0}^{M-1}(f_{n,m}E)W_{\mu_2}^{ms}\right]. \end{aligned}$$

Since $(q_{n,m}E)W_{\mu_2}^{ms} = (q_{n,m}W_{\mu_2}^{-ms})E$, we obtain

$$O_{p,s} = \sum_{n=0}^{N-1} W_{\mu_1}^{np}\left(\sum_{m=0}^{M-1} q_{n,m}W_{\mu_2}^{ms}\right) + \sum_{n=0}^{N-1} W_{\mu_1}^{np}\left(\sum_{m=0}^{M-1} f_{n,m}W_{\mu_2}^{-ms}\right)E.$$

It should be noted that

$$W_{\mu_2}^{-ms} = W_{\mu_2}^{m(M-s)} = W_{-\mu_2}^{ms}.$$

Thus, to calculate the two-side ODFT, we can perform the two-side QDFT of the quaternion image $q_{n,m}$ and one two-side QDFT of the quaternion image $f_{n,m}$. The first two-side QDFT is calculated with the kernel (W_{μ_1}, W_{μ_2}), and the second one with the kernel (W_{μ_1}, W_{μ_2}), or $(W_{\mu_1}, W_{-\mu_2})$. Let $Q_{p,s}$ be the two-side QDFT of the image $q_{n,m}$ and let $F_{p,s}$ be the two-side QDFT of $f_{n,m}$, with the kernel (W_{μ_1}, W_{μ_2}) both. Then,

$$O_{p,s} = Q_{p,s} + F_{p,M-s}E.$$

If $F_{p,s}$ is the two-side QDFT of $f_{n,m}$, with the kernel $(W_{\mu_1}, W_{-\mu_2})$, then

$$O_{p,s} = Q_{p,s} + F_{p,s}E.$$

The inverse two-side ODFT is calculated by the similar formula

$$o_{n,m} = \frac{1}{NM}\sum_{p=0}^{N-1}\sum_{s=0}^{M-1} W_{\mu_1}^{-np}(O_{p,s}W_{\mu_2}^{-ms}) = \sum_{p=0}^{N-1} W_{\mu_1}^{-np}\left(\sum_{s=0}^{M-1} O_{p,s}W_{\mu_2}^{-ms}\right) \quad (6.76)$$

when $n = 0:(N-1)$ and $m = 0:(M-1)$.

We also can consider the following definition of the two-side ODFT:

$$O_{p,s} = \sum_{m=0}^{M-1}\sum_{n=0}^{N-1}(W_{\mu_1}^{np}o_{n,m})W_{\mu_2}^{ms} = \sum_{n=0}^{M-1}\left(\sum_{n=0}^{N-1} W_{\mu_1}^{np}o_{n,m}\right)W_{\mu_2}^{ms}, \quad (6.77)$$

where $p = 0:(N-1)$ and $s = 0:(M-1)$. For this transform, the column-row method can be used. Along each column (when m is fixed) the N-point ls-ODFT is calculated, and then along each row (when n is fixed) of the obtained data, the M-point rs-ODFT is calculated. Thus, since the operation of multiplication for octonions is not associative, the row-column and column-row two-side ODFTs are different.

We consider a discrete color image $f_{n,m}$ in the RGB color space, which is transformed into imaginary part of quaternion numbers, by encoding the red, green, and blue components of the color image,

$$f_{n,m} = 0 + (ir_{n,m} + jg_{n,m} + kb_{n,m}).$$

The zero real part of the quaternion image make it pure quaternion, and we could use the grayscale component in the real part

$$\begin{aligned} f_{n,m} &= a_{n,m} + (ir_{n,m} + jg_{n,m} + kb_{n,m}) \\ &= \frac{r_{n,m} + g_{n,m} + b_{n,m}}{3} + (ir_{n,m} + jg_{n,m} + kb_{n,m}). \end{aligned} \quad (6.78)$$

There are different ways to compose or transfer color images into octonion space and, then, introduce the concept of the octonion image. We consider one of such ways, by using the RGB color model. Let $N \times M$ be the size of the $f_{n,m}$ image components, i.e., $n = 0:(N-1)$ and $m = 0:(M-1)$, and let N and M be even numbers. We consider that the image is the color image transformed to the quaternion space as shown in Eq. (6.78). This quaternion image with four components can be transformed to octonion space, by composing the following "octonion image":

$$\begin{aligned} o_{n,m} &= f_{n,m} + f_{n+1,m}E \\ &= a_{n,m} + ir_{n,m} + jg_{n,m} + kb_{n,m} + Ea_{n+1,m} + Ir_{n+1,m} \\ &\quad + Jg_{n+1,m} + Kb_{n+1,m} \\ &\quad n = 0 : (N/2 - 1), m = 0 : (M - 1). \end{aligned} \tag{6.79}$$

This octonion image $o_{n,m} = (f_{n,m}, f_{n+1,m})$ is of size $(N/2) \times M$.

As an example, the octonion "tree image" is illustrated in Fig. 6.27. Here, the original 256×256 color "tree" image in the RGB model is presented as the "octonion" image with its eight components of size 128×256 each. The grayscale image $a_{n,m}$ of the color image that is the real component of the octonion image $o_{n,m}$ is shown in Fig. 6.27(a). The three components, i.e., the (i, j, k)-part of the imaginary part of the octonion image $o_{n,m}$ is shown in Fig. 6.27(b), as the color image in the RGB model. In part Fig. 6.27(c), the imaginary E-components of $o_{n,m}$ is shown, as a grayscale image. The color image in Fig. 6.27(d) represents the (I, J, K)-part of the imaginary part of the octonion image $o_{n,m}$. The images in Fig. 6.27(a,b) represent the first quaternion image $q_{n,m}$ and the two images in Fig. 6.27(c,d) represent the second quaternion image $f_{n,m}$ of the octonion image $o_{n,m}$.

For the "tree" octonion image, Fig. 6.28 shows the 2-D QDFT, $Q_{p,s}$, of the image $g_{n,m}$ in part (a), and the 2-D QDFT, $F_{p,M-s}$, of the image $f_{n,m}$ in part (b). These two 2-D QDFTs define the 2-D ODFT of the octonion image $o_{n,m}$. The magnitude of the 2-D ODFT of the image $o_{n,m}$ is shown in Fig. 6.28(c). All transforms are cyclically shifted to the center.

The "octonion image" can also be composed as

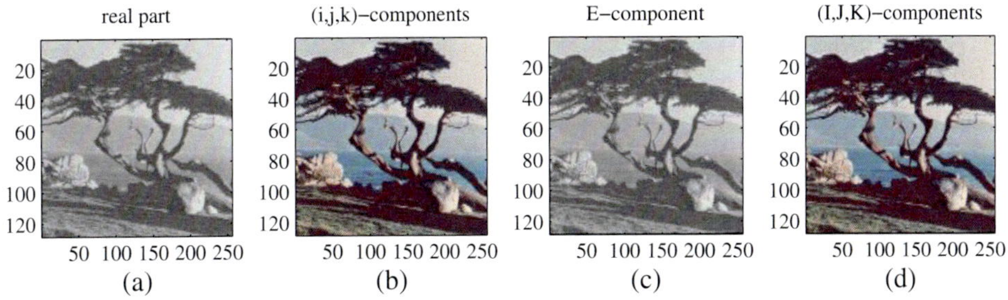

Figure 6.27 (a) The grayscale component, (b) the (i, j, k)-component color image, (c) the E-component, and (b) the (I, J, K)-component color image of the 128×256 octonion "tree" image.

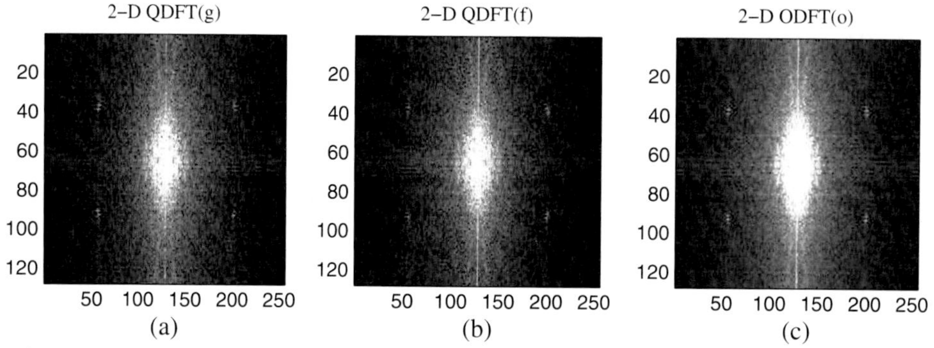

Figure 6.28 (a) The 2-D QDFT of the quaternion image $g_{n,m}$, (b) the 2-D QDFT of the quaternion image $f_{n,m}$, and (c) the 2-D ODFT of the octonion "tree" image $o_{n,m}$. (All transforms are shown in the absolute scale.)

$$\begin{aligned} o_{n,m} &= f_{n,m} + f_{n,m+1}E \\ &= a_{n,m} + ir_{n,m} + jg_{n,m} + kb_{n,m} + Ea_{n,m+1} + Ir_{n,m+1} \\ &\quad + Jg_{n,m+1} + Kb_{n,m+1}, \\ &\quad n = 0 : (N-1),\ m = 0 : (M/2-1). \end{aligned} \tag{6.80}$$

The octonion image $o_{n,m} = (f_{n,m}, f_{n,m+1})$ is of size $N \times (M/2)$.

The color image can also be divided by two halves, and each halve can be transformed to quaternion space and two quaternion images $q_{n,m}$ and $f_{n,m}$ are composed, which together result in an octonion image. In this simple model, two parts q and f are not much correlated, in comparison to the model described above.

6.12 2-D Right-Side ODFT

The 2-D right-side ODFT (rs-ODFT) is defined as follows:

$$O_{p,s} = \sum_{n=0}^{N-1}\left(\sum_{m=0}^{M-1} o_{n,m} W_{\lambda_2;\,M}^{ms}\right) W_{\lambda_1;\,N}^{np} = \sum_{n=0}^{N-1}\sum_{m=0}^{M-1}\left(o_{n,m} W_{\lambda_2;\,M}^{ms}\right) W_{\lambda_1;\,N}^{np}, \tag{6.81}$$

where $p = 0 : (N-1)$ and $s = 0 : (M-1)$. Here, we also consider that the pure unit octonion numbers λ_1 and λ_2 are quaternions μ_1 and μ_2, respectively. This transform can be considered as the row-column 2-D rs-ODFT. The calculation of this transform first is performed by rows, and then by columns. The inverse 2-D rs-ODFT is calculated by

$$o_{n,m} = \frac{1}{NM} \sum_{p=0}^{N-1} \left(\sum_{s=0}^{M-1} O_{p,s} W_{\mu_1;N}^{-np} \right) W_{\mu_2;M}^{-ms},$$

where $n = 0 : (N-1)$ and $m = 0 : (M-1)$.

We can also consider the column-row 2-D rs-ODFT,

$$O_{p,s} = \sum_{m=0}^{M-1} \left(\sum_{n=0}^{N-1} o_{n,m} W_{\mu_1;N}^{np} \right) W_{\mu_2;M}^{ms} = \sum_{m=0}^{M-1} \sum_{n=0}^{N-1} \left(o_{n,m} W_{\mu_1;N}^{np} \right) W_{\mu_2;M}^{ms}, \quad (6.82)$$

where $p = 0 : (N-1)$ and $s = 0 : (M-1)$.

6.12.1 Tensor transform-based 2-D rs-ODFT

The octonion image $o_{n,m}$ can be described in tensor representation as the set of 1-D octonion signals and the 2-D rs-ODFT can be calculated by 1-D ODFTs of these signals, which we call the splitting-signals. For simplicity of calculations, we consider the case when the size of the image is $N \times N$ and two pure quaternions μ_1 and μ_2 are equal. The 2-D rs-ODFT is

$$O_{p,s} = \sum_{n=0}^{N-1} \sum_{m=0}^{N-1} \left(o_{n,m} W_{\mu}^{ms} \right) W_{\mu}^{np}, \quad p, s = 0 : (N-1), \quad (6.83)$$

where $\mu_2 = \mu_1 = \mu$ and $W_{\mu}^{ms} = W_{\mu;N}^{ms}$. Considering the first and second quaternion components of the image

$$o_{n,m} = q_{n,m} + f_{n,m} E,$$

we can write the above definition of the 2-D rs-ODFT as

$$O_{p,s} = \sum_{n=0}^{N-1} \sum_{m=0}^{N-1} \left(q_{n,m} W_{\mu}^{ms} \right) W_{\mu}^{np} + \sum_{n=0}^{N-1} \sum_{m=0}^{N-1} \left[\left(f_{n,m} E \right) W_{\mu}^{ms} \right] W_{\mu}^{np}.$$

In the first double sum, the term

$$\left(q_{n,m} W_{\mu}^{ms} \right) W_{\mu}^{np} = q_{n,m} \left(W_{\mu}^{ms} \right) W_{\mu}^{np} = q_{n,m} W_{\mu}^{ms+np}.$$

Using the properties of multiplication of the quaternions with the unit E, we can write the addend in the second double sum as

$$\left[(f_{n,m}E)W_{\mu}^{ms}\right]W_{\mu}^{np} = \left[(f_{n,m}W_{\mu}^{-ms})E\right]W_{\mu}^{np} = \left[(f_{n,m}W_{\mu}^{-ms})W_{\mu}^{-np}\right]E$$
$$= \left[(f_{n,m}(W_{\mu}^{-ms}W_{\mu}^{-np})\right]E = \left[f_{n,m}W_{\mu}^{-(ms+np)}\right]E,$$

or

$$\left[(f_{n,m}E)W_{\mu}^{ms}\right]W_{\mu}^{np} = \left[f_{n,m}W_{\mu}^{m(N-s)+n(N-p)}\right]E = \left[f_{n,m}W_{-\mu}^{(ms+np)}\right]E.$$

Thus,

$$O_{p,s} = \sum_{n=0}^{N-1}\sum_{m=0}^{M-1} q_{n,m}W_{\mu}^{ms+np} + \sum_{n=0}^{N-1}\sum_{m=0}^{M-1}\left[f_{n,m}W_{-\mu}^{(ms+np)}\right]E. \tag{6.84}$$

Also, we can write

$$O_{p,s} = \sum_{n=0}^{N-1}\sum_{m=0}^{M-1} q_{n,m}W_{\mu}^{ms+np} + \sum_{n=0}^{N-1}\sum_{m=0}^{M-1}\left[f_{n,m}W_{\mu}^{m(N-s)+n(N-p)}\right]E. \tag{6.85}$$

Thus,

$$O_{p,s} = Q_{p,s} + F_{N-p,N-s}E,\ p,s = 0:(N-1). \tag{6.86}$$

In tensor representation, the discrete octonion image $o_{n,m}$ is described by a set of 1-D splitting-signals of length N each,

$$\chi : \{o_{n,m}\} \to \{o_{T_{p,s}} = \{o_{p,s,t}; t = 0:(N-1)\}\}_{(p,s)\in J_{N,N}}.$$

Each octonion splitting-signal defines the 2-D rs-ODFT at the frequency-points of the set

$$T_{p,s} = \{(kp \bmod N, ks \bmod N); k = 0, 1, \ldots, (N-1)\}.$$

The octonion splitting-signal is the sum of two quaternion splitting-signals

$$o_{T_{p,s}} = q_{T_{p,s}} + f_{T_{p,s}}E. \tag{6.87}$$

These two splitting-signals are the splitting-signals of the quaternion images $q_{n,m}$ and $f_{n,m}$, respectively. The construction of such splitting-signals is described in detail in Section 6.6. The first quaternion splitting-signal $q_{T_{p,s}} = \{q_{p,s,t};\ t = 0:(N-1)\}$ is calculated by

$$q_{p,s,t} = \sum_{(n,m)\in V_{p,s,t}} q_{n,m} = (q_e)_{p,s,t} + i(q_i)_{p,s,t} + j(q_j)_{p,s,t} + k(q_k)_{p,s,t} \tag{6.88}$$

and the second quaternion splitting-signal $f_{T_{p,s}} = \{f_{p,s,t};\ t = 0 : (N-1)\}$ is calculated by

$$f_{p,s,t} = \sum_{(n,m)\in V_{p,s,t}} f_{n,m} = (f_e)_{p,s,t} + i(f_i)_{p,s,t} + j(f_j)_{p,s,t} + k(f_k)_{p,s,t}. \quad (6.89)$$

The summation is by the set

$$V_{p,s,t} = \{(n,m) \in X;\ np + ms = t \bmod N\},\ \ t = 0 : (N-1).$$

The main equality in Eq. (6.84) can be written as

$$O_{p,s} = \sum_{t=0}^{N-1} q_{p,s,t} W_{\mu}^{t} + \left(\sum_{t=0}^{N-1} f_{p,s,t} W^{t}_{-\mu}\right) E, \quad (6.90)$$

or

$$O_{p,s} = \sum_{t=0}^{N-1} q_{p,s,t} W_{\mu}^{t} + \left(\sum_{t=0}^{N-1} f_{p,s,t} W_{\mu}^{-t}\right) E. \quad (6.91)$$

The octonion splitting-signals are periodic, i.e., $o_{p,s,t+N} = o_{p,s,t}$. Therefore, as in the case of quaternion images, the splitting-signal $o_{T_{p,s}}$ defines all values of the 2-D rs-ODFT at frequency-points of the set $T_{p,s}$,

$$O_{kp \bmod N, ks \bmod N} = \sum_{t=0}^{N-1} o_{p,s,t} W_{\mu}^{kt},\ \ k = 0 : (N-1). \quad (6.92)$$

Since $o_{p,s,t} = q_{p,s,t} + f_{p,s,t}E$, the above equation can be opened as

$$O_{kp \bmod N, ks \bmod N} = \sum_{t=0}^{N-1} q_{p,s,t} W_{\mu}^{kt} + \left(\sum_{t=0}^{N-1} f_{p,s,t} W_{\mu}^{-kt}\right) E,\ \ k = 0 : (N-1).$$

Thus, the 1-D rs-QDFT of the splitting-signal

$$\{q_{p,s,t};\ t = 0 : (N-1)\} \rightarrow Q_k = \sum_{t=0}^{N-1} q_{p,s,t} W_{\mu}^{kt},\ \ k = 0 : (N-1),$$

defines the first four components of the 2-D rs-ODFT of the octonion image $o_{n,m}$ at the frequency-points of the set $T_{p,s}$. The 1-D rs-QDFT of the splitting-signal $\{f_{p,s,t};\ t = 0 : (N-1)\}$ defines the next four components of the 2-D rs-ODFT of $o_{n,m}$ at $T_{p,s}$. This rs-QDFT can be written in two forms as

$$\sum_{t=0}^{N-1} f_{p,s,t} W_{\mu}^{-kt} = \sum_{t=0}^{N-1} f_{p,s,t} W_{-\mu}^{kt} = \sum_{t=0}^{N-1} f_{p,s,t} W_{\mu}^{(N-k)t}.$$

It is the N-point rs-QDFT with the kernel $W_{-\mu}^{kt}$. This transform can also be calculated by the N-point rs-QDFT with the kernel W_{μ}^{kt} following the permutation $p \to (N-p)$,

$$\{f_{p,s,t};\ t = 0 : (N-1)\} \to F_k = \sum_{t=0}^{N-1} f_{p,s,t} W_{\mu}^{kt} \to F_{(N-k) \bmod N}.$$

Thus, the 2-D rs-ODFT at the frequency-points of the set $T_{p,s}$ can be calculated as

$$O_{kp \bmod N, ks \bmod N} = Q_k + F_{N-k},\ k = 0 : (N-1). \tag{6.93}$$

If we define the octonion splitting-signals as

$$o_{T_{p,s}} = q_{T_{p,s}} + f_{T_{N-p,N-s}} E, \tag{6.94}$$

then, the 2-D rs-ODFT at the sets $T_{p,s}$ can be calculated as

$$O_{kp \bmod N, ks \bmod N} = Q_k + F_k E,\ k = 0 : (N-1).$$

Here, F_k stands for the 1-D QDFT of the splitting-signal $f_{T_{N-p,N-s}}$.

In tensor representation, the calculation of the 2-D rs-ODFT requires twice operations used for the 2-D rs-QDFT. In the considered two cases, when N is prime and a power of two, the $N \times N$-point 2-D rs-ODFT requires $2(N+1)$ and $3N$ N-point 1-D rs-QDFTs, respectively.

6.13 2-D Left-Side ODFT

The 2-D left-side ODFT (ls-ODFT) of the octonion image $o_{n,m}$ is defined as

$$O_{p,s} = \sum_{n=0}^{N-1} W_{N;\lambda_1}^{np} \left(\sum_{m=0}^{M-1} W_{M;\lambda_2}^{ms} o_{n,m} \right) = \sum_{n=0}^{N-1} \sum_{m=0}^{M-1} W_{N;\lambda_1}^{np} \left(W_{M;\lambda_2}^{ms} o_{n,m} \right), \tag{6.95}$$

where $p = 0 : (N-1)$ and $s = 0 : (M-1)$. The calculation of this transform is performed first by the M-point ls-ODFT along the rows of the image and then by the N-point ls-ODFTs along the columns. We will call this transform the row-column 2-D ls-ODFT. The numbers λ_1 and λ_2 are considered to be pure quaternions μ_1 and μ_2, respectively.

The 2-D inverse ls-ODFT is calculated by

$$o_{n,m} = \frac{1}{NM}\sum_{p=0}^{N-1}\sum_{s=0}^{M-1} W_{N;\mu_1}^{np}\left(W_{M;\mu_2}^{ms} O_{p,s}\right),$$

where $n = 0 : (N-1)$ and $m = 0 : (M-1)$.

The 2-D ls-ODFT of the octonion image $o_{n,m}$ can also be defined as

$$O_{p,s} = \sum_{m=0}^{M-1} W_{M;\mu_1}^{ms}\left(\sum_{n=0}^{N-1} W_{N;\mu_2}^{np} o_{n,m}\right) = \sum_{m=0}^{M-1}\sum_{n=0}^{M-1} W_{M;\mu_1}^{ms}\left(W_{N;\mu_2}^{np} o_{n,m}\right), \quad (6.96)$$

where $p = 0 : (N-1)$ and $s = 0 : (M-1)$. The calculation of this transform is performed first by the N-point ls-ODFTs of the columns of the image and, then, by the M-point ls-ODFTs of the rows. This is the column-row 2-D ls-ODFT.

In both definitions, the 2-D ls-ODFTs can be calculated by two 2-D QDFTs. To show this fact, we consider the definition of the transform in Eq. (6.95) and presentation of the octonion image by its two quaternion components $o_{n,m} = q_{n,m} + f_{n,m}E$. The 2-D ls-ODFT of the octonion image $o_{n,m}$ can be written as

$$O_{p,s} = \sum_{n=0}^{N-1}\sum_{m=0}^{M-1} W_{N;\mu_1}^{np}\left[W_{M;\mu_2}^{ms}(q_{n,m} + f_{n,m}E)\right],$$

or

$$O_{p,s} = \sum_{n=0}^{N-1}\sum_{m=0}^{M-1} W_{N;\mu_1}^{np}\left(W_{M;\mu_2}^{ms} q_{n,m}\right) + \sum_{n=0}^{N-1}\sum_{m=0}^{M-1} W_{N;\mu_1}^{np}\left[W_{M;\mu_2}^{ms}(f_{n,m}E)\right]. \quad (6.97)$$

Due to the property $q_1(q_2E) = (q_2q_1)E$, the following equalities hold for the addend in the second double sum:

$$W_{M;\mu_2}^{ms}(f_{n,m}E) = \left(f_{n,m}W_{M;\mu_2}^{ms}\right)E,$$

and then,

$$W_{N;\mu_1}^{np}\left[\left(f_{n,m}W_{M;\mu_2}^{ms}\right)E\right] = \left[\left(f_{n,m}W_{M;\mu_2}^{ms}\right)W_{N;\mu_1}^{np}\right]E.$$

Thus, if we denote by $Q_{p,s}$ the 2-D ls-QDFT of the quaternion image $q_{n,m}$ and by $F_{p,s}$ the 2-D rs-QDFT of the quaternion image $f_{n,m}$, we obtain

$$O_{p,s} = Q_{p,s} + F_{p,s}E = Q_{p,s}^{(ls)} + F_{p,s}^{(rs)}E,$$

where $p = 0 : (N - 1)$ and $s = 0 : (M - 1)$.

6.13.1 Tensor transform and the 2-D ls-ODFT

When considering the tensor representation of the octonion image $o_{n,m}$, the same set of 1-D octonion splitting-signals describes the right-side and left-side 2-D ODFTs. This unique representation of the image allows for calculating both transforms. For the two-side 2-D ODFT, such representation does not have a place or may be defined in a different way. The tensor representation with its fast algorithms for 2-D DFT was derived from the mathematical structure of the transform for traditional images, or complex 2-D images, and, then, used for the quaternion and octonion images. On this line from the real images till the octonion images, there is no place for the two-side 2-D ODFT. The application of the tensor representation for the 2-D ls-ODFT can be described similarly to the 2-D rs-ODFT case.

We consider the case when two pure quaternions μ_1 and μ_2 are equal to μ. The 2-D ls-ODFT is

$$O_{p,s} = \sum_{n=0}^{N-1}\sum_{m=0}^{M-1} W_{N;\mu}^{np}\left(W_{M;\mu}^{ms} o_{n,m}\right), \tag{6.98}$$

where $p = 0 : (N - 1)$ and $s = 0 : (M - 1)$. As is shown above, the 2-D ls-ODFT of the octonion image $o_{n,m} = q_{n,m} + f_{n,m}E$ can be written as

$$O_{p,s} = \sum_{n=0}^{N-1}\sum_{m=0}^{M-1} W_{\mu}^{np}\left(W_{\mu}^{ms} q_{n,m}\right) + \left[\sum_{n=0}^{N-1}\sum_{m=0}^{M-1} \left(f_{n,m} W_{\mu}^{ms}\right) W_{\mu}^{np}\right] E,$$

where $W_{\mu}^{np} = W_{N;\mu}^{np}$. The multiplication is associative for quaternion numbers and we can write that

$$\begin{aligned} O_{p,s} &= \sum_{n=0}^{N-1}\sum_{m=0}^{M-1} \left(W_{\mu}^{np} W_{\mu}^{ms}\right) q_{n,m} + \left[\sum_{n=0}^{N-1}\sum_{m=0}^{M-1} f_{n,m}\left(W_{\mu}^{ms} W_{\mu}^{np}\right)\right] E \\ &= \sum_{n=0}^{N-1}\sum_{m=0}^{M-1} W_{\mu}^{np+ms} q_{n,m} + \left(\sum_{n=0}^{N-1}\sum_{m=0}^{M-1} f_{n,m} W_{\mu}^{ms+np}\right) E. \end{aligned} \tag{6.99}$$

In tensor representation, the discrete octonion image $o_{n,m}$ is described by a set of 1-D splitting-signals of length N each,

$$o_{T_{p,s}} = q_{T_{p,s}} + f_{T_{p,s}} E, \tag{6.100}$$

which are described in Eqs. (6.87)–(6.89). Equation (6.99) can thus be written as

$$O_{p,s} = \sum_{t=0}^{N-1} W^{t}_{\mu} q_{p,s,t} + \left(\sum_{t=0}^{N-1} f_{p,s,t} W^{t}_{\mu}\right) E. \tag{6.101}$$

The components of the octonion splitting-signals are periodic, i.e., $o_{p,s,t+N} = o_{p,s,t}$. Therefore, as in the case of the quaternion images, the octonion splitting-signal $o_{T_{p,s}}$ defines all values of the 2-D rs-ODFT at the frequency-points of the set $T_{p,s}$,

$$O_{kp \bmod N, ks \bmod N} = \sum_{t=0}^{N-1} W^{kt}_{\mu} q_{p,s,t} + \left(\sum_{t=0}^{N-1} f_{p,s,t} W^{-kt}_{\mu}\right) E, \quad k = 0 : (N-1). \tag{6.102}$$

Thus, the 1-D ls-QDFT of the splitting-signal $\{q_{p,s,t};\ t = 0 : (N-1)\}$

$$Q_k = Q^{(ls)}_k = \sum_{t=0}^{N-1} W^{kt}_{\mu} q_{p,s,t}, \ k = 0 : (N-1),$$

defines the first four components of the 2-D ls-ODFT of the octonion image $o_{n,m}$ at the frequency-points of the set $T_{p,s}$. The 1-D rs-QDFT of the splitting-signal $\{f_{p,s,t};\ t = 0 : (N-1)\}$,

$$F_k = F^{(ls)}_k = \sum_{t=0}^{N-1} f_{p,s,t} W^{kt}_{\mu}, \quad k = 0 : (N-1),$$

defines the next four components of the 2-D ls-ODFT of $o_{n,m}$ at the same set $T_{p,s}$. Thus, the 2-D ls-ODFT at the frequency-points of the set $T_{p,s}$ can be calculated as

$$O_{kp \bmod N, ks \bmod N} = Q_k + F_k E, \ k = 0 : (N-1). \tag{6.103}$$

The calculation of the 2-D ls-ODFT requires twice operations used for the 2-D ls-QDFT. In the considered two cases, when N is prime and a power of two, the $N \times N$-point ls-ODFT requires $2(N+1)$ and $3N$ N-point rs-QDFTs, respectively.

References

[1] T. A. Ell, "Quaternion-Fourier transforms for analysis of 2-dimensional linear time-invariant partial differential systems," *Proc. the 3nd IEEE Conference on Decision and Control*, **1**(4), 1830–1841, San Antonio, Texas (1993).

[2] S. J. Sangwine, "Fourier transforms of colour images using quaternion, or hypercomplex, numbers," *Electronics Letters*, **32**(21), 1979–1980 (1996).

[3] S. J. Sangwine, "The discrete quaternion-Fourier transform," *Proc. IPA97*, **443**, 790–793 (1997).

[4] S. J. Sangwine, T. A. Ell, J. M. Blackledge, and M. J. Turner, "The discrete Fourier transform of a color image," *Proc. Image Processing II Mathematical Methods, Algorithms and Applications*, 430–441 (2000).

[5] S. J. Sangwine and T. A. Ell, "Hypercomplex Fourier transforms of color images," *Proc. IEEE Intl. Conf. Image Process.*, **1**, 137–140 (2001).

[6] E. Hitzer, "Quaternion Fourier transform on quaternion fields and generalizations," *Advances in Applied Clifford Algebras*, **17**(3), 497–517, May 2007.

[7] E. Hitzer, "Directional uncertainty principle for quaternion Fourier transforms," *Advances in Applied Clifford Algebras*, **20**(2), 271–284 (2010).

[8] M. H. Yeh, "Relationships among various 2-D quaternion Fourier transforms," *IEEE Signal Processing Letters*, **15**, 669–672 (2008).

[9] Z. Yang and S.-I. Kamata, "Hypercomplex polar Fourier analysis for color image," in Proc. *International Conference on Image Processing (ICIP)*, 2117–2120 (2011).

[10] A. M. Grigoryan and S. S. Agaian, "Retolling of color imaging in the quaternion algebra," *Applied Mathematics and Sciences: An International Journal (MathSJ)*, **1**(3), 23–39 (2014).

[11] A. M. Grigoryan and S. S. Agaian, "Alpha-rooting method of color image enhancement by discrete quaternion Fourier transform," [9019-3], *Proc. SPIE*, **9019**, 901904 (2014) [doi: 10.1117/12.2040596].

[12] A. M. Grigoryan, J. Jenkinson, and S. S. Agaian, "Quaternion Fourier transform based alpha-rooting method for color image measurement, and enhancement," *SIGPRO-D-14-01083R1, Signal Processing*, **109**, 269–289 (2015) [doi: 10.1016/j.sigpro.2014.11.019].

[13] J. W. Cooley and J. W. Tukey, "An algorithm the machine computation of complex Fourier series," *Math. Comput*, **9**(2), 297–301 (1965).

[14] L. Rabiner and R. W. Schaefer, *Speech Signal Processing*, Prentice Hall, Englewood Cliffs, New Jersey (1983).

[15] R. E. Blahut, *Fast Algorithms for Digital Signal Processing*, Addison-Wesley, Reading, Mass (1985).

[16] A. M. Grigoryan, "An algorithm of the two-dimensional Fourier transform," *Izvestiya VUZov SSSR, Radioelectronica*, USSR, Kiev, **27**(10), 52–57 (1984). (Available in http://fasttransforms.com/Art-USSR-papers/Art_IzvuzovSSSR1984.pdf)
[17] A. M. Grigoryan, "An optimal algorithm for computing the two-dimensional discrete Fourier transform," *Izvestiya VUZ SSSR, Radio-electronica*, USSR, Kiev, **29**(12), 20–25 (1986). (Available in http://fasttransforms.com/Art-USSR-papers/Art_IzvuzovSSSR1986.pdf)
[18] A. M. Grigoryan, "New algorithms for calculating discrete Fourier transforms," *Journal Vichislitelnoi Matematiki i Matematicheskoi Fiziki*, AS USSR, Moscow, **26**(9), 1407–1412 (1986). (Available in http://fasttransforms.com/Art-USSR-papers/Art_JVMATofASUSSR1986.pdf)
[19] A. M. Grigoryan and M. M. Grigoryan, "Two-dimensional Fourier transform in the tensor presentation and new orthogonal functions," *Avtometria, AS USSR Siberian section*, Novosibirsk, (1), 21–27 (1986).
[20] A. M. Grigoryan, "An algorithm for computing the two-dimensional Fourier transform of equal orders," Proc. IX All-Union Conference on Coding and Transmission of Information, USSR, Odessa, **2**, 49–52 (1988). (Available in http://fasttransforms.com/Art-USSR-papers/Art_AllUnConf1988.pdf)
[21] A. M. Grigoryan, "An algorithm for computing the discrete Fourier transform with arbitrary orders," *Journal Vichislitelnoi Matematiki i Matematicheskoi Fiziki*, AS USSR, Moscow, **30**(10), 1576–1581 (1991). (Available in http://fasttransforms.com/Art-USSR-papers/Art_JVMATofASUSSR1991.pdf)
[22] A. M. Grigoryan, "2-D and 1-D multi-paired transforms: Frequency-time type wavelets," *IEEE Trans. on Signal Processing*, **49**(2), 344–353 (2001).
[23] A. M. Grigoryan and S. S. Agaian, "Split manageable efficient algorithm for Fourier and Hadamard transforms," *IEEE Trans. Signal Processing*, **48**(1), 172–183 (2000).
[24] A. M. Grigoryan and S. S. Agaian, *Multidimensional Discrete Unitary Transforms: Representation, Partitioning, and Algorithms*, Marcel Dekker Inc., New York (2003).
[25] A. M. Grigoryan and M. M. Grigoryan, *Brief Notes in Advanced DSP: Fourier Analysis with MATLAB*, CRC Press, Taylor, and Francis Group, Boca Raton, Florida (2009).
[26] A. M. Grigoryan and S. S. Agaian, "Tensor transform-based quaternion Fourier transform algorithm," *Information Sciences*, **320**, 62–74 (2015).
[27] A. M. Grigoryan and S. S. Agaian, "2-D Left-side quaternion discrete Fourier transform fast algorithms," *Proc. IS&T International*

Symposium, 2016 Electronic Imaging: Algorithms and Systems XIV, San Francisco, California (2016).

[28] N. Ranganadh, A. M. Grigoryan, and B. D. Tushara, "Implementation of FFT by using MATLAB: SIMULINK on Xilinx Virtex-4 FPGAs: Performance of a paired transform based FFT," *International Journal of Advanced Computer Research*, **3**(2), 72–78 (2013).

[29] N. Ranganadh and M. Guravaiah, "Performance case study of Grigoryan FFT over Cooley-Tukey FFT using TMS DSP processors," *International Journal of Advanced Computer Research*, **2**(4), 452–457 (2012).

[30] R. Narayanam, P. Parimal, and A. M. Grigoryan, "Performances of Texas Instruments DSP, and Xilinx FPGAs for Cooley-Tukey, and Grigoryan FFT algorithms," *SOURCE Journal of Engineering and Technology*, **1**(2), 83 (2011).

[31] R. C. Gonzalez and R. E. Woods, *Digital Image Processing*, 2nd Edition, Prentice Hall, Upper Saddle River, New Jersey (2002).

[32] A. M. Grigoryan and K. Naghdali, "On a method of paired representation: Enhancement and decomposition by series direction images," *Journal of Mathematical Imaging and Vision*, **34**(2), 185–199 (2009).

[33] A. M. Grigoryan, "Solution of the problem on image reconstruction in computed tomography," *Journal of Mathematical Imaging and Vision*, **54**(1), 35–63 (2016).

[34] A. M. Grigoryan and N. Du, "2-D images in frequency-time representation: Direction images, and resolution map," *Journal of Electronic Imaging*, **19**(3), 033012 (2010) [doi: 10.1117/1.3483906].

[35] A. M. Grigoryan and N. Du, "Principle of superposition by direction images," *IEEE Trans. on Image Processing*, **20**(9), 2531–2541 (2011).

[36] A. M. Grigoryan and M. M. Grigoryan, *Image processing: Tensor transform, and discrete tomography with MATLAB*, CRC Press Taylor, and Francis Group (2012).

[37] A. M. Grigoryan, "Image reconstruction from finite number of projections: Method of transferring geometry," *IEEE Trans. on Image Processing*, **22**(12), 4738–4751 (2013).

[38] A. M. Grigoryan, "Another Step towards Successful Tomographic Imaging in Cancer: Solving the Problem of Image Reconstruction," in *Computer-Aided Cancer Detection, and Diagnosis: Recent Advances*, (J. Tang and S. Agaian, Editors), SPIE Press, Ch. 11, 295–338 (2014) [doi: 10.1117/3.1002311.ch11].

[39] S. S. Agaian, H. G. Sarukhanyan, K. O. Egiazarian, and J. Astola, *Hadamard Transforms*, SPIE Press, Bellingham, Washington (2011) [doi: 10.1117/3.890094].

[40] S. C. Pei, J. J. Ding, and J. H. Chang, "Efficient implementation of quaternion Fourier transform, convolution, and correlation by 2-D complex FFT," *IEEE Trans. Signal Process.*, **49**, 2783–2797 (2001).

[41] T. A. Ell and S. J. Sangwine "Hypercomplex Fourier transforms of color images," *IEEE Trans. Image Process.*, **16**(1), 22–35 (2007).

[42] D. E. Dudgeon and R. M. Mersereau, *Multidimensional Digital Signal Processing*, Prentice Hall: Englewood Cliffs, NJ (1984).

[43] R. C. Staunton, *Hexagonal Sampling in Image Processing*, in Advances in Imaging, and Electron Physics, **107**, 231–307 (1999).

[44] R. M. Mersereau, "The processing of hexagonally sampled two-dimensional signals," *Proc. IEEE*, **67**(6), 930–949 (1979).

[45] D. P. Petersen and D. Middleton, "Sampling, and reconstruction of wave-number-limited functions in *N*-dimensional Euclidean spaces," *Information and Control*, **5**, 279–323 (1962).

[46] D. H. Hubel, *Eye, Brain, and Vision*, Scientific American Books, New York (1988).

[47] A. L. Nel, "Hexagonal image processing," *Proc. Communications and Signal Processing, COMSIG 1989*, Southern African Conference on, 109–113 (1989).

[48] A. B. Watson and A. J. Ahumada Jr., "A hexagonal orthogonal-oriented pyramid as a model of image representation in visual cortex," *Biomedical Engineering, IEEE Transactions*, **36**(1), 97–106 (1989).

[49] R. C. Giles, P. R. Kotiuga, and M. Mansuripur, "Parallel micro-magnetic simulations using Fourier methods on a regular hexagonal lattice," *Magnetics, IEEE Transactions*, **27**(5), 3815–3818 (1991).

[50] G. E. Rivard, "Direct fast Fourier transform of bivariate functions," *IEEE Trans. ASSP*, **25**, 250–252 (1977).

[51] R. M. Mersereau, "A two-dimensional FFT for hexagonally sampled data", in Digital Signal Processing, ed. V. Cappellini and A. G. Constantinides, Academic Press, 93–101 (1980).

[52] M. R. Mersereau and T. C. Speake, "A unified treatment of Cooley–Tukey algorithms for the evaluation of the multidimensional DFT," *IEEE Trans. Acoustic, Speech, and Signal Processing*, **29**(5), 1011–1018 (1981).

[53] A. Guessoum and R. M. Mersereau, "Fast algorithms for the multidimensional discrete Fourier transform," *IEEE ASAP*, **34**(4), 937–943 (1986).

[54] I. Her, "Geometric transformations on the hexagonal grid," *IEEE Trans. Image Processing*, **4**(9), 1213–1222 (1995).

[55] J. L. Zapata and G. X. Ritter, "Fast Fourier transform for hexagonal aggregates", *Journal of Mathematical Imaging and Vision*, **12**(3), 183–197 (2000).

[56] A. M. Grigoryan, "Efficient algorithms for computing the 2-D hexagonal Fourier transforms," *IEEE Transaction on Signal Processing*, **50**(6), 1438–1448 (2002).

[57] A. M. Grigoryan, S. S. Agaian, and E. R. Dougherty, "Splitting of the 2–D Hexagonal DFT," *Proc. Pattern Recognition and Information Processing PRIP2001*, Belarus (2001).

[58] A. M. Grigoryan, S. S. Agaian, and E. R. Dougherty, "Splitting of the 2–D convolution: Optimal linear filtration," *Proc. ICICS'99, The Second International Conference on Information, Communications and Signal Processing*, Singapore (1999).

Chapter 7
Color Image Enhancement and QDFT

This chapter presents methods for color image enhancement based on the discrete quaternion Fourier transform. We choose the quaternion Fourier transform, because it is well-suited for color image processing applications, it processes all three color components (red, green, blue) simultaneously, it captures the inherent correlation between the components, it does not generate color artifacts or blending, and finally, it does not need an additional color restoration process. Also we describe a measure to evaluate the quality of the enhanced color images [1]–[5]. Many experimental results show that the α-rooting enhancement based on the quaternion Fourier transform outperforms other enhancement methods such as the Fourier transform based α-rooting algorithm and the multi-scale retinex. On top, the new methods not only provide true color fidelity for poor quality images but also average the color components to gray value for balancing colors. These methods can be used to enhance edge information and sharp features in images, as well as for enhancing low contrast images. The proposed algorithms are simple to apply and design, which makes them very practical in image enhancement.

Digital image enhancement has a long history for more than 50 years. It has remained one of the most active research areas in image processing and computer vision. Recently, several image enhancement algorithms, and many applicable systems have been exploited [3, 4]. Current research in image enhancement covers such wide topics as algorithms based on the human visual system, histograms with hue preservation, JPEG-based enhancement for the visually impaired, and histogram modification techniques [5]. Two major classifications of image enhancement techniques can be defined: the spatial domain enhancement and the transform domain enhancement. Digital image enhancement is a powerful tool for many image-processing applications, when the critical details are not seen clearly enough. The purpose of image enhancement is to improve a digital image quality and to support the human perception [7]–[9]. The Fourier transform plays important role in image processing and is used in different stages of processing such as filtering, coding, recognition, and restoration analysis. The methods of the Fourier transform were applied for the Hartley, Hadamard, cosine, and other transforms [3, 6, 29]. Transform methods of image enhancement are based on manipulation with all or a part of spectral components of the transform. We focus on the well-known method of α-rooting [10], although other methods, including the log-α-rooting, modified unsharp masking [11], enhancement by entropy [12] and methods based on wavelet transforms [16]–[20] are also used in image enhancement.

Many traditional methods of image enhancement were applied for processing color images. The Fourier transform-based method is one of these methods, which in many cases is applied to each color plane-component of the image separately. In other words, the color image is considered as a triplet of separate 2-D grayscale images and each of these images represents red, green, or blue component of the color. Quaternion numbers of Hamilton's [21] were used in Ell's work [22], and after that time much attention was given to the transformation of the color components to the imaginary subspace of quaternions, "imaginary part" of which consists of three components. We use the well-known α-rooting method of image enhancement in the subspace of quaternions. For that, the 2-D discrete quaternion Fourier transforms (QDFT) [21] are used to perform the α-rooting of color images. The color images are considered in the RGB format, and three color components (r, g, b) are placing into the three imaginary parts (i, j, k) of quaternion number space. The preliminary results show that the application of the 2-D QDFTs plus α-rooting method result in high quality color images; therefore, these quaternion transforms can be effectively used for enhancing color images together with the traditional Fourier transform-based method.

7.1 Transform-Based Image Enhancement

The basic idea behind the frequency domain methods consists in computing a discrete unitary transform (DUT) of the image, for instance the 2-D discrete Fourier transform (2-D DFT), manipulating the transform coefficients by an operator **M**, and, then, performing the inverse transform, as shown in Fig. 7.1.

Transform-based image enhancement methods include techniques such as alpha-rooting, weighted α-rooting, modified unsharp masking, and filtering, which are all motivated by the human visual response [3, 9], [11]–[12]. The main advantages of transform-based image enhancement techniques are a low

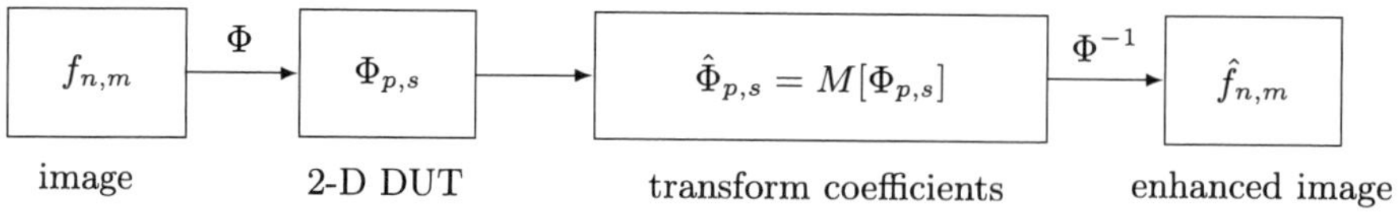

Figure 7.1 Block-diagram of the transform-based image enhancement.

complexity of computations, high quality of enhancement, and the critical role of fast unitary transforms in digital image processing. The transformation Φ in the above diagram is the Fourier transform, and other transforms can also be used. We consider the case when $\mathbf{M}$ is an operator of magnitude and the enhancement is described by

$$\{f_{n,m}\} \to \{F_{p,s} = |F_{p,s}|e^{-j\vartheta_{p,s}}\} \to \mathbf{M} \circ \mathrm{F} \to \{\hat{F}_{p,s} = M[|F_{p,s}|]e^{-j\vartheta_{p,s}}\} \to \{\hat{f}_{n,m}\}. \tag{7.1}$$

In the α-rooting method of image enhancement [10], the magnitude of the Fourier transform of the image is transformed as

$$|F_{p,s}| \to M[|F_{p,s}|] = |F_{p,s}|^{\alpha},$$

for each sample (p, s); a value α is taken from the interval $(0, 1]$. The coefficients of the Fourier transform are multiplied by coefficients $C(p,s) = |F_{p,s})|^{\alpha-1}$. This is one of the simplest parameterized method that is used effectively in image enhancement. The "best" value of α maybe adjusted interactively by the user, or can be found automatically as discussed in the next section.

7.2 Quantitative Measure of Image Enhancement

To measure the quality of images and select optimal processing parameters, we consider the described in [1] quantitative measure of image enhancement that is based on the ratio of the maximum and minimum of intensity in the logarithm scale.

The image enhancement measure estimation (EME) is based on the idea of calculating a visibility image which is defined as the ratio of maximums and minimums. This image enhancement measure can be used for selecting the best parameters for image enhancement by the Fourier transform, as well as other unitary transforms Φ. The measure is defined as follows (a review of the quantitative enhancement measure can be found in Refs. [2]–[6] and [30]–[34]). A discrete image $\{f_{n,m}\}$ of size $N_1 \times N_2$ is divided by $k_1 k_2$ blocks of size $L_1 \times L_2$ each, where integers $L_i = [N_i/k_i]$, $i = 1, 2$. The quantitative measure of enhancement of the image processed by Φ transform, $M_{\mathbf{a};\Phi} : \{f_{n,m}\} \to \{\hat{f}_{n,m}\}$, is defined by

$$EME_{\mathbf{a},\Phi}(\hat{f}) = EME_{L_1,L_2;\mathbf{a},\Phi}(\hat{f}) = \frac{1}{k_1 k_2}\sum_{k=1}^{k_1}\sum_{l=1}^{k_2} 20 \ln\left[\frac{\max_{k,l}(\hat{f})}{\min_{k,l}(\hat{f})}\right], \tag{7.2}$$

where $\max_{k,l}(\hat{f})$ and $\min_{k,l}(\hat{f})$ respectively are the maximum and minimum of the image $\hat{f}_{n,m}$ inside the (k,l)th block, and $\mathbf{a}$ is a parameter, or a vector parameter of the enhancement algorithm. $EME_{\mathbf{a},\Phi}(\hat{f})$ is called a measure of enhancement, or measure of improvement of the image f with respect to the transform Φ.

In calculation of the EME function in Eq. (7.2), the blocks with zero $\min_{k,l}(\hat{f})$ can be skipped or the number 1 can be added to the image $\hat{f}$, to avoid zero values of $\hat{f}$ for $\ln(\hat{f})$ function. In the (k,l)th block, the ratio in the logarithm

$$r_{k,l} = \ln\left[\frac{\max_{k,l}(\hat{f})}{\min_{k,l}(\hat{f})}\right] = \ln[\max_{k,l}(\hat{f})] - \ln[\min_{k,l}(\hat{f})]$$

shows the range of intensity of the image $\hat{f}$ in the logarithm scale. Therefore, $EME_{\mathbf{a}}(\hat{f})$ determines the average-block scale of intensities in the image.

We define a parameter $\mathbf{a}_0$ such that $EME_{\mathbf{a}_0;\Phi}(\hat{f}) = \max_{\mathbf{a}}\{EME_{\mathbf{a};\Phi}(\hat{f})\}$ to be the best (or optimal) Φ-transform-based image enhancement vector parameter. Experimental results show that the Fourier transform can be considered as the optimal, when compared with the cosine, Hartley, Hadamard, and other transforms. Therefore, the enhancement measure $EME_{\Phi}(\hat{f})$ will be considered with respect to the Fourier transform, i.e., when $\Phi = \mathcal{F}$. When Φ is the identity transformation, I, the EME of $\hat{f} = f$ is called the enhancement measure of the image f, i.e., $EME(f) = EME_I(f)$.

As an example, Fig. 7.2 shows the image in part (a), along with two curves of the measure of enhancement in part (b), which were calculated by using 5×5 and 7×7 blocks. The image enhancement by the α-rooting method has been parameterized by α varying in the interval (0.35,1). Both the curves have pikes of maximum at $\alpha_0 = 0.93$ that corresponds to the best visual estimation

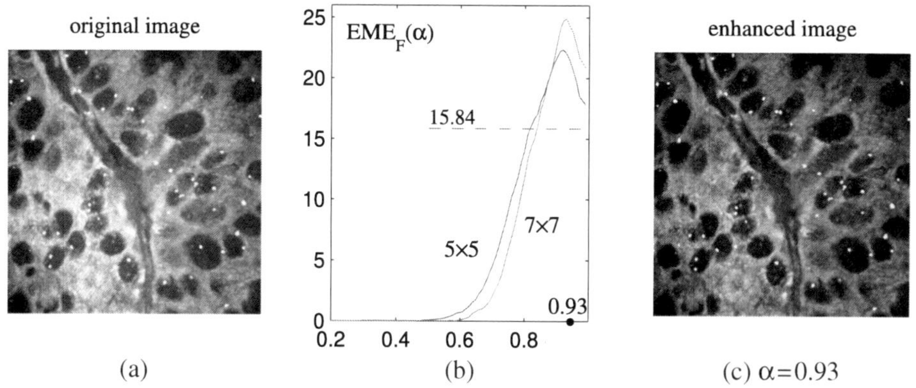

Figure 7.2 Parameterized image-enhancement based on the Fourier transform by the α-rooting method. (a) FISH image, (b) two curves $EME(\alpha)$, and (c) the image enhanced by 0.93-rooting method.

of enhancement. The α-rooting enhancement of the original image when $\alpha = 0.93$ is illustrated in c. The enhancement equals $EME_{0.93}(\hat{f}) - EME(f) = 25.03 - 15.84 = 9.19$.

When processing color images, to select values of α for image enhancement, we can analyze the enhancement functions EME_α for all three channels, which we call $E_r(\alpha) = EME_\alpha(\hat{f}_R)$, $E_g(\alpha) = EME_\alpha(\hat{f}_G)$, and $E_b(\alpha) = EME_\alpha(\hat{f}_B)$. The measure of image enhancement, $EME(\alpha)$, as defined in Eq. (7.2), is calculated in the spatial domain, after image has been processed by transformation Φ, or the Fourier transformation in our case. Instead of calculating three functions of *EME*, we need consider a new enhancement measure of the color image.

7.2.1 Quaternion transform-based image enhancement

The quaternion Fourier transform-based method of image enhancement can be used for color images after mapping them into the imaginary subspace of quaternion numbers, and by manipulating the transform coefficients by an operator **M**, and performing then the inverse 2-D QDFT, as shown in Fig. 7.3.

α-rooting Algorithm:

(Step 1) Perform the 2-D QDFT of the color image.

(Step 2) Multiply the transform coefficients, $Q_{p,s}$, by some quaternion factors.

(Step 3) Perform the inverse 2-D QDFT.

Therefore the use of the 2-D QDFT in α-rooting adopted to the case of color images is much promised. The α-rooting image enhancement technique can be used to enhance edge information and sharp features in images, as well as for enhancing even low contrast images. As an example, Fig. 7.4 shows the graphs of the calculated enhancement measure functions (EME) for the red, green, and blue channels of the color image in part (a). The color image is in Fig. 7.4(b). The color image enhancement by the 0.85-rooting with the 2-D QDFT is shown in Fig. 7.4(c). We also consider another example. Figure 7.5 in part (a) shows the enhancement measure functions (EME) for three channels. The original color image is shown in Fig. 7.4(b) and the results of the 0.82- and 0.94-rootings by the 2-D QDFT are shown in Fig. 7.4(c,d). To illustrate the performance of the color image enhancement algorithm, we will generalize the known EME measure of image enhancement.

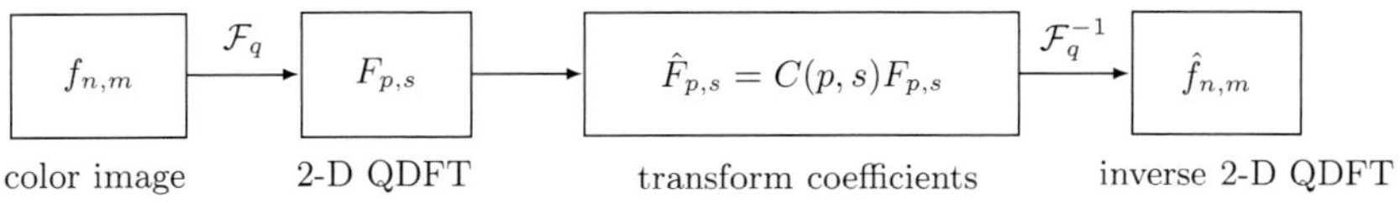

Figure 7.3 Block-diagram of the quaternion Fourier transform-based image enhancement.

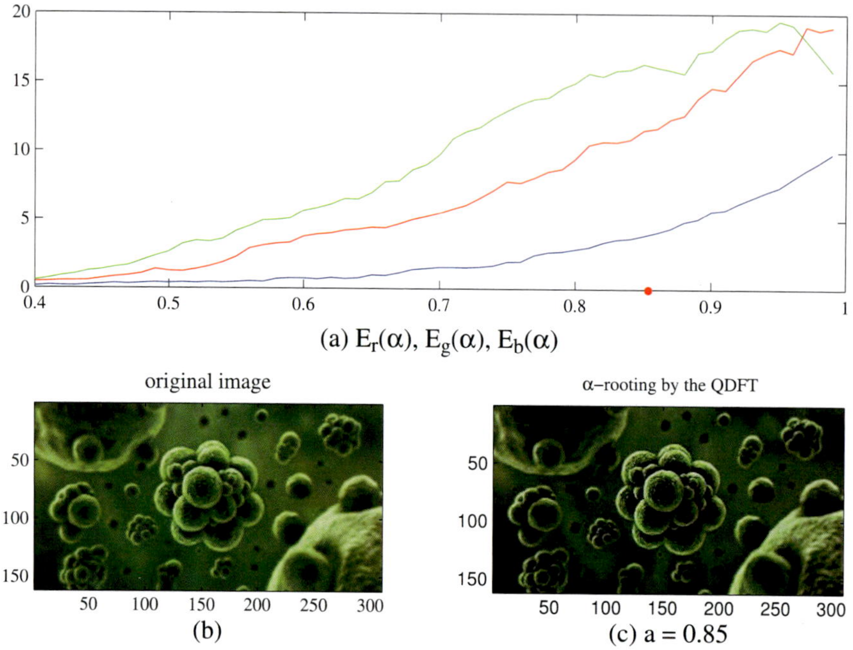

Figure 7.4 (a) Three enhancement functions (EME) for color channels of (b) the color image. (c) The 0.85-rooting by the 2-D QDFT.

7.3 New Color Image Quality Measure

This section introduces a simple color image enhancement metric. The measure of quality of images can be used for selecting optimal processing parameters for image enhancement. Note that the measure of image enhancement, $EME(\alpha)$, is calculated in the spatial domain after processing the image in the frequency domain. Instead of calculating three functions $EME_a(\hat{f}_R)$, $EME_a(\hat{f}_G)$, and $EME_a(\hat{f}_B)$, we propose the following enhancement measure [36, 40] of the color image $\hat{f} = (\hat{f}_R, \hat{f}_G, \hat{f}_B)$:

$$EMEC(\hat{f}) = \frac{1}{k_1 k_2} \sum_{k=1}^{k_1} \sum_{l=1}^{k_2} 20 \log_{10} \left[\frac{\max_{k,l}\{\hat{f}_R, \hat{f}_G, \hat{f}_B\}}{\min_{k,l}\{\hat{f}_R, \hat{f}_G, \hat{f}_B\}} \right]. \tag{7.3}$$

The image is in the RGB color space, i.e., $\hat{f} = (\hat{f}_R, \hat{f}_G, \hat{f}_B)$.

Thus, the maximum and minimum values of the image $\hat{f}$ in the (k,l)-th block are calculated as $\max(\hat{f}) = \max(\hat{f}_R, \hat{f}_G, \hat{f}_B)$ and $\min(\hat{f}) = \min(\hat{f}_R, \hat{f}_G, \hat{f}_B)$. In the block where the minimum is zero, the ratio $\log_{10} \max(\hat{f})/\min(\hat{f})$ is counted as $\log_{10} \max(\hat{f})$, of 1 is added to the image. This concept of enhancement

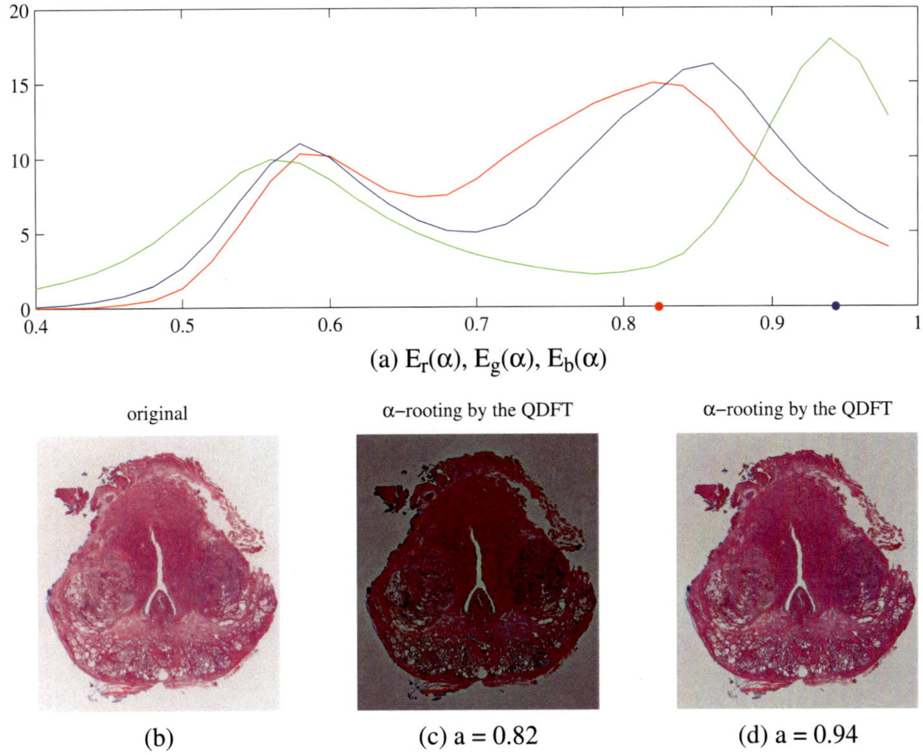

Figure 7.5 (a) Enhancement functions (EME) for all three channels of (b) the color image. (c) The 0.82- and (d) 0.94-rooting by the 2-D QDFT.

measure, EMEC, can be used when processing the image by the traditional 2-D DFT, as well as with other transforms including the 2-D QDFT.

To select the vector parameter α for color image enhancement, one can take the values with high enhancement measure. Figure 7.6 shows such an enhancement in part (b) together with the original image in part (a). The enhancement measure function for this color image is shown in part (c). Preliminary results show that the presented algorithm out-performs other enhancement methods such as the traditional Fourier transform-based alpha-rooting and the multi scale retinex (de facto standard) algorithms. The enhancement measure $EMEC(\hat{f})$ becomes the vector function $EMEC_{\alpha_r,\alpha_g,\alpha_b}(\hat{f})$.

For the quaternion images $\hat{f} = (\hat{f}_A, \hat{f}_R, \hat{f}_G, \hat{f}_B)$ with the real part $\hat{f}_A \neq 0$, the enhancement measure of the processed quaternion image $\hat{f} = (\hat{f}_A, \hat{f}_R, \hat{f}_G, \hat{f}_B)$ can be defined in the following way [37]–[39]:

$$EMEC(\hat{f}) = \frac{1}{k_1 k_2} \sum_{k=1}^{k_1} \sum_{l=1}^{k_2} 20 \log_{10} \left[\frac{\max_{k,l}\{\hat{f}_A, \hat{f}_R, \hat{f}_G, \hat{f}_B\}}{\min_{k,l}\{\hat{f}_A, \hat{f}_R, \hat{f}_G, \hat{f}_B\}} \right]. \qquad (7.4)$$

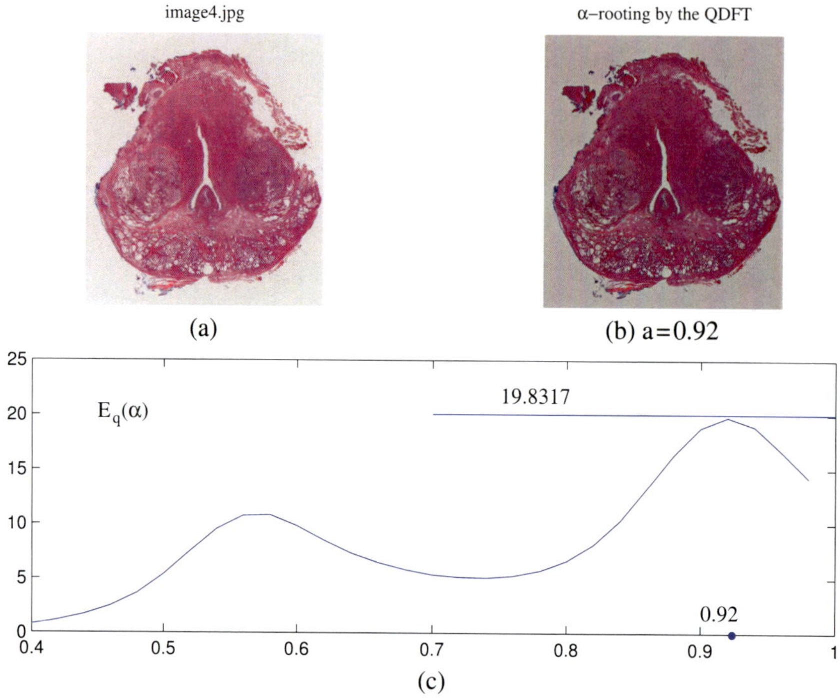

Figure 7.6 (a) The color image of size 1001 × 901, (b) the 0.92-rooting by the 2-D QDFT, and (c) the enhancement function (EMEC).

The color enhancement measure EMEC can be used to estimate enhancement in the RGB color space and other color models [36]. In the XYZ space, when $\hat{f} = (\hat{f}_X, \hat{f}_Y, \hat{f}_Z)$, the enhancement measure is defined as follows:

$$E_q(\alpha) = EMEC_a(\hat{f}) = \frac{1}{k_1 k_2} \sum_{k=1}^{k_1} \sum_{l=1}^{k_2} 20 \log_{10} \left[\frac{\max_{k,l}\{\hat{f}_X, \hat{f}_Y, \hat{f}_Z\}}{\min_{k,l}\{\hat{f}_X, \hat{f}_Y, \hat{f}_Z\}} \right]. \tag{7.5}$$

As an example, Fig. 7.7 shows such an enhancement of the same image in the XYZ color space in part (a). The measure of enhancement of the image is 19.9913, when $\alpha = 0.90$, and for the original image is 10.1933 (which is the $\alpha = 1$ case). The image enhanced by the optimal α-rooting when $\alpha = 0.92$ is shown in Fig. 7.7(b) and its transform into the RGB space in Fig. 7.7(c). The enhancement measure of the image in Fig. 7.7(b) for this α is 20.8884 and the measure of this image in the RGB color space is 20.97.

The optimal enhancement of the image in the RGB color space is $EMEC = 19.8317$ and after processing this image in the XYZ color space, the enhancement measure is larger, 20.97.

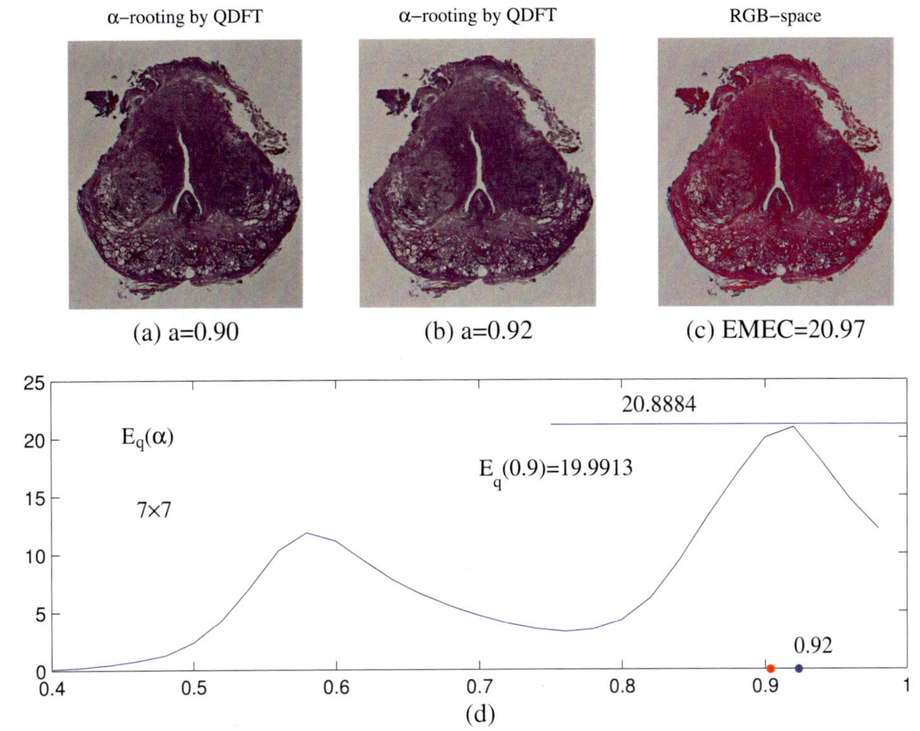

Figure 7.7 (a) The 0.90-rooting and (b) the 0.92-rooting by the 2-D QDFT in the XYZ color space, (c) its transformation to the RGB color space, and (d) the enhancement function *EMEC*(α) in the XYZ color space.

7.4 Enhancement of Images by Colors

In this section, we describe a few methods of color image enhancement when separately processing three colors in the known RGB model. First we consider the color "tree" image of size 256×256, which is shown in Fig. 7.8(a), along with its grayscale image components, which are red, green, and blue channels in Fig. 7.8(b,c,d), respectively. The red component is an image of high quality and many details of the "tree" image can be observed. In the green, the foreground looks well, but not the scene behind the tree. The blue color component is not a high quality image, when compared with other two channels. Small details and edges of the tree and other objects in the image do not have much of blue color.

To apply the concept of the α-rooting to the color image like this "tree" image, each of three color channels, or only one, can be processed. As an example, Fig. 7.9 shows the red channel of the image in part (a), along with the 0.88-rooting enhancement in part (b). The curve of the enhancement function $EME(\alpha)$ is shown in Fig. 7.9(c). The enhancement measure of the original red channel is 9.59, i.e., $EME(1) = 9.59$, as shown by the horizontal

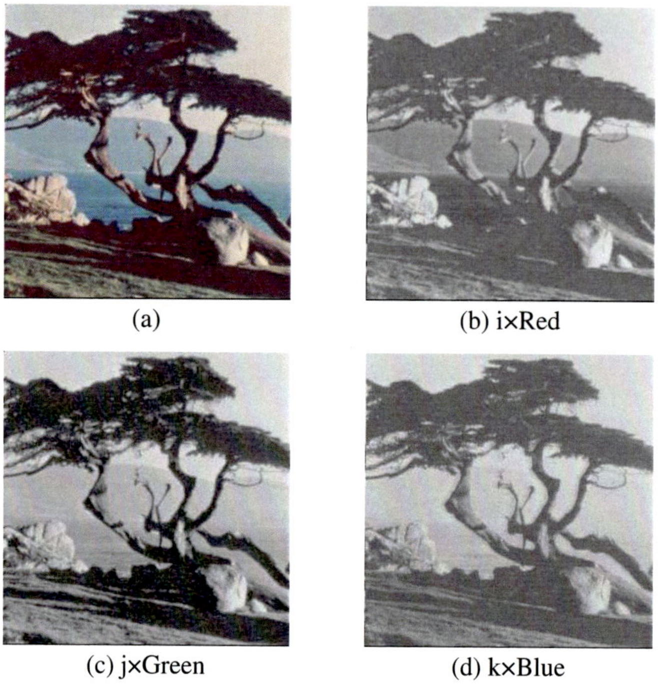

Figure 7.8 (a) Color "tree" image in RGB format, and (b) red, (c) green, and (d) blue channels of the image.

line in the figure. The enhancement measure exceeds this level for values of parameter $\alpha > 0.82$, and the maximum value is at the point 0.88. The block size $L_1 \times L_2$ for the function EME is 7×7. This block size is used for calculating the enhancement measure for all images in this chapter.

Figure 7.10 shows the green channel of the image in part (a), along with the 0.96-rooting enhancement in part (b). The enhancement function $EME(\alpha)$ is shown in part (c). When $\alpha > 0.93$ the enhancement measure exceeds the level of 16.92 which is the enhancement measure of the original green channel. The maximum value of the measure EME is at the point $\alpha = 0.96$ and equals 31.06. The blue channel of the "tree" image is shown in Fig. 7.11(a), and the 0.90-rooting enhancement of this channel in part Fig. 7.11(b). The enhancement measure of the blue channel is 9.58, which is very close to such number for the red channel. The enhancement function $EME(\alpha)$ for the blue channel is shown in Fig. 7.11(c). The maximum value of this measure EME is at the point $\alpha = 0.90$ and equals 29.95.

Figure 7.12 shows the color image in part (a), along with the results of image processing by only one channel, red in part (b), green in part (c), and blue in part (d), by the corresponding "optimal" α-rooting method of

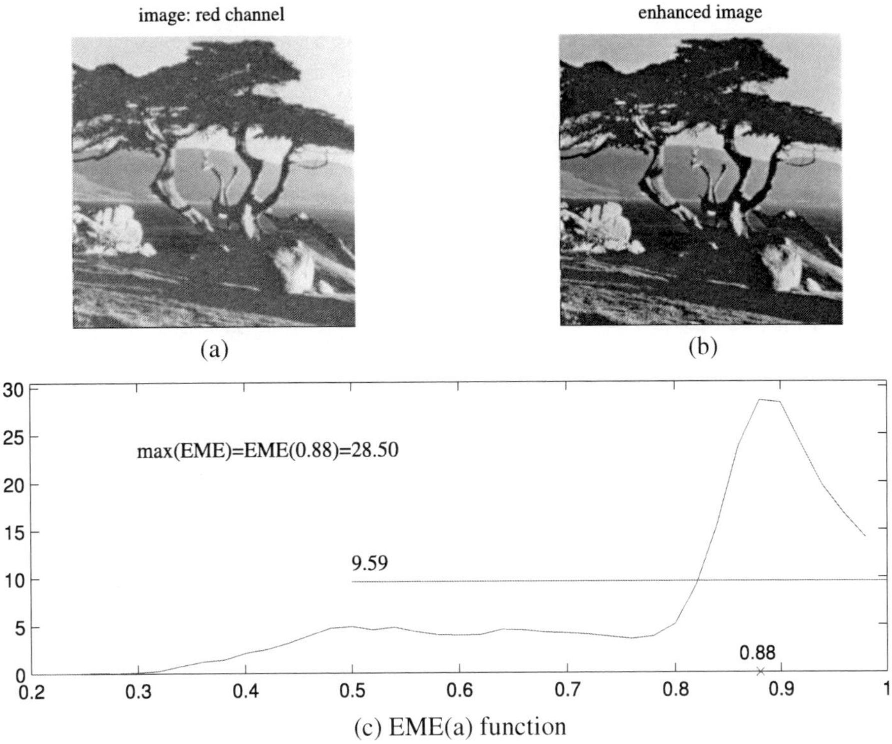

Figure 7.9 (a) Red channel of the image, and (b) 0.88-rooting enhancement, and (c) the function $EME(\alpha)$.

enhancement. When processing the red channel, the enhanced color image added more green on the grass and more blue on the water, sky, and the scene far behind the water. There are more red colors appeared when processing only the green channel, and yellow colors after processing the only blue channel. When comparing these three images, one can see that the image in part Fig. 7.12(b) is of the highest quality. These images show that the processing only one channel may change much the color palette of the image.

By taking the average of the "optimal" processed three channels, which can be written as $(R_{0.88}, G, B)$, $(R, G_{0.96}, B)$, and $(R, G, B_{0.90})$, we obtain the image $2/3(R, G, B) + 1/3(R_{0.88}, G_{0.96}, B_{0.90})$. Figure 7.13 shows the original "tree" image in part (a), along with the average of three images enhanced by one channel only. It was expected that processing all three channels will result an image of a quality better than in Fig. 7.13(b). For such comparison, the whole processed image $(R_{0.88}, G_{0.96}, B_{0.90})$ is shown in Fig. 7.14(b), together with the original image in Fig. 7.14(a). As shown above, the vector of values of *EME*s for original three channels equals $[9.59, 16.92, 9.58]$ and this vector

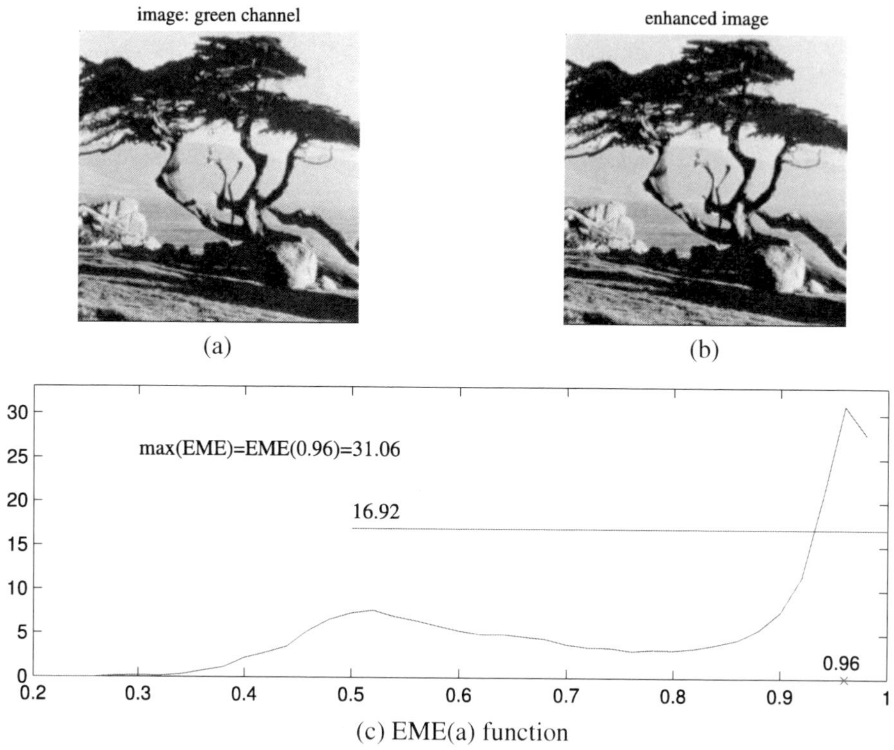

Figure 7.10 (a) Green channel of the image, and (b) 0.96-rooting enhancement, and (c) the function $EME(\alpha)$.

EME is reduced to $[28.50, 31.06, 29.95]$. We call such color image processing the $[0.88, 0.96, 0.90]$-rooting enhancement. One can notice that the enhanced is the image with more green color component, which has the high enhancement measure. Other vector parameters of α for channels can also be used, to enhance the colors in different degrees.

The optimal enhancement of each channel can be used to compose the grayscale image from the processed color image as $G = (R + G + B)/3$. Figure 7.15 shows the grayscale "tree" image in part (a), which is calculated by averaging three channels. The grayscale image calculated from the image in Fig. 7.13(b) is shown in Fig. 7.15(b). One can observe the enhancement of this image. The measure of enhancement EME for the grayscale image in Fig. 7.15(a) equals 13.18 and 17.30 for the image in Fig. 7.15(b).

Now, we describe the methods for color image enhancement when three channels processed together, after transferring the color image into quaternion space. For that, we apply the concept of the two-side 2-D QDFT together with the α-rooting. The results of the color image enhancement by the 2-D QDFT will be compared with the above described methods when separately processing the color channels.

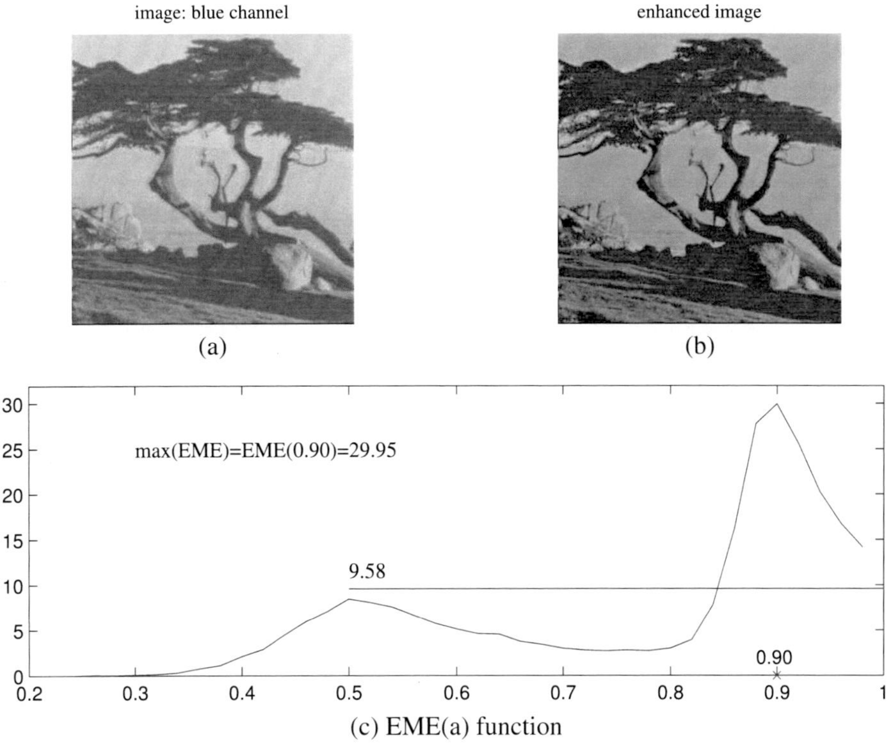

Figure 7.11 (a) Blue channel of the image, and (b) 0.90-rooting enhancement, and (c) the function *EME*(α).

7.5 2-D Quaternion DFT in Image Enhancement

A quaternion number has four components, a real part and three imaginary parts, which naturally coincides with the three components, R(ed), G(reen), and B(lue) of a color pixel. Therefore, a discrete color image $f(n,m)$ in the RGB color space can be transformed into an imaginary part of quaternion numbers form by encoding the red, green, and blue components of the RGB value as a pure quaternion (with zero real part):

$$f(n,m) = R(n,m)i + G(n,m)j + B(n,m)k.$$

The advantage of using quaternion based operations to manipulate color information in an image is that we do not have to process each color channel independently, but rather, treat each color triple as a whole unit. We believe that, by using quaternion operations, higher color information accuracy can be achieved because a color is treated as an entity. Another advantage of the use of quaternion-type representation is that a color image can be treated as a vector field or the hypercomplex Fourier transforms can handle color image pixels as vectors and thus offer scope to process color images holistically;

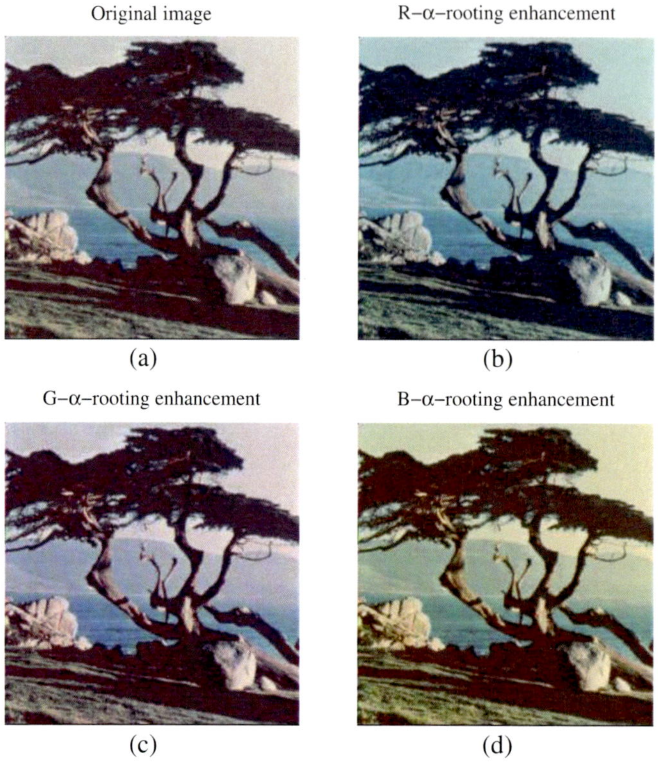

Figure 7.12 (a) The image, and the 0.88-, 0.96-, and 0.90-rooting enhancements of the image by only (b) red, (c) green, and (d) blue channels.

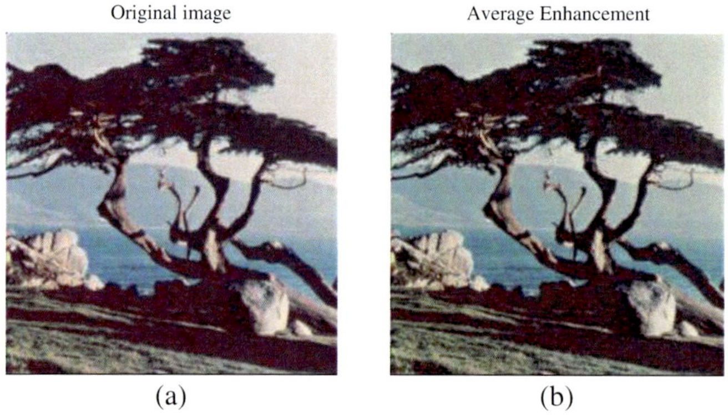

Figure 7.13 (a) The "tree" image and (b) the average of 0.88-, 0.96-, and 0.90-rooting enhancement of three channels.

rather than as separated luminance and chrominance, or separate color space components (example: red, green, blue) [27].

The quaternion multiplication is not commutative and the definition of the 2-D QDFT is not unique [23, 25, 26]. The color image $f_{n,m}$ is considered of

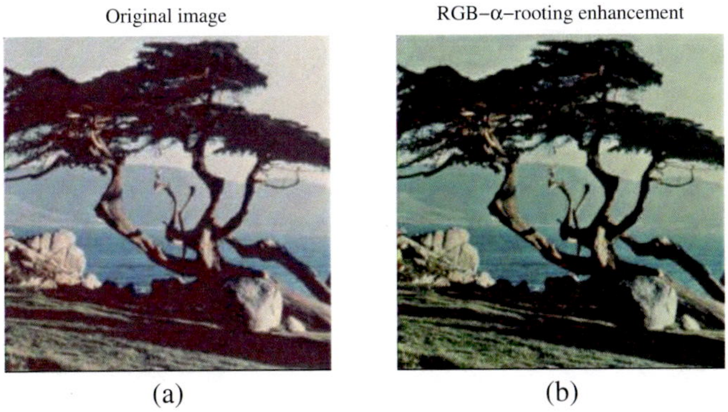

Figure 7.14 (a) The "tree" image and (b) [0.88, 0.96, 0.90]-rooting enhancement by three channels.

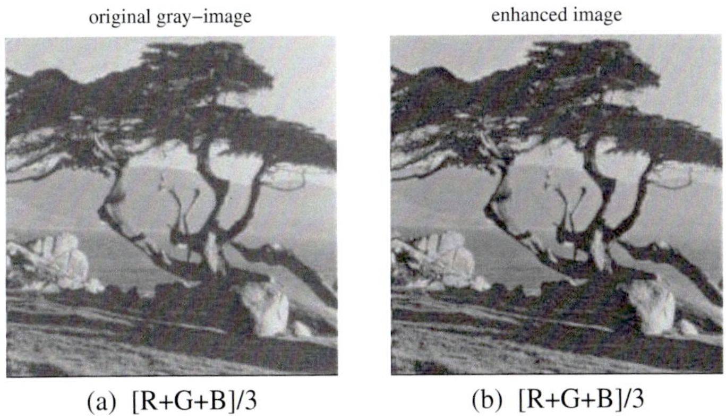

Figure 7.15 (a) The grayscale "tree" image and (b) the grayscale image being the sum of "optimal" α-rooting enhancement of three channels.

the size $N \times M$. Different QDFT can be used in image enhancement, and we consider the two-side QDFT [24]:

$$F_{p,s} = \sum_{n=0}^{N-1} \sum_{m=0}^{M-1} W_j^{np} f_{n,m} W_k^{ms} = \sum_{n=0}^{N-1} W_j^{np} \left(\sum_{m=0}^{M-1} f_{n,m} W_k^{ms} \right), \tag{7.6}$$

where $p = 0 : (N-1), s = 0 : (M-1)$. The basis functions are defined by the exponential coefficients

$$W_j = \cos(2\pi/N) - j\sin(2\pi/N), W_k = \cos(2\pi/M) - k\sin(2\pi/M).$$

Figure 7.16 shows the magnitude of the 2-D 256×256-point QDFT, $|F_{p,s}|$, of the "tree" image in part (a), together with the magnitude of the 2-D 256×256-point DFT in part (b). The magnitude of the 2-D DFT is defined as

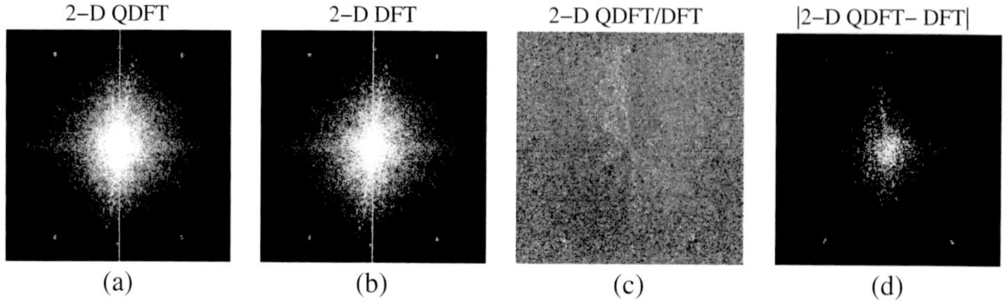

Figure 7.16 Magnitudes of (a) the 2-D QDFT and (b) the 2-D DFT of the color "tree" image, (c) the ratio and (d) difference of magnitudes.

the square root of the sum $|DFT(R)|^2 + |DFT(G)|^2 + |DFT(B)|^2$. The ratio of these magnitudes is shown in part (c), and the difference of magnitudes in part (d).

The inverse 2-D two-side QDFT is calculated by the similar formula

$$f_{n,m} = \frac{1}{NM}\sum_{n=0}^{N-1}\sum_{m=0}^{M-1} W_j^{-np} F_{p,s} W_k^{-ms} = \sum_{p=0}^{N-1} W_j^{-np}\left(\sum_{s=0}^{M-1} F_{p,s} W_k^{-ms}\right),$$

when $n = 0{:}(N-1)$ and $m = 0{:}(M-1)$. The 2-D QDFT can be calculated by using the column-row algorithm over each image component. In the $N = M$ case, this calculations requires $4(3N)$ real valued N-point DFTs. When N is a power of two, the number of operations of multiplication in the real valued N-point DFT equals $[N/2(\log_2 N - 3) + 2]/2$ [28]. The total number of such operations for the 2-D QDFT equals $3N^2(\log_2 N - 3) + 12N$. For a pure quaternion image $f_{n,m}$ the number of these operations is reduced to $9/4N^2(\log_2 N - 3) + 9N$.

7.5.1 Examples and experimental results

We consider a few examples of color image processing by α-rooting. The magnitude $|F_{p,s}|$ of the 2-D QDFT is modified as $|F_{p,s}|^\alpha$, and then the inverse 2-D QDFT is calculated. To select values of α for image enhancement, first we analyze the enhancement functions EME_α for all three channels, which we call $E_r(\alpha) = EME_\alpha(\hat{f}_R)$, $E_g(\alpha) = EME_\alpha(\hat{f}_G)$, and $E_b(\alpha) = EME_\alpha(\hat{f}_B)$. As an example, Fig. 7.17 shows these functions for the "tree" image in part (a). The EME functions for the red and blue channels reach maximums, 29.3324 and 31.2091, at points close to $\alpha = 0.90$. The 0.90-rooting of the image is shown in Fig. 7.17(b). The maximum 30.5665 of the EME for the green channel occurs at point 0.97. One can see that, at point $\alpha = 0.95$, three curves of EME functions intersect. This is a point of equal enhancement for all three channels. The 0.95-rooting of the "tree" image is shown in Fig. 7.17(c).

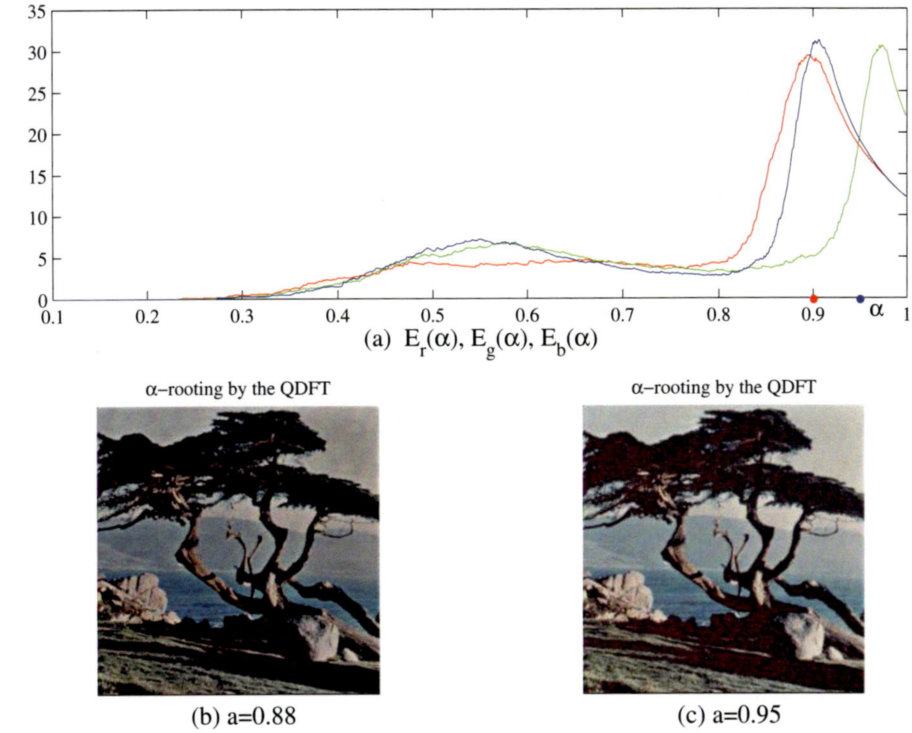

Figure 7.17 (a) Enhancement functions for three channels of the "tree" image. The α-rooting by the 2-D QDFT for (b) $\alpha = 0.90$ and (c) $\alpha = 0.95$.

We now consider the image enhancement by using the measure of image enhancement $EMEC(\alpha)$ defined in Eq. (7.4). Figure 7.18 shows the curve of the function $E_q(\alpha)$ for the color "tree" image in part (a). The maximum of this function is at point $\alpha = 0.9630$. The corresponding α-rooting of the image is shown in Fig. 7.18(b).

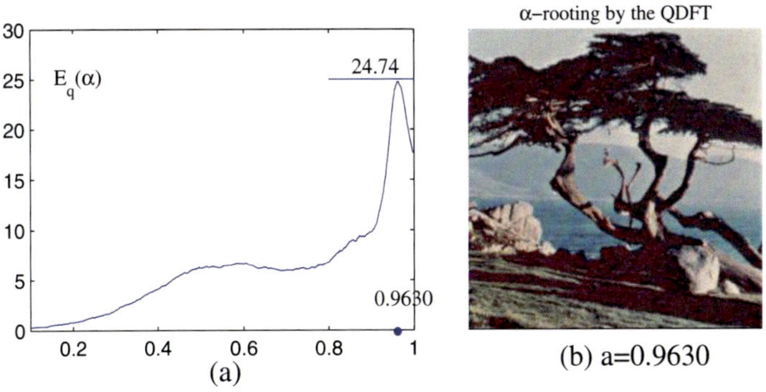

Figure 7.18 (a) Enhancement function of the "tree" image and (b) the 0.9630-rooting by the 2-D QDFT.

A vagueness in choosing optimal values of α when separately working with three color channels is replaced with the unique value of this parameter when working with the color image in quaternion field. Similar results can be obtained when enhanced many other color images. As an example, Fig. 7.19 shows the color image in part (a), along with the α-rooting for $\alpha = 0.964$ in part (b). The enhancement measure is given in Fig. 7.19(c); the function $E_q(\alpha)$ reaches its maximum at point 0.964.

Figure 7.20 shows the "room" image of size 322 × 212 in part (a), with the result of enhancement by the retinex method in part (b) (both images were taken from NASA's web page). The image after applying the histogram equalization to each color channel is shown in Fig. 7.20(c), and the α-rooting image enhancement for $\alpha = 0.96$ is shown in Fig. 7.20(d). Since the image in its right part is very dark, the α-rooting was implemented first on the negative image as follows:

$$f_{n,m} \to g_{n,m} = (255 - f_{n,m}) \to [\alpha - \text{ rooting}] \to \hat{g}_{n,m} \to \hat{f}_{n,m} = (255 - \hat{g}_{n,m}). \tag{7.7}$$

The above examples show that the α-rooting by the 2-D QDFT is an effective method for enhancing color images in the given RGB model of color images. The new enhancement measure *EMEC* allows for automatically

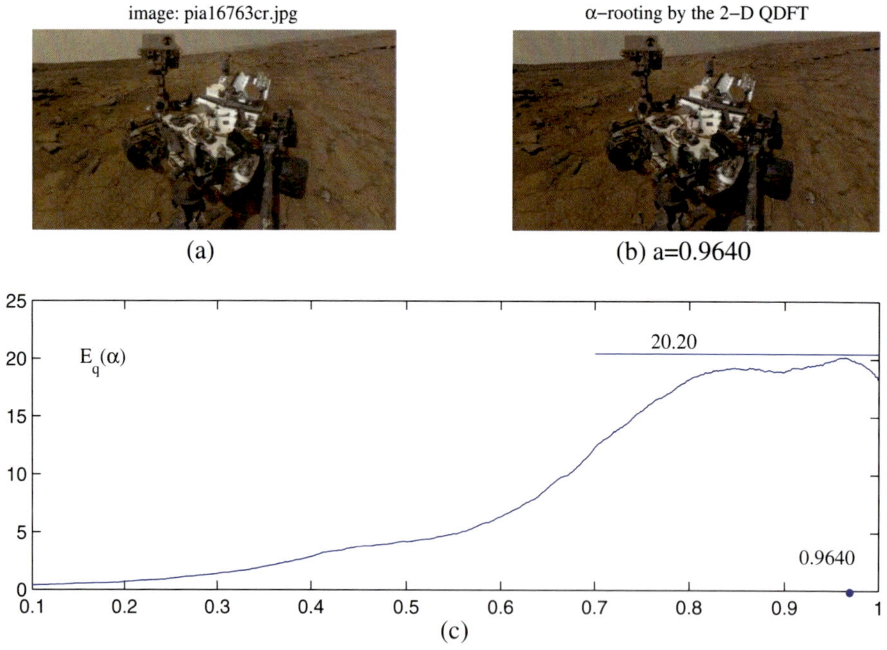

Figure 7.19 (a) The 466 × 248 × 3-point color "roller" image, (b) 0.964-rooting of the image by the 2-D QDFT, and (c) the graph of the function $E_q(\alpha)$.

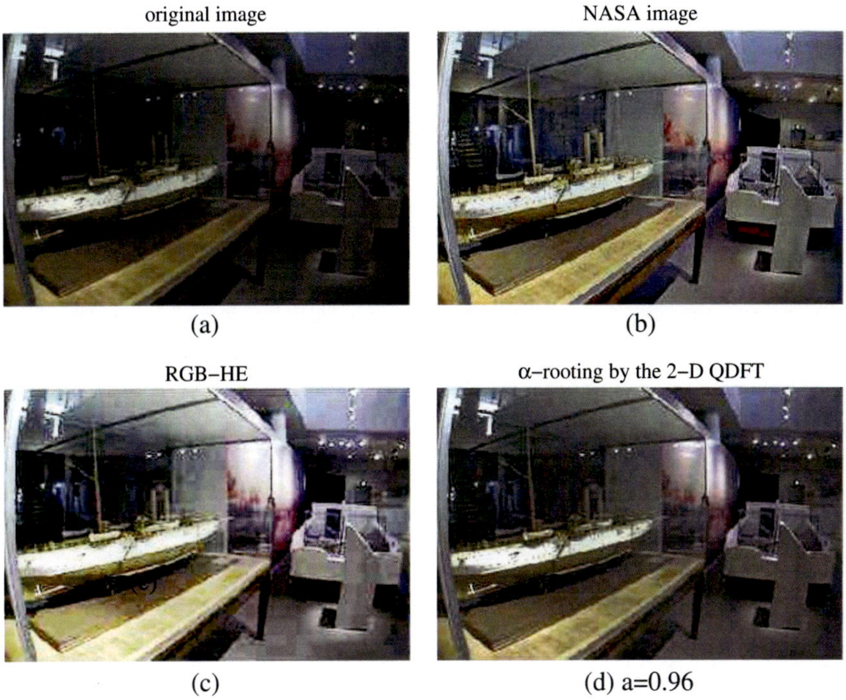

Figure 7.20 (a) The "room" image, (b) the retinex enhancement, (c) the three channel histogram equalization, and (d) the 0.96-rooting by the 2-D QDFT.

selection of values of α in image enhancement. The new methods are well-suited for color image processing applications because all three color components (R,G,B) are processed simultaneously.

References

[1] S. S. Agaian, K. Panetta, and A. M. Grigoryan, "A new measure of image enhancement," *Proc. IASTED Int. Conf. Signal Processing & Communication*, Marbella, Spain, 19–22 (2000).

[2] S. S. Agaian, K. Panetta, and A. M. Grigoryan, "Transform-based image enhancement algorithms," *IEEE Trans. on Image Processing*, **10**(3), 367–382 (2001).

[3] A. M. Grigoryan and S. S. Agaian, *Multidimensional Discrete Unitary Transforms: Representation, Partitioning and Algorithms*, Marcel Dekker Inc., New York (2003).

[4] A. M. Grigoryan and S. S. Agaian, "Transform-based image enhancement algorithms with performance measure," *Advances in Imaging and Electron Physics, Academic Press*, **130**, 165–242 (2004).

[5] S. S. Agaian, B. Silver, and K. A. Panetta, "Transform coefficient histogram-based image enhancement algorithms using contrast entropy," *Image Processing, IEEE Transactions on*, **16**(3), 741–758 (2007).

[6] A. M. Grigoryan and M. M. Grigoryan, *Brief Notes in Advanced DSP: Fourier Analysis With MATLAB*, CRC Press, Taylor and Francis Group, Boca Raton, New York (2009).

[7] W. M. Morrow, R. B. Paranjape, R. M. Rangayyan, and J. E. L. Desautels, "Region-based contrast enhancement of mammograms," *IEEE Trans. on Medical Imaging*, **11**(3), 392–406 (1992).

[8] T. L. Ji, M. K. Sundareshan, and H. Roehrig, "Adaptive image contrast enhancement based on human visual properties," *IEEE Trans. on Medical Imaging*, **13**(4), 573–586 (1994).

[9] P. Zamperoni, "Image enhancement," *Advanced in Image and Electron Physics*, **92**, 1–77, 1995.

[10] J. H. McClellan, "Artifacts in alpha-rooting of images," *Proc. IEEE Int. Conf. Acoustics, Speech, and Signal Processing*, 449–452 (1980).

[11] Y. Zhou, K. Panetta, S. Agaian, and H. Jia, "Nonlinear unsharp masking for mammogram enhancement," *Information Technology in Biomedicine, IEEE Trans. on*, **15**(6), 918–928 (2011).

[12] E. A. Silva, K. A. Panetta, and S. S. Agaian, "Quantify similarity with measurement of enhancement by entropy," *Proc. SPIE*, **6579**, 65790U (2007) [doi: 10.1117/12.720087].

[13] S. Hecht, "The visual discrimination of intensity and the Weber-Fechner law," *The Journal of General Physiology*, 235–267 (1924).

[14] R. C. Gonzalez and R. E. Woods, *Digital Image Processing*, 2nd Edition, Prentice Hall, Upper Saddle River, New Jersey (2002).

[15] W. K. Pratt, *Digital Image Processing*, 3rd Edition, John Wiley & Sons, Inc. (2001).

[16] D.-Y. Tsai, Y. Lee, M. Sekiya, S. Sakaguchi, and I. Yamada, "A method of medical image enhancement using wavelet analysis," *Proc. ICSP'02, IEEE*, 723–726 (2002).

[17] S. DelMarco and S. Agaian, "The design of wavelets for image enhancement and target detection," *Proc. SPIE*, **7351**, 735103 (2009) [doi: 10.1117/12.816135].

[18] A. Grigoryan and S. Agaian, "Tensor form of image representation: enhancement by image-signals," *Proc. SPIE*, **5014** (2003) [doi: 10.1117/12.473063].

[19] F. T. Arslan and A. M. Grigoryan, "Fast splitting α-rooting method of image enhancement: Tensor representation," *IEEE Trans. on Image Processing*, **15**(11), 3375–3384 (2006).

[20] A. M. Grigoryan and N. Du, "2-D images in frequency-time representation: Direction images and resolution map," *Journal of Electronic Imaging*, **19**(3), 033012 (2010) [doi: 10.1117/1.3483906].

[21] W. R. Hamilton, *Elements of Quaternions*, Logmans, Green and Co., London (1866).
[22] T. A. Ell, "Quaternion-Fourier transforms for analysis of 2-dimensional linear time-invariant partial differential systems," *Proc. the 3nd IEEE Conference on Decision and Control*, **1**(4), 1830–1841, San Antonio, Texas, (1993).
[23] S. J. Sangwine, "Fourier transforms of colour images using quaternion, or hypercomplex, numbers," *Electronics Letters*, **32**(21), 1979–1980 (1996).
[24] S. J. Sangwine, "The discrete quaternion-Fourier transform," *IPA97*, Conference Publication, (443), 790–793 (1997).
[25] S. J. Sangwine, T. A. Ell, J. M. Blackledge, and M. J. Turner, "The discrete Fourier transform of a color image," *Proc. Image Processing II Mathematical Methods, Algorithms and Applications*, 430–441 (2000).
[26] M. H. Yeh, "Relationships among various 2-D quaternion Fourier transforms," *IEEE Signal Processing Letters*, **15**, 669–672 (2008).
[27] S. J. Sangwine and T. A. Ell, "Hypercomplex Fourier transforms of color images," *in Proc. IEEE Intl. Conf. Image Process.*, **1**, 137–140 (2001).
[28] A. M. Grigoryan and S. S. Agaian, "Split manageable efficient algorithm for Fourier and Hadamard transforms," *IEEE Trans. Signal Processing*, **48**(1), 172–183 (2000).
[29] A. M. Grigoryan, "2-D and 1-D multi-paired transforms: Frequency-time type wavelets," *IEEE Trans. on Signal Processing*, **49**(2), 344–353 (2001).
[30] E. Wharton, S. Agaian, and K. Panetta, "A logarithmic measure of image enhancement," *Proc. SPIE*, **6250**, 62500P (2006) [doi: 10.1117/12.665693].
[31] E. Wharton, S. Agaian, and K. Panetta, "Comparative study of logarithmic enhancement algorithms with performance measure," *Proc. Electronic Imaging* (2006).
[32] E Wharton, K Panetta, and S. Agaian, "Human visual system based similarity," *Proc. Systems, Man and Cybernetics, SMC 2008. IEEE International Conference on*, 685–690 (2008).
[33] E. Wharton, K. Panetta, and S. Agaian, "Human visual system based similarity metrics," *IEEE System, Man and Cybernetics Conference*, Singapore (2008).
[34] S. Agaian, "Visual morphology," *Proc. SPIE*, **3646** (1999) [doi:10.1117/12.341081].
[35] K. A. Panetta, C. Gao, and S. S. Agaian, "No reference color image contrast and quality measures," *IEEE Transactions on Consumer Electronics*, **59**(3), 643–651 (2013).
[36] A. M. Grigoryan and S. S. Agaian, "Retolling of color imaging in the quaternion algebra," *Applied Mathematics and Sciences: An International Journal (MathSJ)*, **1**(3), 23–39 (2014).

[37] A. M. Grigoryan and S. S. Agaian, "Alpha-rooting method of color image enhancement by discrete quaternion Fourier transform," *Proc. SPIE*, **9019**, 901904 (2014) [doi: 10.1117/12.2040596].

[38] A. M. Grigoryan, J. Jenkinson, and S. S. Agaian, "Quaternion Fourier transform based alpha-rooting method for color image measurement and enhancement," *SIGPRO-D-14-01083R1, Signal Processing*, **109**, 269–289 (2015) [doi: 10.1016/j.sigpro.2014.11.019].

[39] A. M. Grigoryan and S. S. Agaian, "Color Enhancement and Correction for Camera Cell Phone Medical Images Using Quaternion Tools," Chapter 4, 77–117. In book: Electronic Imaging Applications in Mobile Healthcare, (J. Tang, S.S. Agaian and J. Tan, Eds.,) *SPIE Press*, Bellingham, Washington (2016) [doi: 10.1117/3.2204748.ch4].

[40] A. M. Grigoryan and S. S. Agaian, "Image Processing Contrast Enhancement," 22p, Wiley Encyclopedia of Electrical and Electronics Engineering, John Wiley & Sons, Inc. (2017).

Chapter 8
Gradients, Face Recognition, Visualization, and Quaternions

It is well known that the human visual system is very sensitive to gradients. An image gradient is a directional change in the intensity or color in the image. Image gradients have been used to create new images with "visualized edges." Moreover, image gradients are very important (or they are central building blocks) in many graphics, computer vision, robotics, imaging systems and visual surveillance applications. For instance, gradients can be used for extracting the useful information (structure, feature, and properties of objects) from images.

In this chapter, we cover the commonly used several gradient operators and introduce the human visual system based several new gradient operators for grayscale images, as well as for color images. New gradient operators in quaternion space, wherein the color images can be transferred, are also described and examples are given. The concepts of Weber-Fisher law-related visibility operators and images are presented and their applications in face recognition are described in Refs. 40 and 44. The extraction of features in the facial image representation is an important task in face recognition systems. Since the most information of the objects is in the contours of objects, different types of gradients are used to extract the edges, contours, and texture of objects. In face images, the important features of the size, shape, and face orientation are in points of the lines of eyes, nose, lips, and cheekbones. We also consider the application of visibility images related to the Michelson contrast, EME measure in face detection, and facial image representation [13, 28], [34]–[46]. The grayscale facial image representation with the local binary patterns (LBP) is described and examples are given. The application of color image gradients and visibility images in extracting features for color face recognition is also presented. For that, the quaternion operation of the 2-D convolution of the quaternion image with the quaternion kernel is described and applied for quaternion gradient operations.

8.1 Correlation Function and Gradients

The correlation is an important concept which is used when comparing images, as well for finding a target object in the image. The *correlation* (or, *cross-correlation*) of two images $f(x, y)$ and $g(x, y)$ is defined by

$$corr_{fg}(x,y) = \int_{-\infty}^{-\infty}\int_{-\infty}^{-\infty} f(\xi - x, \eta - y)g(\xi, \eta)d\xi d\eta.$$

In the discrete case of images $f_{n,m}$ and $g_{n,m}$, the correlation is defined by

$$corr_{fg}(n,m) = \sum_k \sum_l f_{k-n,l-m}g_{k,l}.$$

If we denote $\bar{f}(x,y) = f(-x, -y)$, then the correlation can be calculated by the convolution as

$$corr_{fg}(x,y) = (\bar{f} * g)(x,y).$$

When $f = g$, the correlation function is called *the autocorrelation* of the image $f(x, y)$. The autocorrelation is real $cor_{ff}(x,y) \geq 0$, and its maximum is achieved at point (0,0), i.e.,

$$cor_{ff}(0,0) \geq cor_{ff}(x,y), \quad \forall(x,y).$$

Therefore, the correlation function also can be considered in the normalized form

$$R_{fg}(x,y) = \frac{cor_{fg}(x,y)}{\sqrt{cor_{ff}(0,0)}\sqrt{cor_{gg}(0,0)}},$$

which yields $0 \leq R_{fg}(x,y) \leq 1$. The images f and g are correlated if the correlation function $R_{fg}(x,y)$ has values close to 1.

In discrete imaging, if $f_{n,m}$ is an object to be found in the image $g_{n,m}$, instead of $f_{n,m}$ its contour image can be used. As an example, we consider the known task of finding the face-target in the dollar bill image $g_{n,m}$. Figure 8.1 shows the target image in part (a), together with its rotated contour image in part (b), which is used when calculating the correlation by the convolution.

The correlation function of these images has the highest value $R_{fg}(n,m)$ at the point $(n,m) = (189,463)$ and the next three maximums at the neighbor points $(701,447), (701,448)$, and (702, 447). The result of processing the image $g_{n,m}$ after calculating the maximums of the correlation function is shown in Fig. 8.2.

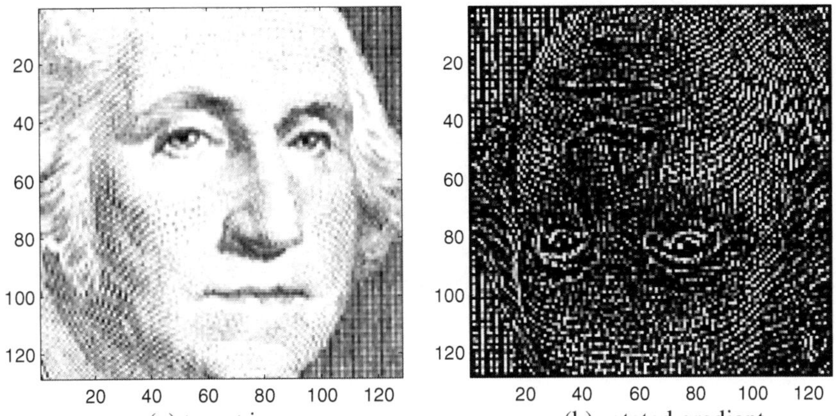

Figure 8.1 (a) The image $f_{n,m}$ and (b) the contour image of the rotation $\bar{f}_{n,m}$.

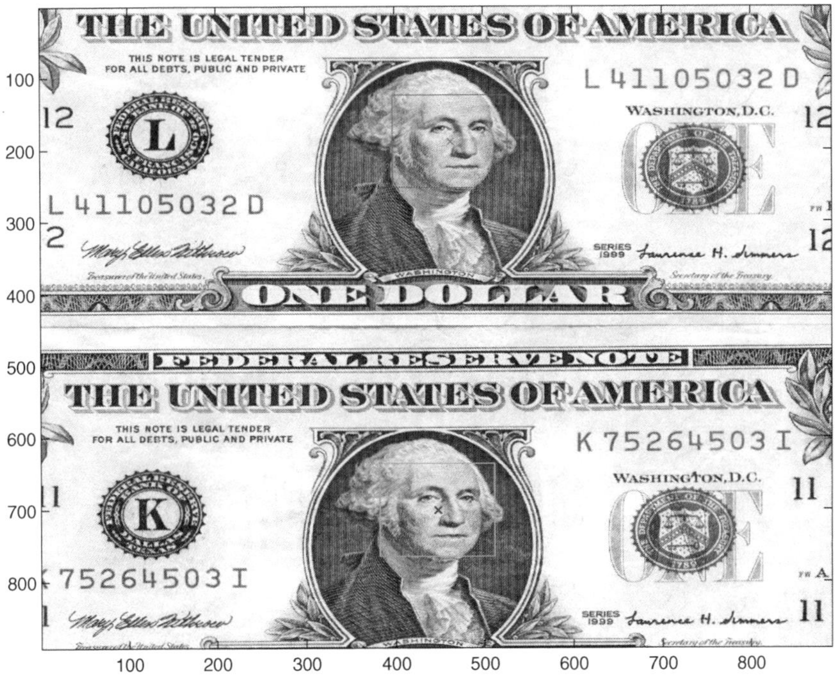

Figure 8.2 The image of two dollars with two marked rectangles at centers of maximums of the correlation function.

The selection of the gradient operators for extracting the edges (or contours) in the face image is a special topic in digital image processing. In the next section, we consider a few gradient operators, which are used in imaging.

8.2 Laplacian Gradient

To detect, extract, and enhance the edges of objects in the image $f(x, y)$, different differential operators of the first and second orders can be used. The points of edges and contours can be defined as points at which jumps of image intensity or the derivatives of intensity are observed. Many gradient operators allow for calculating the angle or orientation of lines and edges of objects in the image.

The first order partial derivatives with respect to x and y are described as

$$G_x = (G_x f)(x, y) = \frac{\partial f(x, y)}{\partial x} \quad \text{and} \quad G_y = (G_y f)(x, y) = \frac{\partial f(x, y)}{\partial y}. \tag{8.1}$$

These operators are directed, because they depend on the edge orientation. If for instance, the point (x, y) is on the line $x \sin\vartheta - y \cos\vartheta - t = 0$, the direction, or angle ϑ of the line can be calculated by Ref. [19]

$$\vartheta = -\arctan(G_x/G_y),$$

or, we can write $\vartheta + \pi/2 = \arctan(G_y/G_x)$. Thus, the ratio of these derivatives shows the direction that is orthogonal to the line of the image. The vector $(\partial f/\partial x, \partial f/\partial y)$ is called *a gradient* of the intensity.

Calculating the square of the gradient, we obtain a non-linear operator

$$G^2 f = \left[\frac{\partial f(x, y)}{\partial x}\right]^2 + \left[\frac{\partial f(x, y)}{\partial y}\right]^2. \tag{8.2}$$

The Laplacian of the image is defined as

$$\nabla^2 f = \frac{\partial^2 f(x, y)}{\partial^2 x} + \frac{\partial^2 f(x, y)}{\partial^2 y} \tag{8.3}$$

and ∇^2 is the linear operator. Both operators G^2 and ∇^2 are isotropic operators, i.e., they are rotation invariant operators. The result of application of an isotropic operator to the rotated image is the same as we apply the rotated operator to the original image.

As is mentioned by Marr and Hildreth [21], to better define the points on edges and make edge detection less sensitive to noise points, it is recommended first to convolute (or smooth) the image with the two-dimension Gaussian function

$$g(x, y) = \frac{1}{2\pi\sigma^2} \exp\left[-\frac{1}{2\pi\sigma^2}(x^2 + y^2)\right],$$

and then to use the Laplacian gradient. The value of the variance σ^2 can be selected, because the convolution removes all small elements in the image, which have sizes smaller than $\sqrt{\sigma^2}$.

8.2.1 Discrete gradient operators

The gradient operation is the 2-D discrete linear convolution with the kernel, or mask $h_{k,l}$ of a small size $(2L_1 + 1) \times (2L_2 + 1)$,

$$G(f)_{n,m} = \sum_{k=-L_1}^{L_1} \sum_{l=-L_2}^{L_2} f_{n+k,m+l} h(k,l), \tag{8.4}$$

where L_1 and L_2 integers ≥ 1.

The simplest discrete analog of the gradients along the X and Y directions are defined with the masks (the centers are underlined)

$$G_x = [-\underline{1} \quad 1] \text{ and } G_y = \begin{bmatrix} 1 \\ -\underline{1} \end{bmatrix}. \tag{8.5}$$

These gradients images are calculated by $(G_x f)_{n,m} = (f_{n,m+1} - f_{n,m})$ and $(G_y f)_{n,m} = (f_{n-1,m} - f_{n,m})$. In this case, we consider that $h_{0,0} = -1$ and $h_{0,1} = 1$, for the 2×1 kernel of the gradient G_x. For the 1×2 kernel of the gradient G_y, $h_{0,0} = -1$ and $h_{-1,0} = 1$.

The discrete analogs of the gradients along the X and Y directions also are the operators with the following masks:

$$G_x = \begin{bmatrix} -1 & +1 \\ -\underline{1} & +1 \end{bmatrix} \text{ and } G_y = \begin{bmatrix} +1 & +1 \\ -\underline{1} & -1 \end{bmatrix}. \tag{8.6}$$

Thus, these gradients images are calculated by

$$\begin{aligned} (G_x f)_{n,m} &= (f_{n-1,m+1} - f_{n-1,m} + f_{n,m+1} - f_{n,m}), \\ (G_y f)_{n,m} &= (f_{n-1,m} - f_{n,m} + f_{n-1,m+1} - f_{n,m+1}). \end{aligned} \tag{8.7}$$

Therefore, the square of the gradient image $G^2_{n,m}$ is calculated by

$$\begin{aligned} (G^2 f)_{n,m} &= (G_x f)^2_{n,m} + (G_y f)^2_{n,m} \\ &= 2\left[(f_{n-1,m+1} - f_{n,m})^2 + (f_{n-1,m} - f_{n,m+1})^2\right]. \end{aligned}$$

As an example, Fig. 8.3 shows the gradient images $G_x(f)$ in part (a), $G_y(f)$ in part (b), the square $(G^2_x + G^2_y)(f)$ in part (c), and the Laplacian $\nabla^2 f$ in part (d).

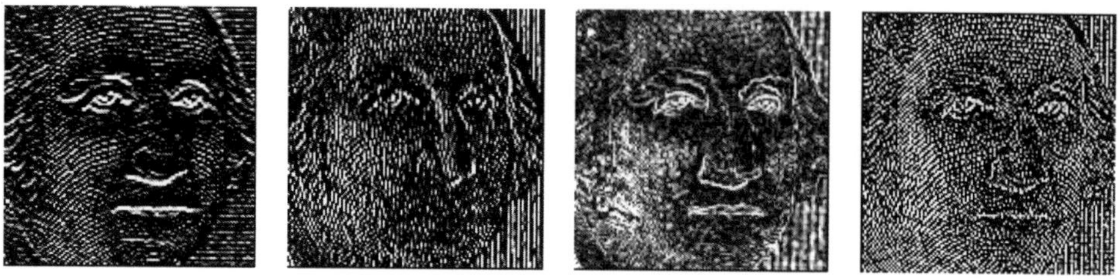

(a) G_x (b) G_y (c) $G_x^2+G_y^2$ (d) Laplacian

Figure 8.3 Images of the gradient operators (a) G_x, (b) G_y, (c) $(G_x^2+G_y^2)$, and (d) ∇^2.

Among other gradient operators, we can mention the Sobel operators, which for the 3×3 window, are defined by the following masks:

$$G_x = \begin{bmatrix} -1 & 0 & +1 \\ -2 & \underline{0} & +2 \\ -1 & 0 & +1 \end{bmatrix} \quad \text{and} \quad G_y = \begin{bmatrix} +1 & +2 & +1 \\ 0 & \underline{0} & 0 \\ -1 & -2 & -1 \end{bmatrix}. \tag{8.8}$$

These masks are convolution kernels in calculation of the second order gradients,

$$\begin{aligned} (G_x f)_{n,m} &= (f_{n-1,m+1} + 2f_{n,m+1} + f_{n+1,m+1}) \\ &\quad - (f_{n-1,m-1} + 2f_{n,m-1} + f_{n+1,m-1}), \\ (G_y f)_{n,m} &= (f_{n-1,m-1} + 2f_{n-1,m} + f_{n-1,m+1}) \\ &\quad - (f_{n+1,m-1} + 2f_{n+1,m} + f_{n+1,m+1}). \end{aligned} \tag{8.9}$$

Figure 8.4 shows the images that correspond to the Sobel gradient operators.

The absolute gradient operator is defined as the image

$$|G(f)_{n,m}| = \sqrt{(G_x f)_{n,m}^2 + (G_y f)_{n,m}^2}$$

or, it can be estimated by $|G(f)_{n,m}| = |(G_x f)_{n,m}| + |(G_y f)_{n,m}|$.

Figure 8.5 shows the images that correspond to the Sobel gradient operators for the "man" image.

The concept of the Laplacian operator in digital imaging is based on the window and coefficients in the mask for the window. If we use the 3×3 cross window, when the neighbors by the horizontal and vertical direction are considered, the following images are calculated:

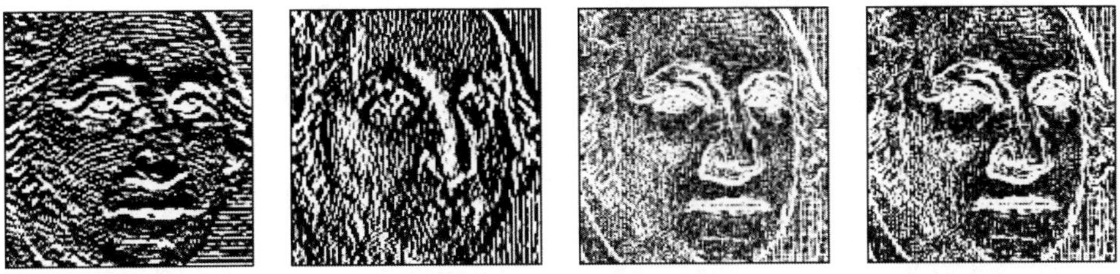

Figure 8.4 Images of the Sobel gradient operators (a) G_x, (b) G_y, (c) $|G_x| + |G_y|$, and (d) $G_x^2 + G_y^2$.

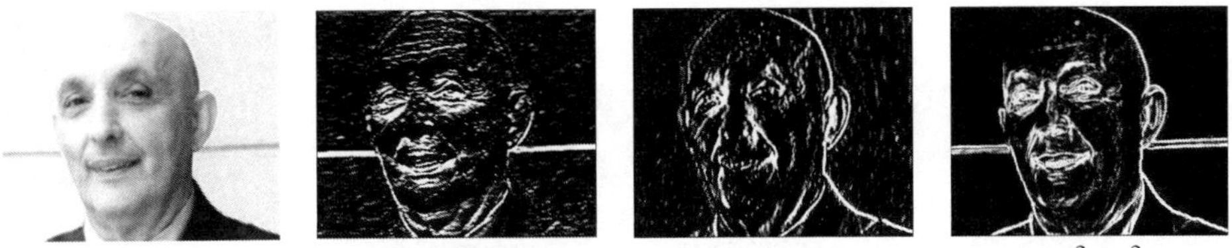

Figure 8.5 The "man" image (a) before and after processing by the Sobel gradient operators (b) G_x, (c) G_y, and (d) $G_x^2 + G_y^2$.

$$\frac{\partial^2 f(x,y)}{\partial^2 x} \to (\nabla_x^2 f)_{n,m} = f_{n,m-1} - 2f_{n,m} + f_{n,m+1},$$
$$\frac{\partial^2 f(x,y)}{\partial^2 y} \to (\nabla_y^2 f)_{n,m} = f_{n-1,m} - 2f_{n,m} + f_{n+1,m}.$$

Therefore, the Laplacian image is calculated by

$$\begin{aligned}(\nabla^2 f)_{n,m} &= (\nabla_x^2 f)_{n,m} + (\nabla_y^2 f)_{n,m} \\ &= \left[f_{n,m-1} + f_{n-1,m} + f_{n,m+1} + f_{n+1,m} - 4f_{n,m}\right].\end{aligned} \tag{8.10}$$

For this Laplacian operator, we use the operation of the 2-D discrete linear convolution with the 3×3 cross window,

$$(\nabla^2 f)_{n,m} = \sum_{k=-1}^{1}\sum_{l=-1}^{1} f_{n+k,m+l} h_+(k,l), \quad h_+ = \frac{1}{4}\begin{bmatrix} 0 & 1 & 0 \\ 1 & \underline{\mathbf{-4}} & 1 \\ 0 & 1 & 0 \end{bmatrix}. \tag{8.11}$$

The Laplacian of the image can also be calculated (namely, estimated) by using the following masks with the *X* and square-windows:

$$h_{\times} = \frac{1}{4}\begin{bmatrix} 1 & 0 & 1 \\ 0 & \underline{\mathbf{-4}} & 0 \\ 1 & 0 & 1 \end{bmatrix} \text{ and } h_s = \frac{1}{20}\begin{bmatrix} 1 & 4 & 1 \\ 4 & \underline{\mathbf{-20}} & 4 \\ 1 & 4 & 1 \end{bmatrix}.$$

Figure 8.6 shows the images that correspond to the Laplacian gradient operators calculated for the above three 3×3 windows.

The Laplacian operator h_s is an eight-neighbor operator. We can also mention the following eight-neighbor Laplacian operators [20]:

$$\frac{1}{8}\begin{bmatrix} 1 & 1 & 1 \\ 1 & \underline{\mathbf{-8}} & 1 \\ 1 & 1 & 1 \end{bmatrix} \text{ and } \frac{1}{8}\begin{bmatrix} 2 & -1 & 2 \\ -1 & \underline{\mathbf{-4}} & -1 \\ 2 & -1 & 2 \end{bmatrix}.$$

8.3 Weber-Fechner Visibility Images

To measure the quality of images and select optimal processing parameters, the EME measure of image enhancement was used in Section 7.2. This measure can be compared with the Weber-Fechner law of the human visual system. Weber stated that the smallest noticeable change in the image intensity, ΔI, is proportional to the original intensity I (before adding the difference), i.e.,

$$\frac{|\Delta I|}{I} = w \approx \text{const.} \tag{8.12}$$

Here w is a number that is called the Weber fraction; it is a constant within narrow limits of intensities. The value of this constant is estimated experimentally; it is assumed that $w \approx 0.015$ [11].

Fechner considered the lower value I_0 of intensity of the retina in the human eye and proposed to use the above law as [12]

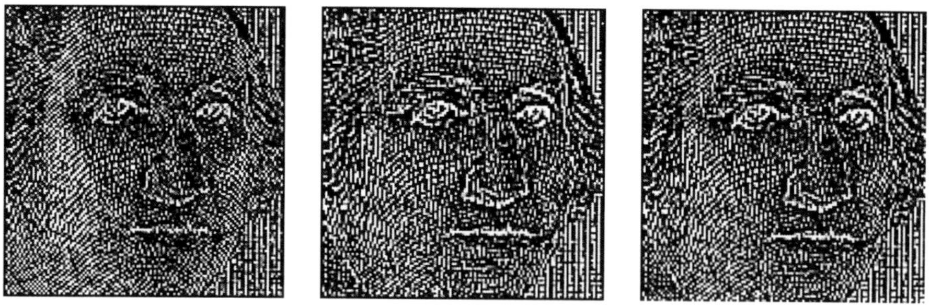

(a) 3×3 W_+ (b) 3×3 W_X (c) 3×3 W_s

Figure 8.6 The Laplacian gradient images calculated for (a) the cross window, (b) the *X* window, and (c) the square window.

$$k\frac{|\Delta I|}{I_0 + I} = w, \tag{8.13}$$

where k is a constant.

To describe this statement in more detail, we consider the grayscale image $f_{n,m}$ of size $N \times M$. The increment of intensities can be described as a result of application of one of the windowed gradient operations which are used in digital image processing, such as the Sobel, Prewitt, Frei-Chen operators [20], or by using the mean operation as

$$\Delta f_{n,m} = f_{n,m} - \text{Mean}_W(f_{n,m}), \tag{8.14}$$

where W is the window of small size, for instance 3×3 or 5×5. In general, the mean operation $\text{Mean}_W(f_{n,m})$ provides the "background" information.

As examples, we can consider the following operators:

$$\text{Mean}_{W_+}(f_{n,m}) = \frac{1}{4}(f_{n-1,m} + f_{n+1,m} + f_{n,m-1} + f_{n,m+1}) \tag{8.15}$$

and

$$\text{Mean}_{W_X}(f_{n,m}) = \frac{1}{4}(f_{n-1,m-1} + f_{n-1,m+1} + f_{n+1,m-1} + f_{n+1,m+1}). \tag{8.16}$$

Also, the eight neighbors of the pixel (n, m) can be considered with the mean operation

$$\text{Mean}_{W_S}(f_{n,m}) = \frac{1}{2}\left[\text{Mean}_{W_+}(f_{n,m}) + \text{Mean}_{W_X}(f_{n,m})\right]. \tag{8.17}$$

These three operators correspond to the 3×3 cross, X, and square-windows

$$W_+ = \begin{bmatrix} 0 & 1 & 0 \\ 1 & 0 & 1 \\ 0 & 1 & 0 \end{bmatrix}, \quad W_X = \begin{bmatrix} 1 & 0 & 1 \\ 0 & 0 & 0 \\ 1 & 0 & 1 \end{bmatrix}, \quad \text{and} \quad W_S = \begin{bmatrix} 1 & 1 & 1 \\ 1 & 0 & 1 \\ 1 & 1 & 1 \end{bmatrix}.$$

These windows are shown in Fig. 8.7 together with the horizontal and vertical bars with the centers marked by the bullets. Other windows of a size greater than 3×3 can also be used in the mean operator.

The Weber law related coefficients at each pixel (n, m) can be written as

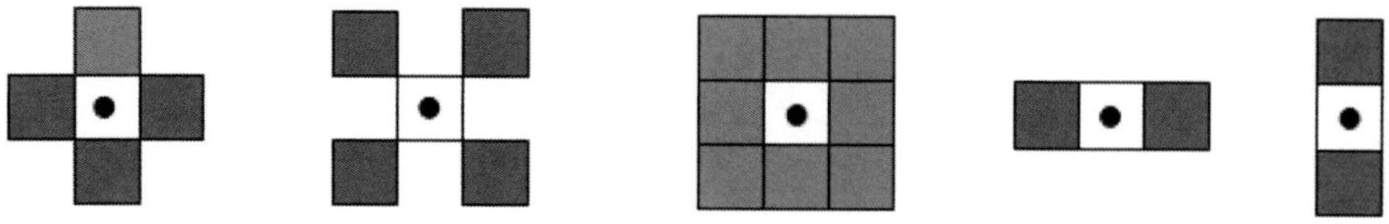

Figure 8.7 Geometrical figures for the symmetric windows: cross, X, square, and two bars.

$$w(n,m) = \frac{|f_{n,m} - \text{Mean}_W(f_{n,m})|}{f_{n,m}}. \tag{8.18}$$

Since the image may have zero values, these coefficients can be calculated by

$$w(n,m) = \frac{|f_{n,m} - \text{Mean}_W(f_{n,m})|}{f_0 + f_{n,m}}, \tag{8.19}$$

where $n = 0:(N-1)$, $m = 0:(M-1)$, and f_0 is a small number, for instance $f_0 = 0.01$ or 0.001. With a scaling coefficient k, these data can be considered as the image

$$w(f)_{n,m} = k\frac{|f_{n,m} - \text{Mean}_W(f_{n,m})|}{f_0 + f_{n,m}}. \tag{8.20}$$

It is assumed that $w(f)_{n,m} = 0$ if $f_{n,m} = 0$.

Thus, the Weber visibility operator is defined as

$$w : f_{n,m} \rightarrow w(f)_{n,m} = k\frac{|f_{n,m} - \text{Mean}_W(f_{n,m})|}{f_0 + f_{n,m}}. \tag{8.21}$$

In calculation of differences in image intensity [Eq. (8.14)], one of the first order gradients is used. Therefore, we call this operator *the first order Weber visibility operator*. The image $w(f)_{n,m}$ is called *the first order Weber visibility image*, or simply *the Weber visibility image*. The second order gradients, including the Laplacian isotropic operator, can also be used when calculating the Weber visibility image. The visibility image is considered as a characteristic of the original image $f_{n,m}$, which describes small noticeable changes in the intensity. In general, the mean operator can also be modified with a parameter α, i.e., we can consider the operator

$$w : f_{n,m} \rightarrow w(f)_{n,m} = k\frac{|f_{n,m} - \alpha \text{Mean}_W(f_{n,m})|}{f_0 + f_{n,m}}, \tag{8.22}$$

where $\alpha > 0$ is not necessarily to be 1.

As an example, Fig. 8.8 shows the 512 × 512 grayscale image in part (a) and the Weber visibility image calculated by using the X-window in part (b).

Below is the script of the MATLAB®-based code "wfvi_image.m" to calculate the Weber visibility image by the 3 × 3 cross-window.

```
% ---------------------------------------------------------
%   call: wfvi_image.m / Art Grigoryan, 02-12-2014
%    Weber-Fechner visibility image calculation
% ---------------------------------------------------------
    Y=imread('peper.tiff','tiff');
    [N M L]=size(Y);                     % 512 x 512 x 3
    X=double(Y(:,:,2));                  % green channel
    h=[0 1 0; 1 0 1; 0 1 0];             % 3 x 3 cross-window
    h=h/sum(sum(h));
    % --- Weber visibility image (WVI)
    WVI=zeros(N,M);
    for n=2:N-1
        n1=n-1; n2=n+1;
        for m=2:M-1
            x1=X(n,m);
            if (x1>0)
                b1=X(n1:n2,m-1:m+1);
                a=sum(sum(b1.*h));
                WVI(n,m)=abs(x1-a)/(.001+x1);
                                              % f_0=0.001
            end
        end
    end
    figure;           colormap(gray(255)); % Fig. 8.8
    subplot(1,2,1); image(X);    axis image;
    subplot(1,2,2); image(400*WVI);    axis image;
% ---------------------------------------------------------
```

Figure 8.9 shows the Weber visibility images calculated with the X-window and square-window in parts (a) and (b), respectively.

The Weber visibility images can be processed in different scales, by using the known monotonic functions, including the logarithm, arctangent, square root, and the 3rt root. The sine function also can be used, since the values of the ratio $\Delta I/I$ are considered to vary from the small number 1/255 to

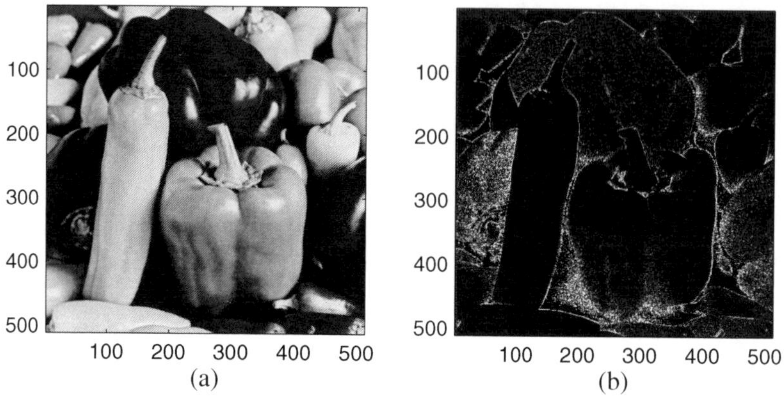

Figure 8.8 (a) The "peppers" image and (b) the Weber visibility image for the 3×3 cross-windowed mean operator.

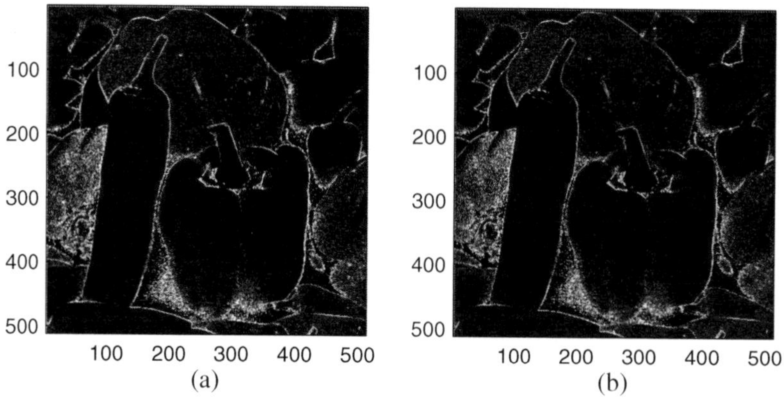

Figure 8.9 The Weber visibility "peppers" images which are calculated by using (a) the X-windowed and (b) the square-windowed mean operators.

maximum 1 in the absolute scale. Figure 8.10 shows four of these functions, and we can notice that the sine and arctangent functions do not differ much in the unit interval.

Therefore, when using the logarithm, sine, arctangent, and square root functions, the corresponding Weber visibility operators are defined as

$$w : f_{n,m} \rightarrow w(f)_{n,m} = k \ln\left[1 + \frac{|f_{n,m} - \alpha \mathrm{Mean}_W(f_{n,m})|}{f_0 + f_{n,m}}\right], \tag{8.23}$$

$$w : f_{n,m} \rightarrow w(f)_{n,m} = k \sin\left[1 + \frac{|f_{n,m} - \alpha \mathrm{Mean}_W(f_{n,m})|}{f_0 + f_{n,m}}\right], \tag{8.24}$$

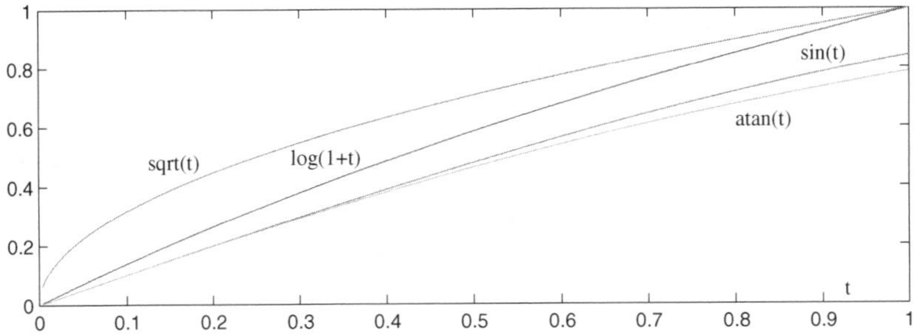

Figure 8.10 The square root, logarithm, sine, and arctangent functions.

$$w : f_{n,m} \to w(f)_{n,m} = k \arctan\left[\frac{|f_{n,m} - \alpha\text{Mean}_W(f_{n,m})|}{f_0 + f_{n,m}}\right], \tag{8.25}$$

and

$$w : f_{n,m} \to w(f)_{n,m} = k\sqrt{\frac{|f_{n,m} - \alpha\text{Mean}_W(f_{n,m})|}{f_0 + f_{n,m}}}. \tag{8.26}$$

As an example, we consider the "cameraman" image shown in Fig. 8.11 in part (a) and its Weber visibility image $w(n, m)$ in part (b), which was calculated by Eq. (8.20) with the 3×3 cross window W.

We call the images defined by Eqs. (8.23)–(8.26) the logarithm-Weber, sine-Weber, arctangent-Weber, and square-Weber visibility images, respectively.

The Weber visibility "cameraman" images with the logarithm, arctangent, and sine functions are shown in Fig. 8.12 in parts (a), (b), and (c), respectively.

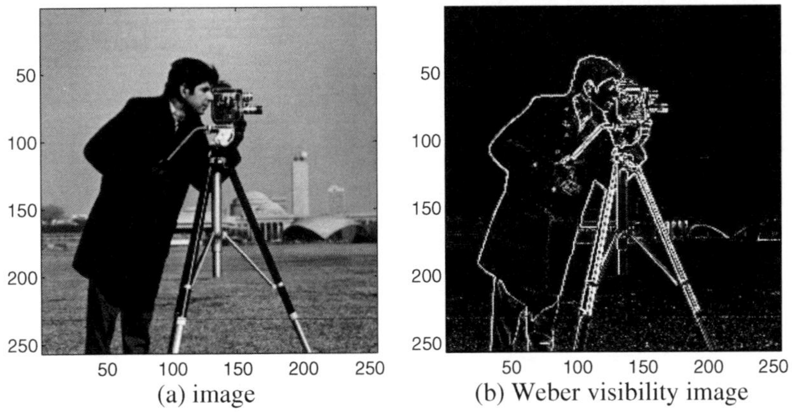

Figure 8.11 (a) The original and (b) the Weber visibility "cameraman" images.

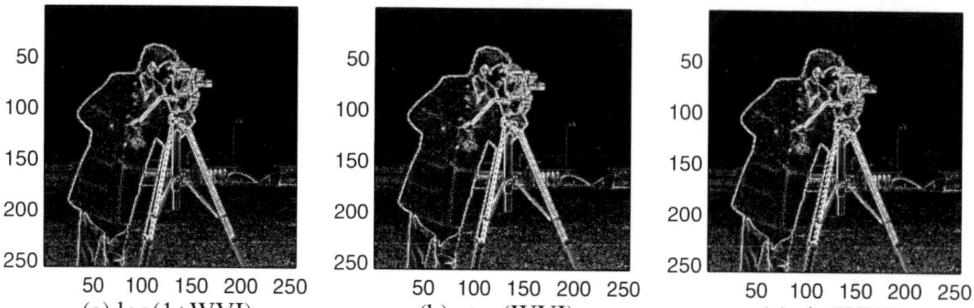

(a) log(1+WVI) (b) atan(WVI) (c) sin(WVI)

Figure 8.12 (a) The logarithm, (b) the arctangent, and (c) the sine Weber visibility "cameraman" images.

Now, we consider the polynomial Weber-law visibility images. In the definition of the Weber law related coefficients in Eq. (8.18), a difference $(J - I)/I$ is used as in the Taylor series of the logarithm function

$$\log(J) = \log(I) + \frac{J - I}{I} - \frac{1}{2}\left(\frac{J - I}{I}\right)^2 + \cdots .$$

Therefore, considering this series, we introduce the following modification of the Weber visibility image $w(n, m)$:

$$w_\alpha(f)_{n,m} = w(f)_{n,m} - \alpha\frac{1}{2}\left[w(f)_{n,m}\right]^2, \tag{8.27}$$

where α is the parameter with values from the interval [0, 1]. If $\alpha = 0$, the image $w_0(n, m) = w(n, m)$. Thus, the new modified Weber visibility operator is defined as

$$A : f_{n,m} \rightarrow A(f)_{n,m} = w_\alpha(f)_{n,m}. \tag{8.28}$$

As examples, Fig. 8.13 shows the polynomial Weber visibility "cameraman" images calculated for $\alpha = 0.05$, 0.075, and 0.125 in parts (a), (b), and (c), respectively.

Another modification of the Weber visibility image $w(n, m)$ was introduced by Grigoryan by using the sine function with a given frequency ω_0,

$$A(f)(n, m) = \sin\left[\omega_0 \cdot w(f)_{n,m}\right]. \tag{8.29}$$

Thus, the parametric sine-Weber visibility operator is defined as

$$A(\omega_0) : f_{n,m} \rightarrow A(f)_{n,m} = \sin\left\{\omega_0\left[\frac{|f_{n,m} - \text{Mean}_W(f_{n,m})|}{f_0 + f_{n,m}}\right]\right\}. \tag{8.30}$$

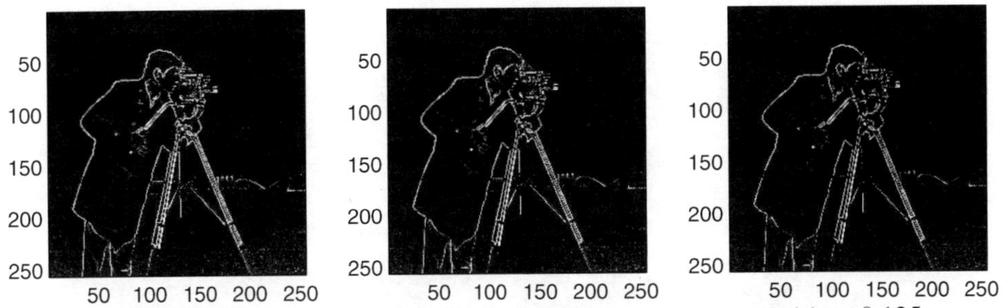

(a) α=0.05 (b) α=0.075 (c) α=0.125

Figure 8.13 The polynomial Weber visibility images for (a) $\alpha = 0.05$, (b) $\alpha = 0.075$, and (c) $\alpha = 0.125$.

As examples, Fig. 8.14 shows the sine-Weber visibility "cameraman" images calculated for the frequencies $\omega_0 = 0.75, 0.5,$ and 0.35 (rad/s) in parts (a), (b), and (c), respectively.

8.4 Image Visualization by the Michelson Contrast

The Michelson contrast definition is based on the ratio of difference of the maximum and minimum to their sum. This concept can be used to construct and analyze the Michelson visibility images in a way similar to the Weber-Fechner law related case. For the image $f_{n,m}$, the visibility image related to the Michelson definition of the contrast is defined in the neighbor W of the pixel (n, m) by the following operator [28]:

$$C : f_{n,m} \rightarrow C(f)_{n,m} = k\frac{\text{Max}_W(f_{n,m}) - \text{Min}_W(f_{n,m})}{\text{Max}_W(f_{n,m}) + \text{Min}_W(f_{n,m})}, \qquad (8.31)$$

where k is a constant. The image $C(f)_{n,m}$ is called *the Michelson visibility image* of $f_{n,m}$.

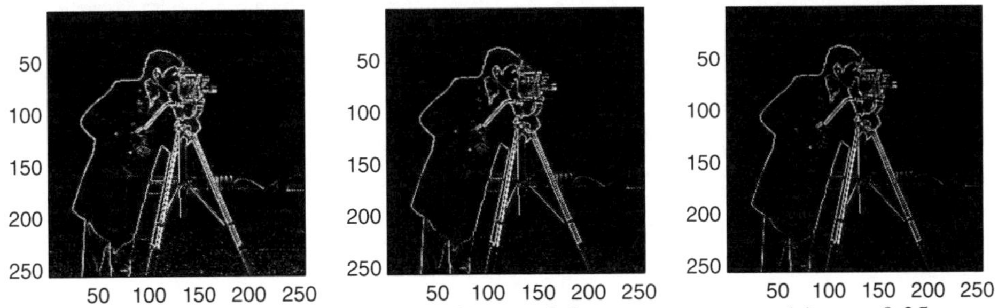

(a) $\omega_0 = 0.75$ (b) $\omega_0 = 0.5$ (c) $\omega_0 = 0.35$

Figure 8.14 The parametric sine-Weber visibility "cameramen" images for (a) $\omega_0 = 0.75$, (b) $\omega_0 = 0.5$, and (c) $\omega_0 = 0.35$.

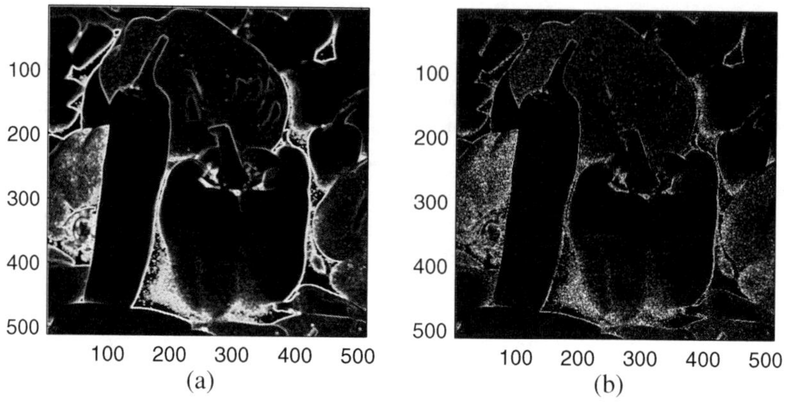

Figure 8.15 (a) The Michelson and (b) the Weber visibility "peppers" images.

As an example, Fig. 8.15 shows the Michelson visibility image for the "peppers" image in part (a). When comparing with the Weber visibility image [which is given in Fig. 8.15(b)], one can notice that $C(f)_{n,m}$ shows the smallest changes in image intensity.

The Michelson visibility operator has various modifications [13], which include the following four Agaian-DelMarco visibility operators:

$$
\begin{aligned}
C_2 : f_{n,m} &\to C_2(f)_{n,m} = k\frac{|\mathrm{Max}_W(f_{n,m}) - 2f_{n,m} + \mathrm{Min}_W(f_{n,m})|}{\mathrm{Max}_W(f_{n,m}) + 2f_{n,m} + \mathrm{Min}_W(f_{n,m})},\\
C_3 : f_{n,m} &\to C_3(f)_{n,m} = k\frac{\sqrt{[(\mathrm{Max}_W(f_{n,m}) - f_{n,m})(f_{n,m} - \mathrm{Min}_W(f_{n,m}))]}}{f_{n,m} + \mathrm{Min}_W(f_{n,m})},\\
C_4 : f_{n,m} &\to C_4(f)_{n,m} = k\frac{\sqrt{[(\mathrm{Max}_W(f_{n,m}) - f_{n,m})(f_{n,m} - \mathrm{Min}_W(f_{n,m}))]}}{\mathrm{Max}_W(f_{n,m}) + \mathrm{Min}_W(f_{n,m})},\\
C_5 : f_{n,m} &\to C_5(f)_{n,m} = k\frac{\sqrt{[(\mathrm{Max}_W(f_{n,m}) - f_{n,m})(f_{n,m} - \mathrm{Min}_W(f_{n,m}))]}}{[(\mathrm{Max}_W(f_{n,m}) + f_{n,m})(f_{n,m} + \mathrm{Min}_W(f_{n,m}))]}.
\end{aligned}
\tag{8.32}
$$

All these operators describe different second-order gradient operations. Therefore, these operators are called *the second order Michelson visibility operators*. The visibility images corresponding to the operators C_k, where $k = 2$, $3, 4$, and 5, are called the k-th Michelson visibility images.

For the "peppers" image, Fig. 8.16 shows the 2nd, 3rd, 4th, and 5th Michelson visibility images in parts (a), (b), (c), and (d), respectively.

The sine-Michelson visibility images for these operators, C_k, are calculated by the operators

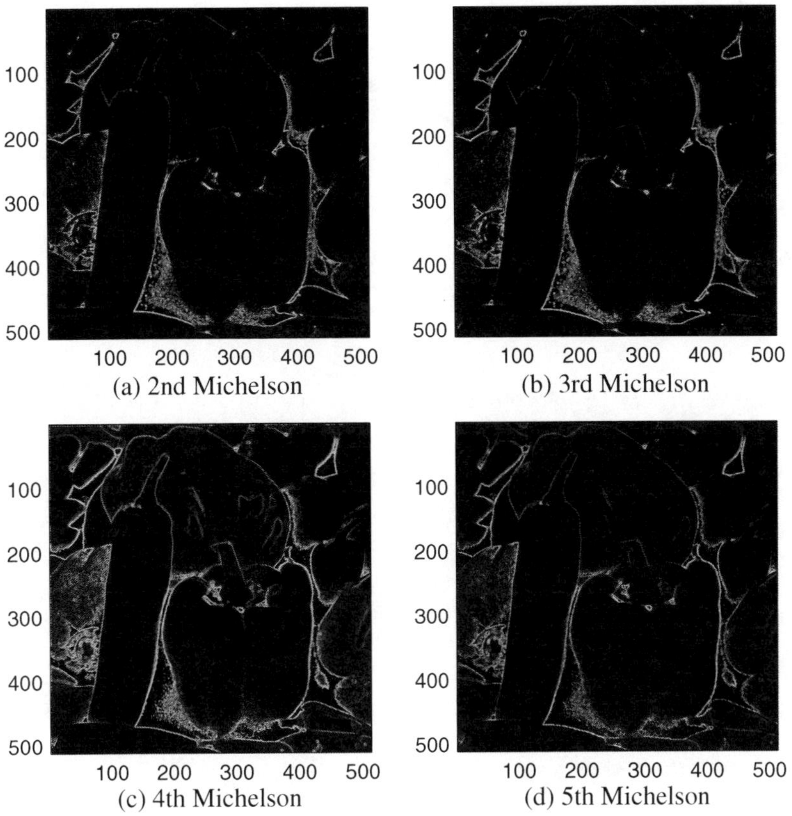

Figure 8.16 Michelson visibility "peppers" images of number (a) 2, (b) 3, (c) 4, and (d) 5.

$$A(\omega_0) : f_{n,m} \to A(f)_{n,m} = \sin\left\{\omega_0 [C_k(f)]_{n,m}\right\}, \quad k = 1 : 4. \tag{8.33}$$

8.5 EME-Type Measures and Visibility

The concept of contrast ratio is an important quality characteristic of the image, which relates to the ability of the human visual system to discern the image brightness and small details, edges, or contours in the image. Values of the image $f_{n,m}$ correspond to the brightness values, which usually are quantized in the integer interval [0, 255]. The concept of the contrast ratio is difficult to describe and define analytically. In projected image systems [47], the American National Standards Institute (ANSI) defines the contrast as "the dynamic video range of a display device as a numeric relationship between the brightest color (typically white) and the darkest color (typically black) that the system is capable of producing." For instance, the contrast ratio of 400:1 means that the brightest white point in the image is 400 times as bright as the black darkest point. A higher contrast ratio therefore means a larger difference in brightness. In any display system, including projection, the higher the number, the better it is. "On the other hand the higher the contrast is, the less perceivable difference an additional increase will make and there will come a point at which no continued increase (in absolute, objective terms) will be perceived by the viewer at all" [48]. The minimum acceptable contrast for normal observers in the ANSI standard is the 3:1 contrast ratio [49].

$EME(f)$ is the enhancement measure of the image $f_{n,m}$, which calculates the average range of intensity of the image in the logarithm scale. Thus, EME of the image is the estimated block-wise by using the logarithm range of the image

$$EME(f) = \text{Mean}\left\{\ln\left[\max_{k,l}(f)\right] - \ln\left[\min_{k,l}(f)\right]\right\}.$$

Here, the max and min operations are operations in the blocks, or in general, in a window W of these blocks, which may be of form other than the square. Therefore, we denote these operations by Max_W and Min_W.

The visibility image related to the EME measure is defined in the neighbor W of the pixel (n, m) by the following EME visibility operator:

$$E : f_{n,m} \to E(f)_{n,m} = k \ln\left[\frac{\text{Max}_W(f_{n,m})}{\text{Min}_W(f_{n,m})}\right], \tag{8.34}$$

where k is a constant. The window W plays the role of the $L_1 \times L_2$ block. If $\text{Min}_W(f_{n,m}) = 0$, the value $E(f)_{n,m} = 0$, or we can add a small number f_0 in the denumerator in the ratio. The image $E(f)_{n,m}$ is called *the EME visibility image* of $f_{n,m}$.

As an example, Fig. 8.17 shows the EME visibility "peppers" image in part (a), which was calculated for the square 3×3-window. The same image in the logarithm scale is shown in part (b), and the Michelson visibility image is given in part (c), for comparison. One more logarithm operation on the

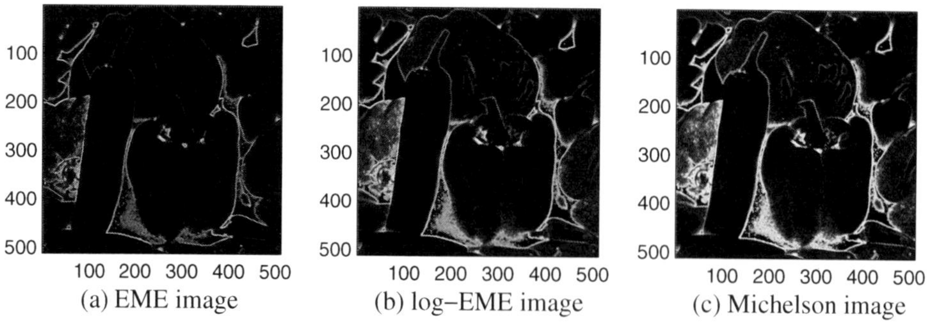

Figure 8.17 The EME measure visibility "peppers" images: (a) the original and (b) in the logarithm scale. (c) The Michelson visibility image.

$\ln[E(f)]$ results in an visibility image similar to the Michelson visibility image. The measure of image enhancement can therefore be used with the logarithm operation as follows:

$$LEME(f) = EME[E(f)] = \frac{1}{k_1 k_2} \sum_{k=1}^{k_1} \sum_{l=1}^{k_2} 20 \ln \left[\frac{\max_{k,l} E(f)}{\min_{k,l} E(f)} \right]. \tag{8.35}$$

For the 3×3 window, the EME measure can be calculated from its visibility image taken at the center-pixels of only not overlapping translated windows (or blocks). The visibility image at these pixels is shown in Fig. 8.18(a). This image is of size $(N/3) \times (M/3)$ after rounding, i.e., 170×170. The EME measure of the original image $f_{n,m}$ is the sum of all values of the visibility image, which equals 10.3076.

In a similar way, the image shown in Fig. 8.18(b) was calculated from the Michelson visibility "peppers" image and then the sum the image values was calculated; it is 3.8926. This calculation is related to the contrast measure, which is called *Michelson visibility measure* (MVM),

$$MVM(f) = \frac{1}{k_1 k_2} \sum_{k=1}^{k_1} \sum_{l=1}^{k_2} \left[\frac{\max_{k,l}(f) - \min_{k,l}(f)}{\max_{k,l}(f) + \min_{k,l}(f)} \right].$$

The Grigoryan-Agaian enhancement measures, which relate to the sine and arctangent visibility images defined in (8.24) and (8.25), are calculated in the following way:

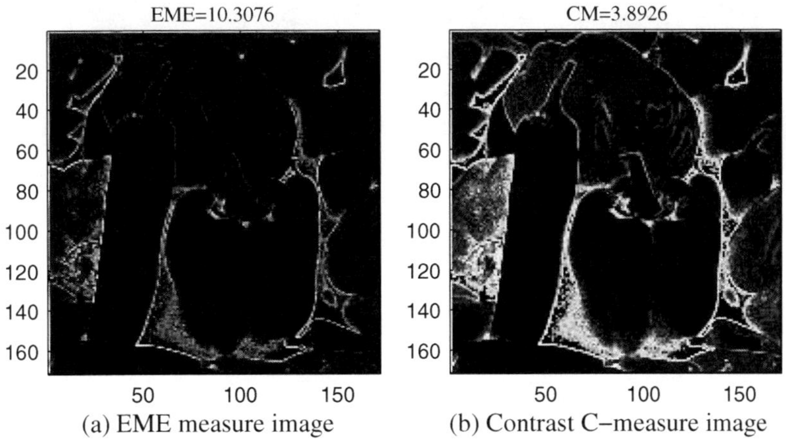

(a) EME measure image (b) Contrast C–measure image

Figure 8.18 The images of (a) the EME and (b) the CEM measures.

$$GAEM(f) = \frac{1}{k_1 k_2} \sum_{k=1}^{k_1} \sum_{l=1}^{k_2} \sin\left\{ \omega_0 \left[1 + \frac{|\text{center}_{k,l}(f) - \alpha \text{mean}_{k,l}(f)|}{f_0 + \text{center}_{k,l}(f)} \right] \right\} \tag{8.36}$$

and

$$GAEM(f) = \frac{1}{k_1 k_2} \sum_{k=1}^{k_1} \sum_{l=1}^{k_2} \arctan\left\{ 1 + \frac{|\text{center}_{k,l}(f) - \alpha \text{mean}_{k,l}(f)|}{f_0 + \text{center}_{k,l}(f)} \right\}. \tag{8.37}$$

Here, $\text{center}_{k,l}(f)$ is the value of the image f in the center of the (k,l)-th block, and $\text{mean}_{k,l}(f)$ is the mean value in this block.

Other enhancement measures that are used in grayscale imaging include the following ones [4, 5, 13, 24, 26, 29]:

(1.) The EME entropy measure

$$EMEE = \frac{1}{k_1 k_2} \sum_{k=1}^{k_1} \sum_{l=1}^{k_2} \left[\frac{\max_{k,l}(f)}{\min_{k,l}(f)} \right] \ln \left[\frac{\max_{k,l}(f)}{\min_{k,l}(f)} \right];$$

(2.) The Agaian-Michelson enhancement measure

$$AME = -\frac{1}{k_1 k_2} \sum_{k=1}^{k_1} \sum_{l=1}^{k_2} 20 \ln \left[\frac{\max_{k,l}(f) - \min_{k,l}(f)}{\max_{k,l}(f) + \min_{k,l}(f)} \right];$$

(3.) The Agaian-Michelson entropy enhancement measure

$$AMEE = -\frac{1}{k_1 k_2} \sum_{k=1}^{k_1} \sum_{l=1}^{k_2} \left[\frac{\max_{k,l}(f) - \min_{k,l}(f)}{\max_{k,l}(f) + \min_{k,l}(f)} \right] \ln \left[\frac{\max_{k,l}(f) - \min_{k,l}(f)}{\max_{k,l}(f) + \min_{k,l}(f)} \right];$$

(4.) The parameterized Agaian-Michelson entropy enhancement measure

$$AMEE(\alpha) = -\frac{1}{k_1 k_2} \sum_{k=1}^{k_1} \sum_{l=1}^{k_2} \alpha \left[\frac{\max_{k,l}(f) - \min_{k,l}(f)}{\max_{k,l}(f) + \min_{k,l}(f)} \right]^{\alpha} \ln \left[\frac{\max_{k,l}(f) - \min_{k,l}(f)}{\max_{k,l}(f) + \min_{k,l}(f)} \right];$$

(5.) The second derivative, or the Panetta-Agaian measure of enhancement

$$PAME = -\frac{1}{k_1 k_2} \sum_{k=1}^{k_1} \sum_{l=1}^{k_2} 20 \ln \left[\frac{|\max_{k,l}(f) - 2\text{center}_{k,l}(f) + \min_{k,l}(f)|}{\max_{k,l}(f) + 2\text{center}_{k,l}(f) + \min_{k,l}(f)} \right];$$

and

(6.) The mean-square-root measure of enhancement

$$MSRME = \frac{1}{k_1 k_2}\sqrt{\sum_{k=1}^{k_1}\sum_{l=1}^{k_2}\frac{|\ln|\text{center}_{k,l}(f) - \text{mean}_{k,l}(f)||}{\ln[\text{center}_{k,l}(f) + \text{mean}_{k,l}(f)]}}.$$

For the AMEE measure, the value of the parameter α is selected from the interval (0, 1). The features of these measures and their advantages in practical applications can be found in Ref. [29].

8.6 Multiplicative Visibility Images

A generalization (or better to say a modification) of the Weber ratio assumes the use of an addition power function as follows:

$$w(\beta) = \left(\frac{|\Delta I|}{I}\right) I^{\beta}, \tag{8.38}$$

where β is the parameter with values of the interval $[-1, 1]$. The $\beta = 0$ case corresponds to the Weber law. The $\beta = 0.5$ case corresponds to the de Vries-Rose law [14, 15]. The $\beta = 1$ case relates to the intensity difference, or gradient. The $\beta = -1$ case relates to the saturation region [16, 17].

Considering the modification for the Weber ratio case, we can introduce the multiplicative visibility operator as

$$v : f_{n,m} \to v(f)_{n,m} = k\left[\frac{|f_{n,m} - \text{Mean}_W(f_{n,m})|}{f_0 + f_{n,m}}\right](f_{n,m})^{\beta}. \tag{8.39}$$

The image $v(f)_{n,m}$ is called *the multiplicative Weber visibility image* (MWVI).

As an example, Fig. 8.19 shows the multiplicative visibility "peppers" images $v(f)_{n,m}$ calculated for the values of β equal 0.25, 0.5, and 0.6 in parts (a), (b), and (c), respectively.

The generalization of the Michelson visibility operator is introduced as

$$c : f_{n,m} \to c(f)_{n,m} = k\left[\frac{\text{Max}_W(f_{n,m}) - \text{Min}_W(f_{n,m})}{\text{Max}_W(f_{n,m}) + \text{Min}_W(f_{n,m})}\right](f_{n,m})^{\beta}, \tag{8.40}$$

where k is a constant. The image $c(f)_{n,m}$ is called *the multiplicative Michelson visibility image.*

The multiplicative EME visibility operator is introduced as

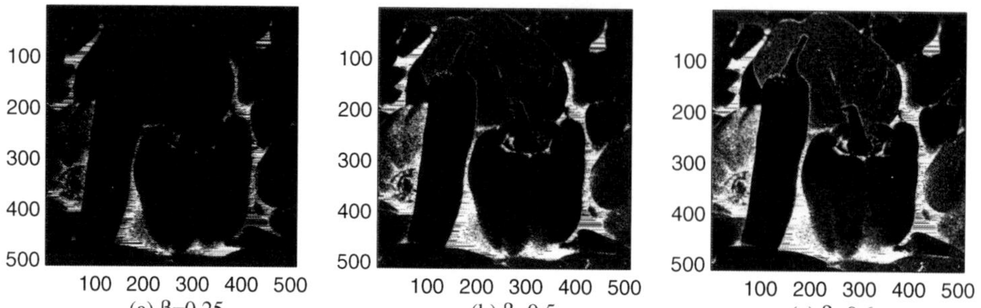

(a) β=0.25 (b) β=0.5 (c) β=0.6

Figure 8.19 The multiplicative Weber visibility "peppers" images for the 3×3 cross-windowed mean operator with (a) $\beta = 0.25$ (b) $\beta = 0.5$, and (c) $\beta = 0.6$.

$$e : f_{n,m} \rightarrow e(f)_{n,m} = k \ln \left[\frac{\mathrm{Max}_W(f_{n,m})}{\mathrm{Min}_W(f_{n,m})} \right] (f_{n,m})^{\beta}, \tag{8.41}$$

where k is a constant. The image $e(f)_{n,m}$ is called *the multiplicative EME visibility image.*

8.7 Two-Window Gradient Operators

The class of gradient operations $\Delta f_{n,m}$ is large. It also contains the operators operating with a few windows of different sizes. For instance, we consider the following two mean operators with square windows W_1 and W_2 of sizes 3×3 and 5×5:

$$\mathrm{Mean}_{W_1} = \frac{1}{9} \begin{bmatrix} 0 & 0 & 0 & 0 & 0 \\ 0 & 1 & 1 & 1 & 0 \\ 0 & 1 & \underline{\mathbf{1}} & 1 & 0 \\ 0 & 1 & 1 & 1 & 0 \\ 0 & 0 & 0 & 0 & 0 \end{bmatrix} \quad \text{and} \quad \mathrm{Mean}_{W_2} = \frac{1}{25} \begin{bmatrix} 1 & 1 & 1 & 1 & 1 \\ 1 & 1 & 1 & 1 & 1 \\ 1 & 1 & \underline{\mathbf{1}} & 1 & 1 \\ 1 & 1 & 1 & 1 & 1 \\ 1 & 1 & 1 & 1 & 1 \end{bmatrix}.$$

The window W_1 is the sub-window of W_2, i.e., $W_1 \subset W_2$. The two-window gradient operator, which is defined as the difference of these mean operators, has the mask

$$\Delta_{W_1, W_2} = \mathrm{Mean}_{W_1} - \mathrm{Mean}_{W_2} = \frac{1}{25} \begin{bmatrix} -1 & -1 & -1 & -1 & -1 \\ -1 & \frac{16}{9} & \frac{16}{9} & \frac{16}{9} & -1 \\ -1 & \frac{16}{9} & \underline{\mathbf{\frac{16}{9}}} & \frac{16}{9} & -1 \\ -1 & \frac{16}{9} & \frac{16}{9} & \frac{16}{9} & -1 \\ -1 & -1 & -1 & -1 & -1 \end{bmatrix}. \tag{8.42}$$

We can also call this operator W_1-in-W_2 *gradient of size* 5×5.

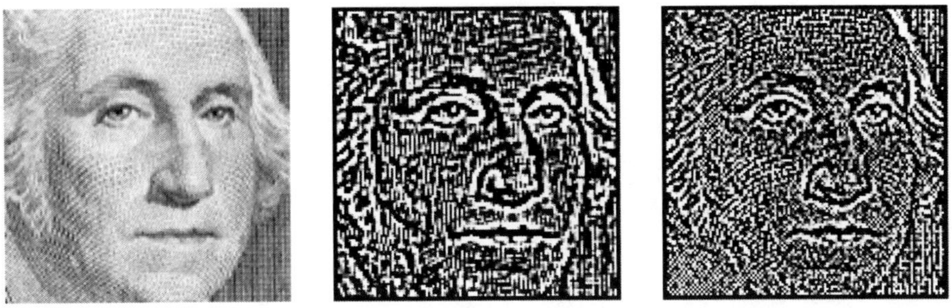

(a) (b) (c)

Figure 8.20 The image (a) before and after processing by the two-window gradient operators with (b) the square-windows and (c) the circle-windows.

As an example, Fig. 8.20 shows the image in part (a) and the gradient image $\Delta_{W_1,W_2}(f)$ in part (b), which was calculated by the above two-window gradient operator. In Fig. 8.20(c), the two-window gradient operator was calculated by the following mask:

$$\Delta_{W_1,W_2} = \text{Mean}_{W_1} - \text{Mean}_{W_2} = \frac{1}{13}\begin{bmatrix} 0 & 0 & -1 & 0 & 0 \\ 0 & -1 & \frac{8}{5} & -1 & 0 \\ -1 & \frac{8}{5} & \underline{\mathbf{\frac{8}{5}}} & \frac{8}{5} & -1 \\ 0 & -1 & \frac{8}{5} & -1 & 0 \\ 0 & 0 & -1 & 0 & 0 \end{bmatrix}. \tag{8.43}$$

This gradient is the difference of the following mean operators with the circle-windows of sizes 3×3 and 5×5,

$$\text{Mean}_{W_1} = \frac{1}{5}\begin{bmatrix} 0 & 0 & 0 & 0 & 0 \\ 0 & 0 & 1 & 0 & 0 \\ 0 & 1 & \underline{\mathbf{1}} & 1 & 0 \\ 0 & 0 & 1 & 0 & 0 \\ 0 & 0 & 0 & 0 & 0 \end{bmatrix} \quad \text{and} \quad \text{Mean}_{W_2} = \frac{1}{13}\begin{bmatrix} 0 & 0 & 1 & 0 & 0 \\ 0 & 1 & 1 & 1 & 0 \\ 1 & 1 & \underline{\mathbf{1}} & 1 & 1 \\ 0 & 1 & 1 & 1 & 0 \\ 0 & 0 & 1 & 0 & 0 \end{bmatrix}.$$

As an example, Fig. 8.21 shows the "man" image in part (a), and the gradient images $\Delta_{W_1,W_2}(f)$ obtained by the above two-window gradient operators in parts (b) and (c).

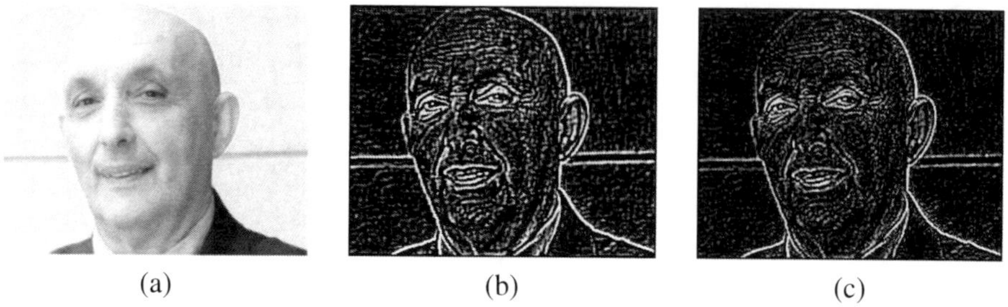

Figure 8.21 The "man" image (a) before and after processing by the two-window gradient operators with (b) the square-windows and (c) the circle-windows.

In another example, Fig. 8.22 shows the "cameraman" image in part (a), and the gradient images $\Delta_{W_1, W_2}(f)$ obtained by the same two gradient operators in parts (b) and (c).

In a similar way, we can construct the gradients operators as the differences of mean operators with windows of sizes 3×3 and 7×7, 5×5 and 7×7, or 5×5 and 9×9. For instance, the two-window gradient operator, which is composed by the 5×5 and 7×7 circle-windows, can be defined as

$$\text{Mean}_{W_1} - \text{Mean}_{W_2} = \frac{1}{13}\begin{bmatrix} 0 & 0 & 0 & 0 & 0 & 0 & 0 \\ 0 & 0 & 0 & 1 & 0 & 0 & 0 \\ 0 & 0 & 1 & 1 & 1 & 0 & 0 \\ 0 & 1 & 1 & \underline{\mathbf{1}} & 1 & 1 & 0 \\ 0 & 0 & 1 & 1 & 1 & 0 & 0 \\ 0 & 0 & 0 & 1 & 0 & 0 & 0 \\ 0 & 0 & 0 & 0 & 0 & 0 & 0 \end{bmatrix} - \frac{1}{25}\begin{bmatrix} 0 & 0 & 0 & 1 & 0 & 0 & 0 \\ 0 & 0 & 1 & 1 & 1 & 0 & 0 \\ 0 & 1 & 1 & 1 & 1 & 1 & 0 \\ 1 & 1 & 1 & \underline{\mathbf{1}} & 1 & 1 & 1 \\ 0 & 1 & 1 & 1 & 1 & 1 & 0 \\ 0 & 0 & 1 & 1 & 1 & 0 & 0 \\ 0 & 0 & 0 & 1 & 0 & 0 & 0 \end{bmatrix}.$$

The mask of this gradient operator is

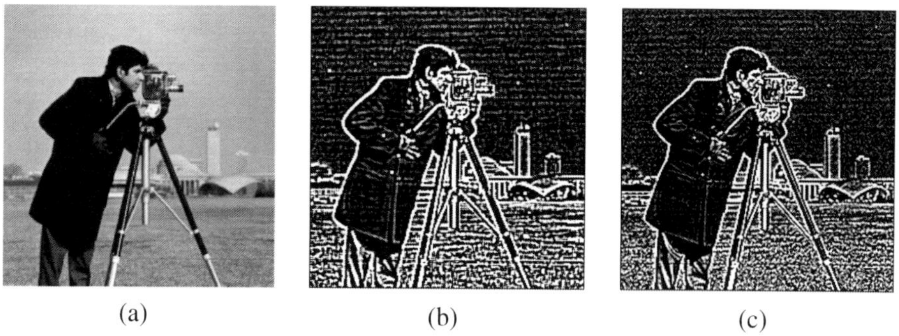

Figure 8.22 The "cameraman" image (a) before and after processing by the two-window gradient operators with (b) the square-windows and (c) the circle-windows.

$$\Delta_{W_1,W_2}=\frac{1}{25}\begin{bmatrix}0&0&0&-1&0&0&0\\0&0&-1&a&-1&0&0\\0&-1&a&a&a&-1&0\\-1&a&a&\underline{a}&a&a&-1\\0&-1&a&a&a&-1&0\\0&0&-1&a&-1&0&0\\0&0&0&-1&0&0&0\end{bmatrix},$$

where $a = 12/13 = 0.9231$.

As examples, Fig. 8.23 shows the gradient images in parts (a) and (b), which are obtained by the above two-window gradient operator for the images of Figs. 8.20 and 8.21, respectively.

The simple case of the W_1-in-W_2 gradient is when the first window W_1 contains only one point, the center, i.e., the mask is the number 1. For example, such a gradient of size 3×3 with the cross-window is defined as

$$\Delta_{1,W_2}=[\underline{\mathbf{1}}]-\frac{1}{5}\begin{bmatrix}0&1&0\\1&\underline{\mathbf{1}}&1\\0&1&0\end{bmatrix}=\frac{1}{5}\begin{bmatrix}0&-1&0\\-1&\underline{\mathbf{4}}&-1\\0&-1&0\end{bmatrix}=-\frac{1}{5}h_{+}.$$

Thus, this gradient up to the coefficient $-1/5$ is the mask h_+ of the Laplacian gradient with the 3×3 cross window, which is defined in Eq. (8.11). The Laplacian gradient $h_\times$ with the 3×3 X-window is similar to the following W_1-in-W_2 gradient:

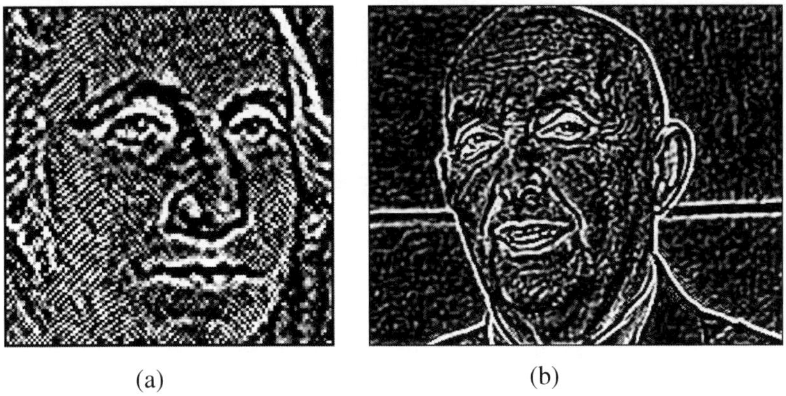

(a) (b)

Figure 8.23 Two gradients images after processing by the two-window gradient operator with the circle-windows W_1 and W_2 of sizes 5×5 and 7×7, respectively.

$$\Delta_{1,W_2} = [\mathbf{1}] - \frac{1}{5}\begin{bmatrix} 1 & 0 & 1 \\ 0 & \mathbf{1} & 0 \\ 1 & 0 & 1 \end{bmatrix} = \frac{1}{5}\begin{bmatrix} -1 & 0 & -1 \\ 0 & \underline{\mathbf{4}} & 0 \\ -1 & 0 & -1 \end{bmatrix} = -\frac{1}{5}h_{\times}.$$

8.8 Facial Image Representation

The process of face recognition includes two parts, the facial representation, to obtain a set of features that are used in the next part, and classification. In this section, we consider the grayscale facial image representation, or description, which is based on the local binary patterns (LBP) over the whole facial image [50]. We describe this representation in terms of simple gradient operators with following composition of the 8-bit LBP image and its histogram, which can be used as the feature in classification of facial images.

The main parts of representation of the grayscale facial image $f_{n,m}$ of size $N \times M$ are shown in the block-diagram of Fig. 8.24.

The algorithm of facial image processing can be described by the following steps.

Stage 1. The visibility image is calculated,

$$f_{n,m} \to V(f)_{n,m}, \quad n = 0:(N-1), m = 0:(M-1).$$

Such an image can be, for instance, the EME visibility image, which is calculated by Eq. (8.34). In this case, $V(f) = E(f)$. The image $V(f)$ can also be the image obtained by applying one of the gradient operators described in the previous sections.

Stage 2. The 2-D Gaussian function is circular convoluted with the visibility image,

$$V(f)_{n,m} \to G(f)_{n,m} = [V(f) \otimes g]_{n,m},$$

where $g_{n,m}$ is the mask of the Gaussian function with the standard deviation σ equal 0.5,

$$g_{n,m} = \frac{1}{K}\exp\left[-\frac{1}{2\pi\sigma^2}(n^2+m^2)\right], \tag{8.44}$$

where the coefficient of normalization equals

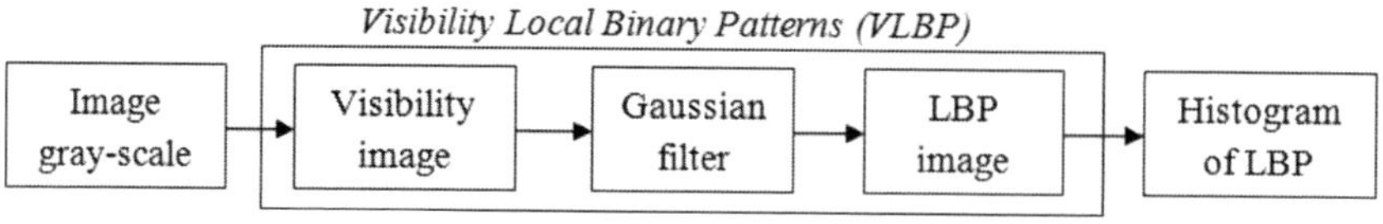

Figure 8.24 The block-diagram of the facial image representation.

$$K=\sum_{n=-L_1}^{L_1-1}\sum_{m=-L_2}^{L_2-1}\exp\left[-\frac{1}{2\pi\sigma^2}(n^2+m^2)\right].$$

The kernel of the Gaussian function is of size $(2L_1+1)\times(2L_2+1)$ which can be smaller or equal to the size of the image. For that, the MATLAB functions “imgaussfilt,” “conv2,” or “imfilter” can be used.

Stage 3. A complex gradient image composition is calculated. In the 3×3-square window W, we consider the following eight gradient operators:

$$A_1=\begin{bmatrix}-1&0&0\\0&\underline{1}&0\\0&0&0\end{bmatrix},A_2=\begin{bmatrix}0&-1&0\\0&\underline{1}&0\\0&0&0\end{bmatrix},A_3=\begin{bmatrix}0&0&-1\\0&\underline{1}&0\\0&0&0\end{bmatrix},A_4=\begin{bmatrix}0&0&0\\0&\underline{1}&-1\\0&0&0\end{bmatrix},$$
$$A_5=\begin{bmatrix}0&0&0\\0&\underline{1}&0\\0&0&-1\end{bmatrix},A_6=\begin{bmatrix}0&0&0\\0&\underline{1}&0\\0&-1&0\end{bmatrix},A_7=\begin{bmatrix}0&0&0\\0&\underline{1}&0\\-1&0&0\end{bmatrix},A_8=\begin{bmatrix}0&0&0\\-1&\underline{1}&0\\0&0&0\end{bmatrix}. \tag{8.45}$$

The center of the masks is in the original point (0,0) and it is underlined. The order of these gradient operators is shown in Fig. 8.25, but it can be changed. These eight neighbor points are called *the sampling points* (SP),

$$SP=\left[(-1,\,-1),(0,\,-1),(1,\,-1),(1,0),(1,1),(0,1),(-1,1),(-1,0)\right]. \tag{8.46}$$

The set of sampling points can also be considered in the following order:

$$SP=\left[(-1,\,-1),(-1,0),(-1,1),(0,\,-1),(0,1),(1,\,-1),(1,0),(1,1)\right].$$

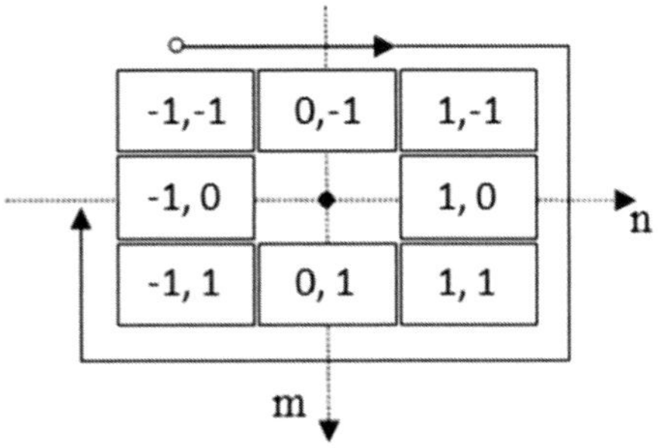

Figure 8.25 The coordinates and the order of eight neighbor sampling points.

These gradient operators are applied on the facial image $f_{n,m}$ and eight binary images are calculated by

$$(B_k)_{n,m} = u\left[A_k(f)_{n,m}\right], \quad k = 1, 2, \ldots, 8, \tag{8.47}$$

where the unit step function is defined as $u[t] = 1$, if $t \geq 0$, and $u[t] = 0$, otherwise. Then, the new image, which is called *the local binary pattern (LBP) image*, is calculated by

$$b_{n,m} = B_1(f)_{n,m} + 2B_2(f)_{n,m} + 2^2 B_3(f)_{n,m} + \cdots + 2^7 B_8(f)_{n,m}. \tag{8.48}$$

The LBP image can also be written in the standard form

$$b_{n,m} = \sum_{r=1}^{8} 2^{r-1} u[f_{n,m} - f_{n+s(k),m+p(k)}], \tag{8.49}$$

where $[s(k), p(k)]$ are the sampling points, which are ordered as shown in Eq. (8.46). The LBP image has the range of 256 integer levels s, from 0 to 255. The histogram of this image, $H(s)$ is considered to be the feature of the facial image $f_{n,m}$, which can be used in face classification.

Example 8.1

Let the values of the image $f_{n,m}$ in the 3×3 window translated to point $(n, m) = (7, 9)$ be

$$I = \begin{bmatrix} 1 & 2 & 4 \\ 2 & \underline{3} & 3 \\ 1 & 4 & 2 \end{bmatrix}, \quad (f_{7,9} = 3).$$

Then, the first gradient results in the following value:

$$A_1(I) = \begin{bmatrix} -1 & 0 & 0 \\ 0 & \underline{1} & 0 \\ 0 & 0 & 0 \end{bmatrix} \circ \begin{bmatrix} 1 & 2 & 4 \\ 2 & \underline{3} & 3 \\ 1 & 4 & 2 \end{bmatrix} = -1 + 3 = 2,$$

and therefore, $u[A_1(I)] = u[2] = 1$. Similarly, we obtain the results of the remaining seven gradient operations, $A_2(I) = 1$, $A_3(I) = -1$, $A_4(I) = 0$, $A_5(I) = 1$, $A_6(I) = -1$, $A_7(I) = 2$, and $A_8(I) = 1$. Therefore,

$$u[A_1(I), A_2(I), \ldots, A_8(I)] = u[2, 1, -1, 0, 1, -1, 2, 1] = (1, 1, 0, 0, 1, 0, 1, 1).$$

The decimal number that corresponds to this binary pattern is

$$1 + 2^1 + 2^4 + 2^6 + 2^7 = 211.$$

It is the value of the LBP image at point (7, 9), i.e., $b_{7,9} = 211$.

As an example, Fig. 8.26 shows the facial image of size 114×94 in part (a), the EME visibility image in part (b), and the visibility image filtered by the 2-D Gaussian function with the standard deviation of 0.5 in part (c). The results of further facial image processing are shown in Fig. 8.27. The LBP image $b_{n,m}$ is shown in Fig. 8.27(a) and the histogram of the image in Fig. 8.27(b).

It should be noted that instead of the operators given in Eq. (8.45), other gradient operators can also be used for calculating the LBP image. For instance, in the 3×3-square window W, we consider the following eight gradient operators (with the scale factor 1/9):

$$A_1 = \begin{bmatrix} -8 & 1 & 1 \\ 1 & \underline{1} & 1 \\ 1 & 1 & 1 \end{bmatrix}, A_2 = \begin{bmatrix} 1 & -8 & 1 \\ 1 & \underline{1} & 1 \\ 1 & 1 & 1 \end{bmatrix}, A_3 = \begin{bmatrix} 1 & 1 & -8 \\ 1 & \underline{1} & 1 \\ 1 & 1 & 1 \end{bmatrix}, A_4 = \begin{bmatrix} 1 & 1 & 1 \\ 1 & \underline{1} & -8 \\ 1 & 1 & 1 \end{bmatrix},$$

$$A_5 = \begin{bmatrix} 1 & 1 & 1 \\ 1 & \underline{1} & 1 \\ 1 & 1 & -8 \end{bmatrix}, A_6 = \begin{bmatrix} 1 & 1 & 1 \\ 1 & \underline{1} & 1 \\ 1 & -8 & 1 \end{bmatrix}, A_7 = \begin{bmatrix} 1 & 1 & 1 \\ 1 & \underline{1} & 1 \\ -8 & 1 & 1 \end{bmatrix}, A_8 = \begin{bmatrix} 1 & 1 & 1 \\ -8 & \underline{1} & 1 \\ 1 & 1 & 1 \end{bmatrix}.$$

This set of eight gradients perform the comparison of the image at sampling points with the mean intensity in the 3×3 window [51]. For instance, the first gradient can be written as

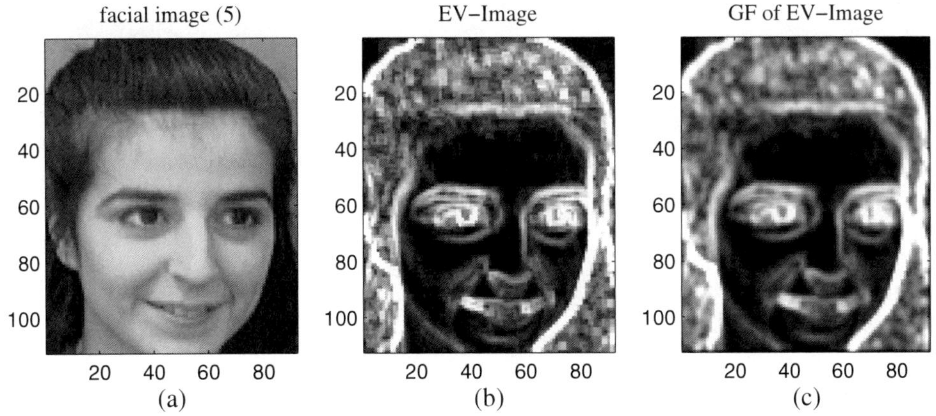

Figure 8.26 (a) The original image and the EME visibility image (b) before and (c) after filtering by the 2-D Gaussian function.

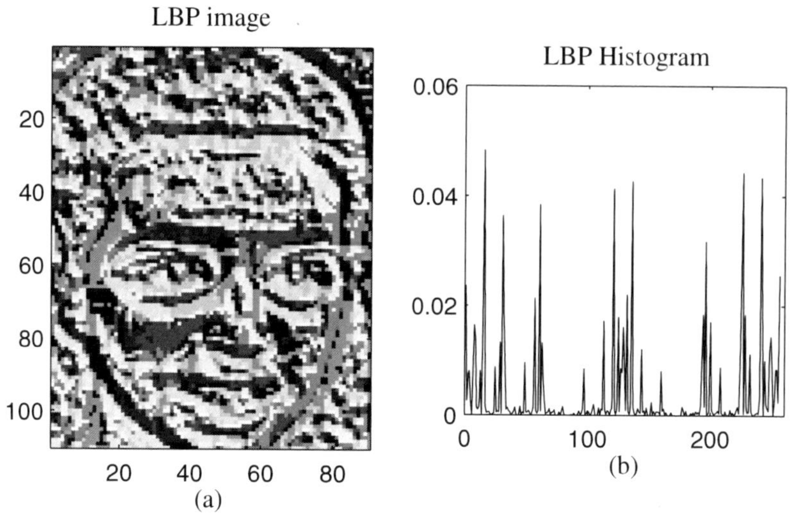

Figure 8.27 (a) The LBP image and (b) the histogram of the image.

$$A_1 = \frac{1}{9}\begin{bmatrix} -8 & 1 & 1 \\ 1 & \underline{1} & 1 \\ 1 & 1 & 1 \end{bmatrix} = \frac{1}{9}\begin{bmatrix} 1 & 1 & 1 \\ 1 & \underline{1} & 1 \\ 1 & 1 & 1 \end{bmatrix} - \begin{bmatrix} 1 & 0 & 0 \\ 0 & \underline{0} & 0 \\ 0 & 0 & 0 \end{bmatrix}.$$

For example, if the values of the image in the 3×3 window are

$$I = \begin{bmatrix} 1 & 2 & 2 \\ 2 & \underline{3} & 3 \\ 1 & 2 & 2 \end{bmatrix}, \quad \text{mean}(I) = \frac{1}{9}(1+2+2+1+4+3+1+2+2) = 2,$$

then, the first gradient results in the following value:

$$\begin{aligned} A_1 \circ I &= \frac{1}{9}\begin{bmatrix} -8 & 1 & 1 \\ 1 & \underline{1} & 1 \\ 1 & 1 & 1 \end{bmatrix} \circ \begin{bmatrix} 1 & 2 & 2 \\ 2 & \underline{3} & 3 \\ 1 & 2 & 2 \end{bmatrix} \\ &= \frac{1}{9}(-8+2+2+2+3+3+1+2+2) = 1. \end{aligned}$$

This value equals $\text{mean}(I) - I_{-1,-1} = 2 - 1 = 1$.

The corresponding LBP image can also be written as

$$b_{n,m} = \sum_{r=1}^{8} 2^{r-1} u[M[f]_{n,m} - f_{n+s(k),m+p(k)}]. \tag{8.50}$$

Here, $M[f]_{n,m}$ is the local mean value of the image in the 3×3 window translated at point (n, m).

We also mention the so-called center-symmetric local binary patterns [52] when only the following four gradient operators are used:

$$A_1 = \begin{bmatrix} 0 & 0 & 0 \\ -1 & \underline{0} & 1 \\ 0 & 0 & 0 \end{bmatrix}, A_2 = \begin{bmatrix} -1 & 0 & 0 \\ 0 & \underline{0} & 0 \\ 0 & 0 & 1 \end{bmatrix}, A_3 = \begin{bmatrix} 0 & -1 & 0 \\ 0 & \underline{0} & 0 \\ 0 & 1 & 0 \end{bmatrix}, A_4 = \begin{bmatrix} 0 & 0 & -1 \\ 0 & \underline{0} & 0 \\ 1 & 0 & 0 \end{bmatrix}.$$

The LBP image $b_{n,m}$ is calculated with fever levels (maximum 16), than in Eq. (8.48), i.e.,

$$b_{n,m} = B_1(f)_{n,m} + 2B_2(f)_{n,m} + 2^2 B_3(f)_{n,m} + 2^3 B_4(f)_{n,m}. \tag{8.51}$$

All gradients described above are defined in the neighborhood of the 3×3 window. Many other gradient operators can be suggested in the neighborhood of the 5×5 or 7×7 window, for constructing LBP images with large number of levels.

8.8.1 Facial image representation with the uniform LBP

The described in the previous sub-section LBP image is composed from different eight-bit patterns of decimals in the interval [0, 255]. The histogram of this image with maximum 256 values can be used as the feature of the facial image. To reduce the number of binary patterns, the LBP image is modified by using a special mapping, or the so called uniform LBP look up table, which is based on the concept of uniform binary patterns [50]. For that, Stage 3 of the complex gradient image composition into the uniform LBP image is accomplished in the following way.

The lookup table (T), which is calculated from the uniform *local binary patterns* is given in Eq. (8.52). Given an integer number $n \in [0, 255]$, its 8-bits local binary pattern (········) is called *uniform*, if there are no less than two bitwise transitions from 0 to 1, or 1 to 0. For instance, for $n = 2$, the command "bitget(2,1:8)" gives the pattern (01000000) with two transitions. The same number of transitions is in the patterns (01100000) and (00011000) of the numbers 6 and 24, respectively. The first not uniform pattern in the table is for $n = 5$, since it equals (10100000) with 3 transitions. The next such number is 9 and "bitget(9,1:8)" is (10010000).

$$T=\begin{bmatrix}
0 & 1 & 2 & 3 & 4 & \bullet & 5 & 6 & 7 & \bullet & \bullet & \bullet & 8 & \bullet & 9 & 10\\
11 & \bullet & \bullet & \bullet & \bullet & \bullet & \bullet & \bullet & 12 & \bullet & \bullet & \bullet & 13 & \bullet & 14 & 15\\
16 & \bullet & \bullet & \bullet & \bullet & \bullet & \bullet & \bullet & \bullet & \bullet & \bullet & \bullet & \bullet & \bullet & \bullet & \bullet\\
17 & \bullet & \bullet & \bullet & \bullet & \bullet & \bullet & \bullet & 18 & \bullet & \bullet & \bullet & 19 & \bullet & 20 & 21\\
22 & \bullet & \bullet & \bullet & \bullet & \bullet & \bullet & \bullet & \bullet & \bullet & \bullet & \bullet & \bullet & \bullet & \bullet & \bullet\\
\bullet & \bullet & \bullet & \bullet & \bullet & \bullet & \bullet & \bullet & \bullet & \bullet & \bullet & \bullet & \bullet & \bullet & \bullet & \bullet\\
23 & \bullet & \bullet & \bullet & \bullet & \bullet & \bullet & \bullet & \bullet & \bullet & \bullet & \bullet & \bullet & \bullet & \bullet & \bullet\\
24 & \bullet & \bullet & \bullet & \bullet & \bullet & \bullet & \bullet & 25 & \bullet & \bullet & \bullet & 26 & \bullet & 27 & 28\\
29 & 30 & \bullet & 31 & \bullet & \bullet & \bullet & 32 & \bullet & \bullet & \bullet & \bullet & \bullet & \bullet & \bullet & 33\\
\bullet & \bullet & \bullet & \bullet & \bullet & \bullet & \bullet & \bullet & \bullet & \bullet & \bullet & \bullet & \bullet & \bullet & \bullet & 34\\
\bullet & \bullet & \bullet & \bullet & \bullet & \bullet & \bullet & \bullet & \bullet & \bullet & \bullet & \bullet & \bullet & \bullet & \bullet & \bullet\\
\bullet & \bullet & \bullet & \bullet & \bullet & \bullet & \bullet & \bullet & \bullet & \bullet & \bullet & \bullet & \bullet & \bullet & \bullet & 35\\
36 & 37 & \bullet & 38 & \bullet & \bullet & \bullet & 39 & \bullet & \bullet & \bullet & \bullet & \bullet & \bullet & \bullet & 40\\
\bullet & \bullet & \bullet & \bullet & \bullet & \bullet & \bullet & \bullet & \bullet & \bullet & \bullet & \bullet & \bullet & \bullet & \bullet & 41\\
42 & 43 & \bullet & 44 & \bullet & \bullet & \bullet & 45 & \bullet & \bullet & \bullet & \bullet & \bullet & \bullet & \bullet & 46\\
47 & 48 & \bullet & 49 & \bullet & \bullet & \bullet & 50 & 51 & 52 & \bullet & 53 & 54 & 55 & 56 & 57
\end{bmatrix}. \quad (8.52)$$

Here, the bullets show the binary patterns which are not uniform. The total number of uniform patterns is 58 which are labeled from 0 to 57. Therefore, for all non-uniform patterns the label 58 is assigned.

```
% -----------------------------------------------------------------
% call: artlbpuni4table.m / Art Grigoryan, Nov. 2, 2015
% similar BitwiseToLBP.m / Stack Overflow Community,
http://stackoverflow.com/
% To calculate the uniform table for LBP with 8 neighbors in
the 3x3 window
% -----------------------------------------------------------------
  function [LbpTable,bitsLabel] = artlbpuni4table
      LbpTable = zeros(1,256);
      bitsLabel = 0;
      for n = 1:256
            bits=bitget(n-1,1:8);
            a=nnz(diff(bits([1:end,1]))) <= 2;
            if (a==1)
                LbpTable(n) = bitsLabel;
                bitsLabel = bitsLabel + 1;
            else
                LbpTable(n) = 58; % not uniform bit pattern
            end
      end
      bitsLabel=bitsLabel+1;
  end
% -----------------------------------------------------------------
```

The 16×16 table in Eq. (8.52) can be calculated from the 256-point row vector, "LbpTable," by using the command

```
T=reshape(LbpTable,16,16)';
```

The uniform LBP table is used to compose the new image by applying this table after each gradient image calculation as

$$\begin{aligned}(g_1)_{n,m} &= T[B_1(f)_{n,m}],\\ (g_2)_{n,m} &= T[(g_1)_{n,m} + B_2(f)_{n,m}],\\ (g_3)_{n,m} &= T[(g_2)_{n,m} + B_3(f)_{n,m}],\\ \cdots &= \cdots\cdots \quad \cdots\cdots \quad \cdots\cdots\\ (g_8)_{n,m} &= T[(g_7)_{n,m} + B_8(f)_{n,m}].\end{aligned} \tag{8.53}$$

The image $(g_8)_{n,m}$ is the uniform LBP image and it has the range of 59 integer levels s, from 0 to 58. Thus, 256 integer levels in the original LBP image have been reduced to only 59 levels in the uniform LBP image. The histogram of this image, $H(s)$ is considered to be the feature of the facial image $f_{n,m}$, which is used in face classification.

To compare the results of the uniform and non-uniform LBP histograms, we consider the facial image of Fig. 8.26. The image $(g_8)_{n,m}$ before and after using the mapping by the uniform LBP table is shown in Fig. 8.28 in parts (a) and (b), respectively. The normalized histogram of the uniform LBP image is given in Fig. 8.28(c).

Figure 8.29 shows another facial image of the same size 114×94 in part (a). The EME visibility image before and after filtering by the 2-D Gaussian function with the standard deviation of 0.5 is shown in Fig. 8.29(b,c).

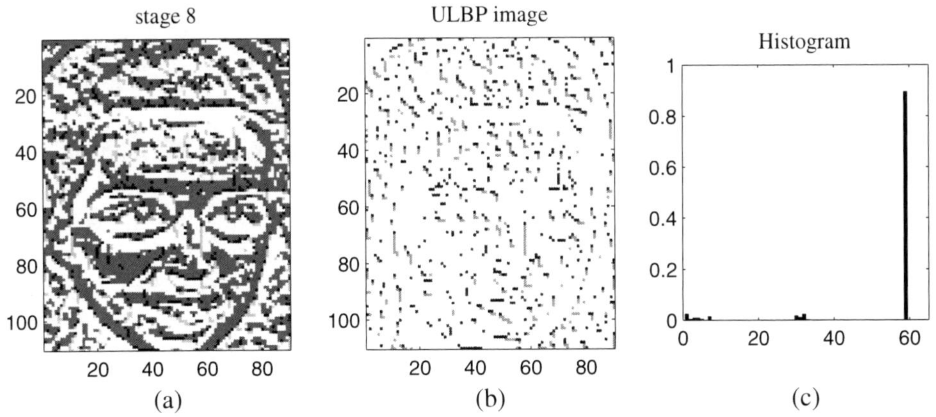

Figure 8.28 (a) The image $(g_8)_{n,m}$ before the mapping, (b) the uniform LBP image, and (c) the histogram of the image.

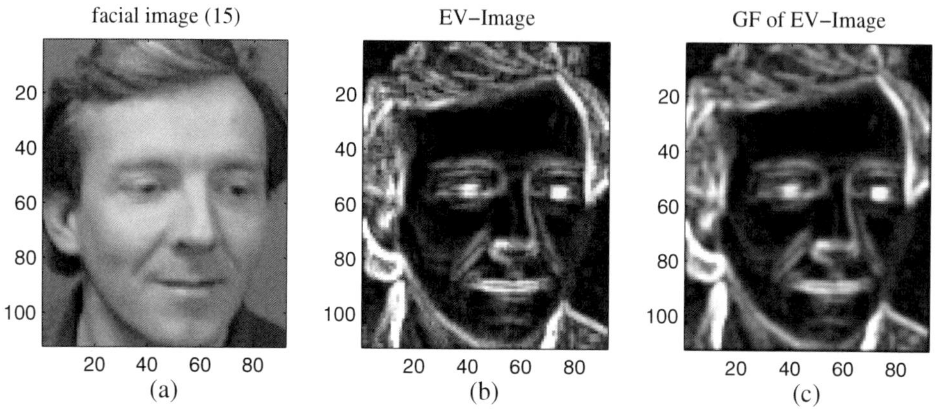

Figure 8.29 (a) The original image and the EME visibility image (b) before and (c) after filtering by the 2-D Gaussian function.

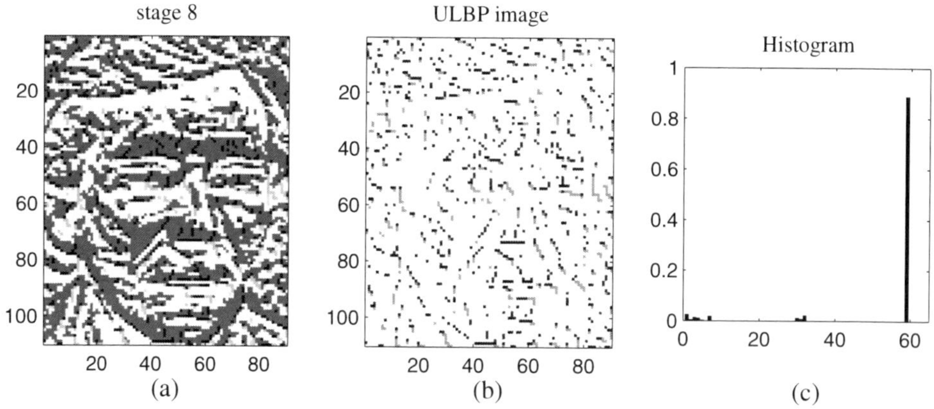

Figure 8.30 (a) The image $(g_8)_{n,m}$ before the mapping, (b) the uniform LBP image, and (c) the histogram of the image.

Figure 8.30 shows the image $(g_8)_{n,m}$ before and after mapping by the uniform LBP table in parts (a) and (b), respectively. The normalized histogram of the uniform LBP image is given in Fig. 8.30(c).

The representation of the facial image by the LBP image and its histogram is calculated over the whole facial image. In face recognition, such concepts can also be calculated separately over the parts of the image, on which the image is divided. In this case, the histogram concatenated from all histograms of the LBP images of the parts is used as a feature of the facial image [53].

8.9 Color Visibility Images

All mentioned concepts of visibility images can be applied component-wise for color images. For example, we consider the EME measure and the RGB color model, wherein the color image $f_{n,m} = (r_{n,m}, g_{n,m}, b_{n,m})$.

The EME visibility color image (EVCI) is defined by the operator

$$e : f_{n,m} \to e(f)_{n,m} = \left[e(r)_{n,m}, e(g)_{n,m}, e(b)_{n,m} \right] \tag{8.54}$$

where the color components of this image are calculated as in Eq. (8.34). Thus,

$$e(x)_{n,m} = k \ln \left[\frac{\mathrm{Max}_W(x_{n,m})}{\mathrm{Min}_W(x_{n,m})} \right]. \tag{8.55}$$

Here, the image $x_{n,m}$ is used for color components $r_{n,m}$, $g_{n,m}$, and $b_{n,m}$.

As an example, Fig. 8.31 shows the color "peppers" image in part (a). The color image composed by the measure EME visibility images (EVI) of three color components of the image is shown in part (b) and the grayscale component of the EVCI in part (c).

The color components of the multiplicative visibility color image (MEVCI) are calculated by

$$e(x)_{n,m} = k \ln \left[\frac{\mathrm{Max}_W(x_{n,m})}{\mathrm{Min}_W(x_{n,m})} \right] (x_{n,m})^{\beta}. \tag{8.56}$$

As an example, we consider the $\beta = 1$ case.

Figure 8.32 shows the MEVCI of the "peppers" image in part (a) and its grayscale component in (b).

We also consider the $\beta = 1$ case with the color "flowers" image that is shown in Fig. 8.33 in part (a). The color image composed by the EME visibility images of three color components is shown in part (b) and the grayscale component of this image in part (c).

Figure 8.34 shows the multiplicative MEVCI of the "flowers" image. The color image is in part (a) and its grayscale image in part (b).

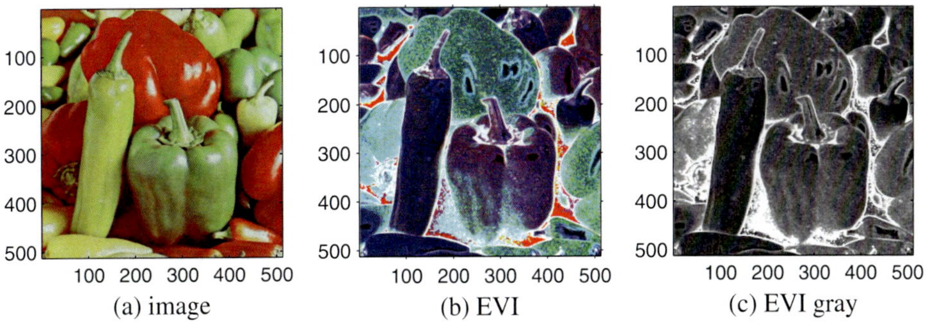

Figure 8.31 (a) The color image, (b) the EVCI, and (c) the grayscale image of the EVCI.

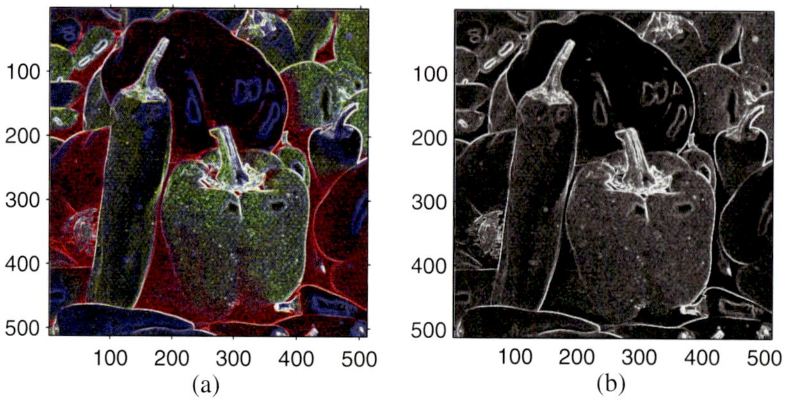

Figure 8.32 (a) The MEVCI of the "peppers" image and (b) the grayscale image of the MEVCI.

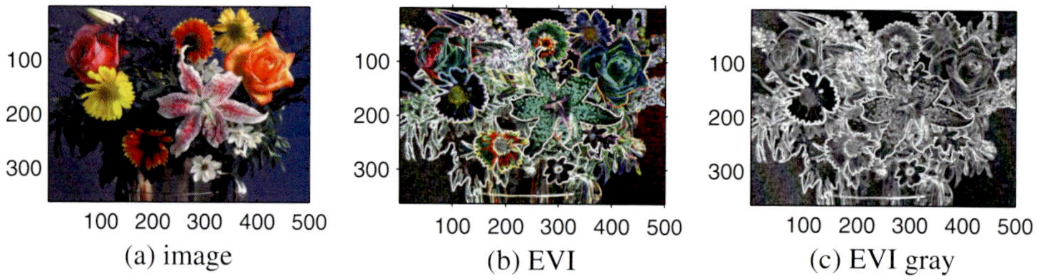

Figure 8.33 (a) The color "flowers" image, (b) the MEVCI, and (c) grayscale image of the MEVCI.

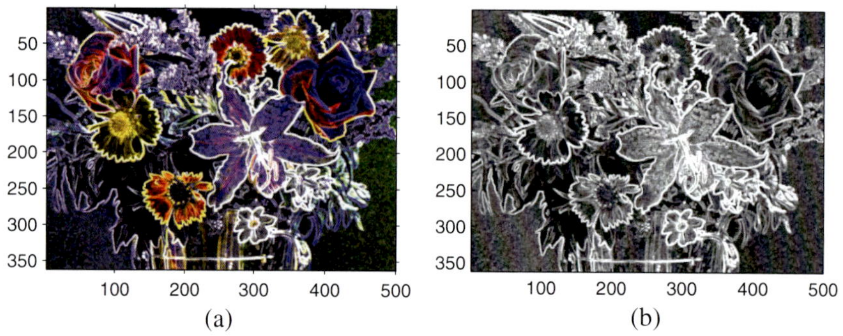

Figure 8.34 (a) The "flower" MEVCI and (b) the grayscale image of this MEVCI.

8.9.1 Other multiplicative visibility color images

The Weber visibility color image (WVCI) is defined by the operator

$$v : f_{n,m} \to v(f)_{n,m} = \left[v(r)_{n,m}, v(g)_{n,m}, v(b)_{n,m} \right], \tag{8.57}$$

where the color components of this image are calculated as in Eq. (8.20). Thus,

$$v(x)_{n,m} = k\left[\frac{|x_{n,m} - \mathrm{Mean}_W(x_{n,m})|}{f_0 + x_{n,m}}\right], \tag{8.58}$$

and the image $x_{n,m}$ is used for color components $r_{n,m}$, $g_{n,m}$, and $b_{n,m}$.

As an example, Fig. 8.35 shows the color "peppers" image in part (a), and the color image composed by the Weber visibility images of three color components, the red, green, and the blue, in part (b). The grayscale component of this WVCI is given in part (c).

The multiplicative WVCI is defined as in Eq. (8.57) with the color components

$$v(x)_{n,m} = k\left[\frac{|x_{n,m} - \mathrm{Mean}_W(x_{n,m})|}{f_0 + x_{n,m}}\right](x_{n,m})^{\beta} \tag{8.59}$$

where $\beta \in [-1.1]$.

The Michelson visibility color image (MVCI) is defined by the operator

$$c : f_{n,m} \to c(f)_{n,m} = \left[c(r)_{n,m}, c(g)_{n,m}, c(b)_{n,m}\right],$$

where the color components of this image are calculated by.

$$c(x)_{n,m} = k\frac{|\mathrm{Max}_W(x_{n,m}) - \mathrm{Min}_W(x_{n,m})|}{\mathrm{Max}_W(x_{n,m}) + \mathrm{Min}_W(x_{n,m})}. \tag{8.60}$$

Here, the image $x_{n,m}$ is used for the color components $r_{n,m}$, $g_{n,m}$, and $b_{n,m}$.

As an example, Fig. 8.36 shows the color "peppers" image in part (a), and the color image composed by the Michelson visibility images (MVI) of three

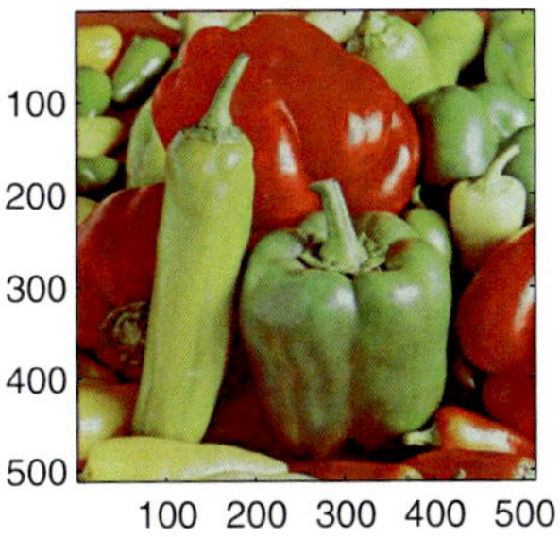

(a) image

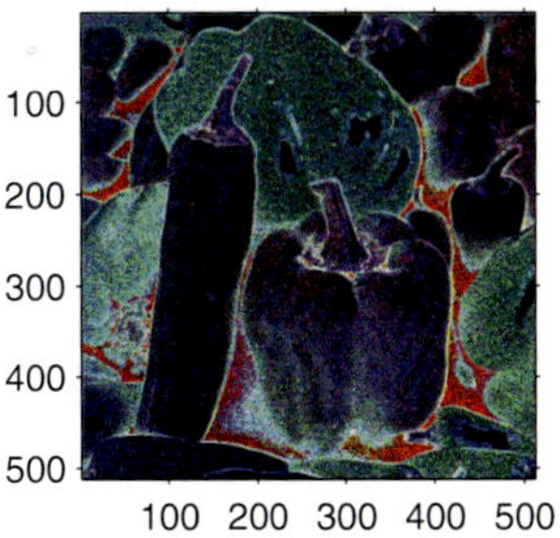

(b) WVI

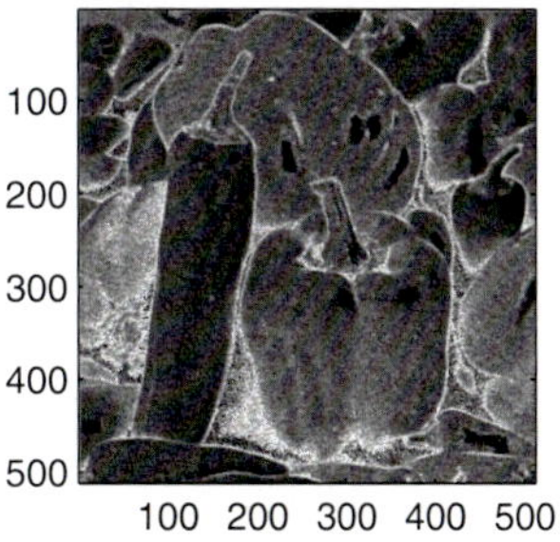

(c) WVI gray

Figure 8.35 (a) The color image, (b) the WVCI, and (c) grayscale image of the WVCI.

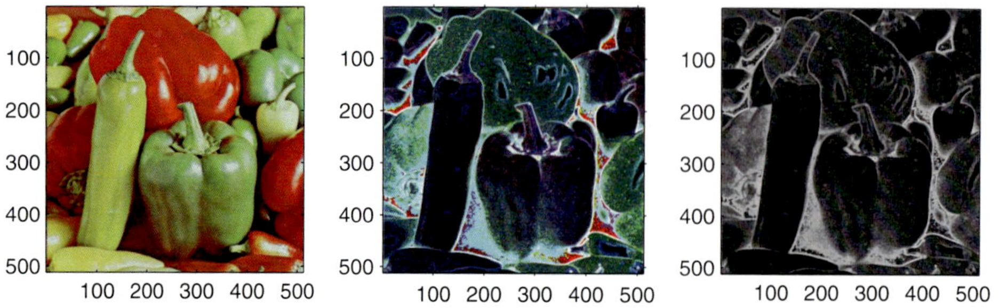

(a) image (b) MVI (c) MVI gray

Figure 8.36 (a) The color image, (b) the MVCI, and (c) grayscale image of the MVCI.

color components in part (b). The grayscale component of the MVCI is shown in part (c).

The multiplicative Michelson visibility color image is defined with the color components that are calculated by

$$c(x)_{n,m} = k\left[\frac{|\mathrm{Max}_W(x_{n,m}) - \mathrm{Min}_W(x_{n,m})|}{\mathrm{Max}_W(x_{n,m}) + \mathrm{Min}_W(x_{n,m})}\right](x_{n,m})^{\beta}. \tag{8.61}$$

Here, the image $x_{n,m}$ is used for the color components $r_{n,m}$, $g_{n,m}$, and $b_{n,m}$.

8.10 Quaternion Image Gradients

The gradient operators can be defined for color images by operating separately each color component, as well as by introducing new gradient operators in quaternion space wherein color images can be transformed. In this section, a few examples of the gradient operators in quaternion imaging are described.

In the RGB model, the color image $f_{n,m} = (r_{n,m}, g_{n,m}, b_{n,m})$ of size $N \times M$ can be represented in the quaternion space as

$$f_{n,m} \to q_{n,m} = a_{n,m} + (ir_{n,m} + jg_{n,m} + kb_{n,m}),$$

where the grayscale image $a_{n,m} = (r_{n,m} + g_{n,m} + b_{n,m})/3$, or $a_{n,m} = 0$. We consider the convolution of this image with a quaternion kernel, or a four-dimensional mask $H_{n,m}$ with the right-side multiplication as

$$(q * H)_{n,m} = \sum_{n_1=-L_1}^{L_1} \sum_{m_1=L_2}^{L_2} q_{n-n_1,m-m_1} H_{n_1,m_1}. \tag{8.62}$$

Here, we assume that the quaternion kernel is of size $(2L_1 + 1) \times (2L_2 + 1) \times 4$, where L_1 and L_2 are small positive integers, and

$$H_{n,m} = (H_e)_{n,m} + i(H_i)_{n,m} + j(H_j)_{n,m} + k(H_k)_{n,m}. \tag{8.63}$$

The real components H_e, H_i, H_j, and H_k are of size $(2L_1 + 1) \times (2L_2 + 1)$ each.

If the components $H_i = H_j = H_k = 0$, each color component of the image $f_{n,m}$ is processing by the same real mask H_e. This is the traditional component-wise color image processing, or the edge detection in color imaging, when H_e is a gradient operator. Not all edges in the color image can be found by processing separately the color components by a single gradient, especially for color images of low contrast. Therefore, we consider the cases of most interest, when the kernel is quaternion with the equal or different components.

8.10.1 1-in-4 Gradients

Let (H_e) be the mask of a real quaternion operator, and let H be the following kernel of the quaternion gradient:

$$H = [1 + i + j + k]H_e. \tag{8.64}$$

In the gradient H, the same mask in used for four components of the kernel H and we call such quaternion gradients the 1-in-4 gradients. As an example, we consider H_e to be the gradient operator G_x or G_y along the X and Y directions, respectively. We call the quaternion operators with masks

$$H_x = [1 + i + j + k]G_x \quad \text{and} \quad H_y = [1 + i + j + k]G_y \tag{8.65}$$

the quaternion gradients along the X and Y directions, respectively. The quaternion images

$$\begin{aligned} H_x(q)_{n,m} &= [1 + i + j + k]G_x(q)_{n,m} \quad \text{and} \\ H_y(q)_{n,m} &= [1 + i + j + k]G_y(q)_{n,m} \end{aligned} \tag{8.66}$$

are called *the quaternion gradient images* along the X and Y directions, respectively. We denote by $A_e(f)$, $A_i(f)$, $A_j(f)$, and $A_k(f)$, components of the quaternion gradient image. In other words,

$$\begin{aligned} \left[A_e(f)\right]_{n,m} &= \left[H(q)_{n,m}\right]_e, \quad \left[A_i(f)\right]_{n,m} = \left[H(q)_{n,m}\right]_i, \\ \left[A_j(f)\right]_{n,m} &= \left[H(q)_{n,m}\right]_j, \quad \left[A_k(f)\right]_{n,m} = \left[H(q)_{n,m}\right]_k. \end{aligned}$$

For instance, let G_x or G_y be the Sobel operators with the 3×3 window, which are given in Eq. (8.8). We denote the corresponding gradients H by H_x and H_y and call them *the quaternion Sobel gradients* along the X and Y directions, respectively.

If we denote the components of the gradient H by $h_e + (ih_i + jh_j + kh_k)$, where $h_e = h_i = h_j = h_k = h$, then the convolution in Eq. (8.62) can be written as

$$(q * H)_{n,m} = \sum_{n_1=-L_1}^{L_1} \sum_{m_1=L_2}^{L_2} q_{n-n_1,m-m_1}(1 + i + j + k)h_{n_1,m_1}. \tag{8.67}$$

In this equation, for each point (n_1, m_1), the product of $(q_e + iq_i + jq_j + kq_k)_{n-n_1,m-m_1} = q_{n-n_1,m-m_1}$ and the gradient can be written in matrix form as

$$(q * H)_{n,m} = \sum_{n_1=-L_1}^{L_1} \sum_{m_1=L_2}^{L_2} \begin{bmatrix} q_e & -q_i & -q_j & -q_k \\ q_e & q_i & q_j & -q_k \\ q_e & -q_i & q_j & q_k \\ q_e & q_i & -q_j & q_k \end{bmatrix} n - n_1, m - m_1 \, h_{n_1,m_1}.$$

As an example, Fig. 8.37 shows the color "peppers" image in part (a). The imaginary part of the quaternion Sobel gradient $H_x(q)$ as a color image is shown in part (b). The real part of the gradient is shown in part (c), i.e., the image calculated from the above equation as

$$(A_e(f))_{n,m} = \sum_{n_1=-L_1}^{L_1} \sum_{m_1=L_2}^{L_2} (q_e - q_i - q_j - q_k)_{n-n_1,m-m_1} h_{n_1,m_1}.$$

One can note that, if $q_e = 0$, this image is the minus average of the color components of the convolution, i.e.,

$$A_e(f) = (q * H)_e = -\frac{1}{3}[q_i * H + q_j * H + q_k * H].$$

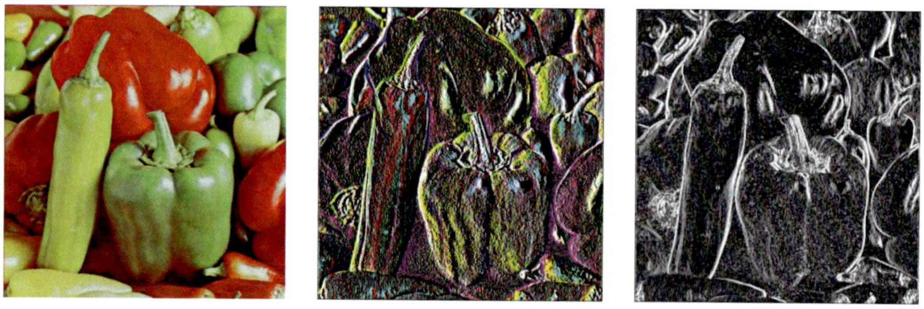

(a) (b) (c)

Figure 8.37 (a) The color "peppers" image $f_{n,m}$, and (b) the imaginary part and (c) the real part $A_e(f)$ of the quaternion Sobel gradient image.

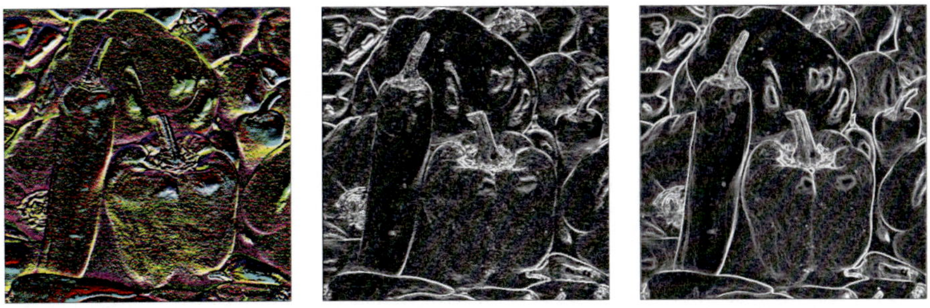

(a) (b) (c)

Figure 8.38 (a) The imaginary part and (b) the real part of the quaternion Sobel gradient $H_y(q)_{n,m}$ of the "peppers" image, and (c) the Sobel gradient image $|H(q)|_{n,m}$.

The result of processing the "peppers" image with the quaternion Sobel gradient H_y is shown in Fig. 8.38. The imaginary part of the gradient as a color image is shown in Fig. 8.38(a) and the real part of the quaternion gradient in Fig. 8.38(b). The sum of these two gradients, namely, their magnitude $|H(q)| = |[H_x(q)]_e| + |[H_y(q)]_e|$ is shown in part Fig. 8.38(c).

We also consider the example with the "flowers" image, which is shown in Fig. 8.39(a). The imaginary part of the quaternion Sobel gradient $H_x(f)$ is shown in Fig. 8.39(b). The real part of the gradient is shown in Fig. 8.39(c).

The result of processing the "flowers" image with the Sobel gradient $H_y(q)$ is shown in Fig. 8.40. The imaginary part of this gradient image as a color image is shown in Fig. 8.40(a) and the real part of the gradient image in Fig. 8.40(b). The Sobel gradient image $|H(q)| = |[H_x(q)]_e| + |[H_y(q)]_e|$ is shown in Fig. 8.40(c).

8.10.2 4-in-4 Gradients

In this section, we consider the quaternion gradients H which are composed by different components, i.e., when the masks h_e, h_i, h_j, and h_k are different.

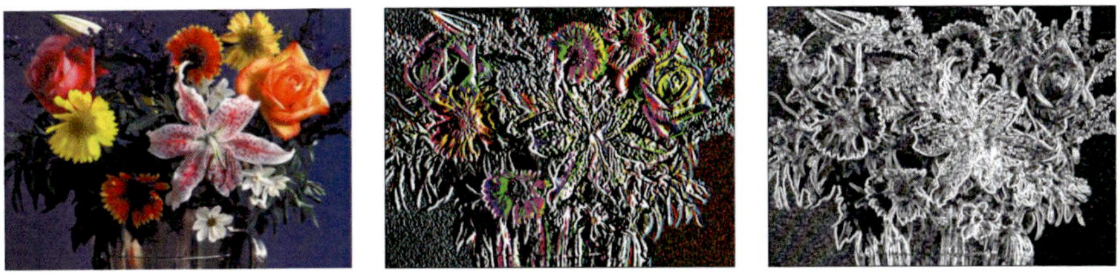

(a) (b) (c)

Figure 8.39 (a) The color "flowers" image $f_{n,m}$, (b) the imaginary part, and (c) the real part $A_e(f)$ of the quaternion Sobel gradient image.

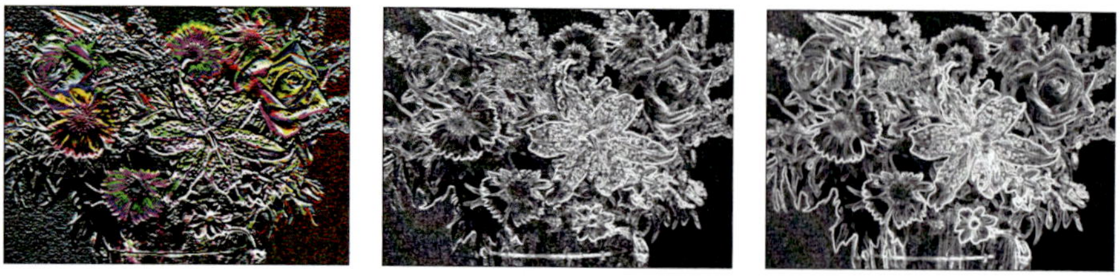

(a) (b) (c)

Figure 8.40 (a) The imaginary part and (b) the real part of the quaternion Sobel gradient $H_y(q)_{n,m}$ of the "flowers" image and (c) the Sobel gradient image $|H(q)|_{n,m}$.

As an example, the quaternion gradient can be composed by the following gradients:

$$h_e = \begin{bmatrix} -1 & 0 & 1 \\ -2 & 0 & 2 \\ -1 & 0 & 1 \end{bmatrix}, h_i = \begin{bmatrix} -1 & 0 & 1 \\ -1 & 0 & 1 \\ -1 & 0 & 1 \end{bmatrix}, h_j = \begin{bmatrix} 1 & 2 & 1 \\ 0 & 0 & 0 \\ -1 & -2 & -1 \end{bmatrix}, h_k = \begin{bmatrix} 1 & 1 & 1 \\ 0 & 0 & 0 \\ -1 & -1 & -1 \end{bmatrix}.$$

The mask of this quaternion gradient with four components can be written as the following 6×6 matrix:

$$[H] = \left[\begin{array}{c|c} h_e & h_i \\ \hline h_j & h_k \end{array}\right] = \left[\begin{array}{ccc|ccc} -1 & 0 & 1 & -1 & 0 & 1 \\ -2 & 0 & 2 & -1 & 0 & 1 \\ -1 & 0 & 1 & -1 & 0 & 1 \\ \hline 1 & 2 & 1 & 1 & 1 & 1 \\ 0 & 0 & 0 & 0 & 0 & 0 \\ -1 & -2 & -1 & -1 & -1 & -1 \end{array}\right]. \tag{8.68}$$

The first two masks, which are written in the first three columns of this matrix, correspond to the Sobel gradients, and the next two masks correspond to the Prewitt gradients along the X and Y-directions, respectively.

As an example, Fig. 8.41 shows the color image in part (a) and the imaginary part of this gradient image as a color image in part (b). The real part of the gradient image is shown in part (c).

In another example shown in Fig. 8.42, the color image of size 500×302 is shown in part (a) and the real part of the quaternion gradient image $H(q)_{n,m}$ in part (b).

Now, we consider the Prewitt compass gradients with directions: West, North, South, and East and compose the quaternion gradient with the mask H corresponding to the following 6×6 matrix:

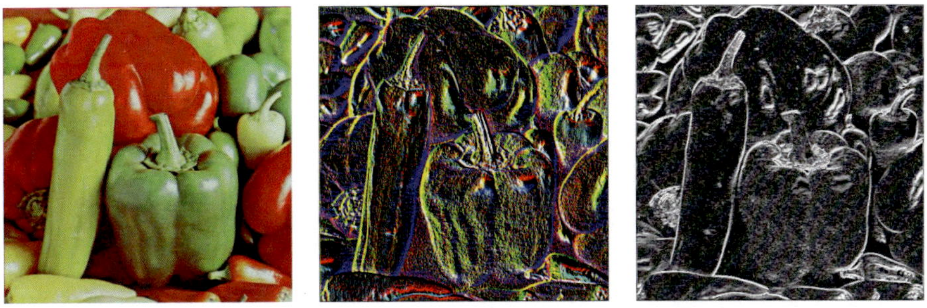

(a) (b) (c)

Figure 8.41 (a) The color "peppers" image, (b) the imaginary part, and (c) the real part of the quaternion gradient image $H(q)_{n,m}$.

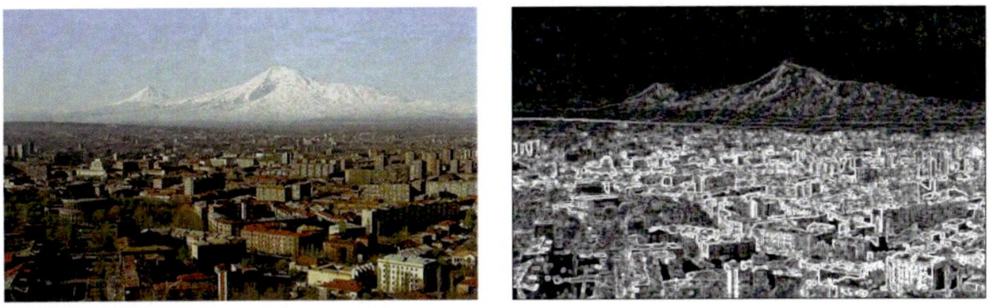

(a) (b)

Figure 8.42 (a) The color "Mountains and city" image and (b) the real part of the quaternion gradient image $H(q)_{n,m}$.

$$[H] = [H_P] = \left[\begin{array}{c|c} h_W & h_N \\ \hline h_S & h_E \end{array}\right] = \left[\begin{array}{ccc|ccc} -1 & 1 & 1 & -1 & -1 & -1 \\ -1 & -2 & 1 & 1 & -2 & 1 \\ -1 & 1 & 1 & 1 & 1 & 1 \\ \hline 1 & 1 & 1 & 1 & 1 & -1 \\ 1 & -2 & 1 & 1 & -2 & -1 \\ -1 & -1 & -1 & 1 & 1 & -1 \end{array}\right]. \quad (8.69)$$

We call the quaternion gradient with this mask *the Prewitt compass quaternion gradient.* As an example, Fig. 8.43 shows the color image in part (a) and the imaginary part of this gradient image as a color image in part (b). The real part of the gradient image is shown in part (c).

For comparison, we calculate *the Robinson 5-level compass quaternion gradient,* which is defined by the Robinson compass gradients with the same directions. The mask of this quaternion gradient is presented by the following 6×6 matrix:

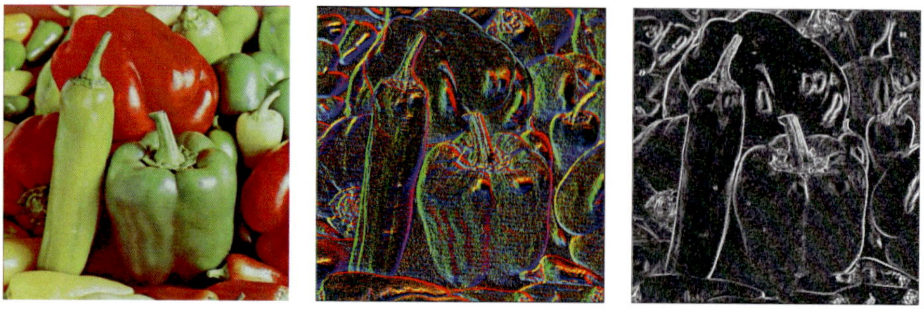

(a) (b) (c)

Figure 8.43 (a) The color "peppers" image, (b) the imaginary part, and (c) the real part of the Prewitt compass quaternion gradient image $H(q)_{n,m}$.

$$[H] = [H_R] = \left[\begin{array}{c|c} h_W & h_N \\ \hline h_S & h_E \end{array}\right] = \left[\begin{array}{ccc|ccc} -1 & 0 & 1 & -1 & -2 & -1 \\ -2 & 0 & 2 & 0 & 0 & 0 \\ -1 & 0 & 1 & 1 & 2 & 1 \\ \hline 1 & 2 & 1 & 1 & 0 & -1 \\ 0 & 0 & 0 & 2 & 0 & -2 \\ -1 & -2 & -1 & 1 & 0 & -1 \end{array}\right]. \tag{8.70}$$

Figure 8.44 shows the imaginary part of the Robinson 5-level quaternion "peppers" gradient image as a color image in part (a). The real part of the gradient image is shown in Fig. 8.44(b). The results of above images show that these quaternion gradients H can be used for edge detection.

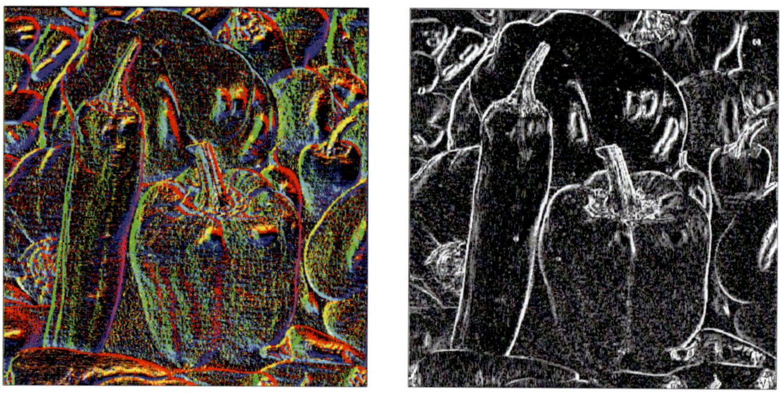

(a) (b)

Figure 8.44 (a) The imaginary part and (b) the real part of the Robinson 5-level compass quaternion gradient "peppers" image $H(q)_{n,m}$.

Thus, the application of quaternion algebra allows for enlarging the class of gradient operators for color imaging. The operation of the quaternion 2-D convolution for the described above gradient operators is considered with the multiplication of the image by the kernel of the gradient from the right. The 2-D quaternion convolution for such gradient operators can also be used with the multiplication by the kernel from the left. These two 2-D convolutions allow for considering new operations of the quaternion window-mean filters and other filters.

8.11 Color Facial Image Representation

The facial color image recognition can be accomplished by analyzing color components separately, or by using the grayscale image calculated as the average of colors in the RGB model. In this section, we consider a representation of the color facial image, to obtain a set of features that can be used in face classification, when the color image is transformed into the quaternion space. The color image can be presented as a full 4-D quaternion image or as a 3-D pure quaternion image, and then can be processed by the quaternion gradient operations or a visibility image can be calculated. Then, the real component of the obtained quaternion image can be used as a grayscale image for color facial image presentation.

The main parts of processing the color facial image $f_{n,m}$ of size $N \times M$ are shown in the block-diagram of Fig. 8.45. The color image is considered in the RGB model. The processing of color images in the XYZ, CMY, CMYK, and other color models is described similar to the RGB model case.

The algorithm of the facial image processing can be described by the following steps.

Stage 1. The color image is transformed to the quaternion image by

$$f_{n,m} \rightarrow q_{n,m} = (a_{n,m}, r_{n,m}, g_{n,m}, b_{n,m}),$$

where the real components $a_{n,m} = (r_{n,m} + g_{n,m} + b_{n,m})/3$, or $a_{n,m} \equiv 0$.

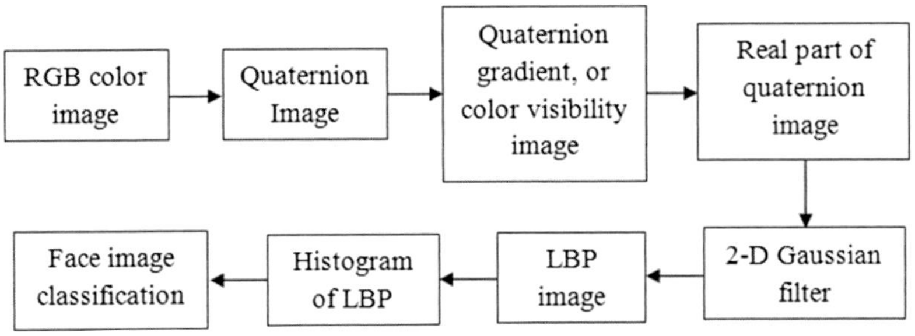

Figure 8.45 The block-diagram of the color facial image processing.

Stage 2. Given a quaternion gradient or image visibility image operator V, a new quaternion image is calculated by

$$q_{n,m} \to V(q)_{n,m}, \quad n = 0 : (N-1), m = 0 : (M-1).$$

The grayscale image, which is the real component of the quaternion image

$$(q_e)_{n,m} = [V(q)_{n,m}]_e,$$

is considered for processing in the next stage.

Stage 3. The 2-D Gaussian function is circular convoluted with the grayscale image,

$$(q_e)_{n,m} \to G(q_e)_{n,m} = (q_e \otimes g)_{n,m},$$

where $g_{n,m}$ is the mask of the Gaussian function with the standard deviation equal 0.5. The value of this parameter can be changed, for instance by 1, depending on the image.

Stage 4. A complex gradient image composition is calculated. In the 3×3-square window W, we consider the following eight gradient operators $A_1, A_2, \ldots, A_8$ which are given in Eq. (8.45),

$$G(q_e) \to A_k[G(q_e)], \quad k = 1, 2, \ldots, 8.$$

Stage 5. Eight binary images are calculated by

$$(B_k)_{n,m} = u\Big\{A_k[G(q_e)]_{n,m}\Big\}, \quad k = 1, 2, \ldots, 8. \tag{8.71}$$

The uniform LBP table is used to compose the new image by applying the uniform LBP table after each gradient image calculation to obtain the images $(g_1)_{n,m}, (g_2)_{n,m}, \ldots, (g_7)_{n,m}$, and $(g_8)_{n,m}$, as described in Eq. (8.53). Thus, we obtain the LBP image that can be written as

$$b_{n,m} = (g_8)_{n,m} = \sum_{r=1}^{8} 2^{r-1} u\big[(q_e)_{n,m} - (q_e)_{n+s(k),m+p(k)}\big], \tag{8.72}$$

where $(s(k), p(k))$ are the sampling points given in Eq. (8.46).

The uniform LBP image is the image $(g_8)_{n,m}$ with maximum of 59 integer levels s, from 0 to 58. The histogram of this image, $H(s)$ is considered to be the feature of the color facial image $f_{n,m}$, which can be used in face classification.

As an example, Fig. 8.46 shows the color facial image of size 313×250 in part (a). The quaternion gradient image $H(q)_{n,m}$ calculated by using the quaternion gradient in Eq. (8.68) is shown in Fig. 8.46(b), and after filtering by the 2-D Gaussian function with the standard deviation of 0.5 in part (c).

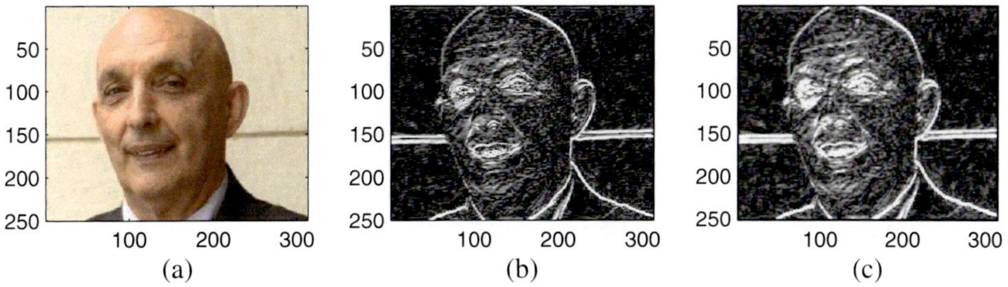

Figure 8.46 (a) The original image and the quaternion gradient image (b) before and (c) after filtering by the 2-D Gaussian function.

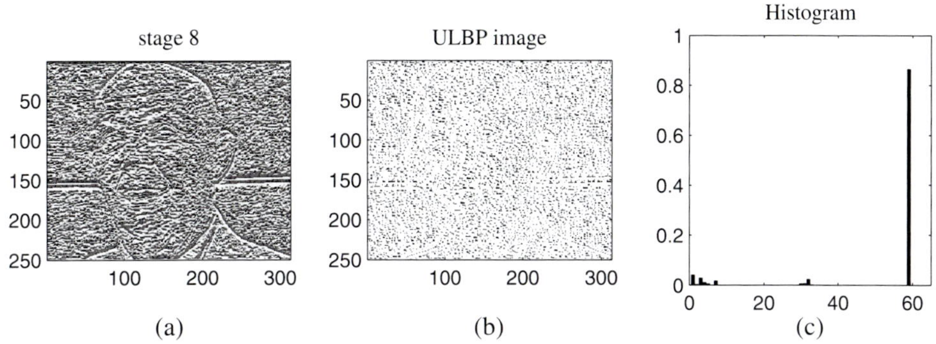

Figure 8.47 (a) The image $(g_8)_{n,m}$ before the mapping, (b) the uniform LBP image, and (c) the histogram of the image.

The results of further facial image processing are shown in Fig. 8.47. The image $(g_8)_{n,m}$ before and after using the mapping by the uniform LBP table is shown in Fig. 8.47(a,b). The normalized histogram of the uniform LBP image is given in Fig. 8.47(c).

References

[1] S. S. Agaian, K. Panetta, and A. M. Grigoryan, "A new measure of image enhancement," *Proc. IASTED Int. Conf. Signal Processing & Communication*, Marbella, Spain, 19–22 (2000).

[2] S. S. Agaian, K. Panetta, and A. M. Grigoryan, "Transform-based image enhancement algorithms," *IEEE Trans. on Image Processing*, **10**(3), 367–382 (2001).

[3] A. M. Grigoryan and S. S. Agaian, *Multidimensional Discrete Unitary Transforms: Representation, Partitioning and Algorithms*, Marcel Dekker Inc., New York (2003).

[4] A. M. Grigoryan and S. S. Agaian, "Transform-based image enhancement algorithms with performance measure," *Advances in Imaging and Electron Physics, Academic Press*, **130**, 165–242 (2004).
[5] S. S. Agaian, B. Silver, and K. A. Panetta, "Transform coefficient histogram-based image enhancement algorithms using contrast entropy," *Image Processing, IEEE Transactions on*, **16**(3), 741–758 (2007).
[6] A. M. Grigoryan and M. M. Grigoryan, *Brief Notes in Advanced DSP: Fourier Analysis With MATLAB*, CRC Press, Taylor and Francis Group, Boca Raton, New York (2009).
[7] W. M. Morrow, R. B. Paranjape, R. M. Rangayyan, and J. E. L. Desautels, "Region-based contrast enhancement of mammograms," *IEEE Trans. on Medical Imaging*, **11**(3), 392–406 (1992).
[8] T. L. Ji, M. K. Sundareshan, and H. Roehrig, "Adaptive image contrast enhancement based on human visual properties," *IEEE Trans. on Medical Imaging*, **13**(4), 573–586 (1994).
[9] P. Zamperoni, "Image enhancement," *Advanced in Image and Electron Physics*, **92**, 1–77 (1995).
[10] E. A. Silva, K. A. Panetta, and S. S. Agaian, "Quantify similarity with measurement of enhancement by entropy," *Proc. SPIE*, **6579**, 65790U (2007) [doi: 10.1117/12.720087].
[11] T. Acharya and A. K. Ray, *Image Processing; Principles and Applications*, A Wiley Interscience Publication, New Jersey (2005).
[12] S. Hecht, "The visual discrimination of intensity and the Weber-Fechner law," *The Journal of General Physiology*, 235–267 (1924).
[13] S. DelMarco and S. Agaian, "The design of wavelets for image enhancement and target detection," *Proc. SPIE*, **7351**, 735103 (2009) [doi: 10.1117/12.816135].
[14] H. L. de Vries, "The quantum character of light and its bearing upon threshold of vision, the differential sensitivity and visual acuity of the eye," *Physica*, **10**, 553–564 (1943).
[15] A. Rose, "The sensitivity performance of the human eye on an absolute scale," *J. Opt. Soc. Am.*, **38**, 196–208 (1948).
[16] G. Buchsbaum "An analytical derivation of visual nonlinearity," *IEEE Trans. on Biomedical Engineering*, **27**(5), 237–242 (1980).
[17] K. A. Panetta, E. J. Wharton, and S. S. Agaian, "Human visual system-based image enhancement and logarithmic contrast measure," *IEEE Transactions on Systems, Man, and Cybernetics, Part B (Cybernetics)*, **38**(1), 174–188 (2008).
[18] R. C. Gonzalez and R. E. Woods, *Digital Image Processing*, 2nd Edition, Prentice Hall, Upper Saddle River, New Jersey (2002).
[19] B. K. P. Horn, *Robot Vision*, The MIT Press, Cambridge (1986).
[20] W. K. Pratt, *Digital Image Processing*, 3rd Edition, John Wiley & Sons, Inc. (2001).

[21] D. Marr and E. Hildreth, "Theory of edge detection," *Proc. R. Boc. Lond. B 207*, Printed in Great Britain, 187–217 (1980).

[22] A. M. Grigoryan and S. S. Agaian, "Split manageable efficient algorithm for Fourier and Hadamard transforms," *IEEE Trans. Signal Processing*, **48**(1), 172–183 (2000).

[23] A. M. Grigoryan, "2-D and 1-D multi-paired transforms: Frequency-time type wavelets," *IEEE Trans. on Signal Processing*, **49**(2), 344–353 (2001).

[24] E. Wharton, S. Agaian, and K. Panetta, "A logarithmic measure of image enhancement," *Proc. SPIE*, **6250**, 62500P (2006) [doi: 10.1117/12.665693].

[25] E. Wharton, S. Agaian, and K. Panetta, "Comparative study of logarithmic enhancement algorithms with performance measure," *Proc. Electronic Imaging* (2006) [doi: 10.1117/12.647489].

[26] E. Wharton, K. Panetta, and S. Agaian, "Human visual system based similarity," *Proc. Systems, Man and Cybernetics, SMC 2008. IEEE International Conference on*, 685–690 (2008).

[27] E. Wharton, K. Panetta, and S. Agaian, "Human visual system based similarity metrics," *IEEE System, Man and Cybernetics Conference*, Singapore (2008).

[28] S. Agaian, "Visual morphology," *Proc. SPIE*, **3646** (1999) [doi:10.1117/12.341081].

[29] K. A. Panetta, C. Gao, and S. S. Agaian, No Reference Color Image Contrast and Quality Measures, *IEEE Transactions on Consumer Electronics*, **59**(3), 643–651 (2013).

[30] A. M. Grigoryan and S. S. Agaian, "Retolling of color imaging in the quaternion algebra," *Applied Mathematics and Sciences: An International Journal (MathSJ)*, **1**(3), 23–39 (2014).

[31] A. M. Grigoryan and S. S. Agaian, "Alpha-rooting method of color image enhancement by discrete quaternion Fourier transform," *Proc. SPIE*, **9019**, 901904 (2014) [doi: 10.1117/12.2040596].

[32] A. M. Grigoryan, J. Jenkinson, and S. S. Agaian, "Quaternion Fourier transform based alpha-rooting method for color image measurement and enhancement," *SIGPRO-D-14-01083R1, Signal Processing*, **109**, 269–289 (2015) [doi: 10.1016/j.sigpro.2014.11.019].

[33] A. M. Grigoryan and S. S. Agaian, "Color Enhancement and Correction for Camera Cell Phone Medical Images Using Quaternion Tools," Chapter 4, 77–117. In book: Electronic Imaging Applications in Mobile Healthcare, (J. Tang, S.S. Agaian, and J. Tan, Eds.,) *SPIE Press*, Bellingham, Washington (2016) [doi: 10.1117/3.2204748.ch4].

[34] S. C. Nercessian, S. S. Agaian, and K. A. Panetta, "A generalized set of kernels for edge and line detection," *Proc. SPIE*, **7245**, 72450U (2009) [doi:10.1117/12.805517].

[35] S. C. Nercessian, S. S. Agaian, and K. A. Panetta, “Multi-scale image enhancement using a second derivative-like measure of contrast,” *Proc. SPIE*, **8295**, 82950Q (2012) [doi: 10.1117/12.906494].

[36] R. Kogan, S. Agaian, and K. A. Panetta, “Visualization Using Rational Morphology and Zonal Magnitude Reduction,” *Proc. SPIE*, **3304** (1998) [doi: 10.1117/12.304595].

[37] E. Wharton, S. Agaian, and K. Panetta, “Adaptive Multi-Histogram Equalization using Human Vision Thresholding,” SPIE Electronic Imaging, (2007) [doi: 10.1117/12.704474].

[38] S. Agaian and A. Almuntashri, “Human visual system-based edge detection using image contrast enhancement and logarithmic ratio,” *Proc. SPIE*, **7708**, 770812 (2010) [doi: 10.1117/12.849609].

[39] K. Panetta, L. Bao, and S. Agaian, “A human visual no-reference image quality measure,” *IEEE Instrumentation & Measurement Magazine*, **19** (3), 34–38 (2016).

[40] Q. Wan, K. Panetta, and S. Agaian, “Autonomous facial recognition based on the human visual system,” *IEEE International Conference on Imaging Systems and Techniques (IST)*, 1–6, (2015).

[41] C. Gao, K. Panetta, and S. Agaian, “Color image attribute and quality measurements,” *Proc. SPIE*, **9120**, 91200T (2014) [doi: 10.1117/12.2050197].

[42] C. Gao et al., “A new color contrast enhancement algorithm for robotic applications,” presented at the IEEE conference on Technologies for Practical Robot Applications (2012).

[43] C. Gao, K. Panetta, and S. Agaian, “No reference color image quality measures,” *Cybernetics (CYBCONF)*, 2013 IEEE International Conference on, 243–248, (2013).

[44] D. Mandal, K. Panetta, and S. Agaian, “Face recognition based on logarithmic local binary patterns,” *Proc. SPIE*, **8655**, 865514 (2013) [doi: 10.1117/12.1000250].

[45] Z. Yicong, S. S. Agaian, and K. Panetta, “Nonlinear Unsharp Masking for Enhancing Suspicious Regions in Mammograms,” Computer-Aided Cancer Detection and Diagnosis: Recent Advances, *SPIE*, (2013) [doi: 10.1117/3.1002311.ch4].

[46] D. Mandal, K. Panetta, and S. Agaian, “Human visual system inspired object detection and recognition,” *IEEE International Conference on Technologies for Practical Robot*, (2012).

[47] ANSI/INFOCOMM 3M-2011 Projected Image System Contrast Ratio, InfoComm International (2011), http://www.infocomm.org/cps/rde/xbcr/infocomm/ANSIINFOCOMM_3M-2011_PISCR_Overview.pdf

[48] TV Contrast Ratio Explained - Part 1; http://www.practical-home-theater-guide.com/contrast-ratio.html

[49] ANSI/HFS 100-1988, American National Standard for Human Factors Engineering of Visual Display Terminal Workstations, Section 6, 17–20 (1988).

[50] T. Ojala, N. Pietikäinen, and T. Maenpaa, "Multiresolution gray scale and rotation invariant texture classification with local binary patterns," *IEEE Trans. Pattern Analysis and Machine Intelligence*, **24**(7), 971–987 (2002).

[51] H. Jin, Q. Liu, H. Lu, and X. Tong, "Face detection using improved LBP under Bayesian framework," *Proc. of the Third Int. Conf. on Image and Graphics (ICIG)*, 306–309 (2004).

[52] M. Heikkilä, M. Pietikäinen, and C. Schmid, "Description of interest regions with center-symmetric local binary patterns," *Patternt Recognition*, **42**(3), 425–436 (2009).

[53] T. Ahonen, A. Halid, and M. Pietikäinen, "Face recognition with local binary patterns," *Proc. Euro. Conf. Computer Vision (ECCV)*, 469–481 (2004).

Chapter 9
Color Image Restoration and QDFT

In this chapter, we consider the model of quaternion signal (or image) degradation when the signal (or image) is convoluted and an additive noise is added. The classical model of such a model leads to the solution of the optimal Wiener filter, where the optimality with respect to the mean square error. The characteristic of this filter can be found in the frequency domain by using the Fourier transform. For quaternion signals and images, the inverse problem is complicated by the fact that the quaternion arithmetic is not commutative. The quaternion Fourier transform does not map the convolution to the operation of multiplication. In this chapter, we analyze the linear model of the signal and image degradation with an additive independent noise and describe the simple optimal filtration of the quaternion signals and images in the frequency domain.

9.1 Problem of Image Restoration

In many applications of signal and image processing when the distortions take place, it is important to solve the problem of signal and image restoration. The linear filtration of a degraded image is the well-known procedure in signal and image restoration [1]–[9]. Methods of optimal restoration, including the Wiener filtration [10]–[13], are used widely in signal and image processing and number of applications continues to grow [14]. These methods were developed for denoising medical images in ultrasound imaging [15], in parameter estimation for linear stochastic systems [16], for edge preserving maximum a posteriori estimation of images in the presence of blur and noise [17], for noise removal from hyper-spectral images [18], in electron microscopy [19], for blind image convolution [20], Bayesian robust optimal filters [21]–[30], and for many other applications [31, 35]–[42]. Many methods of image filtration were applied for processing the color images. The Fourier transform-based method is one of these methods, which in many cases is applied to each color plane-component of the color image separately. In other words, the color image is considered as a triplet of separate 2-D grayscale images and each of these images represents the red, green, or blue component of the color.

The existent variety of methods for solving problem of image restoration, including the classical Fourier transformation, are described by the convolution type integral:

$$\int_X h(y-x)o(x)dx = u(y), \quad y \in Y, \tag{9.1}$$

with a given function $h(x)$ called the kernel of the equation. Here $o = o(x)$ is an unknown function of two variables of a certain plane X (true image plane), and $u = u(y)$ is the given function, defined in another plane Y (possible image plane).

Our problem is in finding the approximate solutions $\hat{o}(x)$ of the true image $o(x)$ from the given data $u(y)$. The right side $u(y)$ of Eq. (9.1) often is not known and instead of it we have a certain (changed) element of the initial data $i(y) = u(y) + n(y)$, where $n(y)$ is the experimental error (noise). In other words, we have the (non-linear) system wherein a noise $n(y)$ is additively added on the output image $u(y)$,

$$\int_X h(y-x)o(x)dx + n(y) = i(y), \quad y \in Y. \tag{9.2}$$

9.2 Classical Model of Image Restoration

We consider a system of image formation that is shown in Fig. 9.1, taking into account an additive noise. The integral Eq. (9.2) can be analyzed in the frequency domain Ω. For the given summable function $f(x)$ on the real line R (or plane R^2), we denote by $\mathcal{F}[f](\omega)$ its Fourier transform at the frequency ω, i.e.,

$$\mathcal{F}[f](\omega) = \int f(x)e^{-j(wx)}dx, \; \omega \in (-\infty, \infty).$$

A similar double integral is used for the Fourier transform of the 2-D function. Owing to the Boreal theorem about multiplication of spectra, Eq. (9.2) takes the form

$$I(\omega) = O(\omega)H(\omega) + N(\omega). \tag{9.3}$$

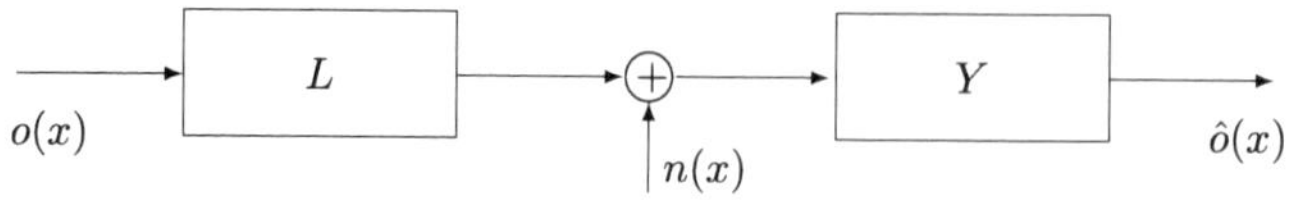

Figure 9.1 Linear model of signal/image restoration with an additive noise.

Here $I(\omega)$ is the Fourier image $\mathcal{F}[i](\omega)$ of the degraded (observed) image formed from the true one $O(\omega) = \mathcal{F}[o](\omega)$ by the linear system L with the transfer function $H(\omega) = \mathcal{F}[h](\omega)$ when an additive noise $N(\omega) = \mathcal{F}[n](\omega)$ acts during the image formation.

9.2.1 Optimal filter

We consider the solution $\hat{O}$ of the convolution Eq. (9.3) which minimizes the standard error ρ^2 and is optimal with respect to the metric

$$\rho^2(\hat{O}, O) = \int_\Omega < |\hat{O}(\omega) - O(\omega)|^2 > d\omega = \min\,, \tag{9.4}$$

where $<\cdot>$ denotes an expected value. We shall consider that all variables $O(\omega)$ and $N(\omega)$ are square integrable functions on Ω, i.e.,

$$\int_\Omega < |O(\omega)|^2 > d\omega < +\infty, \quad \int_\Omega < |N(\omega)|^2 > d\omega < +\infty\,.$$

The error is defined as the expected error across the frequency domain. Assuming the stochastic integral in the first line exists in the mean-square sense, the error is itself a random variable and the existence of the integral permits us to bring the expectation inside the integral, so that the integral becomes deterministic [25].

The indicated process is called *optimal*, or *Wiener, filtration* of image and the corresponding filter is called *the Wiener filter.* As is well known, the response function of the filter equals

$$Y_{Win}(\omega) = \frac{\bar{H}(\omega)}{|H(\omega)|^2 + \phi_{N/O}(\omega)}, \quad \omega \in \Omega. \tag{9.5}$$

Here $\phi_{N/O}(\omega)$ denotes the ratio noise-signal $\phi_N(\omega)/\phi_O(\omega)$, where $\phi_N(\omega)$ and $\phi_O(\omega)$ are spatial spectral densities (SSDs) of the restored and true images, i.e., $\phi_N(\omega) =< |N(\omega)|^2 >$ and $\phi_O(\omega) =< |O(\omega)|^2 >$. The value $\bar{H}(\omega)$ is the complex conjugate to $H(\omega)$. The optimal linear filter depends on both the object signal and degradation. The error for the optimal linear filter is

$$\min_{\hat{O}=YI} \rho^2(\hat{O}, O) = \int_{R^2} \frac{\phi_N(\omega)}{|H(\omega)|^2(\omega) + \phi_{N/O}(\omega)} d\omega. \tag{9.6}$$

Now, we derive the optimal filter in the case when $< O(\omega)N(\omega) >= 0$, i.e., the variables $O(\omega)$ and $N(\omega)$ are independent. We have the following Lagrangian in the integral in Eq. (9.4):

$$\begin{aligned}\mathcal{L} &=< |O(\omega) - Y(\omega)[O(\omega)H(\omega) + N(\omega)]|^2 > \\ &= \phi_O(\omega) + Y(\omega)\bar{Y}(\omega) < |O(\omega)H(\omega) + N(\omega)|^2 > \\ &\quad - \bar{Y}(\omega)[\phi_O(\omega)\bar{H}(\omega)] - Y(\omega)\phi_O(\omega)H(\omega).\end{aligned} \tag{9.7}$$

The minimizing condition

$$\begin{aligned}\frac{\partial \mathcal{L}}{\partial \bar{Y}} &= Y(\omega) < |O(\omega)H(\omega) + N(\omega)|^2 > -\phi_0(\omega)\bar{H}(\omega) = 0 \\ &= Y(\omega)[\phi_O(\omega)|H(\omega)|^2 + \phi_N] - \phi_0(\omega)\bar{H}(\omega) = 0\end{aligned} \tag{9.8}$$

results in the filter with the response function of Eq. (9.5).

Comparing the obtained image estimate in the proposed model with the inverse filter, $Y_{Inv}(\omega) = 1/H(\omega)$,

$$\hat{O}_{Inv}(\omega) = Y_{Inv}(\omega)\hat{I}(\omega) = O(\omega) - N_O(\omega)/H(\omega), \tag{9.9}$$

we obtain the following errors of approximations:

$$\rho^2(O, \hat{O}_{Win}) = \int \frac{\phi_N(\omega)}{|H(\omega)|^2 + \phi_{N/O}(\omega)} dw \le \rho^2(O, \hat{O}_{Inv}) = \int \frac{\phi_N(\omega)}{|H(\omega)|^2} dw. \tag{9.10}$$

In general, the error of the $O(\omega)$ restoration by a linear filter $Y(\omega)$ is

$$\rho^2(\hat{O}, O) = \int_\Omega [|1 - Y(\omega)H(\omega)|^2\phi_O(\omega) + |Y(\omega)|^2\phi_N(\omega)]d\omega. \tag{9.11}$$

In the considered model, $\hat{O}(\omega)$ is the best approximation of the true image. The Wiener filtration is successfully used in the various problems of image processing and has different modifications [6, 7, 12, 14].

As an example, the Wiener filtration of the 1-D degraded signal is illustrated in Fig. 9.2. The original signal of length 512 was convoluted with the window $h = \{\ldots 0, 1, 2, \underline{3}, 2, 1, 0, ..\}/9$ and the additive noise with the normal distribution $\mathcal{N}(1, 4)$ was added to the signal. Here, $h(0) = 3$.

Figure 9.3 shows the spectral characteristics of the 1-D optimal Wiener filter. The magnitude of the frequency characteristic H_p is shown in Fig. 9.3(a) and the magnitude of the Wiener filter in Fig. 9.3(b). Both characteristics are plotted in the decibel scale, $|X| \to -20\log_{10}(|X|)$. The ratio of noise-to-signal ϕ_N/ϕ_O is given in part c. The mean-root-square errors (MRSE) of convolution and the degraded signal are 0.6974 and 0.7223, respectively. The error of the optimal filtration of this signal is 0.3065.

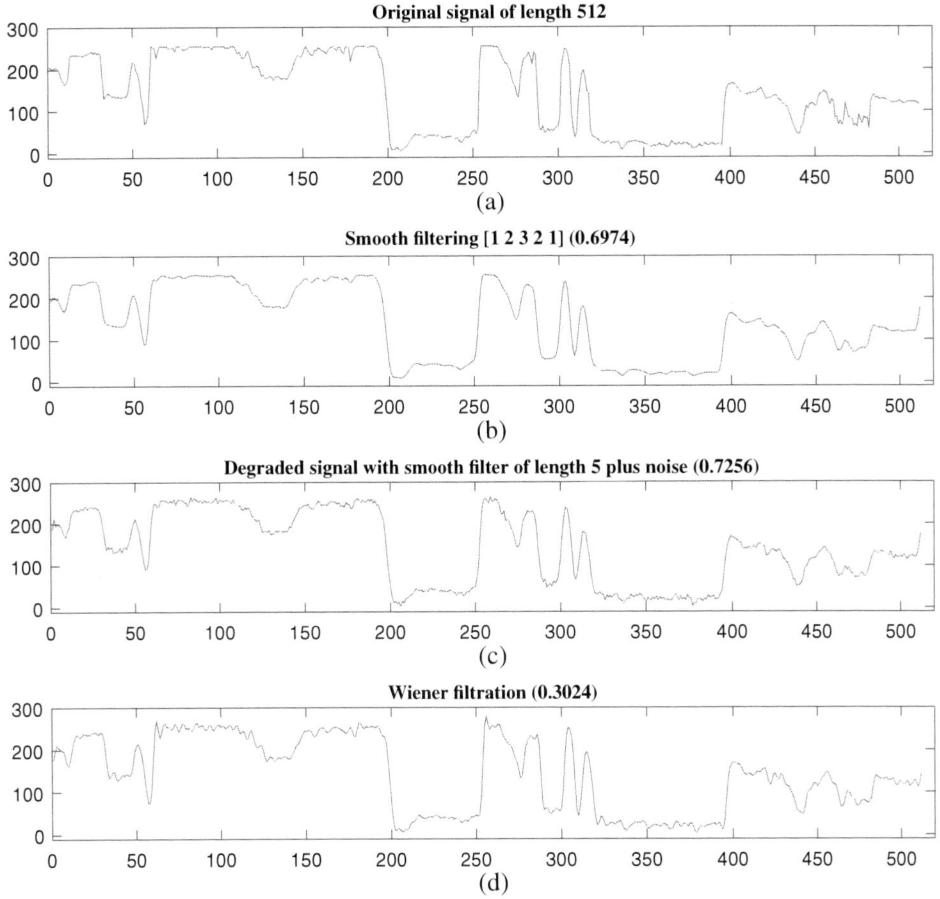

Figure 9.2 (a) Original signal, (b) smooth signal, (c) noisy signal, and (d) signal processed by the 1-D Wiener filter.

For the 2-D case, Fig. 9.4 shows the result of the optimal filtration of the degraded "Miki-Anoush-Mini" image in the model blur + noise. The original image is shown in Fig. 9.4(a), and the circular convolution of the image with the 3×3 window

$$h = \frac{1}{15}\begin{bmatrix} 1 & 2 & 1 \\ 2 & \underline{3} & 2 \\ 1 & 2 & 1 \end{bmatrix}, \quad \left[h(0,0) = \frac{3}{15} = \frac{1}{5}\right],$$

is shown in Fig. 9.4(b). The noisy image in Fig. 9.4(c) is obtained by adding a small by amplitude random noise distributed as $\mathcal{N}(0,4)$. The optimal restoration of the image with error 0.0132 is given in Fig. 9.4(d). The MRS errors of blurring is 0.0353 and adding noise is 0.0361.

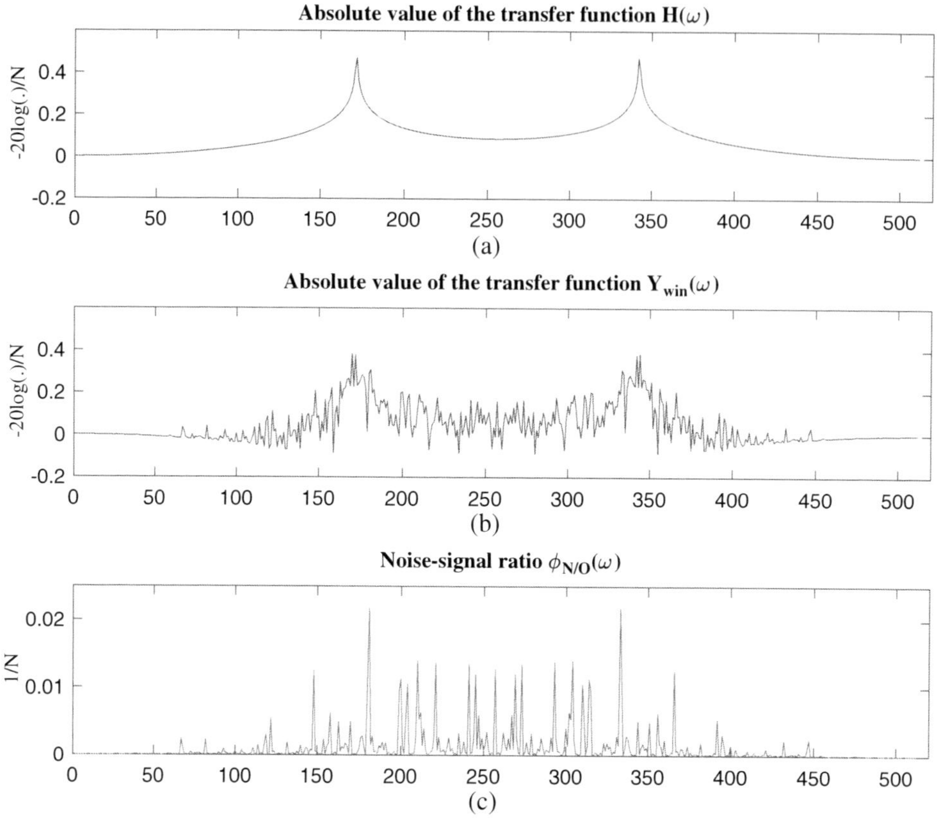

Figure 9.3 Spectral characteristics of (a) the window *h*, (b) the 1-D optimal Wiener filter, and (c) the noise-to-signal ratio ϕ_N/ϕ_O.

Figure 9.5 shows the magnitude of the frequency characteristic $H_{p,s}$ in part (a), and the magnitude of the Wiener filter in part (b). Both characteristics are plotted in the decibel scale.

We consider the RGB color model that is used widely in color imaging [3, 50]. A quaternion number has four components, a real part and three imaginary parts, which naturally coincides with the three components, R(ed), G(reen), and B(lue) of a color pixel. Therefore, a discrete color image $f(n,m)$ in the RGB color space can be transformed into imaginary part of quaternion numbers, by encoding the red, green, and blue components of the RGB value as a pure quaternion (with zero real part):

$$f_{n,m} = 0 + (r_{n,m}i + g_{n,m}j + b_{n,m}k).$$

In this additive model, colors are produced by adding components $C = rR + gG + bB$ and reproduce a broad range of colors. Practically, the color is expressed as the triplet (R,G,B), each component of which can vary

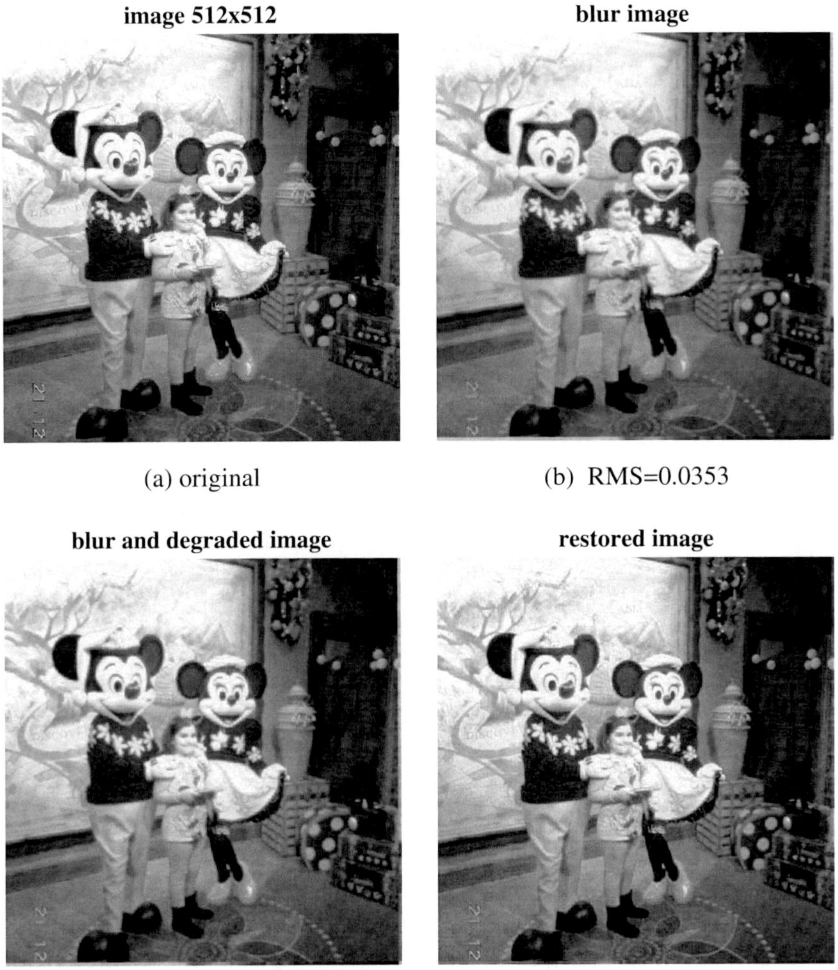

(a) original (b) RMS=0.0353

(c) RMS=0.0361 (d) RMS=0.0132

Figure 9.4 (a) Original 512 × 512 image, (b) blur image, (c) noisy image, and (d) image processed by the 2-D Wiener filter.

from zero to a defined maximum value, for instance 1. The advantage of using quaternion based operations to manipulate color information in an image is that we do not have to process each color channel independently, but rather, treat each color triple as a whole unit. By using quaternion operations, higher color information accuracy can be achieved because the color is treated as an entity [51, 52].

The quaternion DFT has fast algorithms in the 1-D and 2-D cases and can be effectively used for color image filtration and enhancement, as shown in Chapter 7. One of the main properties of the traditional discrete Fourier transform is the calculation of the linear convolution as the multiplication. This is an unique property of the classical Fourier transform, which makes it very

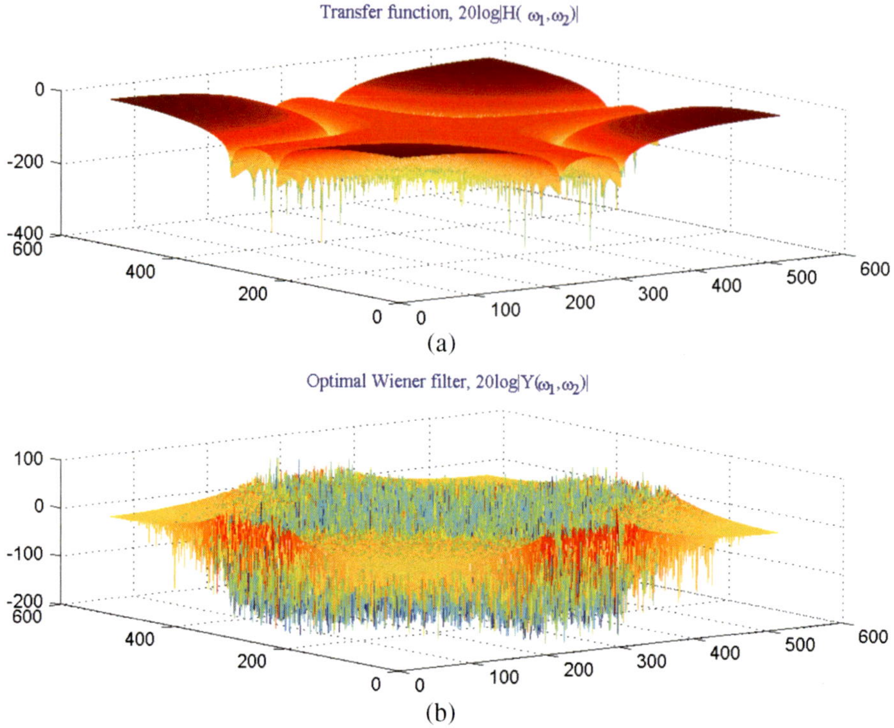

Figure 9.5 (a) The response function of the blurring and (b) the response function of the 2-D Wiener filter.

useful in solutions of many problems including signal and image filtration [7, 8], enhancement [46]–[44], and reconstruction [6], [42]–[49]. The quaternion DFT in the considered above definition, as well as in many other definitions of such transform, does not posses this property. As an example, Fig. 9.6 shows separately the four components of the quaternion signal $f_n = a_n + ib_n + jc_n + kd_n$ of length 512 are shown in the first column of the figure. This signal was convoluted with the quaternion sequence g_n each component of which is the filter $h_n = [1, 2, \underline{3}, 2, 1]/9$, i.e., $g_n = (1 + i + j + k)h_n$. The result of the inverse QDFT of the product of 1-D QDFTs of signals f_n and g_n is shown in the second column of the figure. One can see that only the first component of the inverse QDFT looks like a convoluted signal, and other three components differ much from their originals.

9.3 Optimal Filtration of Color Images

In this section, we consider the convolution in the quaternion space of discrete signals and images. For simplicity of calculations, we consider the 1-D case; the 2-D case can similarly be described.

Let f_n be the quaternion signal of length N,

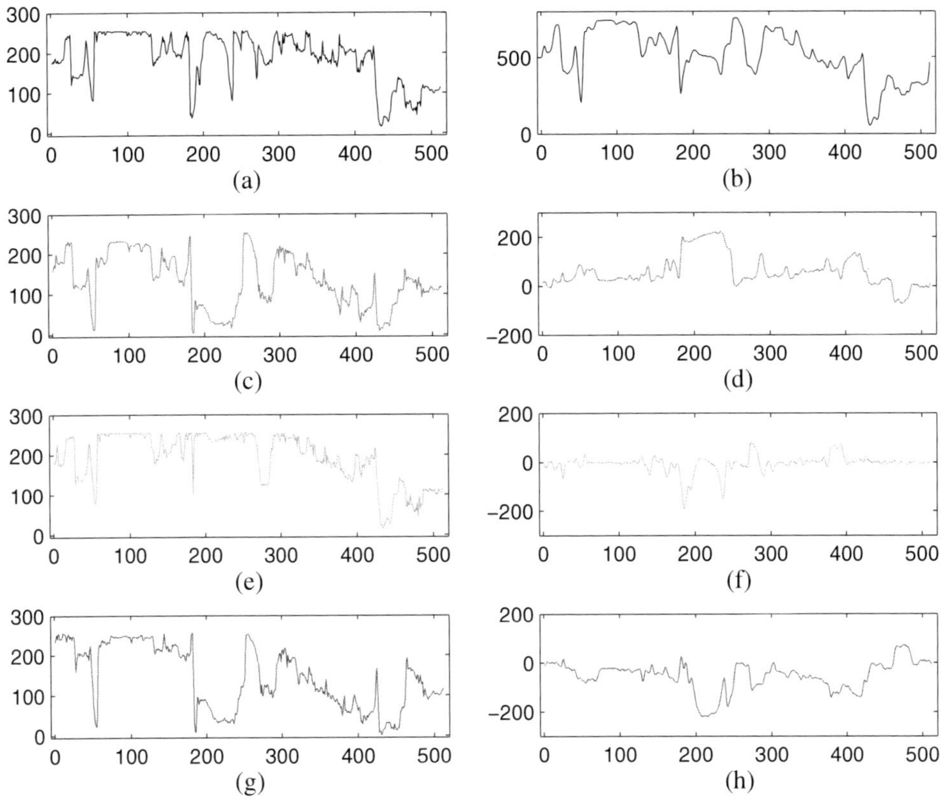

Figure 9.6 Four components of the quaternion signal before (a,c,e,g) and after (b,d,f,h) the convolution.

$$f_n = [(f_1)_n, (f_2)_n, (f_3)_n, (f_4)_n] = (f_1)_n + i(f_2)_n + j(f_3)_n + k(f_4)_n,$$

where $n = 0:(N-1)$.

We consider the signal in matrix representation, by introducing the following four matrices:

$$E = \begin{bmatrix} 1 & 0 & 0 & 0 \\ 0 & 1 & 0 & 0 \\ 0 & 0 & 1 & 0 \\ 0 & 0 & 0 & 1 \end{bmatrix}, I = \begin{bmatrix} 0 & -1 & 0 & 0 \\ 1 & 0 & 0 & 0 \\ 0 & 0 & 0 & 1 \\ 0 & 0 & -1 & 0 \end{bmatrix},$$

$$J = \begin{bmatrix} 0 & 0 & -1 & 0 \\ 0 & 0 & 0 & -1 \\ 1 & 0 & 0 & 0 \\ 0 & 1 & 0 & 0 \end{bmatrix}, K = \begin{bmatrix} 0 & 0 & 0 & -1 \\ 0 & 0 & 1 & 0 \\ 0 & -1 & 0 & 0 \\ 1 & 0 & 0 & 0 \end{bmatrix}$$

with the multiplications shown in Table 9.1.

Table 9.1 Table of multiplications of quaternion matrices.

	E	I	J	K
E	E	I	J	K
I	I	–E	–K	J
J	J	K	–E	–I
K	K	–J	I	–E

We consider the following model of signal degradation. Let h_n be the sequence that we call *the impulse response characteristic* of a linear system,

$$h_n = [(h_1)_n, (h_2)_n, (h_3)_n, (h_4)_n],$$

which is presented in the matrix form as

$$h_n = (h_1)_n E + (h_2)_n I + (h_3)_n J + (h_4)_n K, \ n = 0 : (N-1). \tag{9.12}$$

We define the concept of the circular linear convolution as

$$g_n = (f * h)_n = (f_1 * h_1)_n E + (f_2 * h_2)_n I + (f_3 * h_3)_n J + (f_4 * h_4)_n K. \tag{9.13}$$

Such defined operation has properties similar to the traditional convolution. This convolution is linear, commutative, and the associative property and distributive property also take place.

The difference between two quaternion signals f_n and g_n can be calculated in the following metric:

$$\begin{aligned} \rho^2(f,g) &= \rho^2(f_1,g_1) + \rho^2(f_2,g_2) + \rho^2(f_3,g_3) + \rho^2(f_4,g_4) \\ &= \sum_{n=0}^{N-1}[(f_1)_n - (g_1)_n]^2 + \sum_{n=0}^{N-1}[(f_2)_n - (g_2)_n]^2 \\ &\quad + \sum_{n=0}^{N-1}[(f_3)_n - (g_3)_n]^2 + \sum_{n=0}^{N-1}[(f_4)_n - (g_4)_n]^2. \end{aligned}$$

To minimize the error between the signals f_n and g_n, it is enough to minimize the errors between their real and three imaginary components,

$$\rho_k^2(f,g) = \rho^2[(f_k)_n, (g_k)_n] = \sum_{n=0}^{N-1}[(f_k)_n - (g_k)_n]^2, \ k = 1:4. \tag{9.14}$$

Now, we consider the model of the signal with the independent noise

$$g_n = (f * h)_n + n_n, \tag{9.15}$$

where the noise is presented in the matrix form as

$$n_n = (n_1)_n E + (n_2)_n I + (n_3)_n J + (n_4)_n K, \quad n = 0 : (N-1).$$

Each component $(n_k)_n$ of the noise is independent for each component of the signal $(f_k)_n$, $k = 1, 2, 3, 4$. The response function of the optimal filter for each component of the degraded quaternion signal is defined as

$$Y_k(\omega) = \frac{\bar{H}(\omega)}{|H(\omega)|^2 + \phi_{N_k/F_k}(\omega)}, \quad \omega \in \Omega, \ k = 1, 2, 3, 4. \tag{9.16}$$

Here, $\phi_{N_k/F_k}(\omega)$ denotes the component noise-signal ratio $\phi_{N_k}(\omega)/\phi_{F_k}(\omega)$, where $\phi_{N_k}(\omega)$ and $\phi_{F_k}(\omega)$ are spatial spectral densities of the restored and original signal component, i.e., $\phi_{N_k}(\omega) = <|N_k(\omega)|^2>$ and $\phi_{F_k}(\omega) = <|F_k(\omega)|^2>$.

The corresponding error of approximation of the kth component of the quaternion signal is

$$\rho^2(F_k, Y_k G_k) = \int_\Omega \frac{\phi_{N_k}(\omega)}{|H(\omega)|^2 + \phi_{N_k/F_k}(\omega)} dw. \tag{9.17}$$

Therefore,

$$\rho^2(f, \hat{f}) = \sum_{k=1}^{4} \rho^2[(f_k)_n, (\hat{f}_k)_n] = \frac{1}{N} \sum_{k=1}^{4} \rho^2(F_k, \hat{Y}_k G_k).$$

As an example in the 2-D case, Fig. 9.7 shows the result of optimal filtering the degraded color "Miki-Anoush-Mini" image in the model blur + noise for each color channel. The RMS error of blurring the image is 0.1009, the error of degraded image is 0.1052. The error of image reconstruction, or $\sqrt{\rho^2(f, \hat{f})} = 0.0419$.

In the quaternion frequency space, these four optimal filters define the following data:

$$(Y)_p = (Y_1)_p E + (Y_2)_p I + (Y_3)_p J + (Y_4)_p K, \quad p = 0 : (N-1).$$

In the case when $\phi_{N_k}(\omega)$ is the same for all components, i.e., $\phi_{N_k}(\omega) = \phi_N(\omega)$, we obtain

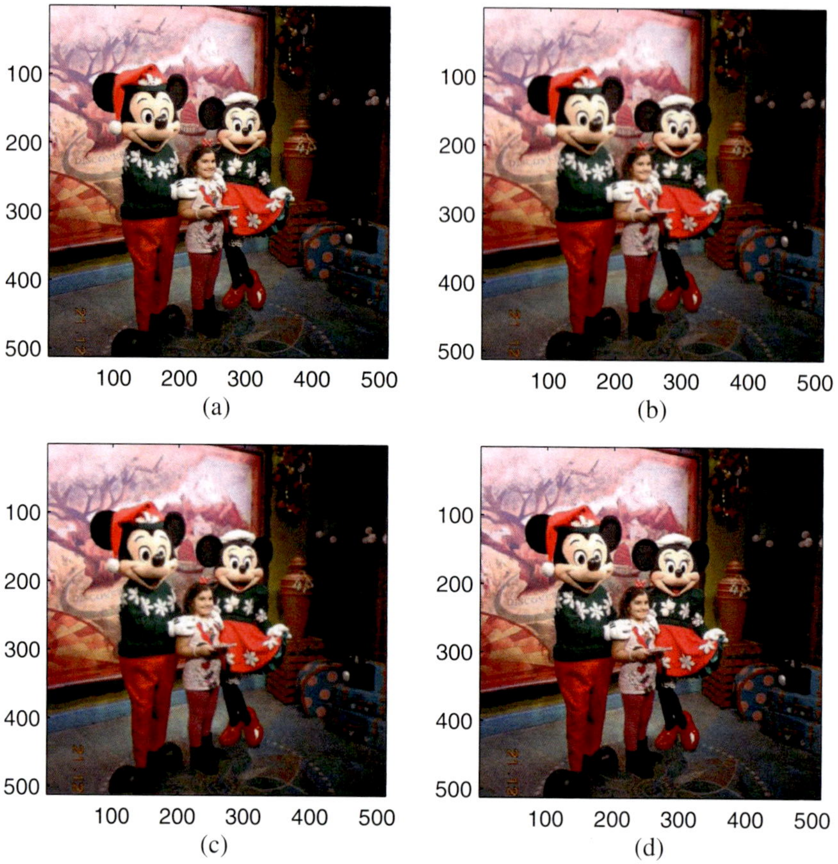

Figure 9.7 (a) True color (512 × 512) image, (b) blur image, (c) noisy image, and (d) image processed by the 2-D optimal filters.

$$(Y)_p = \frac{\bar{H}_p}{|H_p|^2 + (\phi_{N/F_1})_p} + \frac{\bar{H}_p}{|H_p|^2 + (\phi_{N/F_2})_p} I + \frac{\bar{H}_p}{|H_p|^2 + (\phi_{N/F_3})_p} J + \frac{\bar{H}_p}{|H_p|^2 + (\phi_{N/F_4})_p} K. \tag{9.18}$$

On other side, the result of component-wise filtration in the quaternion frequency space is

$$(\hat{F})_p = (Y_1)_p(G_1)_p + (Y_2)_p(G_2)_p I + (Y_3)_p(G_3)_p J + (Y_4)_p(G_4)_p K, \tag{9.19}$$

where $p = 0 : (N-1)$. It is clear that $(\hat{F})_p \neq (\hat{Y})_p(\hat{G})_p$, because the convolution is not transformed into the operation of multiplication when using the quaternion discrete Fourier transform. The inverse QDFT of the product $(\hat{Y})_p(\hat{G})_p$ is a signal that is a new operation (useful or not) of the degraded signal or image, and what is defined in Eq. (9.18) is a "filter." To show this fact, we consider the 2-D case. Figure 9.8 shows the true image in part (a) and the inverse 2-D QDFT of $(\hat{F})_{p,s}$ in part (b), when the real component is zero, i.e., $(G_1)_p = 0$, $p = 0:(N-1)$.

For the 2-D case of the quaternion image, we can consider and process the grayscale and color images as one unit [50],

$$f_{n,m} = [(f_1)_{n,m}, (f_2)_{n,m}, (f_3)_{n,m}, (f_4)_{n,m}], \quad n, m = 0:(N-1),$$

where

$$(f_1)_{n,m} = \frac{r_{n,m} + g_{n,m} + b_{n,m}}{3}$$

and

$$(f_2)_{n,m} = r_{n,m}, \quad (f_3)_{n,m} = g_{n,m}, \quad (f_4)_{n,m} = b_{n,m}.$$

In this chapter, we analyzed the model of quaternion signal degradation when the signal is convoluted and a random additive noise is added. The classical model of such a model leads to the solution of the optimal Wiener filter, where the optimality with respect to the mean square error. For quaternion signals, the inverse problem is complicated by the fact that the quaternion arithmetic is not commutative. The quaternion Fourier transform does not map the convolution to the operation of multiplication. It should be mentioned that the optimal filter in the quaternion frequency space is not direct representation of the filters designed for each color channel separately.

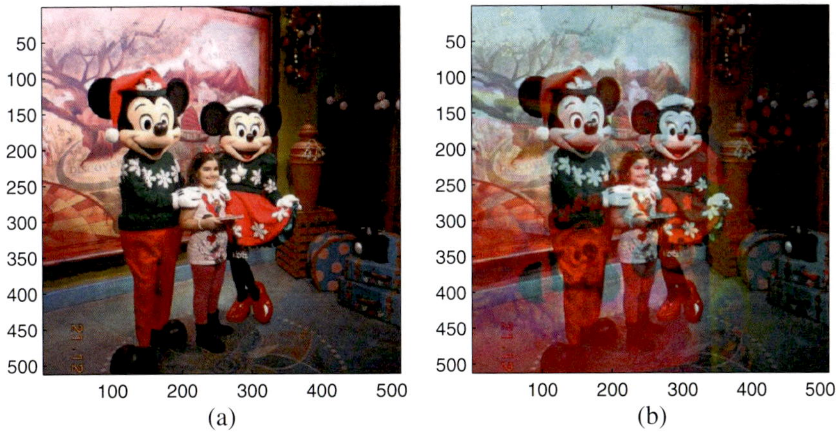

Figure 9.8 (a) True color (512 × 512) image and (b) the inverse 2-D QDFT of the data $(\hat{F})_{p,s}$.

In order to design the optimal quaternion filter for quaternion degraded signals and color images and use the concept of the 2-D QDFT, new effective methods should be found (see Ref. [53]).

References

[1] A. K. Jain, *Fundamentals of Digital Image Processing*, Prentice Hall, Inc. Englewood Cliffs, NJ (1989).

[2] L. Rabiner and B. Gold, *Theory and Application of Digital Signal Processing*, Prentice-Hall Inc. (1975).

[3] R. C. Gonzalez and R. E. Woods, *Digital Image Processing*, 2nd Edition, Prentice Hall, Upper Saddle River, New Jersey (2002).

[4] A. M. Grigoryan and S. S. Agaian, "Preprocessing tools for computer-aided cancer imaging systems," Chapter 2, 23–78, in "Computer-Aided Cancer Detection and Diagnosis: Recent Advances" (Editors: J. Tang and S. Agaian), *SPIE*, (2014) [doi: 10.1117/3.1002311.ch2].

[5] S. S. Agaian, "Advances and problems of the fast orthogonal transforms for signal-images processing applications (part 1)," *Pattern Recognition, Classification, Forecasting*. Yearbook, The Russian Academy of Sciences, Nauka, Moscow, **3**, 146–215 (1990).

[6] A. M. Grigoryan and S. S. Agaian, *Multidimensional Discrete Unitary Transforms: Representation, Partitioning and Algorithms*, Marcel Dekker Inc., New York (2003).

[7] A. M. Grigoryan and M. M. Grigoryan, *Brief Notes in Advanced DSP: Fourier Analysis With MATLAB*, CRC Press, Taylor and Francis Group, London, New York (2009).

[8] A. M. Grigoryan, "Fast paired method of a 1-D cyclic convolution realization," US Patent 7,346,498 (2008).

[9] A. M. Grigoryan, S. S. Agaian, and J. T. Astola, "Quasilinear spaces: a general framework for nonlinear filtering," *Proc. SPIE*, **2424** (1995) [doi: 10.1117/12.205218].

[10] N. Wiener, *The Extrapolation, Interpolation, and Smoothing of Stationary Time Series with Engineering Applications*, John Wiley and Sons, Inc., New York (1949).

[11] C. W. Hesltrom, "Image restoration by the method of least squares," *J. Opt. Soc, Amer.*, **57**, 297–303 (1967).

[12] H. C. Andrews and B. R. Hunt, *Digital Image Restoration*, Prentice Hall, Inc. Englewood Cliffs, NJ (1977).

[13] W. K. Pratt, "Generalized Wiener filter computation techniques," *Trans. IEEE. Comput. C-21*, 636–641 (1972).

[14] A. M. Grigoryan, E. R. Dougherty, and S. S. Agaian, "Optimal Wiener and homomorphic filtration: Review," *Signal Processing*, **121**, 111–138 (2016) [doi: 10.1016/j.sigpro.2015.11.006].

[15] L.-H. Juang and M.-N. Wu, "Image noise reduction using Wiener filtering with pseudo-inverse," *Measurement*, **43**(10), 1649–1655 (2010).
[16] M. Basin and A. Loukianov et al., "Mean-square filtering for uncertain linear stochastic systems," *Signal Processing*, **90**(6), 1916–1923 (2010).
[17] D. Humphrey and D. Taubman, "A filtering approach to edge preserving MAP estimation of images," *Image Processing, IEEE Transactions on*, **20**(5), 1234–1248 (2011).
[18] D. Letexier and S. Bourennane et al., "Nonorthogonal Tensor Matricization for Hyperspectral Image Filtering," *IEEE Geoscience and Remote Sensing Letters*, **5**(1), 3–7 (2008).
[19] P. A. Penczek, "Resolution measures in molecular electron microscopy," *Methods Enzymol*, **482**, 73–100, 2010.
[20] D. Kundur and D. Hatzinakos, "A novel blind deconvolution scheme for image restoration using recursive filtering," *IEEE Trans. on Signal Processing*, **46**(2), 375–390 (1998).
[21] A. M. Grigoryan and E. R. Dougherty, "Bayesian robust optimal linear filters," *Signal Processing*, **81**(12), 2503–2521 (2001).
[22] A. M. Grigoryan and E. R. Dougherty, "Robustness of optimal binary filters," *Journal of Electronic Imaging*, **7**(1), 117–126 (1998) [doi: 10.1117/1.482633].
[23] E. R. Dougherty, *Random Processes for Image and Signal Processing*, SPIE/IEEE Presses, Bellingham 1999.
[24] A. M. Grigoryan and E. R. Dougherty, "Design and analysis of robust binary filters in the context of a prior distribution for the states of nature," *Journal of Mathematical Imaging and Vision 11*, 239–254 (1999).
[25] A. M. Grigoryan and E. R. Dougherty, "Optimization of linear filters under power-spectral-density stabilization," *IEEE Trans. Signal Processing*, **49**(10), 2292–2300 (2001).
[26] A. M. Grigoryan, *Robustness of optimal binary filters: Analysis and design*, Texas A&M University, 1999.
[27] E. R. Dougherty, A. M. Grigoryan, J. Barrera, and N. S. T. Hirata, "Binary filter design: optimization, prior information, and robustness," *Mathematical modeling and estimation techniques in computer vision* (1998).
[28] A. M. Grigoryan and E. R. Dougherty, "Designing robust binary filters," *Proc. SPIE*, **3304** (1998).
[29] E. R. Dougherty and A. M. Grigoryan, "Robustness of optimal binary filters for sparse noise," Optical Science, Engineering and Instrumentation, 130–137 (1997) [doi: 10.1117/12.290275].
[30] E. R. Dougherty and A. M. Grigoryan, "Robustness of optimal binary filters for sparse noise [3167-12]," *Proc. SPIE*, **3167** (1997) [doi: 10.1117/12.290275].

[31] A. M. Grigoryan and S. S. Agaian, "New models and nonlinear filters of image restoration," *Proc. SPIE*, **2662** (1996) [doi: 10.1117/12.235824].

[32] A. M. Grigoryan and S. S. Agaian, "Split manageable efficient algorithm for Fourier and Hadamard transforms," *IEEE Trans. on Signal Processing*, **48**(1), 172–183 (2000).

[33] A. M. Grigoryan, "2-D and 1-D multi-paired transforms: Frequency-time type wavelets," *IEEE Trans. on Signal Processing*, **49**(2), 344–353 (2001).

[34] S. S. Agaian, H. G. Sarukhanyan, K. O. Egiazarian, and J. Astola, *Hadamard Transforms*, SPIE Press, Bellingham, Washington (2011) [doi: 10.1117/3.890094].

[35] A. M. Grigoryan and M. Hajinoroozi, "A novel method of filtration by the discrete heap transforms," *Proc. SPIE*, **9019**, 901903 (2014) [doi: 10.1117/12.2039982].

[36] A. M. Grigoryan and M. Hajinoroozi, "Image and audio signal filtration with discrete Heap transforms," *Applied Mathematics and Sciences: An International Journal (MathSJ)*, **1**(1), 1–18 (2014).

[37] A. M. Grigoryan and S. P. K. Devieni, "New method of signal denoising by the paired transform," *Applied Mathematics and Sciences: An International Journal (MathSJ)*, **1**(2), 1–20 (2014).

[38] S. DelMarco and S. Agaian, "The design of wavelets for image enhancement and target detection," *Proc. SPIE*, **7351**, 735103 (2009) [doi: 10.1117/12.816135].

[39] K. Panetta, Y. Zhou, S. Agaian, and H. Jia, "Nonlinear unsharp masking for mammogram enhancement," *IEEE Transactions on Information Technology in Biomedicine*, **15**, 918–928 (2011).

[40] S. C. Nercessian, K. A. Panetta, and S. S. Agaian, "Non-linear direct multi-scale image enhancement based on the luminance and contrast masking characteristics of the human visual system," *IEEE Transactions on Image Processing*, **22**(9), 3549–3561 (2013).

[41] K. Panetta, S. Agaian, Z. Yicong, and E. J. Wharton, "Parameterized logarithmic framework for image enhancement systems," *Man, and Cybernetics, Part B: Cybernetics, IEEE Transactions on*, **41**(2), 460–473 (2011).

[42] A. M. Grigoryan, "Representation of the Fourier transform by Fourier series," *Journal of Mathematical Imaging and Vision*, **25**(1), 87–105 (2006).

[43] A. M. Grigoryan and S. S. Agaian, "Tensor form of image representation: enhancement by image-signals," *Proc. SPIE*, **5014** (2003) [doi: 10.1117/12.473063].

[44] F. T. Arslan and A. M. Grigoryan, "Fast splitting α-rooting method of image enhancement: Tensor representation," *IEEE Trans. on Image Processing*, **15**(11), 3375–3384 (2006).

[45] A. M. Grigoryan and N. Du, "Principle of superposition by direction images," *IEEE Trans. Image Processing*, **20**(9), 2531–2541 (2011).

[46] A. M. Grigoryan and K. Naghdali, "On a method of paired representation: Enhancement and decomposition by series direction images," *Journal of Mathematical Imaging and Vision*, **34**(2), 185–199 (2009).

[47] A. M. Grigoryan and M. M Grigoryan, *Image Processing: Tensor Transform and Discrete Tomography with MATLAB*, CRC Press Taylor and Francis Group, New York (2012).

[48] A. M. Grigoryan and N. Du, "2-D images in frequency-time representation: Direction images and resolution map," *Journal of Electronic Imaging*, **19**(3), (033012), 1–14 (2010) [doi: 10.1117/1.3483906].

[49] A. M. Grigoryan, "Another Step towards successful tomographic imaging in cancer: Solving the problem of image reconstruction," Chapter 11, 295–338, in "Computer-Aided Cancer Detection and Diagnosis: Recent Advances" (Editors: J. Tang and S. Agaian), *SPIE*, (2014) [doi: 10.1117/3.1002311.ch11].

[50] A. M. Grigoryan and S. S. Agaian, "Retooling of color imaging in the quaternion algebra," *Applied Mathematics and Sciences: An International Journal (MathSJ)*, **1**(3), 23–39 (2014).

[51] A. M. Grigoryan and S. S. Agaian, "Alpha-rooting method of color image enhancement by discrete quaternion Fourier transform," *Proc. SPIE*, **9019**, 901904 (2014) [doi: 10.1117/12.2040596].

[52] A. M. Grigoryan, J. Jenkinson, and S. S. Agaian, "Quaternion Fourier transform based alpha-rooting method for color image measurement and enhancement," *SIGPRO-D-14-01083R1, Signal Processing*, **109**, 269–289 (2015).

[53] A. M. Grigoryan and S. S. Agaian, "Systems and Methods for Fast Quaternion Convolution with Multimedia," US Patent Application, Attorney Docket Number: 292007–8240 (2017).

Index

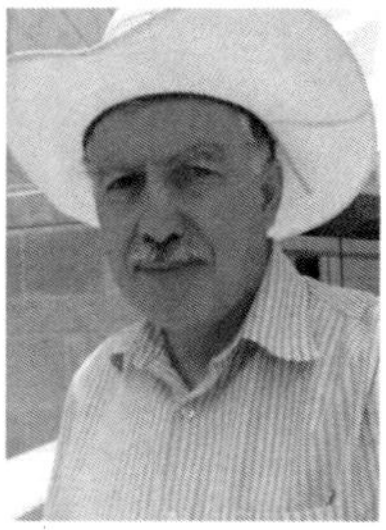

Dr. Artyom M. Grigoryan received the MSs degrees in mathematics from Yerevan State University (YSU), Armenia, USSR, in 1978, in imaging science from Moscow Institute of Physics and Technology, USSR, and in electrical engineering from Texas A&M University, USA, and Ph.D. degree in mathematics and physics from the YSU. He was a senior researcher with the Department of Signal and Image Processing at Institute for Problems of Informatics and Automation, and Yerevan State University, National Academy Science of Armenia. In 1996–2000 he was a Research Engineer with the Department of Electrical Engineering, Texas A&M University. In December 2000, he joined the Department of Electrical Engineering, University of Texas at San Antonio, where he is currently an Associate Professor. He is the author of five books, ten book-chapters, four patents, and many journal papers and specializing in the theory and application of fast multi-dimensional Fourier transforms, elliptic and quaternion Fourier transforms, tensor and paired transforms, unitary heap transforms, design of robust linear and nonlinear filters, image enhancement, encoding, and encryption, computerized tomography, and processing biomedical images.

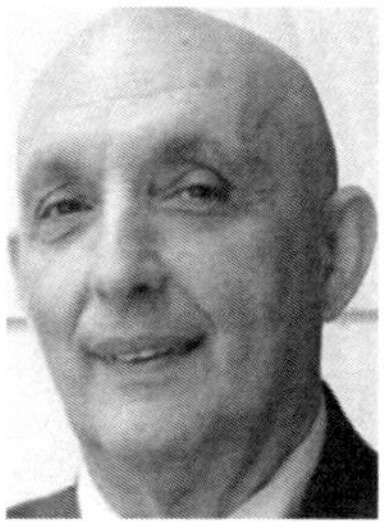

Dr. Sos S. Agaian is Peter T. Flawn Professor of Electrical and Computer Engineering at the University of Texas, San Antonio, and Adjunct Professor at the Graduate School and University Center, City University of New York, and Professor at the Health Science Center, San Antonio. Dr. Agaian received his M.S. degree (summa cum laude) in mathematics and mechanics from Yerevan University, Armenia, the Ph.D. degree in math and physics from the Steklov Institute of Mathematics, Russian Academy of Sciences, and the Doctor of Engineering Sciences degree from the Institute of the Control System, Russian Academy of Sciences. He has authored more than 600 scientific papers, 9 books, and holds 15 patents. He is a Fellow of the International Society for Photo-Optical Instrumentations Engineers, Fellow of the IEEE, Fellow the Science Serving Society (AAAS), and Fellow Imaging Sciences and Technology (IS&T). He also serves as a foreign member of the Armenian National Academy. He is the recipient of MAEStro Educator of the Year, sponsored by the Society of Mexican American Engineers. Dr. Agaian has invented technologies that have been adopted across multiple disciplines, including the US government, and commercialized by industry. He is an Editorial Board Member of the Journal of Pattern Recognition and Image Analysis and an Associate Editor for several journals, including the Journal of Electronic Imaging (SPIE, IS&T) and the System Journal (IEEE). His research interests are multimedia processing, information security, artificial intelligence, computer vision, 3D imaging sensors, Fusion, and biomedical and health informatics.